# HISTORY

OF THE

# UNITED STATES MARINE CORPS

BY

RICHARD S. COLLUM,

*Major U. S. M. C.*

NEW YORK:

L. R. HAMERSLY CO.

1903.

TO THE

CITIZENS OF THE UNITED STATES

THIS WORK IS RESPECTFULLY DEDICATED; WITH A DESIRE THAT THE SERVICES OF THH

UNITED STATES MARINE CORPS

MAY BE INTELLIGENTLY APPRECIATED,
AND THAT THE NATION MAY RECOGNIZE
THE DEBT IT OWES TO THE

OFFICERS AND ENLISTED MEN,

WHO, IN ALL THE TRYING TIMES IN OUR COUNTRY'S HISTORY
HAVE NOBLY DONE THEIR DUTY.

"From the establishment of the Marine Corps to the present time it has constituted an integral part of the Navy, has been identified with it in all its achievements, ashore and afloat, and has continued to receive from its most distinguished commanders the expression of their appreciation of its effectiveness as a part of the Navy." —*Report of House Committee on Naval Affairs; Thirty-ninth Congress, Second Session.*

# PREFACE.

A custom has prevailed throughout the armies of Europe to keep regular record of the services and achievements of their regiments and corps.

This principle has not obtained in our own country, hence a great difficulty arises in presenting an authentic account of the services of any branch of our establishment either military or naval.

This defect has long been felt in relation to the Corps of which this historical record is designed to treat.

An attempt was made by the author in 1874 to present a true and faithful record, but owing to circumstances beyond his control the original idea was changed during his absence from the country.

It is now thought that the proper time has arrived for an edition which will present a record of events that will be acceptable to the Corps and to the nation.

Impressed with this consideration, and influenced by a strong attachment to the welfare and reputation of the Corps in which he has the honor to serve, the author feels persuaded that an impartial account of the services of the United States Marines will not only reflect an additional lustre on its distinguished character, but encourage a spirit of emulation, which is the strongest impulse to great and gallant actions.

Although he may have failed in accomplishing that object with the ability due its importance, he trusts that the fidelity and zeal which has been exercised in the present compilation will obtain for him the indulgent consideration of those who are aware of the difficulties with which he has had to contend.

He desires to express his obligation to his brother officers who have kindly rendered him their cordial assistance, without which this record would have been incomplete.

In narrating these events the information obtained has been compiled from official reports, "Cooper's Naval History," "Hamersly's Record of Living Officers," and "Hamersly's Naval Encyclopedia."

Although this record must fail to do the United States Marines ample justice, the author trusts that sufficient has been said to prove that their distinguished repute is well merited, and that they have not, under any circumstances, tarnished their motto—

"Semper Fidelis."

# CONTENTS.

INTRODUCTION.

## CHAPTER XV.



## CHAPTER XVI.



## CHAPTER XVII.



## CHAPTER XVIII.



## CHAPTER XIX.



## CHAPTER XX.



## CHAPTER XXI.



## CHAPTER XXII.



## CHAPTER XXIII



## CHAPTER XXIV.

CHAPTER XLVIII.



CHAPTER XLIX.



CHAPTER L.



CHAPTER LI.



APPENDIX.

# INTRODUCTION.

The Antiquity of Marines—The Royal Marines of Great Britain—The American Marines.*

THE employment of infantry as part of the regular complement of vessels of war was common to the Phœnicians and to all the maritime states of Greece at least five centuries before the commencement of the Christian era. In the earlier period of history it was not so. When vessels were no larger than pentekonters,—open boats pulling fifty oars,—perhaps up to the time of biremes, the warriors were the oarsmen. But as naval science progressed, and the size of the vessels increased, there gradually sprang up distinct classes, which together made up the *personnel* of the navies about 500 B. C.,—the rowers, the seamen proper, who had the general management of the vessel and sails, and Marines, or fighting men. Marines are especially mentioned in the account of the battle of Lâdé, in the time of Darius, king of Persia, about 497 B. C. The Ionian Greeks, being in a state of revolt, had their fleet drawn up at Lâdé, a small island lying off Miletus, where it was discovered by the Persians. In the battle which ensued, the Samian and Lesbian squadrons deserted the cause, the people of those isles having been won over by Persian emissaries, and were followed by others. "Of those who remained and fought," says Herodotus, "none were so rudely handled as the Chians. They had furnished to the common fleet over one hundred ships, having each of them forty armed citizens on board, and those picked men. Scorning to follow the base examples of the traitors, they fought desperately, till, overcome by numbers, they were obliged to seek safety in flight." The same author, in estimating the strength of the Persian naval force which accompanied the army of Xerxes to Greece, observes that each vessel had on board, besides native soldiers, thirty fighting men, who were either Persians, Medes, or Sacans.

At the dawning of the day of the battle of Salamis, the men-at-arms of the Greek fleet were assembled on shore, and speeches were made to

*This chapter was kindly furnished the author, in 1874, by Captain, now Rear-Admiral, S. B. Luce, U.S.N.

them. "The best of all was that of Themistocles, who, throughout, contrasted what was noble with what was base, and bade them in all that came within the range of man's nature always to make choice of the nobler part." These men-at-arms, or soldiers, which formed part of the complement of the Greek trireme, were called *Epibatæ*, a word all authorities agree in rendering into English by the word Marines. The largest number of Marines found aboard each of the "swift ships" —that is, the regular men-of-war, as distinguished from transports—at this period was forty. Plutarch gives the number on board each Greek trireme at the battle of Salamis as eighteen, four of whom were archers and the rest heavy-armed. During the Poloponnesian War, the average number on board the Athenian trireme was ten. It may not be out of place to give here one of the many incidents of the battle of Salamis, as an illustration of the valor and mode of fighting of the *Epibatæ:* "A Samothracian vessel bore down on an Athenian and sunk it, but was attacked and crippled immediately after by one of the Eginetan squadron. Now, the Samothracians were expert with the javelin, and aimed their weapons so well that they cleared the deck of the vessel which had disabled their own, after which they sprang on board and took it." (Herodotus viii. 90.)

In the account of the battle of Platea, where Mardonius, the great military leader of the Persians, was slain, and the choice of the Persian troops routed, there is an instance related where the military and naval training are curiously blended in the person of the Athenian Sôphanes. "He wore," says Herodotus, "an iron anchor, fastened to the belt which secured his breastplate by a brazen chain; and this, when he came near the enemy, he threw out, to the intent that when they made their charge it might be impossible for him to be driven from his post. As soon, however, as the enemy fled, his wont was to take up his anchor and join the pursuit." Another account states that the anchor was simply a device upon his shield. But in either event the anchor would indicate that he had served afloat. Thucydides makes frequent mention of *Epibatæ*. When the Athenian strategus Demosthenes was operating near Leucas with thirty ships, he landed his forces, and had, besides the army, " the three hundred *Epibatæ* from his own ships," making ten for each trireme. Rawlinson, in translating the word *Epibatæ*, explains that it means "the armed portion of the crew, corresponding to our (English) Marines." So also Dr. Dale, in his translation of "Thucydides," renders the word *Epibatæ* as "the heavy-armed soldiers who served on board ship, answering to our Marines." The eminent Greek scholar, Dr. Arnold, takes the same view. The learned historian of Greece, Mr. Grote, speaks of *Epibatæ* as Marines, and observes that "though not forming a corps permanently distinct, they correspond in function to the English Marines." In the statement that they did not form a distinct corps, Mr. Grote seems to differ from other authorities.

Boeckh, probably one of the very best authorities on the antiquities of Athens, who is so freely quoted by Mr. Grote in his history of Greece and referred to by Dr. William Smith and Rich in their dictionaries of Roman and Greek antiquities, in speaking of matters concerning the Athenian navy, remarks that "The crews of the swift triremes consisted of two descriptions of men,—of the soldiers or Marines appointed to defend the vessels, who were also called *Epibatæ*, and of the sailors. These *Epibatæ* were entirely distinct from the land soldiers, such as the *hoplitæ, peltastæ* and cavalry, and belonged to the vessel." ("Boeckh's Economy of Athens," vol. i., page 373.) They had, moreover, their own officers, called *tricrarchoi*.

During the naval supremacy of Rome, the quinquireme only, as a rule, was admitted into the line of battle. Polybius gives the number of rowers in these vessels as three hundred, and that of the Marines (*clasiarii milites*) as one hundred and twenty. In regard to the manner of fighting, it may be observed that the *Epibatæ* used arrows and darts at a distance, spears and swords in close combat, and as ships increased in size they added *balistæ,* and *turres,* or turrets (*naves turritæ*), and fought from them as from castles on land.

In the earlier period of naval history, when opposing fleets, drawn up in parallel lines, closed at once and decided the issue of the battle by a hand-to-hand contest, the number of Marines aboard each vessel was as large as could be accommodated. But when, in the process of time, military science became better understood, and its principles came to be applied to the management of fleets, naval tactics grew to be an art in which the Athenians rendered themselves pre-eminently successful. Skilful manœuvres and evolutions performed with wonderful celerity and precision took the place of the old style of rushing headlong into battle. To maim and disable an enemy without receiving any injury in return was now the evidence of an expert *tricrarch* and a well-drilled crew; this feat once accomplished, and the enemy's vessel reduced to a helpless wreck upon the water, the survivors were easily despatched by the Marines. Hence, with the improved tactics came a reduction of the number of Marines allowed to a Greek trireme, the precise number being finally settled at ten, as already stated, the entire complement of the trireme being two hundred.

The true functions of the Marine of modern times were little understood, his great value to the navy to which he belonged little appreciated, till within a recent period. According to their own writers, the Corps of Marines of Great Britain was originally instituted in 1664; and, curiously enough, partly with a view to forming a nursery of seamen for the fleet. The privates were encouraged to qualify as able seamen, and were allowed every opportunity of doing so.*

"It having been found necessary on many occasions to embark a number of soldiers on board our ships of war, . . it was judged expedient to appoint certain regiments for that service who were trained to the different modes of sea-fighting, and also made useful in some of those manœuvres of a ship where a great many hands were required. These, from the nature of their duty, were distinguished as marine soldiers, or Marines." ("Grose's Military Antiquities of the English Army," vol. i.) They were expected to be more or less familiar with the duties of seamen, and it is still more curious to observe that this total misapprehension as to the nature of the duties of Marines was transmitted to the colonies in America, and found definite expression in the first act of the Continental Congress establishing a Marine Corps.

In 1740 three additional regiments were raised in America and assembled at New York. All the officers, excepting the captains of companies, who were colonists nominated by the provinces, were appointed by the crown, and Colonel Spottswood, of Virginia, was colonel-commandant of the whole.†

In 1760 the strength of the Corps being increased to eighteen thousand three hundred and fifty-five men, application was made for an increase of the number of field-officers, etc.

Nicolas, in his history of the Royal Marine forces, gives very full accounts of all the battles in which that Corps participated, but although the British Marine had in many a hard-fought action acquitted himself with credit, yet on no occasion had he proved himself of more value to the navy to which he belonged, or reflected more honor on his

---

*"Royal Marines," by T. Smith, R.M. Colburn's *United Service Magazine* for May, and following numbers, 1874.

†The history of the Royal Marines is not without its page of romance. One Hannah Snell, of Worcester, England, it appears, fought in the ranks as a Marine. She belonged to the guard of the Swallow, one of the squadron under Admiral Boscawen, was distinguished for bravery, wounded twelve times in various actions, and was finally discharged without her sex being discovered. This brings to mind the story of the lady whose devotion to the fortunes of one William Taylor induced her to follow that hero to sea. She assumed the name of Richard Carr, behaved with great gallantry, and, notwithstanding the secret of her sex being revealed, she was actually promoted to be first lieutenant of the "Thunder" bomb. (See "Universal Songster," London, 1827, vol. i., page 65.)

Corps, than during the momentous period covering the great mutinies at Spithead, the Nore, and Bantry Bay. The seamen of the fleet fully understanding the advantage, in the stand they had taken against the government, of engaging the Marines as their allies, took care to include them in their demand for redress of abuses. In their answer to the Lords of the Admiralty, dated on board the Queen Charlotte at Spithead, April 19, 1797, they say, "And as a further proof of our moderation, and that we are actuated by a true spirit of benevolence towards our brethren, the Marines, who are not noticed in your lordships' answer," etc. But neither had they, in their first statement of grievances, beginning, "We, the seamen of His Majesty's Navy," alluded, even remotely, to the Marines, showing that the alliance was an after-thought. The Marines were mentioned, however, in the petition to Parliament. But although their "brethren the Marines" doubtless had their own grievances, yet they, as a rule, remained true to their duty, and by their steadfast courage and good discipline suppressed more than one mutiny. One remarkable instance among many of a similar character is related of the crew of the Impétueux, Captain Sir Edward Pellew, afterwards Lord Exmouth. The mutineers had in vain attempted to win over the Marines, and Sir Edward, finding he had them on his side, at once took such energetic measures that the mutiny was put down. The Marine guard of the Castor frigate summarily suppressed a mutiny on board that ship December, 1801. In the same year the mutiny in the squadron in Bantry Bay broke out; this was also put down by the aid of the Marines. On each occasion the good conduct of the Marines was recognized in the official report of the affair, and it was probably due to the reputation for loyalty then acquired under the most trying circumstances, rather than their conduct in battle, that gave rise to the complimentary order of the Admiralty, of April 29, 1802, conveying His Majesty's directions "that the Corps shall be styled Royal Marines." The last serious mutiny necessary in this connection to notice occurred on board the Excellent, 74, in the West Indies.

Lord Hood, in general orders, dated "Blenheim, Barbadoes, December 30, 1802," says, "The commander-in-chief (as well as the members of the court-martial) is highly sensible of the active exertions of the officers of H. M. ship Excellent, in quelling the late mutiny on board that ship; and also the officers, non-commissioned officers, and private Marines belonging to the said ship: who, by their firmness in resisting the attempt to seduce them from their duty, and in opposing men in actual mutiny, have increased, if possible, the high character the Corps has so justly acquired," etc., etc. It is evident that the worth of the

British Marine had begun to be acknowledged, and yet it was not till 1837 that the "iniquitous system"* of rewarding distinguished naval officers by appointing them to the sinecures of generals and colonels of marines was abolished.

The value of the Royal Marine as exhibited in his loyalty during a series of mutinies of ships' companies, extending over a period of five years,—or from 1797 to 1802,—some of which shook the throne of Great Britain to its very foundation, was found to be in the total absence of sympathy between him and the seaman; an inimical feeling, perhaps, which the appointment of naval officers, of whatever class, to positions in the Corps had a decided tendency to break down. Hence it may be assumed as a general proposition, subject to little qualification, that the value of a Marine Corps to a navy, or of a Marine guard to a ship, is in direct proportion to the thoroughness of its military training, its *esprit de corps,* and the strict observance of that line of demarcation which separates the military from the naval element.

Happily for us, our seamen have never been driven into mutiny, nor does it seem possible that such a deplorable condition of affairs could be brought about as would cause an absolute antagonism between them and the Marines. Still, as the legally established "sharpshooters" and necessary military element of our sea-forces, our Marine Corps should jealously guard its integrity as a military body "par excellence."

In 1804 an artillery company was attached to each of the three divisions of the Royal Marine Corps, to supply the service of the bomb-vessels, and in time of peace to drill the whole of the Marines in gunnery. But they were soon available for other purposes, and on the outbreak of the War of 1812, a large body of the Marine Artillery, with a field battery and rocket equipment, accompanied the battalions of Marines then formed for service in America. Later, Sir Howard Douglass complimented the Marine Artillery as being "either a corps of good infantry, of scientific bombardiers, or expert field artillerymen, well constituted, thoroughly instructed, and ably commanded." After fluctuating in numbers through several years, the Corps was so augmented that in 1859 it numbered three thousand officers and men, and was formed into a separate division, with its head-quarters at Fort Cumberland.

The military education of the officers of the Royal Marine Artillery is thorough, and such as could be heartily wished were introduced somewhat into our own Marine Corps. The cadet Marines have to study for two years, more or less, during which they are expected to acquire

*Nicolas.

a competent knowledge of arithmetic, algebra, geometry, plane trigonometry, the use of the sextant, fortification, English history, and French. To this may be added a practical course in naval gunnery. If, on obtaining his commission, the young marine officer is selected to qualify for the artillery, he must be prepared at the end of a year to pass an examination in analytical trigonometry, differential and integral calculus, conic sections, statics and dynamics, hydrostatics and steam, besides being required to have an increased knowledge of fortification. The men are volunteers from the light infantry divisions, possessing specified qualifications as to age, height, intelligence and character.

The course of training, which is, with a few exceptions, common to both officers and men, is very comprehensive: it includes the usual infantry drills and musketry instruction; the exercise of field guns and rockets, with such field-battery movements as are of real practical importance; the service of heavy ordnance, including heavy guns, howitzers, and sea and land service mortars; the naval great gun exercise; mounting and dismounting ordnance, with and without machines; the various methods of slinging and transporting ordnance; knotting and splicing and fitting gun gear; use of pulleys, etc.; a laboratory course, including use and preparation of tubes, rockets, and fuses, making up cartridges, manufacture of port fires, signal lights, rockets, and explosive compounds; a course of practical gunnery, comprising instruction in the nature and uses of the various kind of guns, howitzers, and mortars, of projectiles, sighting ordnance, use of red-hot shot, and such matters connected with the theory of projectiles as may have a practical application.*

The United States Marine Corps has well sustained the high reputation for steadfast courage and loyalty which has been handed down to it from the days of Themistocles. But like their modern prototypes of Great Britain, they have felt the want of proper appreciation. In the resolution of Congress of November 10, 1775, to raise two battalions to be called "first and second battalions of American Marines," it was enjoined that "no person be enlisted into said battalion but *such as are good seamen,* or so acquainted with maritime affairs as to be able to serve with advantage by sea:" clearly showing that our legislators of that day, at least, had little conception of the nature of a properly organized Marine Corps. Unfortunately, the erroneous ideas expressed in that act seem more or less to have been transmitted to the present time; for, strange as it may appear, the Corps, while it has

*"Military Schools and Course of Instruction in the Science and Art of War." Henry Barnard, LL.D. New York, 1872.

its own peculiar organization, is yet without any regimental organization. It would seem to be to the interest of our Marine Corps that it should be brigaded,—the full number allowed by law being three thousand and seventy-four,† though practically reduced by the amount of the annual appropriation,—and companies, or parts of companies, with their proper officers detailed for duty afloat. This course, in connection with the educating of the young Marine officers at West Point, would soon bring our Marine Corps up to that high military standard which it is quite safe to say all naval as well as all Marine officers desire to see it attain.‡

†Act of July 25, 1861.

‡The officers of the Corps, since 1881, are appointed from the graduates of the Naval Academy, a course advocated by the progressive members of the Corps, many years before the passage of the law.—AUTHOR.

# HISTORY

OF THE

# UNITED STATES MARINE CORPS.

## CHAPTER I.

**1740-1777.**—Colonial Marines—Organization of the Corps—First Regular Cruisers—Vessels ordered to be Constructed—Commodore Hopkins appointed Commander-in-Chief—The First Ensign—Descent on New Providence—Alfred, Cabot, and Glasgow—Lexington and Edward—Appointments in the Corps—The Sachem—Doria and Racehorse—Uniform—Adoption of the American Flag—Fox and Hancock.

1740.—The first authentic record of Marines in America bears the date of 1740. Early in that year three additional regiments were raised, when the royal standard was displayed at New York as the port to which any volunteer Marine was to repair. It was supposed that the native Americans were better calculated for the service in this climate than the Europeans, and they were clothed in a manner which was considered well adapted for their duties. The colonels, lieutenant-colonels, and other commissioned officers were appointed by the crown, except the captains of the companies, who were nominated by the American provinces. Their uniform was camlet coats, brown linen waistcoats, and canvas trousers. Colonel Spotiswood, of Virginia, was colonel commandant of the whole.

1775.—On the 8th of June the Continental Congress resolved "that the compact between the crown and the people of Massachusetts Bay is dissolved," and on the 10th of November, before a single vessel of the Navy was sent to sea, the Corps was organized by the following resolution:

*Resolved,* That two battalions of Marines be raised consisting of one colonel, two lieutenant-colonels, two majors, and other officers, as usual in other regiments; that they consist of an equal number of privates with other battalions; that particular care be taken that no persons be appointed to offices, or enlisted into said battalions, but such as are good seamen, or so acquainted with maritime affairs as to be able to serve to advantage by sea when required; that they be enlisted and commissioned to serve for and during the present war with Great Britain and the colonies, unless dismissed by order of Congress; that they be distinguished by the names of the First and Second Battalions of American Marines.

Later in the same month, another resolution was adopted providing against filling the Corps for the army.

But slight record can be found of the services of the Marines during the first year of the existence of the Corps. Men were not over-plenty, either for military or naval duty, and several months elapsed before the Corps was fully organized, equipped, and ready for duty. From the hour of their earliest organization, however, the Marines made a noble record for themselves and for their country. In the early naval fights they played a most important part, amply proving their devotion to the cause in which they were enlisted, and furnishing conclusive evidence of the wisdom of the legislation which called their Corps into existence.

The first regular cruisers that ever got to sea were the sloop Hornet and schooner Wasp, which sailed in November from Baltimore to join the squadron under Commodore Hopkins in the Delaware.

This passage, however, cannot be called a cruise. Cooper says that the honor properly belongs to Captain John Barry, of Philadelphia, who commanded the Lexington, a brig of fourteen guns, which must have left the capes of the Delaware late in January or early in February, 1776, and her orders were to cruise to the southward.

On the 13th of December Congress directed thirteen ships of war to be built. On the 22d of the same month Congress passed a resolution declaring Esek Hopkins commander-in-chief, and appointed officers for all the vessels then in service.

The equipment of all vessels that could be produced was going on in the autumn of 1775, and the first ensign ever shown by a regular man-of-war was hoisted in the Delaware on board the Alfred, by Paul Jones, some time about the last of December. What that ensign was is not positively known, but it is thought to have been a device representing a pine-tree, with a rattlesnake, about to strike, coiled at its root, and bearing the motto, "Don't tread on me." It is certain that such a flag was used at the commencement of the Revolution, and on board of some of the vessels of war, though whether this was the flag borne by the Alfred is not quite so clear. Most of the privateers of the period either wore the arms of the colony from which they sailed, and by which they were authorized to cruise, or they also showed devices of their own, according to the conceits of the different captains and owners.

**1776.**—Commodore Hopkins was directed to proceed to the southward, with a view to act against the naval force, which was then ravaging the coast of Virginia, under Lord Dunmore. The squadron had gone into the bay and rendezvoused under Cape Henlopen early in February. It consisted of the Alfred, Columbus, Doria, Cabot, Providence, Hornet, Wasp, and Fly, despatch vessel. Commodore Hopkins sailed with this force on the 17th of February. Nothing of importance occurred, and in due time the squadron reached Abaco, in the Bahamas, the place of rendezvous. Here Commodore Hopkins determined to make a descent on New Providence, where it was understood a considerable amount of military stores had been collected. A battalion of three hundred men, Marines and landsmen, under the command of Captain Nichols, the senior Marine officer of the service, was put into ten sloops with the hope of surprising the place. An alarm having been given, however, the Providence and Wasp were sent in to cover the landing. This duty was handsomely performed, and Captain Nichols captured the forts by assault and obtained complete command of the place in the course of the afternoon and following morning. A hundred cannon and a large quantity of stores fell into the hands of the Americans, although the governor had succeeded in removing a large quantity of powder during the night. On this occasion, the first that had ever occurred in the regular American Navy, "the Marines under Captain Nichols behaved with a spirit and steadiness that have distinguished the Corps from that hour down to the present moment." On as close an examination of this action as our means will allow, the following officers of the Corps appear to have been attached to the squadron: Alfred, Captains Nichols and Robert Mullin; Columbus, Captain Andrew Porter; Providence, First Lieutenant Peregrine Brown.

On the 6th of April, in the action between the Alfred and Cabot and the British 20-gun frigate Glasgow, the list of casualties on the two American vessels amounted to twelve seamen and eleven marines killed and wounded.

On the 17th of April the Lexington, Captain Barry, encountered the Edward, an armed tender of the Liverpool, and, after a severe action of an hour, captured her. The Lexington had four of her crew killed and several wounded, among the latter being Sergeant Haslit of the Marines. Lieutenant Abel Morgan commanded the Marines.

During the year 1776 the organization of the Corps was further perfected, and the following officers were appointed June 25:

| | | | |
|---|---|---|---|
| Samuel Nichols | Major. | Franklin Read | First Lieutenant. |
| Andrew Porter | Captain. | Peregrine Brown | First Lieutenant. |
| Joseph Hardy | Captain. | Thomas Pownal | First Lieutenant. |
| Samuel Shaw | Captain. | James McClure | Second Lieutenant. |
| Benjamin Deane | Captain. | William Gilmore | Second Lieutenant. |
| Robert Mullin | Captain. | Abel Morgan | Second Lieutenant. |
| John Stuart | Captain. | Hugh Morgan | Second Lieutenant. |
| Daniel Henderson | First Lieutenant. | Richard Harrison | Second Lieutenant. |
| David Lowe | First Lieutenant. | | |

Other appointments during the same year succeeded as follows: July 22, Captains, George J. Osborn and Richard Palmer; First Lieutenant, Stephen Meade; Second Lieutenants, Nathaniel Thwing and Benjamin Thompson; July 24, First Lieutenant, Abraham Van Dyke; August 24, First Lieutenant, Louis de la Valette; September 6, Captain William Nicholson.

That there were Marines in the squadron of Commodore Hopkins, is known from the fact of their having been landed at New Providence, where they were the assailing force; but even the greater portion of the sea officers employed on that occasion had merely letters of appointment, and it is to be presumed that such was also the case with Captains Nichols, Mullin, Porter, and Lieutenant Brown. Captain Nichols was undoubtedly the senior officer of the Corps at New Providence, although the first regular roster published by authority places him at the head of the Corps as major. But this, it will be observed, bears the date of June 25, and Captain Nichols was promoted for his gallantry.

On the 6th of July, two days after the Declaration of Independence, the Sachem, ten guns, Captain Robinson, sailed from the Delaware. When a few days out she fell in with an English letter-of-marque and captured her, after a sharp contest. Both vessels suffered severely in killed and wounded. Some time in August, the Doria, while off the western end of Porto Rico, encountered an English vessel of war. The enemy commenced the action by firing a broadside, which was immediately returned by the Doria. After a severe engagement of two hours the Englishman surrendered. The prize proved to be the Racehorse, twelve guns, Lieutenant Jones, who was mortally wounded, together with a large number of his crew. The casualties on board the Doria amounted to twelve. Lieutenant Daniel Henderson commanded the Marines.

By this time the whole coast was alive with adventures, scarcely a week passing that did not give rise to some incident which exhibited great heroism. Whenever an enemy's cruiser appeared, or attempted to land, skirmishes ensued, and from contemporary accounts the information obtained warrants the statement that the Corps was fully represented, and its officers and men shared in the affairs of personal gallantry displayed on the coast of New England, the Chesapeake, and the coast of the Carolinas.

The following was the uniform for the officers of the Corps, by order of the "Marine Committee," dated Philadelphia, September 5, 1776: "A green coat faced with white, round cuff, slashed sleeves and pockets, with buttons round the cuff, silver epaulette on the right shoulder, skirts turned back, buttons to suit the facings; white waistcoat and breeches edged with green, black gaiters and garters;" green shirts were ordered for the men, "if they can be procured."

1777.—On the 14th of June of this year, Congress first adopted the stars and stripes as the national flag.

On the 27th of June an action was fought off the banks of Newfoundland between the British 32-gun frigate Fox and the American 32-gun frigate Hancock. After an engagement of two hours, which resulted in great loss to the enemy, the Fox surrendered. The officers of the Corps attached to the Hancock were Captain Benjamin Deane, First-Lieutenant Thomas Pownal.

---

## CHAPTER II.

1777.—Capture of Philadelphia—The First Attack by the British Fleet on the Defences of the Delaware—Success of Commodore Hazelwood in repulsing the Enemy—The Advance Squadron of the Enemy succeeds in opening the Channel.

In July, General Howe, the commander of the British forces, occupied Staten Island, and preparations were made for an attack on Philadelphia. On the 23d of the same month a powerful armament, consisting of thirty-six battalions of infantry and a strong detachment of artillery, sailed from Sandy Hook and arrived off the entrance of the Delaware on the 31st. Finding, however, that the Americans had obstructed the navigation of the river, the fleet abandoned the attempt, and proceeded up the Chesapeake on the 14th of August, and the enemy landed at Elk Ferry on the 25th. On the 3d of September the British began their march, but Washington, who had returned from New Jersey by forced marches with ten thousand men, lost no opportunity of harassing the enemy, without bringing on a general engagement. Several actions, however, took place, notably the battles of Brandywine and Germantown, before the British entered Philadelphia, resulting in great loss to them, but their superiority of numbers prevailed, and General Howe entered the city on the 26th of September. The enemy immediately commenced the erection of batteries on the Delaware front, but so prompt and energetic were the officers of the little fleet, that before the batteries could be completed two frigates and some small galleys came up from Mud Island and attacked the lower battery of four guns, anchoring within five hundred yards of the enemy's position. At ten o'clock on the morning of the 10th of September our vessels commenced a heavy cannonade and soon silenced the battery. The tide ebbing, however, one of the frigates, the Delaware, took the ground and was soon in a defenceless condition. She was therefore obliged to strike her colors, and was taken possession of by a detachment of British Marines.

The first effort of the brothers Howe, so soon as Philadelphia was captured, was to open the navigation of the Delaware River, so that the transports containing provisions, stores, and other necessities for the British army might be carried by water to it. Both the British commanders and Washington looked to this matter with the same deep interest.*

There were three fortifications below Philadelphia:

First, Fort Mifflin, on Mud Island, upon the Pennsylvania side of the river, just below where the Schuylkill empties into it, seven miles below the city.

Opposite to Fort Mifflin, and about nineteen hundred yards—the width of the river here—from it, on the Jersey shore, was Fort Mercer, at Red Bank.

The third fortification was Billingsport, four or five miles lower down the river, on the Jersey side.

Two rows of *chevaux-de-frise* were sunk opposite to Billingsport, and four rows between Fort Mifflin and Fort Mercer. In the narrow and difficult channel behind Fort Mifflin, and between it and Province Island, no obstructions had been placed. In the channel there, as up to that time generally known, no vessel of the draft of water made by even the smallest one of the British men-of-war could pass. Between the north end of Hog Island and the south end of Mud Island *one* ship was sunk.

The moment that victory on the Brandywine opened the gates of Philadelphia to the enemy, the British fleet was in motion down the Chesapeake, around Cape Charles, and up the Delaware, in order that it might give, by opening that great watercourse, completeness to the capture. The admiral cast anchor off Chester, on the 6th of October, but Captain Hammond with an advance squadron—the Roebuck, Pearl, Camilla, and Liverpool—had moved up the river as the army advanced towards Philadelphia, and on the 1st of October was off Billingsport. He had informed General Howe that if a sufficient land force could be sent to New Jersey to attack the fortifications at that point, *he* would take upon himself the task of removing the *chevaux-de-frise* which Billingsport covered.

---

* "Without the free navigation of the Delaware, I am confident that General Howe will never remain in Philadelphia.—*Washington.*

On the 1st of October the general accordingly sent Colonel Sterling with two regiments of grenadiers from Chester. They landed about five miles below the fort, now garrisoned by one hundred and fifty men, and on the 2d began their march towards it. Colonel Bradford was there in time to bring off to Fort Mifflin the little garrison and the military stores, Bradford himself remaining until fired upon by the advancing army. Entered by the British on the 2d of October, it was evacuated by them on the 5th. It uncovered the *chevaux-de-frise* opposite to the place, and Captain Hammond went to work with his squadron in redemption of his pledge to remove these obstructions. He had hardly got his squadron at work before Commodore Hazelwood, who was in command of our fleet,* went at it, and attacked the ships so sharply that they were obliged to desist from their efforts and to fall down to Chester.

Captain Hammond, however, on the 25th of October, succeeded, not without much difficulty, being continually harassed by our galleys, fire-rafts, and other armed craft, in opening a sufficient channel for the larger ships.

The plan of the brothers Howe was now developed. It was to attack simultaneously Forts Mifflin and Mercer by land and water; and, as respected Fort Mifflin, to begin their attack by land from batteries previously erected on the few spots of hard ground at the junction of the Delaware and Schuylkill, and along the western banks of the narrow channel behind Fort Mifflin.

## CHAPTER III.

Contemplated Capture by the Combined British Army and Navy of Forts Mercer and Mifflin—The Attempt made October 22, 1777—Attack on that Day on Fort Mercer repulsed with Heavy Loss to the Assailants—The Navy rendering Great Aid—No Serious Attack on Fort Mifflin—Two British Vessels run aground, and on the 23d are set on Fire—Their Magazines explode.

By the 20th of October, a passage for the British vessels being now opened through the lower *chevaux-de-frise,* the brothers Howe prepared to attempt the capture of the forts.

The general had been able to erect on the hard grounds of the rear island a redoubt which enfiladed a principal battery of Fort Mifflin.

*Commodore Hazelwood was born in England in 1726; died at Philadelphia, March 1, 1800. He was promoted commodore in the Pennsylvania State Navy, September 6, 1777. At a later period he was appointed one of the "Commissioners of Purchase" in Philadelphia.

General Howe ordered also a body of troops to be landed for "forcing the redoubt," as the British called Fort Mercer at Red Bank.

In addition to the erection of the battery on Province Island intended to play on the rear of Fort Mifflin, and, as Admiral Howe tells us, *on the galleys* and other armed craft, the squadron from the British fleet had come up the river. It consisted of the Augusta, Roebuck, Liverpool, Pearl, Merlin, and Isis.

The rest of the plan is thus disclosed in Lord Howe's official despatch:

"It was *intended* that the Vigilant should pass through a shallow and very confined channel on Hog Island and the Pennsylvania shore, to arrive and act upon the rear and less defensible part of the work; and the circumstances of the navigation not admitting of a more serious attack, . . . a diversion was proposed to be made at the same time by the Isis and Augusta in the eastern or main channel of the river, as well for engaging the attention of the enemy at Fort Island and the redoubt as to restrain the motion of the galleys and other armed crafts, which had retired under the works at Red Bank when they discovered the danger which they would be exposed to from our batteries on the western shore."

Such was the *plan*. It included, as will be seen, an attack by the Vigilant on the rear of Fort Mifflin. But even an attack could not be made on this fort. And while on Fort Mercer an attack was made, it was one which ended only in disaster.

At five o'clock on the afternoon of the 22d of October a force of Hessians under Count Dunop attacked the fort with great spirit. Having carried the outworks, his troops had now to storm the interior intrenchments. To reach these, the attacking force had to place itself where it would be exposed to the fire of our vessels. The opportunity was not missed by Commodore Hazelwood, the commander of the fleet. Coming as near as possible to the fort, he poured in a hot and incessant fire on the attacking party. The assailants retired, but only to find in their slower and disordered retreat a still more deadly fire from the fleet. The whole advance, attack, and retreat lasted but forty-five minutes; the loss was estimated at not less than four hundred, including Count Dunlop and many of the best officers.

The loss in the garrison amounted to one captain, five sergeants, one fifer, and seven privates killed; one ensign, two sergeants, and twenty privates wounded.

As respected Fort Mifflin, the plan failed on the point by which success was expected to be accomplished,—an attack by the Vigilant on the rear and less defensible part of the work.

On the evening of the 22d the Augusta and Merlin grounded some distance below the second line of *chevaux-de-frise.* At an early hour the next morning empty transports were making their way up from the fleet, and other preparations made for lightening the grounded vessels. It was too late. The Navy and Fort Miffin were upon them. A portion of the Navy could not at once be got into action. The British squadron lay below the stockades; and while certain vessels of our fleet could pass through these to make an attack, care was required in bringing down others. However, twelve galleys and two floating batteries got into action at once. Soon all got in and the action became general. The Augusta, though grounded, had her broadside constantly playing on them; the Roebuck, the two frigates, and their galleys giving good support. After some hours' fighting, a tremendous noise was heard. A volume of thick smoke was seen ascending to the heavens. The magazine of the Augusta exploded, and that of the Merlin soon followed. These successive explosions were so disastrous to the British fleet that only three boat-loads of men were saved from the two well-manned vessels.

This conjoint success by sea and land was, with reason, regarded as an event of the time. It encouraged the friends of the Revolution everywhere, and *dis*couraged the many, who were disposed to put themselves under the protection of the British troops enjoying themselves in Philadelphia.

---

## CHAPTER IV.

The Second Attack on Fort Mifflin—Success of the British—A Portion of the State Fleet saved—The Continental Fleet burnt by its Commander—The Treason of Captain Robert Whyte—Marines attached to the Pennsylvania Navy.

THE new attack on Fort Mifflin began on the 5th of November, from land batteries placed as follows:

Two batteries on Providence Island, having two 32-pounders, one 18-pounder, one 32-pounder.

One battery on Carpenter's Island, having six 24-pounders, one 8-inch howitzer, one 8-inch mortar.

One Battery on Carpenter's Island, having one 8-inch howitzer, one 7-inch mortar.

One battery on Carpenter's Island, having one 15-inch mortar.

One battery on the mainland, having two 12-pounders, one 18-pounder.

On the night of the 14th the Vigilant and Fury got under way, and at the break of day sailed in on the swelling tide behind Hog Island,

and by the time that broad daylight of the 15th had arrived, anchored themselves within a hundred yards of the fort, and at such a point as enabled them to aid in the most effective manner the batteries on the two islands.

The British squadron which sailed up on the morning of the 15th was,—

| | |
|---|---|
| Somerset, five hundred men | 64 guns. |
| Isis, three hundred and fifty men | 50 guns. |
| Roebuck, two hundred and eighty men | 44 guns. |
| Pearl, two hundred and twenty-two men | 32 guns. |
| Liverpool | 28 guns. |
| Cornwallis, galley | 1 gun. |
| Vigilant, one hundred and fifty men | 16 guns. |
| Fury | 3 guns. |

The other vessels of the British fleet which remained below were the flagship Eagle, sixty-four guns; Experiment, fifty guns; Renown, fifty guns; Apollo, thirty-two guns; Zebra, sixteen guns; and the Camilla, twenty guns.

Our own little Navy, which faced this mighty armament, was as follows:

STATE FLEET.

*Galleys.*—Bulldog, Burke, Camden, Congress, Dickinson, Experiment, Effingham, Franklin, Hancock, Ranger, Warren, Washington, Witch.

| | |
|---|---|
| One carrying | one 32-pounder. |
| Two carrying each | one 24-pounder. |
| Ten carrying each | one 18-pounder. |

*Half-Galleys.* Argus, Basilisk, Brimstone, Dragon, Eagle, Fame, Firebrand, Hawk, Hornet, Lion, Porcupine, Racehorse, Repulse, Resolution, Salamander, Terror, Thunder, Tormentor, Viper, Vulture, Wasp.

| | |
|---|---|
| Twenty-one carrying each | one 4-pounder. |

*Floating Batteries.*—Arnold and Putnam.

| | |
|---|---|
| One carrying | twelve 18-pounders. |
| One carrying | ten 18-pounders. |

*Provincial Ship.*—Montgomery.

| | |
|---|---|
| One carrying | fourteen 18-pounders,<br>sixteen cohorns,<br>eight swivels. |

*Armed Schooner.*—Delaware.

*Schooner Galley.*—Lydia.

One Carrying . . . . . . { in bow, two 18-pounders, in stern, two 9-pounders.

*Brig Galley.*—Convention.

One carrying in bow, two 18-pounders; in stern, two 9-pounders.

*Fire-Ships.*—Hecla, Stromboli.

*Fire-Brig.*—Volcano.

*Fire-Brigantine.*—Blast.

*Fire-Sloop.*—Ætna.

*Accommodation Sloops.*—Defiance, Hetty, Industry.

*Ammunition Sloop.*—Sally.

*Sloops.*—Liberty, Speedwell.

*Shallop.*—Black Duck.

CONTINENTAL FLEET.

| | |
|---|---|
| Andria Doria, brig | Fourteen 6-pounders. |
| Hornet, sloop | Twelve 9-pounders. |
| Racehorse, sloop | Ten guns. |
| Fly, sloop | Six guns. |
| Wasp, schooner | Ten 9-pounders. |
| Independence, sloop | Ten 9-pounders. |
| Sachem, sloop | Ten 9-pounders. |
| Mosquito, sloop | Four 9-pounders. |

The fleet at this period was never fully manned; besides, half of our larger vessels were short of cannon. With regard to half our fleet, therefore, at this critical moment, its vessels were no better than "painted ships upon a painted ocean."

The Delaware fleet was under the command of Commodore Hazelwood. After twelve hours of uninterrupted fighting the garrison was evacuated at midnight of the 15th, but not until the blockhouses were reduced to ruins, the palisades beaten down, the barracks battered in every part, most of the guns dismounted, and a large part of the garrison killed. On the fleet twenty-eight men were killed and twelve wounded; while of the twelve galleys in action only one was not shattered by shot. The fire-rafts were almost all destroyed by the tempestuous weather.

It now became necessary to determine in what manner the vessels of the fleet should be disposed of, in order to prevent their capture by the enemy. A council of war was accordingly held on board the Speedwell, on November 19th. It was concluded to endeavor to pass the city through the Jersey or Eastern channel. Commodore Hazelwood got thirteen galleys and nine armed boats under way at three o'clock on the morning of the 20th of November. Aided by the darkness and the fog, he was carried by the "morning flood" alone, in about an hour and a half, through the Jersey channel past the city, and by ten A.M. he was at anchor off Burlington.

The Continental fleet, however, under Captain Robinson, was not so fortunate. On the morning of the 21st, Captain Robinson, being unable to escape up the river, set fire to the remaining vessels to prevent their capture by the enemy. As far as can be ascertained, the following officers of the corps were engaged in these operations: Captain Andrew Porter, Captain Richard Palmer, First Lieutenant Daniel Henderson, First Lieutenant Stephen Meade.

From the fact that the channel at the rear of Fort Mifflin, in the opinion of the British admiral, would float armed vessels, and that he from the beginning meant to operate by means of it, the question naturally arises, *"How came it that the channel at the rear of the fort was left open?"*

In raising this question it must be recollected that great pains were taken to obstruct the navigation of the channel of the Delaware in *front* of Fort Mifflin.

*Whence came this confidence of the British admiral and this purpose?*

*Is it possible that Fort Mifflin was betrayed?*

In a letter from Wilmington, Delaware, dated November 10, 1777, and published in Revington's *Royal Gazette,* at New York, then in possession of the British, the writer refers to the successful opening, just then previously made by Captain Hammond, of the obstructions in the river at Billingsport; he says,—

"The Roebuck has raised three parts of the first tier in the face of the Rebel's whole fire. *The man who laid the same is now in His Majesty's service."*

The laying of the *chevaux-de-frise* at Billingsport was committed to Colonel Robert Smith, of Chester county, Pennsylvania, Captain Robert Wythe and Mr. Samuel Morris, of Philadelphia.

Both Smith and Morris are well-known characters in the history of Pennsylvania, and the loyalty of both at all times to the cause of independence is undeniable. The only person, therefore, to whom this imputation can apply is Captain Wythe.

In the year 1787, ten years after the siege, an Englishman, George Grien, in a note to a translation which he made and then published in London, of "Chastellux's Travels in North America," makes this statement:

"The person principally employed in sinking the *chevaux-de-frise,* and in securing the passage of the river, was one Wythe, who is supposed to have left the river open designedly, as he afterwards went over to the enemy and distinguished himself by every act of hostile virulence against his country."

"Who, then, was this Wythe, a great actor in the events of America, if these British accounts are true; not yet, indeed, equal in infamy with Arnold, but equal in baseness, and more than equal in success."

"Public records show that prior to December 14, 1779, Captain Wythe was attainted of treason by the State of Pennsylvania; that as early as April 23, 1782, he was in the British naval service, commanding a royal cutter."*

Marines attached to the following vessels of the Pennsylvania State Navy 1776 to 1777:

*Arnold Battery.* Captain Thomas Forest, two sergeants, one drummer, and twenty-seven privates.

*Bull Dog.*—Two sergeants, two drummers, one fifer, and fifty-one privates.

*Burke.*—Two sergeants, one drummer, and twenty-six privates.

*Camden.*—Two sergeants, two drummers, one fifer, and thirty-five privates.

*Chatham.*—Two sergeants and thirty-four privates.

*Congress.*—First Lieutenant Hugh Montgomery, two sergeants, two corporals, two drummers, one fifer, and thirty-nine privates.

*Brig Convention.*—First Lieutenant Douglas Spencer, two sergeants, one drummer, and forty-one privates.

*Dickinson.*—Captain John Rice, two sergeants, two corporals, one drummer, two fifers, and sixty-nine privates.

*Effingham.*—First Lieutenant James Kirkpatrick, two sergeants, two corporals, two drummers, two fifers, and fifty-nine privates.

*Experiment.*—First Lieutenant James Glasgow, two sergeants, one drummer, one fifer, and fifty-one privates.

*Franklin.*—Second Lieutenant Samuel Snowden, two sergeants, three drummers, one fifer, and fifty-six privates.

*General Greene.*—Captain Robert Caldwell, Lieutenants John Hambright, Jr., Jonas Mackey, three sergeants, two corporals, one drummer, one fifer, and thirty privates.

*Hancock.*—First Lieutenant William Martin, two sergeants, two drummers, one fifer, and sixty privates.

*Flag-Ship Montgomery.*—Captain William Brown, First Lieutenant James Morrison, Second Lieutenant Thomas Caldwell, two sergeants, two corporals, two drummers, and fifty-eight privates.

---

*"Colonel William Bradford, the Patriot Printer of 1776." By John William Wallace.

## CHAPTER V.

1778-80.—Second Descent on New Providence—Ranger and the Drake—Capture of a British Privateer—Capture of the British Sloop Jason, Schooner Hibernia, and Five Transports—The Expedition to Castine—The Squadron under Commodore Paul Jones—Trumbull and Watt—Alliance—Congress and Savage—The Hyder Ally.

THE naval operations of the year opened with a gallant exploit, achieved by the United States sloop Providence, Captain Rathburne. This vessel carried only 4-pounders and is said to have had a crew of but fifty sailors and a sergeant's guard of Marines on board. Notwithstanding this trifling force, Captain Rathburne made a descent on the Island of New Providence, at the head of twenty-five Marines and sailors. He was joined by a few American prisoners, less than thirty, and, while a privateer of sixteen guns, with a crew of fifty men, lay in the harbor, he seized the forts, got possession of the stores, and obtained command of the place. All the vessels in port, six in number, fell into his hands, and an attempt of the armed populace to overpower him was suppressed by a menace to burn the town. Captain Rathburne spiked the guns of the fort, removed all the ammunition and small-arms, burned two of his prizes, and sailed with the remainder without leaving a man behind him. On the 10th of April, Paul Jones sailed from Brest on a cruise in the Irish Channel. On the 24th he arrived off Carrickfergus in the roads of which the Drake, sloop of war, Captain Burden, a vessel equal in force to the Ranger, lay anchored. The commander of the British vessel sent out an officer, in one of his boats, to ascertain the character of the stranger. By skilful handling, the officer in charge of the boat was decoyed alongside and taken. In a short time the English ship was seen under way. The Ranger now filled and stood off the land with a view to draw her enemy more into the channel, where she lay to, in waiting for her antagonist. The tide being unfavorable, the Drake worked out of the roads slowly, and night was approaching before she drew near the Ranger. The Ranger gave the first broadside, and the battle was fought running free, under easy canvas. It lasted an hour and four minutes, when the Drake called for quarter, her ensign being already down. The loss on the English ship amounted to forty. Her captain and first lieutenant were both desperately wounded, and died shortly after the engagement. The Ranger suffered much less, having Lieutenant Wallingford of the Marines and one man killed and six wounded.

In the manuscript diary of William Jenison, Lieutenant of Marines on board the United States ship Boston, then just arrived on the coast of France, there is noted the following requisition under date:

ON BOARD THE BOSTON. April 11th, 1778.

Wanted for the use and service of the Marines belonging to this ship:

40 green coats faced with white,
40 white waistcoats, and
40 white breeches.

The buttons of the whole to be a plain white. Coats to be open-sleeved, and a belt for every waistcoat.

In behalf of the Captain of Marines,
WILLIAM JENISON, *Lieutenant of Marines.*

It appears from this that the buttons for the Marine uniform at that date were not gilt, but were covered with plain white, like the facings.

Under date of May 13, we find "Regimentals for the Marines, an *uniform ordered by Congress,*" was brought on board and distributed to *officers* and men.

1779.—On the 18th of April the United States ships Warren, Captain J. B. Hopkins, Queen of France, Captain Olney, and Ranger, Captain Simpson, sailed from Boston on a cruise. This squadron captured a British privateer of fourteen guns, and later, off Cape Henry, the cruisers came up with nine sail, seven of which they captured. The vessels taken were the Jason, twenty guns, and the schooner Hibernia, eight guns. In addition to these vessels the brigs Patriot, Prince Frederick, Bachelor John, and schooner Chance, all laden with stores, fell into the hands of the Americans. Among the prisoners were twenty-four British officers on their way to join their regiments at the South. The following officers of Marines served in this squadron: Warren, Captain William Nicholson; Queen of France, First Lieutenant Abraham Van Dyke; Ranger, Second Lieutenant Benjamin Thompson.

Captain John Adams, of the Boston, states in his diary, that the uniforms of the Marines at this period were changed from green to red, the facings being the same in both uniforms.

Adams, who had no love for Paul Jones, writes in his diary, dating from Port Louis, L'Orient, Thursday, May 13, 1799:

"After dinner walked out with Captain Jones and Landais to see Jones's Marines dressed in the English uniform, red and white; a number of very active and clever sergeants and corporals are employed to teach them the exercises and manœuvres and marches, etc.; after which Jones came on board our ship. This is the most ambitious and intriguing officer in the American Navy. Jones has art and secrecy, and aspires high. You see the character of the man in his uniform, and that of his officers and Marines *variant from the uniform established by Congress,*—golden buttons for himself, *two* epaulets; Marines in red and white instead of green. Eccentricities and irregularities are to be expected from him. They are in his character, they are visible in his eyes. His voice is soft and still and small; his eye has keenness and mildness and softness in it."

In the spring of 1779 the British leaders conceived the idea that it was essential to their interests that a military post be established on the coast of Maine. With that object in view, an expedition was fitted out from Halifax, Nova Scotia, for the invasion of that district. About one thousand men were detached from the Seventy-fourth and Eighty-second Regiments of Royal Artillery, and placed under the command of Brigadier-General McLane; the ships North, Nautilus, and Albany, each of sixteen guns, were used as convoys, Captain Mowat in command. This was not the first time that Mowat had invaded the shores of Maine, for it was but a few years before that he had deliberately burned the town of Portland, then called Falmouth, to the ground. His name was still one of terror to the people of that vicinity, and it was a source of great sorrow to them, on the 16th of June, that this officer again landed on their shores. Castine, near the entrance of the Penobscot River, was selected as the spot upon which to establish the post, and the erection of a fort was immediately begun. Batteries were erected upon the banks of the river, and the ships of war were advantageously disposed of in the harbor. These dispositions having been made, the invasion was considered accomplished. But the people of Massachusetts would not quietly submit to allow the enemy's foot to rest undisturbed on their territory,—Maine being but a district of Massachusetts at that time. The spirit which led to the belief, since the evacuation of Boston, that the British dared not invade Massachusetts, still prevailed, and when the news of the landing at Castine was received at Boston, the excitement was intense. It was immediately resolved to send an armed fleet to the scene. The Legislature of Massachusetts was called upon, and, the appeal being favorably received, orders were issued for detaching a force of fifteen hundred men from the State militia, and the Board of War was authorized to engage armed ships for the service. Application to Congress for aid was also successful, and the fleet was rapidly fitted out. The United States frigate Warren, of thirty-two guns, was loaned for the expedition; she carried two hundred and fifty men, under the command of Captain Saltonstall, who was also made commander-in-chief. Congress also granted the use of two sloops-of-war, the Diligence and the Providence, of fourteen and twelve guns respectively, then lying in Boston harbor; the former vessel carried ninety men, and was under the command of Captain Brown, and the latter seventy men, Captain Hacker.

The government of Massachusetts furnished three vessels, as follows: the Tyrannicide, of fourteen guns, Captain Cathcart, ninety men; the Active, sixteen guns, Captain Hallet, one hundred men; the Hazard, fourteen guns, Captain Williams, ninety men. Thirteen vessels were also hired, by the government of Massachusetts, from citizens of Boston, Salem, and Newbury, and fitted up for the expedition. The Hampden, Captain Salter, of Portsmouth, New Hampshire, carried twenty guns and one hundred and thirty men; the Pallas, Captain Johnson, fourteen guns and eighty men; the Sky-Rocket, Captain Burke, sixteen guns and one hundred and twenty men; the Black Prince, eighteen guns, one hundred and sixty men; the Nector, Captain Carnes, eighteen guns, one hundred and fifty men; the Monmouth, Captain Ross, twenty guns, one hundred and sixty men; the Hunter, Captain Brown, twenty guns, one hundred and eighty men; the Putnam, Captain Waters, twenty guns, one hundred and seventy men; the Vengeance, Captain Thomas, eighteen guns, one hundred and ninety men. The Revenge, of eighteen guns, and the Defence, of fourteen, carried one hundred and fifty and eighty men respectively, but the names of their captains are not known. There was also a schooner of eight guns, carrying forty men, whose captain's name has not been preserved. A remarkable fact showing the enthusiasm which existed in the cause is that no less than thirty masters of merchant ships in Newburyport volunteered as common seamen on board the Vengeance, and, in deference to the superior age and experience of one of these, Captain Thomas, Captain Dennis relinquished the command of the ship, himself acting as lieutenant.

The command of the land forces was given to Brigadier-General Lovell, an officer whose superior merit had gained him a great reputation for courage and skill. Unfortunately, however, some delay had prevented a portion of the militia from arriving at Townsend, the place of rendezvous in Maine, at the appointed time, and upon the arrival of the fleet of armed ships and transports punctually on the day named, the general found but nine hundred of the militia instead of the promised fifteen hundred. This was a great disappointment to all; but for numerous reasons the greatest despatch was necessary, and delay would have chafed the spirits of the brave men who were ready for work, so that, after careful consideration of the subject, it was decided in a council of war to proceed at once. The militia therefore embarked on board the transports which had been provided for them, and, accompanied by the armed vessels with three hundred Marines under the command of Captain Osborn, sailed from Townsend, arriving at the place of destination on the 25th of July.

So well, however, had the British defences been arranged that it was not until the 28th that a landing was effected. This was at last accomplished by the Marines and the militia, the place chosen for a landing being at the foot of a precipitous height which rose abruptly and almost perpendicularly from the river. The advance of the British force was stationed in a wood on the top of this very eminence. According to General Lovell's previous orders, the main body of the seamen put off from the ships immediately after the troops had landed. The word to advance was then given, and the intrepid body of Marines and militia, closely followed by the seamen, met their own number of Scotch regulars, and with such unabated ardor and zeal was the battle pushed that the enemy was driven back within their works. But this step was only gained at a great sacrifice to the patriots, for several officers of merit were slain and of the Marines and militia about one hundred were killed and wounded. Many acts of heroic courage on the part of the Marines, seamen, and militia signalized the affair, reflecting the highest credit upon all. Upon reconnoitering, however, General Lovell found that the enemy had intrenched themselves so securely, and their defences were of such strength, that it would be worse than useless to attack them in their stronghold until reinforcements could be obtained. Neither was Commodore Saltonstall willing to attempt the destruction of the British ships lying in the harbor, even after the batteries had been captured. A council of war was called, and it was decided to send for reinforcements and to suspend operations until they should arrive.

Help, however, did not arrive as soon as it was expected, and in the mean time the British had also sent for assistance, which was despatched promptly; and on the 14th of August a British fleet from New York, under the command of Sir George Collier, sailed into the harbor. This fleet comprised the Raisonnable, of sixty-four guns, and the frigates Blonde, Virginia, Greyhound, Galatea, and Camilla. This arrival was a great blow to the Americans, who saw at once that the project of driving out the British must be abandoned, flight being the only recourse left them. Captain Saltonstall immediately communicated with General Lovell, who was on shore with his troops, at one o'clock in the morning, and preparations were at once made to evacuate the position which they held. Before dawn the little army was on board the transports and sailing up the river, the only direction open to them. But their flight was immediately discovered by the enemy, who started in pursuit. The enemy soon got near enough to use their chase guns, and the fire was returned by the Americans. It was undoubtedly the wish of Captain Saltonstall to reach the shallow waters before he was overtaken; but finding this impracticable, he ran his ship on shore and set her on fire. Others followed his example, and most of the vessels were destroyed, though three or four fell into the hands of the enemy. The loss on the Warren was thirty killed and wounded, while the loss among the troops amounted to one hundred. The following officers of the Corps were engaged in this expedition: Captains Osborn, Palmer, and Nicholson; Lieutenants Thuing, McClure, and Gilmore.

On the 14th of August the squadron under the command of Commodore Paul Jones, consisting of the Bonhomme Richard, Alliance, Pallas, Cerf, and Vengeance, sailed from the anchorage under the Isle of Groix, off L'Orient. On the 23d of September, in the combat between the Richard and the Serapis, an unusual number of lives were lost, though no regular authentic report appears to have been given on either side. In the action the Marines suffered out of proportion to the rest of the crew. Out of one hundred and thirty-seven Marines, forty-nine were killed or wounded. On board of the Richard the detachment was commanded by Lieutenant-Colonel Stack; his subordinate officers were Lieutenants Macarthy and O'Kelly. On the Alliance were Captain Park, First Lieutenant Thomas Ellenwood, and Second Lieutenant James Warren.

1780.—The first action of moment that occurred this year between any United States vessel and the enemy has the reputation of being one of the most hotly and obstinately contested combats of the war. On the 2d of June the Trumbull, twenty-eight guns, under the command of Captain James Nicholson, while cruising in latitude 35° 54′, longitude 66° W., made a strange sail to windward. Making sail in pursuit, the Trumbull was soon within a hundred yards of her antagonist, which had in the mean time shown English colors. The English ship fired the first broadside and the action began in earnest. For two hours and a half the vessels lay nearly abeam of each other, giving and receiving broadsides without intermission. At no time were they half a cable's length asunder, and more than once the yards nearly interlaced. Twice was the Trumbull set on fire by the wads of her opponent, and once the enemy suffered in the same way. At last the fire of the Englishman slackened sensibly until it nearly ceased. At this juncture the main- and mizzen-top-masts of the Trumbull went over the side, and, in spite of every effort to secure them, spar after spar came down until nothing was left but the fore-mast. The enemy being in a crippled condition speedily made off on her proper course, but before she was out of sight her main-top-mast also was seen to fall. The enemy proved to be the Watt, a larger and stronger ship and carrying more men than her opponent. The Trumbull lost thirty-nine in killed and wounded, among the former of whom were two of her lieutenants. Captain Andrew Porter of the Corps was attached to this vessel.*

1781.—In February, Captain John Barry, in command of the Alliance, sailed from Boston for France. On the outward passage she captured the Alert. On her return, in company with the Marquis de Lafayette, she captured the Mars, twenty-six guns, and the Minerva, ten guns. After this success the Alliance parted company with her consort and the prizes, and continued to cruise until the 28th of May, when two vessels of the enemy, a sloop of sixteen guns and a brig of fourteen, were encountered. After a severe action of several hours, during which Captain Barry was severely wounded, the British vessels hauled down their colors. Captain Park and Lieutenants Ellenwood and Warren were the officers of the Corps on board the Alliance.

In the engagement on the 6th of September between the Congress and the British sloop-of-war Savage, which resulted in the capture of the latter vessel, the Marines were commanded by Second Lieutenant Richard Harrison.

1782.—In March, 1782, Captain Barney, in command of the Hyder Ally, fitted out by the State of Pennsylvania, with a crew of one hundred and ten seamen and Marines, captured in Delaware Bay the British ship General Monk, eighteen guns, after a hotly-contested combat. This action has been justly deemed one of the most brilliant that ever occurred under the American flag. It was fought in the presence of a vastly superior force that was not engaged; and the ship taken was, in every essential respect, superior to her conqueror.

The termination of the first war in which America was engaged as a separate nation was proclaimed on the 11th of April, after a struggle that had endured for seven years and ten months.

At no period of the naval history of the world is it probable that Marines were more important than during the War of the Revolution. In many instances they preserved the vessels to the country by suppressing the turbulence of their ill-assorted crews, and the effect of their fire not only then, but in all the subsequent conflicts, under those circumstances in which it could be resorted to, has been singularly creditable to their steadiness and discipline. The history of the Navy,

*Captain Porter, although at this time attached to the army as captain of artillery, was detailed at his own request for duty on the Trumbull.

even at that early day, as well as in these latter times, abounds with instances of the gallantry and self-devotion of this body of soldiers. The Marine incurs the same risks from disease and tempests, undergoes the same privations, suffers the same hardships, and sheds his blood in the same battles as the seamen, and the nation owes him the same rewards. Necessity makes him the subordinate, but nations ought never to overlook the important moral and political truth, that the brightest lessons they can teach are those of justice; and no servant of the public should pass a youth of toil and danger, without the consciousness of possessing a claim to a certain and honorable reward, that is dependent only on himself. That this reward was as unwisely as it was unfairly withheld for many years, from *all* connected with the Navy, is acknowledged, and in no instance was this injustice more signally denied than to the Marine Corps."

It remains only to say that the Navy of the Revolution and, consequently, the Corps of Marines, like the army, was disbanded at the termination of the struggle, literally leaving nothing behind it but the recollections of its services and sufferings.

---

## CHAPTER VI.

**1798-1800.**—Commencement of the New Navy—The Organization of the Corps—Marine Camp near Philadelphia—The Quasi French War—Death of Washington.

On the 30th of April, 1798, a regular Navy Department was formally created, and in June, Benjamin Stoddart, of Georgetown, D. C., was made the first secretary. Down to this moment the old treaty of alliance, formed between France and the United States, during the War of the Revolution, was legally in existence; but Congress by law abrogated this treaty on the 7th of July, 1798, on the plea that it had been repeatedly disregarded by France, and that the latter country continued, in the face of the most solemn remonstrances, to practice a system of predatory warfare on the commerce of the country.

On the 11th of July, 1798, an act was approved "for the establishing and organizing a 'Marine Corps.'"

This act provided that, in addition to the military establishment, there should be raised and organized a Corps of Marines, to consist of one major, four captains, sixteen first lieutenants, twelve second lieutenants, forty-eight sergeants, forty-eight corporals, thirty-two drums and fifes, and seven hundred and twenty privates, including the marines already enlisted, or authorized to be raised for the naval armament. The corps was to be formed into as many companies or detachments as the President should direct, with a proper distribution of the commissioned and non-commissioned officers and musicians to each company or detachment. The pay and subsistence of the officers, non-commissioned officers, musicians, and privates was fixed as follows: To a major, fifty dollars per month and four rations per day; to a captain, forty dollars per month and three rations per day; to a first lieutenant, thirty dollars per month and three rations per day; to a second lieutenant, twenty-five dollars per month and two rations per day; and to the non-commissioned officers, musicians, and privates, conformably to the act entitled "An act providing a naval armament," as should be fixed by the President. The President was authorized to continue the enlistment of Marines, until the corps should be complete; and to appoint the commissioned officers, whenever, in the recess of the Senate, an appointment should be necessary. The enlistments were to be for a term of three years, subject to discharge by the President, or by the ceasing or repeal of the laws providing for the naval armament. If the Marine Corps, or any part of it, should be ordered by the President to do duty on shore, and it should become necessary to appoint an adjutant, paymaster, quartermaster, sergeant-major, quartermaster-sergeant, and drum and fife-major, or any of them, the major or commander of the Corps was authorized to appoint such staff-officer or officers from the line of subalterns, sergeants, and musicians respectively, who should be entitled, during the time they should do such duty, to the same extra pay and emoluments which were allowed by law to officers acting in the same capacities in the infantry.

It was further provided that the detachments of the Corps of Marines, thus authorized, should be made in lieu of the respective quotas of Marines, which had been established or authorized for the frigates and other armed vessels and galleys which should be employed in the service of the United States. The President was authorized to detach and appoint such of the officers of this Marine Corps to act on board the frigates and any of the armed vessels of the United States respectively as he should from time to time judge necessary, anything in the act "providing a naval armament" to the contrary notwithstanding. The officers, non-commissioned officers, privates and musicians were to take the same oath and be governed by the same rules and articles of war as were prescribed for the military establishment of the United States and by the rules for the regulation of the Navy, according to the nature of the service in which they should be employed, and should be entitled to the same allowance, in case of wounds or disabilities, according to their respective ranks, as were granted by the "Act to ascertain and fix the military establishment of the United States." It was provided that "the non-commissioned officers, musicians, seamen, and marines, who are or shall be enlisted into the service of the United States, and the non-commissioned officers and musicians who are or shall be enlisted into the army of the United States, shall be and they are hereby exempted, during the term of service, from all personal arrests for any debt or contract." It was specially provided that the Marine Corps established by this act should, at any time, be liable to do duty in the forts and garrisons of the United States, on the seacoast, or any other duty on shore, as the President, at his discretion, should direct.

The following is a list of the officers of the Corps in 1798:

| | |
|---|---|
| MAJOR COMMANDANT. | Date. |
| William W. Burrows | 12 July, 1798. |
| CAPTAINS. | |
| Daniel Carmick | 3 August, 1798. |
| Lemuel Clark | 3 August, 1798. |
| George Memminger | 3 August, 1798. |
| Franklin Wharton | 3 August, 1798. |
| FIRST LIEUTENANTS. | |
| Richard Harwood | 10 July, 1798. |
| S. W. Geddis | 26 July, 1798. |
| John Hall | 2 August, 1798. |
| Philip Edwards | 3 August, 1798. |
| James McKnight | 3 August, 1798. |
| James James | 9 August, 1798. |
| William Cammack | 1 September, 1798. |
| Jonathan Church | 5 September, 1798. |
| Bartholomew Clinch | 5 September, 1798. |
| Diamond Colton | 5 September, 1798. |
| James Tallman | 5 September, 1798. |
| Reuben Lilly | 9 September, 1798. |
| H. A. Williams | 28 September, 1798. |
| David Stickney | 23 October, 1798. |
| Benjamin Strother | 31 October, 1798. |
| D. S. Wynkoop | 16 November, 1798. |
| SECOND LIEUTENANTS. | |
| William Amory | 25 July, 1798. |
| John L. Lewis | 9 August, 1798. |
| John Maine | 12 August, 1798. |
| Henry Caldwell | 2 September, 1798. |
| Anthony Gale | 2 September, 1798. |
| Robert Rankin | 2 September, 1798. |
| Josiah Reddick | 5 September, 1798. |
| John Darley | 8 November, 1798. |
| Newton Keene | 22 December, 1798. |

The following letter from Major W. W. Burrows, Commandant, illustrates the condition of the Corps in 1798:

MARINE CAMP, NEAR PHILADELPHIA,
September 9, 1798.

LIEUTENANT JOHN HALL, Charleston:

DEAR SIR,—I hasten to answer yours of the 17th ult., having already answered your last of the 27th. You may enlist as many men as you can and as many drummers and fifers as possible. I do not care what country the drummers and fifers are of, but you must be careful not to enlist more foreigners than as one to three natives. You can make use of blacks and mulattoes while you recruit, but you cannot enlist them. If you can get any smart lads you can have them taught and enter them as drummers and fifers, but we do not attend to the enlisting of them by companies, as they will never go, only by detachments.

Marines will be wanted very shortly at Charleston, and I hope you will be successful in raising your men. In your instructions you are ordered not to enlist any men under 5 feet 6 inches, but I have thought proper to do away that order, as it is not so essential to have marines of that size, so that if they are not too diminutive you may enlist of any size. I shall be very happy to hear of your success, and you must know, from the affection I bore your father, how dear you are to me.

Your ob't sv't, W. W. B.,
*M. C.*

No commissions are issued yet, but you will see your name in some of the Philadelphia papers of 3d or 4th September, 1798.

The first ship that ever got to sea since the present organization of the Navy was the Ganges, a purchased Indiaman. This vessel was armed and equipped as a twenty-four, and was put under the command of Captain Richard Dale, who was ordered to sail on the 22d of May. On the 12th of June it was determined to instruct "such of the cruisers as might hereafter go to sea," to capture all French cruisers that were hovering on the coast with hostile views on the American commerce. In obedience to these instructions the Constitution captured the French privateer schooner Le Croyable, while the United States and the Delaware early in the autumn captured the privateers Sans Pareil and Jaloux. First Lieutenant John Hall served on board the United States, and First Lieutenant Philip Edwards was attached to the Delaware.

1799.—On the 9th of February, the Constitution, Commodore Truxtun, captured the French frigate Insurgente after a close contest of an hour. The French vessel lost seventy men, killed and wounded, while the Constitution had only three men wounded. First Lieutenant Bartholomew Clinch of the corps was attached to the Constitution.

GENERAL ORDERS TO THE OFFICERS OF THE NAVY AND MARINES.

The President, with deep affliction, announces to the Navy and to the Marines, the death of our beloved fellow-citizen, George Washington, Commander of our armies, and late President of the United States, but rendered more illustrious by his eminent virtues, and a long series of most important services, than by the honors which his grateful country delighted to confer upon him.

Desirous that the Navy and Marines should express, in common with every other description of American citizens, the high sense which all feel of the loss our country has sustained in the death of this great and good man, the President directs that the vessels of the Navy, in our own and foreign ports, be put in mourning for one week, by wearing their colors half mast high; and that the officers of the Navy and of Marines wear crape on the left arm, below the elbow, for six months.

BEN. STODDART.

NAVY DEPARTMENT, 20th December, 1799.

1800.—In the close and hard-fought action between the Constitution and the French vessel La Vengeance on the 1st of February, the detachment of Marines suffered severely: out of fourteen men killed and twenty-five wounded, the detachment suffered a loss exceeding one-fourth. *This engagement lasted five hours within pistol-shot.* The Marines were commanded by Lieutenant Clinch.

Commodore Talbot, in command of the West India Squadron, ascertained that a valuable French letter of marque was lying in Port Platte, a small harbor on the Spanish side of the island of St. Domingo, he therefore determined to attempt cutting her out. An American sloop called the Sally was found on the coast, and the Commodore accordingly threw a party of seamen and Marines into the Sally, giving the command to the First Lieutenant of the Constitution, Mr. Hull, with Captain Daniel Carmick in command of the Marines. The Sally's movements were so well timed as to permit her to arrive off the harbor's mouth at the proper hour. The Sandwich was lying with her broadside bearing on the approach, and there was a battery close by to protect her. Lieutenant Hull in the Sally ran the Sandwich aboard, and carried her without the loss of a man. At the same moment, Captain Carmick landed with the Marines, entered the battery, and spiked the guns. In the actions between the Enterprise and the Flambeau, the Experiment and La Diane, the detachments of Marines shared the honors and losses on those occasions.

The short and irregular struggle with France ended by the treaty of peace which was ratified by the Senate on the 3d of February, 1801.

## CHAPTER VII.

**1801-1805.**—War with Tripoli—The Squadron under Commodore Dale—Capture of the Tripoli by the Enterprise—Commodore Preble assigned to the Command of the Squadron before Tripoli—Loss of the Philadelphia—Destruction of the Philadelphia by Decatur—Actions before Tripoli—The Expedition to Derne—Assault and Capture of the City.

As early as 1800, the Pacha of Tripoli, Jessuf Caramalli, who had deposed his brother Hamet, accused the American government of having bribed the subordinates of Tunis at a higher price than it had him. He therefore demanded of the President, in unequivocal language, a substantial tribute, and informed the American consul that he would wait six months for a present in money, and that if it did not arrive within that time, he would formally declare war against the United States.

No response having been made to this demand, the flag-staff of the American consulate was cut down on the 14th of May, 1801, and war was proclaimed.

A squadron was fitted out and rendezvoused in Hampton Roads, consisting of the President, Captain J. Barron; Philadelphia, Captain S. Barron; Essex, Captain Bainbridge; and Enterprise, Lieutenant-Commander Sterrett. Captain Dale was ordered to the command, who selected the President as his flag-ship. The following officers of the Corps were attached to the squadron: President, First Lieutenant John Heath; Philadelphia, Lieutenant William Osborne; Essex, Lieutenant Gamble; Enterprise, Lieutenant E. S. Lane. The squadron arrived at Gibralter on the 1st of July. On the 1st of August the Enterprise, while running for Malta, encountered the polacre-rigged ship Tripoli, of fourteen guns, and after an action of three hours, within pistol-shot, the Tripolitan surrendered, having suffered a loss of fifty killed and wounded.

**1803.**—The war was carried on, in a desultory way, with success during 1802, by the squadron under Commodore Morris, but it was not until 1803, when Commodore Morris was recalled, that the government realized the necessity of a formidable squadron. Commodore Preble was therefore assigned to the command of the squadron, consisting of the following ships: The Constitution, Philadelphia, Argus, Siren, Nautilus, Vixen, and Enterprise.

On the 31st of October, the Philadelphia, while near Tripoli, discovered a sail in-shore and to windward, standing for Tripoli. Sail was made to cut her off. The Philadelphia, however, unfortunately struck a reef about the distance of a league from the shore. The Tripolitan vessel doubled the edge of the shoal and reached Tripoli in safety. The firing had brought out nine gunboats, which now appeared turning to windward. Every effort was made to lighten the ship, but without avail, as the vessel remained perfectly immovable. After a consultation with his officers, Captain Bainbridge lowered his colors. Before this was done, however, the magazine was drowned, and holes were bored in the ship's bottom. Everything was taken from the officers and men, many of them being stripped half naked. At ten o'clock at night the prisoners were landed and conducted to the presence of the Pacha. The entire crew of the Philadelphia, including officers and Marines, numbered three hundred and fifteen. The ship was finally partially repaired, her guns remounted, and she was moored off the town near the castle of the Pacha.

1804.—The subsequent fate of the Philadelphia is familiar to all. Contemporaneous history relates in what manner, on the 16th of February, the gallant Decatur with a few chosen companions, in the ketch Intrepid, with the harbor filled with light cruisers, gunboats, and galleys, and surrounded by forts and batteries of the enemy, assaulted and burned the Philadelphia. The following Marines volunteered on this occasion: Sergeant Solomon Wren, Corporal Duncan Mansfield, Privates James Noble, John Quinn, Isaac Campfield, Reuben O'Brien, William Pepper, and J. Wolsfrandoff.

When the commander of the American forces had assembled his force before Tripoli, on the 25th of July, it consisted of the Constitution, Siren, Argus, Scourge, Vixen, Nautilus, Enterprise, two bomb-vessels, and six gunboats. The American force numbered one thousand seamen and Marines, exclusive of the officers, while the enemy numbered twenty-five thousand.

At half-past one, on the 3d of August, the bombardment of Tripoli began. In the desperate hand-to-hand fighting which occurred between our small gunboats and those of the enemy, one notable instance deserves to be mentioned. Decatur having boarded one of the enemy's gunboats, encountered the captain of the Tripolitan vessel. While the two leaders were engaged in a desperate struggle, an enemy raised his sword to cleave the skull of Mr. Decatur while he was occupied by his antagonist. A Marine of the Enterprise interposed an arm to save him. The blow was intercepted, but the limb was severed to a bit of skin.

Lieutenant Trippe, of the Vixen, in gunboat No. 6, boarded his enemy concealed by the smoke. The boats, however, separated by the shock of collision, leaving Lieutenant Trippe, Mr. J. D. Henley, and nine Marines on board the Tripolitan. Here, also, the commanders singled each other out, and a severe personal combat occurred while the work of death was going on around them. The Tripolitan officer, however, was killed, and the boat surrendered. While Mr. Trippe was hard pressed by his antagonist, a Turk aimed a blow at him from behind; but just before the latter struck, Sergeant Meredith, of the Marines, passed a bayonet through his body.

The blockade of Tripoli was maintained by different vessels during 1804-5.

It has been related that Jussuf Caramelli, the reigning pacha, was a usurper, having deposed his elder brother Hamet. The latter had escaped from the regency, and, after passing a wandering life, he had taken refuge among the Mamelukes of Egypt. It had often been suggested that the deposed prince might be made useful in carrying on the war against the usurper. Mr. Eaton, the consul at Tunis, who had been a captain in the army, came to America and prevailed upon the government to support his views. Commodore Barron, therefore, sent the Argus, with General Eaton, to Alexandria, where he arrived on the 26th of November. On the 29th General Eaton, accompanied by Lieutenant O'Bannon of the Marines, proceeded to Cairo. The Viceroy of Egypt granted permission for the Prince of Tripoli to pass out of the country unmolested.

**1805.**—The Marines in the expedition were from the Argus, and consisted of Lieutenant O'Bannon, one sergeant, and six privates, who were relied upon to preserve discipline, and by their example animate the hearts of the mercenaries employed. In a letter to Mr. Smith, Secretary of the Navy, dated at Alexandria, February 13, 1805, General Eaton wrote: "Those provinces in our possession will cut off from the country and turn into our own channel a source of provisions, and will open a free intercourse with the interior of the country. I have requested of the commodore for this purpose one hundred stand of arms, with cartridges and two field-pieces, with trains and ammunition; *and also a detachment of one hundred Marines, if necessary, to lead a coup de main.*"

From General Eaton's journal it appears that on Sunday, March 3, 1805, the force under his command left Alexandria on its march to Derne, taking the route across the Desert of Barca. Included in the force were but nine Americans, Lieutenant O'Bannon, Mr. Peck, one sergeant, and six privates of the Marine Corps. Including the footmen and camel-drivers, the whole force numbered about four hundred. This caravan consisted of one hundred and seven camels and a few asses. After marching two hundred miles, eighty mounted warriers joined the pacha. Provisions had been reduced to hard bread and rice. From Alexandria to that point there was not a living stream or rivulet or spring of water. A few days later, forty-seven tents of Arabs joined them, with their families and movables. In this detachment were one hundred and fifty warriors on foot. On March 30, General Eaton wrote: "From Alexandria to this place, we have experienced continual altercations, contentions, and delays among the Arabs. They have no sense of patriotism, truth, or honor; and no attachment where they have no prospect of gain, except to their religion, to which they are enthusiasts. Poverty makes them thieves, and practice renders them adroit in stealing."

April 9 he wrote: "Advanced ten miles. Good water. In the cistern were two dead men; probably murdered by Arabs. Obliged to drink the water, however." On the following day a courier arrived from Derne. April 14, at 4 o'clock P.M., they reached Bonda. But their astonishment was great to find at this port "not the foot-trace of a human being, nor a drop of water." The next morning the Argus, Captain Hull, arrived, and on the 17th the sloop Hornet arrived with provisions. At this point the remainder of the detachment of Marines now landed and joined Lieutenant O'Bannon. On the 18th the march was resumed. On the 24th they marched fifteen miles over mountainous and broken ground, covered with herbage and beautiful red cedars, "the first resemblance of a forest tree," wrote General Eaton, "we have seen during a march of nearly six hundred miles."

Arriving before Derne, on the morning of the 26th, terms of amity were offered the Bey, on condition of allegiance and fidelity. The flag of truce was sent back with this laconic answer, "My head or yours." On the 27th the assault, so novel for Americans to be engaged in, in the other hemisphere, was commenced. The Hornet, Lieutenant Commandant Evans, having run close in, and anchored with springs on her cables, within pistol-shot of a battery of eight guns, opened fire. The Nautilus lay at a little distance to the eastward, and the Argus still farther in the same direction, the two latter firing on the town and battery. In about an hour the enemy were driven from the work, when all the vessels directed their guns at the beach, to clear the way for the advance of the party on shore. The enemy made an irregular but spirited defense, keeping up a heavy fire of musketry, as the assailants advanced, from behind houses and walls. At half-past three, however, Lieutenant O'Bannon and Mr. Mann, midshipman, stormed the principal work, hauling down the Tripolitan ensign, and, *for the first time in the history of the country, hoisting that of the republic on a fortress of the Old World.* The enemy were driven out of the work with so much precipitation that he left his guns loaded and even primed. The guns were immediately turned upon the town, and Hamet Caramelli, having made a lodgement on the other side, so as to bring the enemy between two fires, the place submitted.

In this affair only fourteen assailants were killed and wounded, General Eaton being among the latter. A detachment, consisting of the Marines from Argus, twenty-four cannoniers, and twenty-six Greeks, including their proper officers, were under the immediate command of Lieutenant O'Bannon. It was with this force that the brave O'Bannon passed through a shower of musketry from the walls of the houses, took possession of the battery, planted the American flag upon its ramparts, and turned its guns upon the enemy. The attack was made by about twelve hundred men, while the place was supposed to be defended by three or four thousand. In his official report of the affair, General Eaton said, "The details I have given of Mr. O'Bannon's conduct need no encomium, and it is believed the disposition of our government to encourage merit will be extended to this intrepid, judicious, and enterprising officer." Lieutenant O'Bannon resigned two years afterwards, without having ever received a brevet for his gallant conduct.

A treaty of peace was signed on the 3d of June, 1805, and the war with Tripoli terminated after an existence of four years.

The following officers of the Corps were attached to the squadron, in the several attacks on the city and harbor of Tripoli in July, August, and September, 1804:

Captain John Hall, Lieutenant Robert Greenleaf, Lieutenant John Johnson. The casualties before Tripoli were: killed, Sergeant John Meredith, Private Nathaniel Holmes.

---

## CHAPTER VIII.

**1809-13.**—Increase of the Corps—The War of 1812—President and Belvidere—The Constitution and Guerrière—United States and Macedonian—Wasp and Frolic—Constitution and Java—Hornet and Peacock—The Essex—Lieutenant Gamble—Attack on the Batteries of Craney Island—Actions on the Lakes.

**1809.**—Early in 1809, the Corps was augmented by an addition of nearly seven hundred men, which probably put this important branch of the Navy on a footing equal to the rest of the service, as it then existed; the entire Corps containing about thirteen hundred men when full.

**1812.**—Owing to grievances that had long been drawing to a head, Congress formally declared war against Great Britain on the 18th of June, 1812.

On the 23d of June, in the engagement between the frigate President and the British frigate Belvidere, Lieutenant Heath of the Marines was severely wounded.

On the 19th of August, at 2 P. M., the Constitution, Captain Hull, encountered the British frigate Guerrière. After a desperate action of two hours, during which time both vessels were within pistol-shot, great damage was done on both sides. Finally the vessels touched and both parties prepared to board. Mr. Aylwyn, the master, and Mr. Bush, the Lieutenant of Marines, were upon the taffrail of the Constitution, to be ready to spring. Mr. Morris, the first lieutenant, was shot through the body, Mr. Aylwyn wounded in the shoulder, and Mr. Bush fell by a bullet through the head. At 7 P. M. the Guerrière surrendered. At 3 P. M. on the next day the Guerrière blew up. The British ship had seventy-nine killed and wounded, while the loss of the Constitution amounted to seven killed and seven wounded, among which was First Lieutenant Bush killed, and Private Francis Mullen wounded. After the fall of Lieutenant Bush, Lieutenant Contee became the commanding officer of the detachment, and was commended for his bravery. Captain Hull in his report to the Secretary of the Navy, says of Lieutenant Bush: *"In him our country has lost a valuble and brave officer.*

This action was the first serious conflict of the war, and was characterized by features which, though novel at the time, became identified with nearly all the subsequent engagements of the contest, showing that they were intimately connected with the discipline and system of the Navy.

In the assault and capture of the British vessels Detroit and Caledonia, on the 8th of October, on Lake Ontario, by Lieutenant Elliott of the Navy, the following officers commanded the Marines: First Lieutenant William H. Truman, Second Lieutenant John Contee.

In the action between the United States and the Macedonian, on October 25, in latitude 29° north, longitude 29° 30′ west, which resulted in the surrender of the latter, the Marines of the United States were commended for good conduct and utmost steadiness under the example of their gallant commanders, Lieutenants William Anderson and J. L. Edwards, the weight of the enemy's fire passing a short distance above their heads. The Marines' loss in this action were: killed, Privates Michael O'Donnell and John Roberts; wounded, mortally, Private John Laton.

The action between the Wasp and the British sloop Frolic resulted in the capture of the latter after a combat of forty-three minutes. The loss on board the Frolic amounted to nearly one hundred, while the American vessel had five men killed and five wounded, two of the latter being Marines.

On the 26th of December the Constitution, Commodore Bainbridge, encountered the British frigate Java, and after a combat which lasted two hours the Java surrendered. The Java was literally picked to pieces by shot, spar following spar, until she had not one left. Her hull was also greatly injured, and her loss in men was twenty-two killed and one hundred and two wounded. Of the Constitution's crew nine were killed, and twenty-five wounded, the Marines, as usual, suffering greatly. Captain Archibald Henderson, Lieutenants W. H. Truman and John Contee, were the officers of the corps attached to the Constitution.

On the 24th of February the Hornet, Captain Lawrence, engaged the Peacock, and in thirty minutes from the time the first shot was fired the British vessel surrendered. In this short encounter, which resulted in the sinking of the Peacock, this vessel had her captain and four men killed, and thirty-three wounded. The detachment of Marines on board the Hornet was commanded by Lieutenant Brownlow.

**1813.**—The Essex, Captain Porter, reached the Marquesas at the close of October, and in the early part of December she was again ready for sea. Shortly after, a fort was constructed on a small conical hill, near the water, where the Seringapatam, Sir Andrew Hammond, and Greenwich were warped close in and moored under its guns The command of this fort was given to Lieutenant John M. Gamble, of the Marines, a spirited and intelligent young officer, who also had command of the three prizes above named. The Essex had no sooner disappeared from Nooaheevah than the savages began to pilfer, and to betray a turbulent disposition. Lieutenant Gamble was compelled to land a detachment, in order to bring the natives to terms by a show of force. Fortunately this object was effected without firing a musket.

On the 12th of April, Mr. Gamble began to rig the Seringapatam and the Sir Andrew Hammond, with the intention of quitting the islands, the long absence of the Essex inducing him to despair of her return. Symptoms of a mutiny showed themselves, and he had all the arms and ammunition carried on board the Greenwich, in which vessel he lived. The mutineers, however, took the Seringapatam on the 7th of May. Mr. Gamble was badly wounded in the foot by a pistol ball during the fight. He made every exertion to get to sea with the Sir Andrew Hammond, but on the 9th the natives made an attack, and one officer and three men were killed, and one other was severely wounded. The situation of those remaining became exceedingly critical, the whole party consisting of only eight individuals, of whom two were badly wounded, one was crippled, and another was just recovering from a serious attack of scurvy. In fact, there were but four men on board the Sir Andrew Hammond fit for duty. With this small force the vessel was put to sea, and, without a chart, Lieutenant Gamble made his way to the Sandwich Islands, which he reached in seventeen days, only, however, to fall into the hands of the enemy. He and his crew were carried on board the Cherub, where they learned of the fate of the Essex. He was carried to Rio Janeiro, whence, in the following August, he was allowed to return to New York. His services won the highest praises of his superiors.

Subsequently, Commodore Porter wrote as follows to Lieutenant Gamble: "Allow me to return to you my thanks for your handsome conduct in bringing the Seringapatam to action, which greatly facilitated her capture, while it prevented the possibility of her escape. Be assured, sir, I shall make a suitable representation of the affair to the Hon. Secretary of the Navy."

Accordingly, Commodore Porter sent the following communication to the department: "Captain Gamble at all times greatly distinguished himself by his activity in every enterprise engaged in by the force under my command, and in many critical encounters by the natives of Madison's Island rendered essential services, and at all times distinguished himself by his coolness and bravery. I therefore do, with pleasure, recommend him to the department as an officer deserving of its patronage. I now avail myself of the opportunity of assuring you that no Marine officer in the service ever had such strong claims as Captain Gamble, and that none have been placed in such conspicuous and critical positions, and that none could have extricated themselves from them more to their honor."

In the action between the British frigate Shannon and the American frigate Chesapeake, which resulted in the capture of the latter vessel, Lieutenant Broom was mortally wounded. Of his detachment eleven were killed and twenty wounded.

On the 14th of June a large force of the enemy, consisting of fourteen sail, came into Hampton Roads, and an attack was expected. On the 20th the enemy's ships weighed, and ascended the tide to the mouth of the James River, where, in the afternoon, they were seen making preparations to send up a large force in boats. As so much depended on the defence of the batteries of Craney Island, Captain Cassin, who commanded the naval force at Norfolk, sent three of the lieutenants of the Constitution, Messrs. Neall, Shubrick, and Sanders, on shore with one hundred seamen and fifty Marines, under the command of Lieutenant Breckinridge. The enemy, on the 22d, attempted to land in front of Craney Island with a large force in barges. Fire was opened on the enemy with great coolness and precision, which resulted in the sinking of three of the barges. Forty prisoners were taken, and the loss in killed and wounded must have been great, as Captain Cassin described the fire from the battery to have been precise. The officers, seamen, and Marines of the Constitution gained great credit for their steadiness, discipline, and spirit.

On the 4th of September, in the action between the Enterprise and the Boxer, five Marines of the former were wounded.

In the attack on Fort George the Marines of the squadron were embodied with the regiment of Colonel Macomb. In this handsome affair the commanding general and the commanding naval officer spoke in the highest terms of the conduct of the seamen and Marines. This occupation of Fort George brought with it an evacuation by the British of the whole Niagara frontier. In the battle of Lake Erie the following officers and non-commissioned officers served in the vessels named:

*Lawrence.*—Lieutenant John Brooks, Sergeant James Tull, Sergeant W. S. Johnson, Corporal Philip Starpley.

*Caledonia.*—Sergeant Joseph Beckley, Sergeant James Artis.

*Scorpion.*—Corporal Joseph Berry.

*Ariel.*—Lieutenant Robert Anderson.

*Trippe.*—Lieutenant James Blair, Corporal John Brown.

*Tigress.*—Corporal William Webster.

*Niagara.*—Captain H. B. Brevoort, Lieutenant John Heddleson, Lieutenant John R. Edwards, Sergeant Jonathan Curtis, Sergent Sanford A. Mason.

The following are the casualties in the squadron:

*Lawrence.*—Killed: Lieutenant John Brooks, Corporal Philip Starpley, Privates Jesse Harland and Abner Williams. Wounded: Privates James Bird, William Burnet, William Baggs, David Christie, Henry Vanpool.

*Niagara.*—Wounded: Lieutenant J. R. Edwards, Sergeant Mason, Corporal Scott, Privates Joshua Traphill, Thomas Miller, John Rumas, George McManomy, George Schofield, Samuel Cochran.

In the action on Lake Champlain with the British fleet on the 11th of September, the following casualties occurred in the Corps:

*Saratoga.*—Killed: Private James Carlisle. Wounded: Privates Benjamin Jackson, Jesse Vanhorn, Joseph Ketter, and Samuel Pearson.

*Eagle.*—Killed: Privates John Wallace, Joseph Heaton, and Robert Stratton. Wounded: Privates John McKinney, Matthew Scriver, George Mannering, Henry Jones, and John McCarty.

*Ticonderoga.*—Killed: Privates Deodrick Think and John Sharp. Wounded: Private John Condon.

*Borer.*—Killed: Private James Day. Wounded: Corporal Ebenezer Cobb.

During the operations on the lakes, the Army co-operating with the Navy, a detachment of Marines under Captain Samuel E. Watson served with great efficiency with the troops under Colonel Scott.

A careful *resume* of the service on the lakes may be found in Cooper's "Naval History," who states that "in this inland contest, while the enemy had been active, bold, and full of resources, impartial judges must award the palm to the Americans."

---

## CHAPTER IX.

**1814-15.**—The War of 1812 Continued—The Chesapeake Flotilla under Captain Barney—The Loss of the Adams in the Penobscot—The Advance upon Washington—The Battle of Bladensburg—Capture of the City of Washington—Unsuccessful Attack by the British on Baltimore—Battle of New Orleans—Capture of the Cyane and Levant by the Constitution—The President and Endymion—Hornet and Penguin.

THE government having equipped a large flotilla to protect the waters of the Chesapeake, the command was given to Captain Joshua Barney, the officer who had so much distinguished himself by the capture of the Monk during the War of the Revolution. On the 1st of June this officer left the Patuxent with the Scorpion, two gunboats, and several large barges, in chase of two schooners. He was closing fast by means of sweeps, when a large two-decker was discovered to the southward, necessitating the retreat of our flotilla up the river to the mouth of St. Leonard's Creek. Captain Barney repulsed three distinct attacks of the enemy on the 8th, 9th and 11th of June, the last repulse being very severe, the English suffering materially. Some small works were now thrown up on the shore to protect the American flotilla, and the blockade by the British continued. In the meantime, Captain Miller, of the Marine Corps, joined the flotilla with a strong detachment of Marines, and a considerable force of militia was collected under Colonel Wadsworth. The enemy was also reinforced. On the 26th an attempt was made by the united force of the Americans to raise the blockade. The cannonade lasted two hours. In this affair the flotilla lost thirteen men in killed and wounded; but it effectually raised the blockade and induced the enemy to be more cautious.

On the 17th of August, in very thick weather, the Adams ran ashore on the Isle of Haute, but was got off by lightening. Captain Morris succeeded in getting her into the Penobscot, in Maine, as high up as Hampden, which is several miles above Castine.

While the Adams lay ready to be hove out, with nothing in her, a strong expedition of the enemy, consisting of troops and vessels of war, entered the river and ascended as high as Hampden. A small force of militia was assembled, and a battery was mounted with the guns of the ship in order to protect her; but the irregular troops giving way and leaving the seamen and Marines exposed in the rear, nothing remained but to set the vessel on fire and to make a retreat. After a fatiguing march of two hundred miles the Marines under Captain Samuel E. Watson reached Portland, where they were ordered to report. It is a fact worthy of record that, although the force of the ship was broken up into small detachments, with orders to make the best of their way to Portland, there were no stragglers, and every squad or detachment reported intact.

On the 16th of August the enemy in the Chesapeake commenced that series of movements, which terminated in his advance upon Washington. Captain Barney having received intelligence that the British were coming up the Patuxent in force, sent an express to the Navy Department for instructions. The answer was to land the men and join the Army that was hurriedly assembling for defence of the coast under General Winder, and, if pressed, to burn the flotilla.

On the 21st news was received that the enemy had landed a force of four or five thousand men at Benedict, and that he was marching in the direction of the Capitol. Captain Barney immediately landed four hundred men, leaving the vessels in charge of Lieutenant Frazier, with orders to set fire to them if attacked, and to join the main body with as little delay as possible. The next day this order was executed, a strong detachment of the enemy's seamen and Marines approaching the flotilla to attack it.

On the 22d, Captain Barney joined an assemblage of armed citizens that was called an army at the wood-yard. The next day he marched into Washington and took up his quarters at the Marine Barracks.

After a good deal of uncertainty concerning the movements of the enemy, it was ascertained he was marching directly on Washington, and that it was intended to fight him at Bladensburg. The Marines and flotilla-men left the yard on the morning of the 24th; they arrived on the battle-ground, and were immediately formed about a mile to the west of Bladensburg, holding the centre of General Winder's position. The actual force of General Winder amounted to three thousand, of whom fourteen hundred were regulars, including the Marines and flotilla-men. The force of the enemy numbered seven thousand, composed of the *elite* of his forces in America, as follows: Eighty-fifth Foot, Fourth and Forty-fourth Regiments, Twenty-first Fusileers, two battalions of Royal Marines, two hundred seamen, and a detachment of guns and rockets with a detachment of marine artillery. At 11.30 A. M., after a short skirmish, when the enemy suffered severely in crossing a bridge, the militia fell back, and the British columns appeared, following the line of the public road. Captain Barney took command of the artillery in person, while Captain Miller, of the Marines, was assigned to the command of the remainder of the two detachments, who were armed as infantry. The Marines now formed in line on the right of the artillery, while the seamen were drawn up a little in the rear and on the right flank of the Marines, that permitted them to fire over the heads of the latter. Although the troops that were falling back did not halt, Captain Barney held his position; and as soon as the enemy began to throw rockets, he opened upon him with a heavy and sharp discharge of round shot and grape. The column of the enemy was staggered, and it immediately gave ground. A second attempt to advance was repulsed in the same manner, when the enemy, who, as yet, had been able to overcome resistance by advancing steadily in column, was obliged to make an oblique movement to his left into some open fields, and to deploy forward. Here he threw out a brigade of light troops, consisting of the Eighty-fifth Regiment, the light companies of the Fourth and Forty-fourth, and the First Battalion of Marines, in open order, and advanced in beautiful style upon the command of Captain Barney, five hundred strong, while the head of a strong column was seen in reserve in a copse in its rear. Captain Miller, with the Marines and that portion of the seamen who acted as infantry, met the charge in the most steady and gallant manner, and after a short conflict drove the British light troops back upon their supporting column. The Marines and seamen manifested the utmost steadiness, and there can be no question, had they beeen properly supported, the Americans would have gained the day. No troops, however, remained in line except those under Captain Barney and Captain Miller. Having been so roughly handled, the enemy made no new attempt to advance directly in front of the Marines and seamen, but, after forcing the troops on their right from the field by a demonstration in that direction, they prepared to turn the rear of Captain Barney. While these movements were going on in front and on the right of the able service, which was gratefuly acknowledged by the commanding general in his official report to the Secretary of War. The following officers of the Corps were engaged: Captain Anthony Gale, Lieutenant H. B. Breckinridge. This failure virtually terminated the warfare in the Chesapeake, the enemy shortly after collecting most of his forces at the south, with a view to make a still more serious attempt at New Orleans.

The command of the naval force at New Orleans had been given to Captain Patterson, one of the young officers who had been a prisoner at Tripoli with Captain Bainbridge.

After a series of successes, the enemy finding himself in command of Lake Borgne, sent up a brigade of troops, under Major-General Keane, which succeeded in entering a bayou, and in landing but a few miles below New Orleans. Here he encamped, after advancing to some hard ground, on the night of the 23d of December, with his left flank resting on the Mississippi. No sooner was the position of the British known to the Americans, than General Jackson marched against them, with all the disposable force he could assemble, making a total of about fifteen hundred men, including a detachment of Marines under Major Carmick, and by a prompt and spirited night attack he saved New Orleans. The movements of the troops on this occasion were preceded by Captain Patterson, who, dropping down abreast of the English bivouac, in the schooner Carolina, opened a most galling fire. Eight Marines were killed and eight wounded on this occasion.

**1815.**—On the 8th of January the whole force of the British Army, under Major-General Sir Edward Packenham, was formed for a general attack on the American works. The enemy attacked at daylight, and were defeated with dreadful slaughter. Their loss was Major-Generals Packenham and Gibbs, two thousand men killed and wounded, and nearly one thousand taken prisoners. On the night of the 18th the English retreated to their boats and embarked, abandoning their attempt altogether. The Marine Corps had its share in this glorious campaign, and it acted with its usual good conduct under Major Carmick, who was wounded.

In his report of the battle of New Orleans, dated 27th January, 1815, Commodore Patterson says, "To Major Daniel Carmick, commanding the Marines of this station, I am indebted for the promptness with which my requisitions on him have been complied with, and the strong desire he has always maintained to further, as far as was in his power, my views."

In the glorious action between the Constitution, Captain Stewart, and the British ships Cyane and Levant, on the 20th of February, which resulted in the capture of the enemy's vessels, the Marines were commanded by Captain Archibald Henderson, who was specially mentioned in the official report of that action.

On the morning of the 15th of January, the President, Commodore Decatur, while near Long Island, fell in with the enemy's ships Majestic, Endymion, Pomona, and Tenedos. All sail was made by the British squadron in chase, which continued throughout the forenoon, the wind becoming lighter and baffling. The Endymion being the nearest ship, closed, and a spirited engagement began between the two vessels which lasted two hours and a half, the greater part of the time in a parallel line of sailing. Both ships kept up a smart fire of musketry by their Marines in the tops. The enemy's vessel was now so far injured that she fell astern, most of her sails having been cut from the yards. At 11 P. M. the Pomona got on the weather bow of the American ship, and poured in a broadside; and as the Tenedos was fast closing on the quarter, and the Majestic was within gun-shot astern, further resistance was useless and the President surrendered. In this long and close cannonade, the President lost twenty-four men killed, and fifty-six wounded. The Endymion had eleven killed and fourteen wounded. Lieutenant Levi Twiggs commanded the Marines of the President, and Commodore Decatur, in his official report says, "Lieutenant Twiggs displayed great zeal, his men were well supplied, and their fire was incomparable."

On the 23d of March the Hornet, Captain Biddle, encountered the British brig Penguin, eighteen guns, and after a hot engagement of twenty-two minutes, the enemy surrendered. The English loss was fourteen killed and twenty-eight wounded, while that of the Hornet was one killed and ten wounded. First Lieutenant W. L. Brownlow was in command of the detachment of Marines. Among the incidents of the fight, it is related that Private Michael Smith, who had served under the gallant Porter, in the Essex when she was captured by the British, received a shot through the upper part of the thigh, which fractured the bone, and nearly at the same moment had the same leg broken immediately above the knee by the spanker-boom of the Hornet, which was carried away by the enemy's bowsprit while afoul of her. In this situation, while bleeding upon the deck, and unable to rise, he was seen to make frequent exertions to discharge his musket at the enemy on the topgallant forecastle of the Penguin. This, however, the gallant fellow was unable to accomplish, and was compelled to submit to be carried below. The combat between the Hornet and the Penguin was the last regular action of the war.

## CHAPTER X.

### 1824.—The Mutiny in the Massachusetts State Prison.

THE account of the mutiny in the Massachusetts State Prison, in 1824, has been familiar to the school-boys of the last fifty years. It was given in authentic form in the New England *Galaxy* in 1828, and has since formed one of the standard selections for school readers. The following account is taken from the original story, omitting such portions as do not have bearing on the action of the Marines who were called in to suppress the outbreak: "Three convicts had been sentenced, under the rules of the prison, to be publicly whipped in the yard, and were confined in the solitary cells, waiting punishment. An officer, whose duty it was to attend to such cases, entered one of the cells to see that everything was right, when the inmate sprang past him, closed the door, and locked him in. He then opened the doors of the cells in which were the other two, and the three, after deliberation, released the officer, and ordered him to approach the door leading into the guard-room, and give the necessary signal that all was right, while they stood ready, on its being opened, to rush through and secure the guard and the arms in the room. The officer resolutely refused. They threatened to kill him; and one of them raised a file, sharpened to a point, with the intent of carrying the threat into execution; but the officer remained firm, telling them that they might take his life, but he would not betray his trust. The other two finally interfered and saved his life, forcing him back into the cell, and locking him in. They then passed into the large dining-hall, a long, dark, and damp room. The alarm had already been given, and the prisoners rushed from the workshops, arming themselves with clubs, knives, hammers, chisels, and every variety of weapon within their reach, forming a band whose strength, vileness, and reckless daring could hardly be equalled. Men of all ages and characters, guilty of every variety of infamous crimes, dressed in the motley and peculiar garb of the institution at that period, and displaying the wild and demoniac appearance that always pertains to imprisoned wretches, were gathered together for the single purpose of preventing the punishment which was to be inflicted on the morrow upon their comrades.

"The acting warden and some other officers of the prison were there at the time, and were, naturally, greatly alarmed at the consequences likely to ensue from the conflict necessary to restore order. They huddled together, but could scarcely be said to consult, as the stoutest of them lost all presence of mind in overwhelming fear. The news spread rapidly through the town, and a subordinate officer, of mild and kind disposition, hurried to the scene, and, calm and collected, went into the midst of the officers. The most equably-tempered and the mildest man in the government, as is usually the case, was, in this hour of peril, the firmest. He instantly, upon his own responsibility, despatched a request to Major Wainwright, commander of the Marines stationed at the Navy-Yard, for assistance, and declared his purpose to enter the hall, and try the force of firm demeanor and persuasion upon the enraged multitude. The other officers exclaimed against an attempt so full of hazard, but in vain. They offered him arms,—a sword and pistol; but he refused them, and said that he had no fear, and in case of danger, arms would be of no service; and alone, with only a little rattan, which was his usual walking-stick, he advanced into the hall to hold parley with the enraged and desperate villains.

"He demanded their purpose in thus coming together with arms, in violation of the prison laws. They replied that they were determined to obtain the remission of the punishment of their three comrades. He said that was impossible; the rules of the prison must be enforced, and they must submit. At the hint of submission, they drew a little nearer together, prepared their weapons for service, and as they were dimly seen in the farther end of the hall by those who observed them from the grating that opened to the day, it was difficult to conceive of a more appalling sight, or one of more moral grandeur, than that of a single man standing within their grasp, and exposed to instant death, if a word or look should add to the already intense excitement.

"The excitement, too, was of the most dangerous kind; it did not break forth in noise and imprecation, but was seen only in the dark looks and the strained nerves, that showed a deep determination. They stated that they expected some would be killed, but that death would be better than such imprisonment; and with look and tone which evinced an indomitable purpose, they declared that not a man should leave the hall alive till the sentence of flogging was remitted. At this period of the discussion their evil passions seemed to be more inflamed, and one or two proposed to kill the officer, who still stood firm, and with more temperate pulse than did his friends, who saw from above but could not avert the danger that threatened him. Just at this moment the officer saw the feet of the Marines filing by the small lights. He knew that it was his only time of escape before a conflict for life should begin. He stepped backward, still urging the prisoners to depart before the officers were driven to the use of fire-arms as a last resort. When within three or four feet of the door it was opened, and closed instantly again as he sprang through, and so was unexpectedly restored to his friends.

"Major Wainwright was requested to order his men to fire down upon the convicts through the little windows, first with powder and then with ball, till they were willing to retreat; but he took a wiser as well as a bolder course. Relying upon the effect which firm determination would have upon men so critically situated, he ordered the door to be again thrown open, and marched in at the head of thirty men, who filed through the passage and formed at the end of the hall, opposite the crowd of criminals grouped together at the other end. He stated that he was empowered to quell the rebellion; that he should not quit that hall alive till every convict had returned to his duty. The latter seemed balancing the strength of the two parties, and replied that some of them were ready to die, and only waited for the attack to see which was the more powerful, swearing that they would fight to the last unless the sentence of flogging was remitted, for they would not submit to any such punishment in the prison.

"Major Wainwright now ordered his Marines to load their pieces, and that they might not be suspected of trifling, each man was told to hold up to view the bullet which he afterwards put into his gun. This only caused a growl of determination, and no one blanched, or seemed disposed to shirk from the foremost exposure. They knew that their numbers would enable them to bear down and destroy the handful of Marines, after the first discharge.

"The Marines were ordered to take aim; their guns were presented; but not a prisoner stirred, except to grasp more firmly his weapon. Still desirous, if possible, to avoid such a slaughter as must follow the discharge of the guns, the major advanced a step or two, and spoke even more firmly than before, urging them to depart. Again, and while looking directly into the muzzles of the guns, which they had seen loaded with ball, they declared their intention of fighting it out. The intrepid officer then took out his watch and told his men to hold their pieces aimed at the prisoners, but not to fire till they had orders. Then, turning to the convicts, he said, 'You must leave this hall. I give you three minutes to decide. If at the end of that time a man remains, he shall be shot dead. I speak no more.' No more tragic situation than this can be conceived; at one end of the hall a fearless multitude of desperate and powerful men waiting for the assault; at the other a little band of well-disciplined Marines, waiting with levelled muskets, and ready on the least motion or sign to begin the carnage, and their tall commander holding up his watch to count the

Major-General CHARLES HEYWOOD,

Commandant U S. M. C. Plate 1

lapse of the three allotted minutes. For two minutes not a person nor a muscle was moved; not a sound was heard in the unwonted stillness of the prison, excepted the labored breathings of the infuriated wretches as they began to pant between fear and revenge. At the expiration of two minutes, during which they had faced the ministers of death with unfaltering eyes, two or three of those in the rear, and nearest to the further entrance, went slowly out; a few more followed the example, dropping out quietly and deliberately; and before half the last minute was gone, every man was struck by the panic, and crowded for exit, and the hall was cleared as if by magic. Thus the steady firmness of moral force, and the strong effect of deliberate determination, cowed the most daring men, and prevented a scene of carnage which would have instantly followed the least precipitancy or exertion of physical force by the officers or their subordinates."

---

## CHAPTER XI.

**1824-35.**—Operations against Pirates—Quallah Battoo—Services in New York.

"THE series of revolutions which wrested from Spain her colonial possessions in Mexico and South America caused serious interference with the commerce of the United States. Vessels were plundered, then sent adrift, burnt, or taken possession of, as fancy or interest dictated; the officers, crews, and passengers were always treated with indignity and violence, very frequenly murdered in cold blood, and, in some instances, their bodies after death were abused with disgusting barbarity, after cruelty had exhausted ingenuity in contrivances of insupportable torture." In the fall of 1821 the government took measures to capture and bring to punishment the offenders.

Although the Marines attached to Commodore Porter's squadron participated in all of the expeditions against pirates off Porto Rico, San Domingo, and Cuba, yet as these resulted in the easy capture and destruction of their strongholds, but slight mention is made of their services during this period.

The next hostile occupation of any of the national vessels was in the expedition against the pirates of Quallah Battoo, undertaken by Commodore John Downes in the frigate Potomac in the year 1832. The cause of this expedition originated during the year 1831, in the capture of the American ship Friendship, which was lying in the harbor of Quallah Battoo, taking a load of pepper. The capture was unprovoked by any acts on the part of the Americans, and was accomplished in a treacherous manner, and by the murder of the mate and two seamen of the ship. By the assistance of other American merchantmen, the captain ultimately recovered his vessel, but she was completely rifled of all valuables, and the object of the voyage was broken up. Commodore Downes was instructed to obtain redress for this special and aggravated wrong. Quallah Battoo is situated on the western shore of the island of Sumatra. It lies entirely open to the sea. The navigation in the vicinity is exceedingly difficult and dangerous; reefs and shoals lining the shore. At the time of the expedition the town, including the neighboring plantations, is said to have had a population of four thousand. It was defended by several small forts, mounted with three or four guns each.

On the 5th of February, 1812, the Potomac arrived on the coast of Sumatra. Commodore Downes disguised the frigate as a merchantman, and then stood in and anchored about three miles from the town. Being convinced that any mere verbal demand for satisfaction would be entirely useless, and only give time for the Malays to prepare better for their defence, he determined to make an attack as soon as possible, in hopes of taking them by surprise, and securing the persons of some of the rajahs, holding whom as hostages he might gain higher terms of satisfaction. Preparations to land the same night were therefore made. Two hundred and fifty seamen and Marines were detailed for the duty, and the whole placed under the command of Lieutenant Irwin Shubrick, the first lieutenant of the ship. The Marines were commanded by Lieutenant Alvin Edson and Lieutenant George H. Terrett. At a sufficient time before daylight, to be unobserved, the boats were hoisted out, and the men stowed away in them, when they quietly and swiftly pulled towards shore. They went in through a heavy surf, but were all safely landed, and in fifteen minutes were formed in their line of march. Lieutenant Edson led the van with the Marines. The first division of seamen was commanded by Lieutenant Pinkham, the second by Lieutenant Hoff, and the third by Lieutenant Ingersoll. A 6-pounder was in charge of Sailing-Master Totten.

When the day dawned the march commenced along the beach. The Malays were anticipating an attack, for their scouts were seen. Lieutenant Hoff was ordered to surround the first fort, which was the most northern one. He left the main body with his division, and soon came to the outworks, consisting of a strong stockade of timber. Within this there was still a citadel, armed with small cannon. As soon as the division approached, the Malays commenced the action by a sharp fire from all arms. They fought with great bravery and steadiness, and were found to be no mean adversaries. Lieutenant Hoff succeeded in breaking through the gate-way of the stockade, but the Malays still stood the attack for two hours within their citadel. Finding it impossible to dislodge them by distant fire, preparations were made to take the citadel by storm. It was necessary to tear up some of the palisades and to form a bridge of them, over which the eager seamen rushed, carrying the work which had so long held them in check, by assault, which the Malays could not withstand.

Lieutenant Edson, with the Marines, attacked the next fort, situated in the rear of the town. There the scenes were repeated, the Malays behaving with the same spirit, and made a long and obstinate resistance. But they could not face the Marines, as with steady discipline they stormed the fort and forced their way into it. Lieutenant Shubrick attacked the principal fort with the first and third divisions, aided by the 6-pounder situated at the southern point of the town, near the beach. Here the fight was long and gallantly maintained on both sides. The gate of the outer defences was soon forced, and the men rushed in, anxious to come to closer quarters with their foes. The stronghold, consisting of an elevated platform mounted with several cannon, still held out, and our men had to stand the fire from it without protection. The ladder to the platform had been removed, and in the attempt to climb up to it one man was killed and several were wounded. At this critical moment, Lieutenant Hoff, with the second division, and Lieutenant Edson, with the Marines, having secured their victories, came up with detachments, and took a position between the fort and the water, from which they poured in a very effective cross-fire. The battle now raged against the devoted Malays with exterminating severity. Although thus closely invested, and the numbers falling under so well-conducted a fire, they yet fought with desperation, and it was not till almost all were killed that their position could be carried.

The town was now fired and much of it was reduced to ashes. The last fort was now revealed by opening its fire. A detachment of Marines and seamen was ordered to capture it. The Malays, as usual, fought fiercely. But success had animated the Americans with even unwonted zeal and impetuosity, and after a short conflict they carried this work by assault.

The action had continued for two hours and a half, and with success on our part at every point attacked. This success was purchased at some cost, though it was comparatively small. Two men were killed, two officers and nine men were wounded. The loss of the enemy must have been very severe, for it was known that at least one hundred and fifty were killed. Among those killed was the rajah who was principally concerned in the capture and plunder of the Friendship.

The casualties among the Marines were as follows:

*Killed.*—Private Benjamin J. Brown.

*Wounded.*—Lieutenant Alvin Edson, Privates Daniel H. Cole (mortally), James A. Huster.

**1835.**—During the great fire of 1835 in New York City, it was found necessary to ask for military aid; a requisition was therefore made for as many Marines as could be spared from the barracks in Brooklyn. First Lieutenant J. G. Reynolds was ordered to report with a detachment to the city authorities for such duty as might be assigned. A strong line of sentinels was posted so as to efficiently guard public and private property, and with such signal success that the gratitude of the city was consequently expressed in a vote of thanks to Lieutenant Reynolds and the Marines under his command.

---

## CHAPTER XII.

**1836-42.**—Potomac at Port Mahon—The Services of the Corps during the Campaigns against the Indians in Florida—The United States Exploring Expedition—The Mutiny on board the Brig Somers.

In 1836, on board of the United States ship Potomac, at Port Mahon, a serious outbreak of the crew occurred, which, but for the promptness of the Marines in quelling it, might have resulted in very serious mutiny, and the action of the Marines was a subject of commendation by the officers of the ship.

Upon the breaking out of Indian hostilities in Georgia, in 1836, at a moment when the Indians were ravaging the country, and the disposable force of the army was inadequate to the emergency, Colonel-Commandant Archibald Henderson of the Marines volunteered his services, together with that portion of the Corps stationed at the different Navy-Yards. In this he was seconded most cheerfully by the officers under his command. Their services were accepted, and their places at the Yards temporarily filled by watchmen from civil life. Before leaving Washington, Colonel Henderson was the recipient of a cane, presented by the officers of the Corps. It was made of a piece of the original timber of the corvette Cyane, captured in 1815 by the frigate Constitution, under Commodore Stewart. The cane had a gold head and silver ferrule, and a sword about nine inches in length. On one side of the blade was the following inscription:

To Colonel Archibald Henderson, Commandant of Marines, commander of the guard on board the United States frigate Constitution in the capture of H. B. M. Corvettes, Cyane and Levant.

On the reverse side:

This cane, being a piece of the original timber of the Cyane, is presented by the officers of the Corps, January 1, 1836.

On the head were the words:

Col. A. Henderson, U. S. Marine Corps.

The *National Intelligencer* of June 2, 1836, says, "The detachment of Marines under the command of Colonel Henderson, which so promptly and handsomely volunteered to go against the Creek Indians, will, we understand, leave here this morning in the Columbia for Norfolk, where they will take passage to Charleston, South Carolina, on their route to the scene of savage warfare. This is another striking evidence of the great value of this arm of the national defence; it has shown itself as prompt to defend its country on the land as on the water, the element in which it was designed, originally, exclusively to act. Upon several occasions during the late war with England, detachments from this brave and highly disciplined Corps covered themselves with unfading laurels by their conduct while serving on land; and in every instance of conflict on the water its bravery and efficiency were attested by the official reports of the actions in which it bore a part. In the present emergency it did not wait even an intimation that its services would be acceptable, but promptly came forth, through its commanding officer, in the first hour of danger, and voluntarily offered to leave its comfortable quarters, and within one week from the offer we see a strong and well-appointed detachment of fine-looking men bidding farewell to families and friends, and taking up the line of march to seek a savage and treacherous foe in a distant land and in an inhospitable climate, to stay the ravages of war, and to protect the innocent, the helpless, and the unoffending. They will carry with them the best and warmest wishes of all, that their success may be commensurate with their bravery and zeal."

The Columbus, Georgia, *Sentinel* of July 1 contained the following: "The first battalion of United States Marines, under the command of Colonel Henderson, is now stationed at Camp Henderson, fifteen miles below Columbus, on the western bank of the Chattahoochee. The battalion arrived at this place on the 23d instant, having left Washington on the 1st, and Augusta on the 10th, marching from Augusta to this place in fourteen days, a distance of two hundred and twenty-four miles. On their arrival here, orders were received from General Scott to proceed to their present station with all possible despatch, and there erect a strong picket work, as a place of deposit for provisions, etc., for the eastern wing of the army. Since their arrival there the officers and men have all enjoyed good health and spirits, and pursue their work with vigilance and promptness. Their location being in the most exposed part of the enemy's country, it is a great privation for them to be confined to the monotonous duties of the camp, though well convinced of the importance of their present work. Their camp has been for two successive nights roused by Indians lurking about, and approaching the picket sentinels within a few yards, when they were fired on and pursuit immediately given, but no traces of them could be found. Last night, after the roll of the drum had ceased, a whoop was distinctly heard up the river, which no doubt was a signal to a party above. On Friday morning last, a negro boy who had escaped from the Indians that morning, and who had been a prisoner some five or six weeks, was brought into camp by Captain Love, of the Georgia volunteers. He stated that a party of twenty or thirty had camped the night before within six or seven miles, and had left that morning for a large swamp not far off, no doubt Cowagee Swamp, and that he saw Jim Henry that day, who advised them to go as soon as possible, and that he had been badly wounded in the shoulder. Captains Twiggs's and Dulany's companies, together with a company of Georgia volunteers, under Captain Love, the whole under the command of Captain Twiggs, were immediately despatched in hopes of overtaking them, but without success. They soon came upon their abandoned camp, found their fires burning and meat cooking, and everything about indicating a sudden departure. A negro man was taken in the vicinity of the camp, who, with great reluctance, gave himself up. He was armed with a musket, twenty balls, and a pocket full of powder. He was much frightened, but seemed determined to communicate as little as possible; but as far as he did tell, he corrobated the statements of the boy. The party took a number of horses belonging to the Indians, and a variety of ornaments, which they in their hurry had left behind. There is but little doubt that the Indians are concentrating somewhere in that vicinity, and will make a desperate effort to cross the river; if they do they may be assured the troops now there will give them warm work."

On Friday, the 24th of June, the second battalion of Marines, under command of Lieutenant-Colonel Freeman, consisting of one hundred and sixty men, reached Milledgeville, and immediately proceeded to their destination, Fort Mitchell. Orders were almost immediately given to proceed to Florida.

On the 16th of October the two battalions, consolidated into one regiment of six companies, arrived at Apalachicola, Florida, from Fort Mitchell, Alabama, on their way to Tampa Bay. The following is a list of the officers of the battalion, which was attached to the army under the command of General Jessup:

Colonel-Commandant: Archibald Henderson. Major and Brevet Lieutenant-Colonel: Samuel Miller. Major and Brevet Lieutenant-Colonel: W. H. Freeman. Staff: Brevet Lieutenant-Colonel Charles R. Broom, Paymaster; Captain Parke G. Howle, Adjutant and Inspector; Captain E. J. Weed, Quartermaster; First Lieutenant George F. Lindsay, Assistant Quartermaster; First Lieutenant F. C. Hall, Assistant Commissary; Surgeon, John A. Kearney, U. S. N.; Passed Assistant Surgeon, George B. McKnight, U. S. A.; Sergeant, Major James Gatchell. Captains: Levi Twiggs, John Harris, James Edelin, William Dulany, T. S. McCawley, Brevet Captain Benjamin Macomber; First Lieutenants, Alvin Edson, H. B. Tyler, L. N. Carter, John G. Reynolds, T. L. C. Watkins, F. N. Armistead, George H. Terrett, W. E. Starke, W. E. Lang, and A. H. Ross. Second Lieutenants: D. D. Baker, L. F. Whitney, E. L. West, Robert C. Caldwell, W. L. Young, Josiah Watson, W. McArdle, and John T. Sprague.

The Marines participated in the arduous campaigns under General Jessup, and received from him the highest commendations. At the battle of the Wahoo Swamp, Colonel B. K. Pierce, commanding the right division, mentions in his official report the gallant conduct of First Lieutenant Andrew Ross of the Marines, who was wounded. This officer subsequently died of his wounds.

1837.—The Marines bore an honorable and highly important part in the battle of Hatchee-Lustee, which began on the 22d of January, 1837. The main body of the army, under command of Major-General Jessup, was put in motion on that day, to attack the Indians and negroes in the strongholds which they were said to occupy on the head-waters of the Ocklawaha. On the 23d, Lieutenant-Colonel Caulfield was detached with his battalion of mounted Alabama volunteers, Captain Harris's company of Marines, and Major Morris's Indian warriors, accompanied by General Jessup's aide, Lieutenant Chambers, to attack Osuchee, a chief of some note, who was reported to have a large Indian force under his command in a swamp on the borders of Ahpopka Lake. The chief was surprised, himself and three warriors killed, and nine Indians and eight negroes captured. On the 27th the enemy was found on the Hatchee-Lustee, in and near the "Great Cypress Swamp," and promptly and gallantly attacked. Lieutenant Chambers, with Price's company of Alabama volunteers, by a rapid charge, succeeded in capturing the horses and baggage of the enemy, with twenty-five Indians and negroes, principally women and children, the men having mostly fled into the swamp. Colonel Henderson, leaving one company with the prisoners and horses, entered the swamp with the remainder of his command, drove the enemy across the Hatchee-Lustee, passed that river under their fire, and drove them into a more dense and difficult swamp, in which they dispersed. A considerable force was ordered to support Colonel Henderson. When the troops reached the point where the colonel had entered the swamp, it was ascertained that he was in rapid pursuit of the enemy, and was believed to be fully able to overcome the force opposed to him. The result of the day's operations was the capture by Colonel Henderson's force of two Indian women and three children, and twenty-three negroes, young and old, and over a hundred ponies, with packs on about fifty of them. All their clothes, blankets, and other baggage were abandoned by the enemy, and either taken or destroyed by Colonel Henderson. In his report of the affair Colonel Henderson says,"The regular troops, both artillery and Marines, displayed great bravery and the most untiring and determined perseverance. The Marines, however, I cannot refrain from mentioning in a particular manner. The killed and wounded show where they were, and render any further comment from me unnecessary."

The *Army and Navy Chronicle* of June 15 contained the following: "The Marine Corps has been so much separated for a year or more past that we have not had it in our power to make any report of the changes that have occurred in the stations of its officers. We are rejoiced to perceive that this Corps has been earning a harvest of fame in Florida, by the alacrity, zeal and ability with which the duties assigned to it have been discharged; and we welcome back to their homes, and to comparative repose, those members of it who have been so long actively engaged in combating the savages." The following order, signed by General Jessup, was issued from the Head-quarters Army of the South, Tampa Bay, May 22d: "The presence of Colonel Henderson being required at the headquarters of his Corps, he will proceed to Washington City and report to the adjutant-general of the army. The major-general commanding would be forgetful of what is due to merit, and would do injustice to his own feelings, were he to omit on the present occasion the expression of the high sense he entertains of the distinguished and valuable services rendered by the colonel. He tenders him his warmest thanks for the able, zealous, and cheerful support he has on every occasion received from him, both in Florida and Alabama; and begs him to accept his best wishes for his future fame and happiness. Captain Howle, adjutant and inspector of the Marine Corps, and Surgeon Kearney, of the Navy, medical director of the army, will accompany Colonel Henderson. The major-general thanks them for the zeal, efficiency, and ability with which they have performed every duty which has devolved upon them since they have been attached to this army, and he assures them that they carry with them his best wishes and kindest regards. Lieutenant-Colonel Miller will command, with his brevet rank, the troops south of the Hillsborough."

Colonel Henderson arrived in Washington about the middle of June, accompanied by Captain Howle, of the Corps, and by Surgeon Kearney, of the Navy. The *National Intelligencer*, in announcing their arrival, said, "We are glad to learn that Colonel Henderson, and the officers who accompanied him, have returned to their families in good health. They have suffered much in common with all with whom they have served, not less from the climate, and the peculiar nature of the country which has been the theatre of the war, than from the necessary hardships of service in so wild and destitute a region. The gallant Corps, which it is the good fortune of Colonel Henderson to command, has always been distinguished wheresoever duty has called it. In the present case the Corps deserves peculiar commendation, from having volunteered in the war in Florida, and having repaired to its theatre, a thousand miles distant, to share in its perils and privations. Its commander deserves the praise of having proven himself worthy of his post, both by his gallantry in the field, and by patience and good example under all difficulties; and he, his officers, and men have most honorably maintained the pledge which they gave to the government and to their country when they first tendered their services. The Corps remains in Florida under the veteran and gallant Colonel Miller, to make further sacrifices and endure fresh hardships, should the obstinacy of the Indians, as now appears too probable, unhappily prolong the war."

The uniform of the Marines for a long time, and until 1839, was a green coat with white or buff facings, an undress frock-coat of the same color, a sword with white ivory cross hilt and brass scabbard, and white leather belts. In 1839 it was changed, to take effect from July 4, 1840, to blue with red facings.

The following list exhibits the character and strength of the naval force employed in the Mosquito Fleet from June, 1838, until August, 1842: Lieutenant John T. McLaughlin, commanding in chief. The squadron consisted of the Wave, Panther, revenue-cutters Campbell and Otsego, with a force of one hundred and sixty men, from June,

1838, until November, 1839; from this date until June, 1841, of the Flirt, Wave, Otsego, and barges Mayo and Harney, and two companies of Marines, in all five hundred and forty-one men; from June, 1841, until August, 1842, of the Flirt, Wave, Otsego, Phœnix, Jefferson, Madison, Van Buren, and barges Mayo and Harney, with one hundred and thirty Marines, making an aggregate force of six hundred and and twenty-two men, of whom sixty-eight were officers; one hundred and forty canoes were employed by the squadron during its service. From 1839 to 1840 the Marines on the ships were commanded by First Lieutenant George H. Terrett, with Lieutenant I. R. Wilson and Robert D. Taylor as his subordinates. From 1840 to 1842 they were commanded by First Lieutenant Thomas T. Sloan, with Lieutenants I. R. Wilson and Robert D. Taylor as his subordinates. In addition to the garrison at Indian Key, in 1842, the Marines garrisoned Fort Dallas.

On the 3d of November, 1841, a detachment of one hundred and fifty Marines and seamen, under command of Lieutenant J. T. McLaughlin, commanding the Mosquito Fleet, made a scout through the Everglades. The scout lasted twenty-two days, and although the Indians fled before them in all directions, allowing no opportunity for fighting, the result of the arduous trip was to gain information of an extensive country which had never before been explored, and to exhibit an imposing force in the heart of a country hitherto deemed impenetrable; the tendency of which was to strengthen the wavering and doubtful Indians in their inclinations to submit. The depot for the Florida squadron, commanded by Lieutenant McLaughlin, was at Indian Key, on the eastern coast.

For several weeks Lieutenant Sloan of the Marines was employed with his command on the coonti grounds between the Miami and New Rivers, with instructions to explore every acre of them. During the exploration, Lieutenant Sloan came upon five distinct settlements of Indians, the first of them within five miles of Fort Dallas, between Little River and Arch Creek. The instant he was discovered signal fires sprang up in every direction, as if by magic, and fields and settlements were as suddenly deserted. Large quantities of ripe corn were gathered and destroyed, and many fields laid waste. A second scout of Marines, under Lieutenant Taylor, in co-operation with Lieutenant-Commanders Marchand and Rogers, was compelled to return abruptly to their post for want of water. The fatigue and privation undergone by this detachment was so great that Private Kingsbury fell in his trail and died from sheer exhaustion. Their patient endurance and cheerful alacrity in the discharge of every duty proved the high state of discipline, both of Marines and seamen.

On the 9th of June, Lieutenant McLaughlin received orders to return to Norfolk with the vessels under his command.

Congress, by a law of the 18th of May, 1836, authorized the equipment and employment of a portion of the national Marines for the purpose of exploring those seas in which the whale fisheries, as well as other branches of commercial enterprise, were pursued. On the 20th of March, 1838, Lieutenant Charles Wilkes, U. S. N., was directed to assume the command of the expedition.

The instructions issued to him by the Navy Department, dated the 11th of August, 1838, indicated the following objects to be aimed at: "To explore and survey the Southern Ocean, having in view the important interest of our commerce embarked in the whale-fisheries, as well as to determine the existence of all doubtful islands and shoals; and to discover and accurately fix the position of those which lie in or near the track pursued by our merchant vessels in that quarter, and which may have hitherto escaped the observation of scientific navigators.

"Although the primary object of the expedition is the promotion of the great interests of commerce and navigation, yet all occasions will be taken, not incompatible with the great purpose of the undertaking, to extend the bounds of science, and to promote the acquisition of knowledge. For the more successful attainment of these objects, a corps of scientific gentlemen for the departments of philology, zoölogy, conchology, geology, mineralogy, and botany, with artists and horticulturists, will accompany the expedition, and are placed under your direction. The hydrography and geography of the various seas and countries you may visit in the route pointed out to you will occupy your special attention; and all the researches connected with them, as well as with astronomy, terrestrial magnetism, and meteorology are confided exclusively to the officers of the Navy, on whose zeal and talents the department confidently relies for such results as will enable future navigâtors to pass over the track traversed by your vessels without fear and without danger."

The track marked out in the instructions required a circumnavigation of the globe, with a divergence quite towards the Southern Pole, and so far towards the north as the westward possessions of the United States extended, with frequent crossings of the Pacific Ocean, and extensive cruisings among its countless islands. This course was very fully and faithfully pursued through a space of three years and ten months.

The vessels composing the squadron were the sloop Vincennes, Lieutenant Wilkes, commanding the expedition; the sloop Peacock,

Lieutenant-Commanding W. L. Hudson; the brig Porpoise, Lieutenant-Commanding Cadwalader Ringgold; store-ship Relief, Lieutenant-Commanding A. K. Long; tenders Sea-Gull and Flying-Fish, commanded by Passed-Midshipmen Reid and Knox.

The squadron sailed from Hampton Roads on the 19th of August for Maderia; then took a southerly course, touched at the Cape de Verde Islands, and arrived at Rio on the 24th of November. On the 6th of January they left Rio; stopped at the Rio Negro, doubled Cape Horn, and entered Orange Harbor.

In the latter part of May the squadron rendezvoused in the harbor of Valparaiso, with the exception of the Sea-Gull, which was lost at sea. Passed-Midshipmen Reid and Bacon, with a crew of fifteen men, were lost by this disaster. The squadron next stopped at Callao. They left Callao on the 12th of July, and arrived at Sydney, New South Wales, on the 28th of November. On the 26th of December the expedition once more turned towards the extreme south, and the vessels were ordered to act independently of each other when arrived in the regions of the designed explorations.

The squadron found constant occupation during the ensuing summer in surveying and exploring duties amid the numerous islands of the Southern Archipelago. New Zealand, the Friendly Islands, and the Feejee group were visited for these purposes.

Though prosecuting a work of peace, and especially desirous of establishing permanent relations of amity with the barbarous tribes of the Pacific, in order to the greater security of commerce, still during its stay in the Feejee group, the expedition was under the necessity of twice putting in exercise its military power. In July a surveying party, under Lieutenant Perry, in a launch, and Mr. Knox, in a first cutter, ran into Sualib Bay for shelter during a storm. In endeavoring to beat out again the cutter ran on a reef at low tide, and it was found impossible to get her off. The natives in great numbers, and well armed, soon collected near, and commenced a serious attack. The ammunition in the cutter being all wet, and the launch being unable to render an effective assistance in repelling the savages, the crew abandoned the boat and went on board the launch. On the 12th the launch rejoined the Vincennes and Peacock. Immediately the schooner and eight boats from the ships, with an extra complement of seamen and Marines, under the immediate command of Lieutenants Wilkes and Hudson, proceeded to inflict merited punishment upon the natives. They entered the bay, and cautiously marched to the principal village, meeting with no resistance. The town, consisting of sixty huts, was burned, and the savages were taught a salutary lesson.

A tragical affair occurred during the latter part of July, at Malolo, an island of the same group. Lieutenant Underwood, with a party, landed on this island during a survey carried on in two of the boats of the squadron, under the command of Lieutenant Alden, for the purpose of ascending a height to look for the Porpoise. He was soon recalled, in consequence of some suspicious movements among the natives, bringing with him a young native, who was retained as a hostage. The next morning the same officer landed for the purpose of procuring provisions. The natives manifested considerable reluctance to trading, and Lieutenant Underwood and party remained some length of time on shore in unsuccessful attempts at negotiation. In the meantime Midshipman Henry was sent to join those on shore. The natives now began to collect in greater numbers, and evidently were not in a friendly mood.

The symptoms of an attack became so manifest that Lieutenant Underwood called his party together and ordered a retreat to the boat. At the same time the hostage suddenly jumped out of Lieutenant Alden's boat and commenced running towards the shore. The report of fire-arms was heard on shore, and it became evident to those in the boats that their companions were engaged in a serious conflict with an overpowering number of savages. They pushed for the land, and as soon as their fire-arms were effective, the savages disappeared in a hasty retreat. On landing they found one seaman badly wounded, and Lieutenant Underwood and Midshipman Henry lying prostrate on the beach. The officers in a few moments expired; and their bodies, rescued from cannibalism, were borne off to the Flying-Fish, which was a few miles distant.

This unprovoked massacre called for stern retribution. Lieutenant-Commandant Wilkes was himself at this time on board the Flying-Fish, and the brig Porpoise was also now in company. There were two towns, Saulib and Arro, situated upon opposite sides of the island. The party which was to attack and destroy these consisted of seventy officers, seamen, and Marines, under the command of Lieutenant-Commanding Ringgold. The party in the boats, intended to cut off escape from the island and to co-operate with the former party, was led by Lieutenant-Commandant Wilkes.

The first party landed without opposition on the southeast point of the island, destroyed the plantations in their course, and crossing over the high land, came in sight of Saulib, situated on the southern shore. Here it was found that the natives were mostly assembled with the intention of defending themselves by all means possible to them; and their preparations and position were not to be despised, even by disciplined troops. The village was surrounded entirely by a strong stockade of cocoa-nut trees, placed a few feet apart, and filled between with close and substantial wicker-work. On the outside of this was a wide ditch, filled with water; and on the inside a dry ditch, in which the defenders were intrenched, while they shot through the loop-holes in the palisade. The savages were very confident in the impregnability of their fortress, for they received the advancing Americans with shouts and expressions of defiance. They possessed quite a number of muskets, as well as their ordinary weapons. Almost the entire population were within the enclosure, and the women and children were as defiant and active in defence as the men. Knowing that an assault must be attended with some loss, even though conducted against undisciplined troops, the commander wisely attacked from such a distance as his means would permit. A sharp contest of about fifteen minutes was maintained, during which a chief and six of the savages were killed, and the houses were fired by a rocket. The natives fled, escaping through a gate leading towards the sea. They were allowed to retreat without further attack. A few of the Americans were wounded, but only one severely. The town was entirely consumed.

The party now marched northward across the island to co-operate with the boat party against Arro. The latter reached Arro first, and without resistance set fire to and destroyed it. While these operations were going on, Lieutenant Emmons chased five canoes, containing about forty native warriors. When overtaken they offered fight and made a resolute resistance. But four of them were finally captured, the other escaping. The next day the whole force from the vessels assembled on a hill, and received the population of the island, who sued for pardon.

On the 11th of August the expedition left the Feejee Islands, and in October was at Honolulu. The Porpoise and Vincennes left Honolulu on the 5th of April, 1841. In twenty-two days these vessels were off the bar at the mouth of the Columbia River. In the summer of 1841 the Peacock, Lieutenant-Commandant Hudson, while at Drummond's Island, one of the Kingsmill group, engaged in a conflict with the natives. A large party, under Lieutenant-Commandant Hudson, was on shore for purposes of curiosity. The natives sought to separate and entice the men in different directions. They also pilfered loose articles, and finally made actual hostile demonstrations. Wishing to avoid a collision, the men were called together and returned to the boats. It was then found that one man was missing. Having waited two days for news of the missing man, it was concluded that he had been treacherously murdered, and Lieutenant-Commandant Hudson determined to inflict severe punishment on the islanders. Eighty seamen and Marines in seven boats under the command of Lieutenant Walker were landed.

The savages, to the number of five hundred and upward, fearlessly awaited their approach on the beach, brandishing their weapons and showing a determination to stand their ground. A parley, with a renewed offer of ransom, was unavailing. A few shots were then fired, bringing down some of the chiefs, and a rocket was discharged into the crowd, immediately followed by a general discharge of musketry, which caused a rapid retreat of the enemy. The council-house and town were reduced to ashes and twelve of the natives were killed. On the 18th of July the Peacock was lost while endeavoring to enter the Columbia River. During the month of June, 1842, the squadron arrived at New York. The Marines attached to this expedition were as follows: Vincennes—Quartermaster's Sergeant Marion A. Stearns; Corporals Joseph Clark, Christian Dobleman, Alexander Ogle; Privates Allshouse, Campbell, Disbrow, Dunbar, Richardson, Roberts, Mackenzie, Nebhert, Rogers, Taylor, and Ward. Peacock—Sergeant Joseph Forbes; Corporals Joseph Carey, Thomas Dewees; Privates Hughes, Riley, Harman, Smith, Marsh, Allman, Pottle, and Crook. Brig Porpoise—Corporal W H. King; Privates Cavanagh, Bateman, Lewellen.

On the first day of December, 1842, Commander Alexander Slidell Mackenzie, U. S. N., in command of the brig Somers, in 17° 34′ 28″ north latitude and 57° 57′ 45″ west longitude, executed, by hanging, Philip Spencer, acting midshipman; Samuel Cromwell, boatswain's mate; and Elisha Small, seaman, for attempted mutiny on the high sea.

At the close of the proceedings of the court-martial Commander Mackenzie stated in his defence that "the Somers had no Marines, a body of men distinct from the crew in organization and feeling, on whom, in ordinary ships of war, the police and discipline greatly depend, and who form a counterpoise and check to the turbulent spirits of the common seamen."

In a letter addressed to the Hon. A. P. Upshur, Secretary of the Navy, Commander Mackenzie also made the following statement in relation to Sergeant Michael H. Garty, who was a passenger, invalided home:

"Of the conduct of Sergeant Michael H. Garty I will only say that it was worthy of the noble Corps to which he has the honor to belong; confined to his hammock by a malady which threatened to be dangerous, at the moment when the conspiracy was discovered, he rose upon his feet a well man.

"Throughout the whole period, from the day of Mr. Spencer's arrest to the day after our arrival, and until the removal of the mutineers, his conduct was calm, steady, and soldier-like. But when his duty was done and health was no longer indispensable to its performance, his malady returned upon him, and he is still in his hammock. In view of this fine conduct, I respectfully recommend that Sergeant Garty be promoted to a second lieutenancy in the Marine Corps. Should I pass without dishonor through the ordeal that probably awaits me, and attain in due time to the command of a vessel entitled to a marine officer, I ask no better fortune than to have the services of Sergeant Garty in that capacity."

*The following tabular statement gives a complete list of the Marines on Sea Service who died in the Florida contest between 1836 and 1842:*

| No. | Name. | Rank. | Place of Death. | Date of Death. | Remarks. |
|---|---|---|---|---|---|
| 1 | Thomas T. Starke | Corporal | Steamer Poinsett, E. F. | December, 1839 | |
| 2 | David Cannon | Corporal | Indian Key | September 4, 1841 | |
| 3 | Frederick Dunn | Fifer | Indian Key | November 4, 1841 | |
| 4 | Henry Elisha | Private | Indian Key | August 21, 1841 | |
| 5 | Joseph H. Root | Private | Indian Key | September 24, 1841 | |
| 6 | James V. Sawyer | Private | Indian Key | October 6, 1841 | Drowned. |
| 7 | Joseph Smith | Private | Indian Key | October 6, 1841 | Drowned. |
| 8 | James J. Ayer | Private | Indian Key | November 11, 1841 | |
| 9 | Thomas Ennis | Private | Indian Key | November 25, 1841 | |
| 10 | Robert Gray | Private | Indian Key | November 20, 1841 | |
| 11 | John Nicholson | Private | Indian Key | December 23, 1841 | |
| 12 | Stephen Schoolcraft | Private | Indian Key | December 9, 1841 | |
| 13 | Jacob Schultz | Private | Indian Key | December 6, 1841 | |
| 14 | William Williams | Private | Indian Key | December 16, 1841 | |
| 15 | Rufus Griffin | Private | Fort Dallas | January 16, 1842 | |
| 16 | L. D. Pierpoint | Private | Fort Dallas | January 20, 1842 | |
| 17 | J. Kingsbury | Private | Fort Dallas | March 14, 1842 | |
| 18 | Thomas Walsh | Private | Schooner Flirt | July 18, 1842 | At Sea. |
| 1 | William Smith | Acting Corporal | Wounded in a skirmish with the Indians, January 6, 1841. | | |

*The following is a list of Officers, non-commissioned Officers, Musicians, and Privates of the Marine Corps who were killed in action, or died of wounds received, or disease contracted during the Florida War:*

| No. | Name. | Rank. | Co. | Place of Death. | Date of Death. | Remarks. |
|---|---|---|---|---|---|---|
| 1 | Andrew Ross | First Lieut. | | Fort Heilman | Dec. 11, 1836 | Wounded Nov. 21, 1836, at Wahoo Swamp. |
| 2 | William Tait | Ord. Sergt. | E | Upper post Sannibar River | Dec. 5, 1837 | Of disease. |
| 3 | Henry Marks | Sergeant | A | Fort Brooke | Nov. 9, 1836 | Of disease. |
| 4 | George King | Sergeant | D | Tampa Bay | April 14, 1838 | Of disease. |
| 5 | Daniel Cunningham | Sergeant | | Hatchee-Lustee Swamp | Jan. 27, 1837 | Wounded in both shoulders. |
| 6 | Leonard Stevens | Sergeant | | Hatchee-Lustee Swamp | Jan. 27, 1837 | Wounded in the thigh. |
| 7 | Charles Pike | Corporal | D | Tampa Bay | April 14, 1838 | Of disease. |
| 8 | Matthew McKinley | Corporal | E | Tampa Bay | April 14, 1838 | Of disease. |
| 9 | Thomas P. Peterson | Drummer | B | Hatchee-Lustee Swamp | Jan. 27, 1837 | Killed in action. |
| 10 | Joel Wright | Private | B | Hatchee-Lustee Swamp | Jan. 27, 1837 | Killed in action. |
| 11 | Daniel Brown | Private | D | Chattahoochee River | Oct. 13, 1836 | Of disease. |
| 12 | John Shillingsford | Private | E | Fort Brooke | Dec. 7, 1836 | Of disease. |
| 13 | John Reardon | Private | C | Fort Brooke | March 14, 1837 | Of disease. |
| 14 | W. W. Vanclef | Private | D | Tampa Bay | Feb. 21, 1837 | Of disease. |
| 15 | Alexander Burke | Private | E | Fort Armstrong | Jan. 20, 1837 | Of disease. |
| 16 | James O'Neil | Private | E | Fort Brooke | March 1, 1837 | Of disease. |
| 17 | William Steel | Private | B | Black Creek | Aug. 11, 1837 | Of disease. |
| 18 | John H. Durant | Private | D | Fort Brooke | June 29, 1837 | Of disease. |
| 19 | John Sweeney | Private | E | Tampa Bay | April 27, 1837 | Of disease. |
| 20 | Michael Sullivan | Private | A | Fort Brooke | July 15, 1837 | Of disease. |
| 21 | J. M. Waalfin | Private | C | Mullet Key | Aug. 22, 1837 | Of disease. |
| 22 | Isaac Elburn | Private | D | Fort Monroe | Oct. 1, 1837 | Of disease. |
| 23 | William T. Henry | Private | D | Fort Dulany | Nov. 17, 1837 | Of disease. |
| 24 | John Jackson | Private | D | Fort Dulany | Nov. 22, 1837 | Of disease. |
| 25 | Richard T. Trask | Private | E | Tampa Bay | April 2, 1838 | Of disease. |
| 26 | John A. Perley | Private | E | Fort Denaud | Feb. 3, 1838 | Of disease. |
| 27 | Thomas Fling | Private | E | Columbus | Sept. 11, 1836 | Of disease. |
| 28 | Peter Foley | Private | | Hatchee-Lustee Swamp | Jan. 27, 1837 | In the arm. |
| 29 | John M. Sullivan | Private | | Hatchee-Lustee Swamp | Jan. 27, 1837 | On the lip. |
| 30 | Thomas Irwin | Private | | Fort Armstrong | | Wounded by a musket ball, by accident. |

## CHAPTER XIII.

**1846-48.**—War with Mexico, I.—Operations on the Pacific Coast.

At no time in our national history has the Marine Corps borne a more conspicuous or more honorable part than during the Mexican War. Even before the news of the formal declaration of war reached the American forces on the frontier, the officers and men of the corps under the orders of their naval superiors, made a record for themselves and for the nation, of brave, unfaltering courage. General Scott stated, at the close of the contest, that he had placed the Marines where the hardest work was to be accomplished, and that he had never found his confidence misplaced.

The annexation of Texas required the defence of that boundary-line which she and the government of the United States claimed as her own. Collisions with the Mexican troops soon occurred, and the formal declaration of war was made on the part of the United States on the 12th of May, 1846, and on the part of Mexico on the 23d of the same month.

Commodore Sloat lay at Mazatlan, in the Savannah, in company with different smaller vessels, closely watching the course of events in the Gulf of Mexico. He had a line of active communication across the continent, by means of different mercantile houses, and on the 7th of June he received, through that channel, the information of the battles of Palo-Alto and Resaca-Palma. Instantly he came to a prompt and wise determination. He sailed on the 8th for the northward, leaving the Warren at Mazatlan to wait for further intelligence. The Savannah reached Monterey on the 2d of July. Here the commodore found the Cyane and Levant, and ascertained that the Portsmouth was at San Francisco. Two hundred and fifty seamen and Marines landed under the immediate command of Captain William Mervine. This force raised the standard of the United States, under a salute of twenty-one guns. Commodore Sloat took the necessary measures to secure his bloodless conquest by garrisoning the town with a force of Marines under Lieutenant W. A. T. Maddox. The officers of the corps engaged on this occasion were Captain Ward Marston, Second Lieutenant W. A. T. Maddox, and Second Lieutenant Henry W. Queen.

Military possession was on July 9 taken of Yerba Buena, and the flag of the United States displayed. A proclamation was issued, calling upon all the residents of that district, agreeably to the laws of the

United States regulating the militia, to enroll themselves into a military company, appoint their own officers, and observe such rules and regulations as should be issued for the maintenance of order, and for the protection of property in Yerba Buena and its immediate neighborhood. A military guard was stationed in possession of the customhouse, under Second Lieutenant Henry B. Watson, of the Marines, who was appointed the military commandant *pro tem.* of all the Marines and militia. He was ordered to remain in military possession as the commander of the Marines and local militia, subject to orders from his superior officers. All the militia organized for the protection of the place were required to repair to the appointed rendezvous, upon such signal as he should designate, properly armed and equipped, and he was directed to make requisition for such arms and ammunition as might be required.

Commodore Stockton sailed in the Congress on the 1st of August for San Pedro. As the ship proceeded south, a landing was made at Santa Barbara, where the flag was shifted and a small force of Marines was left in possession. The Congress arrived off San Pedro on the 6th of August. On the 13th the commodore made a junction with the battalion of volunteers under Major Fremont, a brigade of Marines and seamen were landed, and the whole force entered Los Angeles without opposition in the course of that day.

Commodore Stockton now determined to organize a civil government for the entire State. At the head of this government he placed Major Fremont, and appointed Lieutenant A. H. Gillespie of the Marines military governor and commandant of the troops. Having effected these arrangements, he returned to the coast on the 2d of September, and proceeded to Monterey, where everything was found tranquil, though rumors were in circulation of an intended rising among the Indians at the north. The Savannah was immediately sent up the coast, where she was soon joined by the Congress off San Francisco.

About the close of the month the Mexicans in the neighborhood, finding that the main force of the Americans was at a considerable distance, laid siege to Los Angeles. The situation became critical, and Lieutenant Gillespie despatched a courier to Monterey to notify the naval forces there of the danger in which his command was placed. Leaving Lieutenant Maddox of the Marines in command at Monterey, the main portion of the naval force started for San Pedro to raise the siege at Los Angeles. Scarcely had they left Monterey when that place was also threatened with assault. Lieutenant Maddox sent a messenger to bring back, if possible, a portion of the force then on its way to San Pedro. By a fortunate combination of circumstances the messenger reached the Congress, then accompanying a transport of troops under Fremont from San Francisco, also ordered to the assistance of Lieutenant Gillespie. A sufficient force to strengthen the position of Lieutenant Maddox and hold it against the Mexicans was sent to Monterey, and the remainder of the command proceeded on the way towards Los Angeles. It proved impossible, however, to reach that place in season to be of any assistance to Lieutenant Gillespie, who was forced to capitulate. He made excellent terms with the Mexican commander, nevertheless, and was soon allowed to join the main force. It was not, however, until after a brave and determined, though unsuccessful, effort had been made by a force of Marines and seamen under Captain Mervine of the Navy to raise the siege. Lieutenant Gillespie shortly afterwards commanded the Marines, who, in connection with the force of seamen under Lieutenant Minor, of the Navy, made a gallant charge on the Mexicans at San Diego, where the enemy was driven back in confusion. About this time the bravery and good conduct of Lieutenant Gillespie were recognized by the Department, and he received a captain's commission. He was shortly afterwards placed in command of an expedition to surprise the enemy in camp at San Bernardo.

Important events at this period followed each other in rapid and almost confusing succession. Before Captain Gillespie's expedition was prepared to start on its arduous work, Brigadier-General Kearney arrived in California to take command of the land forces. Captain Gillespie's force was immediately ordered to make a junction with Kearney's dragoons. Before the junction could be made, Kearney made an attack on the enemy at San Pasqual, and received a crushing defeat, the Americans losing heavily in killed and wounded, Kearney himself being among the latter. The rapid march of Commodore Stockton's force to San Gabriel and the overwhelming defeat of the enemy at that point followed. The officers of the Marine Corps who participated in these operations were Captain Gillespie, Lieutenant J. Zeilin, Lieutenant J. C. Cash, Lieutenant W. A. T. Maddox, and Lieutenant James Wiley.

On the 29th of December, in consequence of a number of Californians having taken up arms against the United States, and taken Lieutenant Bartlett, United States Navy, six seamen, and others, prisoners, an expedition, consisting of a detachment of Marines, under Lieutenant Tansil, a field-piece in charge of Passed-Midshipman D. Jones, the San José mounted volunteers, under Captain Weber, and the San Francisco mounted volunteers, under Captain William

M. Smith, and Assistant Surgeon M. Duvall, acting surgeon and aid, the whole under the command of Captain Ward Marston, United States Marine Corps, amounting in all to one hundred and one officers and men, left San Francisco in pursuit of the enemy. On the 2d of January, 1847, when within about seven miles of the mission of Santa Clara, and distant some fifty miles from San Francisco, the enemy were discovered, at first few in number, but increasing to about one hundred and thirty men, all mounted on the best horses. This force was driven back about two miles, when the Americans found their advance impeded by a deep marsh, and in attempting to cross it, the Marines and those in front were over their knees in mud. While in this position, the enemy attempted to drive down about five hundred wild cattle to break the ranks, and some twenty-five Californians came within range of the gun and opened fire on the Americans. At this time Captains Smith and Marston distinguished themselves by securing the spare horses of the enemy who became frightened. Captain Marston ordered the Marines to open on the cattle, and the gun to open on the Californians, and Lieutenant Tansil and Passed-Midshipman D. Jones displayed great activity in executing this order, which resulted in driving the Californians and their auxiliaries from the field. That evening a flag of truce was sent in by the Californians, asking, on the part of Colonel Sanches, their commander, an interview with the commanding officer of the Americans, each to be accompanied by two officers. Accordingly, the next morning Captain Marston, accompanied by Captain Smith and Surgeon Duvall, proceeded to the appointed place of meeeting. Colonel Sanches then laid before Captain Marston a series of complaints of abuses which the Californians had experienced, and also professed a perfect willingness on their part to surrender and retire to their homes, if they could be assured of future protection. As Captain Marston had no authority to treat with them, he proposed an armistice till he could communicate with the commander of the district and receive his instructions, which was readily agreed to, and a messenger despatched to San Francisco. On the afternoon of the armistice, the Americans were reinforced by the arrival of Captain Maddox's company of Monterey mounted volunteers. An answer was received on the 6th of January, saying the only conditions with which the Californians could be treated were for them to lay down their arms, surrender their prisoners and horses, and retire to their homes, all of which were complied with on the following day. In the action of the 2d of January, the Americans had two wounded,—a Marine and one of Captain Weber's men. The Californians admitted that eleven of their men were hit in their clothing, but were uninjured, which they attributed to the justice of their cause and the interposition of the Virgin Mary, and for which they celebrated mass in the church of Santa Clara. It was ascertained afterwards that their loss was four killed and five wounded.

1847.—In a general order, issued February 1, 1847, Commodore Shubrick said, "For disinterested conduct, the company of mounted volunteers, under Lieutenant Maddox of the Marine Corps, acting as captain, is tendered the thanks of the commander-in-chief, and will without doubt, receive applause and due recompense from the general government."

In September, 1847, Lieutenant Maddox was appointed military commandant of the Middle Department by Commodore Stockton. Subsequently, in a letter to General Henderson, Commodore Stockton wrote: "It is my duty as well as pleasure to inform you that Lieutenant Maddox was in command of the Middle Department in California and at Monterey when the insurrection broke out in that country, and that his zeal and gallantry were conspicuous in his efforts to suppress it. It would be no more than justice if his services there should be acknowledged by a brevet."

In November, 1847, the Dale, Commander Selfridge, was sent to Guaymas to relieve Commander Montgomery, in the Portsmouth, the latter vessel joining the main squadron. The flag was now flying at La Paz, in the province of Lower California, protected by a detachment of one hundred and ten men of a regiment of New York volunteers, under the command of Lieutenant-Colonel Burton. In the month of November it was also hoisted at San José, in the same province, while the flag-ship was off that place, and Lieutenant Charles Heywood, with four passed midshipmen and twenty Marines, was placed in command of this post. He was also furnished with a 9-pounder carronade, and some fire-arms to be loaned in an emergency to friendly Californians. He posted his force in an old mission-house, situated on the higher portion of the town.

Commodore Shubrick on the 10th of November appeared off Mazatlan, a town of about eleven thousand inhabitants, occupied by Colonel Telles, with a force of twelve hundred men. On the morning of the 11th a force of six hundred seamen and Marines landed and captured the place, Colonel Telles with his whole force retreating on the approach of the Americans. The following officers of the Corps were present: Independence, Lieutenant W. W. Russell; Congress, First Lieutenant J. Zeilin; and a detachment from the Cyane under Sergeant Forrest (the remainder of the guard being on duty at San José).

The Dale arrived at Guaymas on the 8th of November, and relieved the Portsmouth from the duty of guarding that post. The enemy's forces in this vicinity had now increased to a considerable amount, being probably not less than a thousand men. Commander Selfridge had not men enough to garrison the town, so that the flag was kept flying on an island under the guns of his ship. On the 17th of November, Commander Selfridge, supposing that there were none of the enemy in the immediate neighborhood, landed at the head of a party of sixty-five Marines and seamen. They had not proceeded far when they were assailed with a volley of musketry from a house. A ball wounded Commander Selfridge severely in the foot, and the command was relinquished to Lieutenant Smith. The latter officer found himself engaged with three hundred of the enemy. The fire of the little force of Marines and seamen was, however, delivered with such rapidity and precision that the enemy soon began to retreat precipitately. The Mexicans evacuated the town in all possible haste. In this creditable action the Americans suffered no injury, with the exception of the wound of the commander; while the loss of the enemy was not less than thirty killed and wounded. The Marines were commanded by Second Lieutenant Robert Tansill.

The Mexicans who had landed at Mulejè, and whom Commander Selfridge had so gallantly driven into the interior, passed through the country, exciting hostility to the American rule, and drawing the disaffected to their ranks. On the 19th of November a large body of the enemy came within a league of the post established at San José, which was occupied by Lieutenant Heywood, with four officers and twenty Marines. He had also organized and armed about twenty volunteers. The position occupied consisted of two houses, the one an old mission-house, which had to be strengthened and repaired to make it defensible, and the other a private house, so closely adjoining that it was necessary to occupy it lest it might fall into the hands of the enemy. In the latter Mr. McLanahan was stationed with twelve volunteers. At the mission a 9-pounder was mounted, and loop-holes left for musketry.

The main body of the enemy took up a position on an eminence about a quarter of a mile distant. It consisted of one hundred and fifty mounted men, and commenced the attack late in the day by a fire from a 6-pounder. About ten o'clock at night, the whole body of the enemy commenced an attack upon the front and rear of both houses. Their fire was vigorous, but was returned as vigorously by the little garrison. Their cannon was posted in front of the main building, where it was replied to by the 9-pounder. The defence was so effectual that the enemy was glad to cover themselves behind the neighboring buildings, and thus the combat was continued throughout the night; and by the morning the enemy retired to their camp. The next night the Mexicans were again in motion. They now concluded to undertake the capture of the mission-house by assault. In executing this movement, a strong party rushed up towards the front of the house, with the intention of forcing their way in and capturing the gun. But a destructive discharge of musketry brought down their leader and several of their number, when they broke and fled. This repulse in front intimidated the party in the rear, who were approaching with ladders to scale the walls, and they were thus easily driven back. The Mexican loss amounted to eight killed and twenty wounded, while the casualties of the garrison amounted to three wounded.

**1848.**—In the early part of the year 1848 San José became the scene of very active military operations. On the 22d of January a small detachment of Lieutenant Heywood's command, consisting of two officers and six men, were captured on the beach by a large party of the enemy's cavalry. The enemy had concentrated his forces, amounting to three or four hundred men, around the position held at San José. The defence of this position rested with a garrison of twenty-seven Marines, ten seamen, and twenty California volunteers. The town was deserted by the inhabitants, and about fifty women and children sought protection in the imperfect fort, and were supported from its scanty supplies. In the latter part of January the Mexicans appeared in force. On the 4th of February they drew their lines quite closely around the fort and fired upon all who showed themselves. From this day until the 14th of the month this large force kept up a constant, harassing fire upon the little garrison. Several sorties were made, but, notwithstanding, the investment of the post became closer and closer, until on the 14th the supply of water began to fail. The garrison also suffered the loss of Mr. McLanahan, who was wounded and died in a few hours. In the afternoon of the 14th the Cyane anchored off the town, and the interest of the battle now turned in this direction.

At the dawn of day, on the 15th, the boats of the Cyane were seen moving towards the shore. They contained one hundred and two officers, seamen, and Marines under Commander Du Pont. As the Mexicans possessed the advantage of the covers along the road, they were able to keep up an annoying flank and rear fire upon the advancing Americans. Still the latter moved steadily on, driving their opponents back at every point, or giving them volleys of musketry to the right and left. Those in the ship were unable to render any assistance, as it was impossible to separate friend from foe. But the garrison could not curb their impatience, and soon sallied out to form a junction with their companions. A party of the enemy were still firing upon the fort, but the charge led by Lieutenant Heywood drove them from the street they occupied and opened the way for the union of the two detachments. Just before this union was effected, the enemy made a final stand and sought by great exertions to maintain his ground. But it all was in vain, and the shout of victory was raised just as Lieutenant Heywood joined the party of his welcome friends. The enemy lost thirty killed and fifteen wounded. The casualties to Lieutenant Heywood's command during the siege were three killed and four wounded.

On the 14th, Lieutenant Stanley of the Dale landed at Guaymas with a force of Marines and seamen, and marched to the attack of the post of Bacochi Vampa. The barracks were situated on a spur of the mountain facing the west, stony and bushy, which rendered a near approach without discovery impossible. As soon as the sentinel hailed them, according to previous arrangement, they charged in double-quick time. But duty and ambition could not take them up to the barracks as fast as fear carried the flying enemy away. They made but one prisoner; several of the others left their arms, ammunition, accoutrements, and stores, and one officer the best part of his uniform.

On the receipt of this information of a provisional suspension of hostilities on the part of the main army in Mexico, Commodore Shubrick entered into a similar arrangement, and withdrew the garrison fiom Mazatlan, leaving a few Marines to guard the works until the establishment of a definite treaty of peace. Until the close of the war, therefore, the duty of the naval force was confined to the maintenance of the positions already held.

The following is an extract from the report of Commodore Shubrick, commanding Pacific Squadron, dated on board the Independence, Mazatlan, February 21: "The Marines have behaved with the fidelity and constancy which characterizes the valuable Corps, and I embrace this opportunity respectfully to recommend that ships coming to this station be allowed as large a complement of these valuable men as possible. The service would be greatly benefited by doubling the number allowed to each ship, and reducing to the same extent, if necessary, the complement of landsmen and ordinary seamen. The want of Marines is strongly felt in all operations on shore."

The following is a list of the officers of the Corps attached to the squadron on the Pacific coast in 1846-48:

Frigate Savannah, Captain Ward Marston, Second Lieutenant Henry W. Queen; frigate Constitution, Second Lieutenant Joseph W Curtis; frigate Congress, First Lieutenant J. Zeilin; sloop Portsmouth, Second Lieutenant Henry B. Watson; sloop Cyane, Second Lieutenant W. A. T. Maddox; Independence, Captain James Edelin, Second Lieutenant William W. Russell; sloop Dale, Second Lieutenant Robert Tansill; First Lieutenant A. H. Gillespie, on special service; Columbus, Captain Henry B. Tyler, First Lieutenant W A. T. Maddox, Second Lieutenant John C. Cash; Ohio, Captain J. L. Hardy, Second Lieutenant James A. Buchanan, Second Lieutenant James Wiley.

---

## CHAPTER XIV.

**1846-48.**—War with Mexico, II.—Naval Operations on the East Coast of Mexico.

When hostilities commenced, Commodore Conner was in command of the Gulf Squadron. Instructions were at once issued to him to employ his command in the blockade of the Mexican ports. These orders were issued on the 13th of May, 1846.

While the Marines connected with the naval force on the west coast were doing such excellent service, the officers and men belonging to the Corps and attached to the squadron were also aiding materially in the work going on in that section. The following is a list of the officers of the Corps attached to the Gulf Squadron in 1846, Commodore David Conner, commander-in-chief: Frigate Cumberland, First Lieutenant D. D. Baker; frigate Raritan, First Lieutenant William Lang; frigate Potomac, First Lieutenant Addison Garland; sloop John Adams, First Lieutenant R. C. Caldwell; sloop St. Mary's, Second Lieutenant John D. Simms. In 1847 the list of officers attached to the squadron was as follows: Frigate Raritan, Captain Alvin Edson, Second Lieutenant George Adams; frigate Potomac, First Lieutenant Addison Garland; sloop Albany, Second Lieutenant William B. Black; sloop John Adams, First Lieutenant Robert C. Caldwell, Second Lieutenant Felix G. Mayson; sloop St. Mary's, Second Lieutenant John D. Simms. The following were the officers attached to the same squadron, Commodore M. C. Perry, commander-in-chief, in 1848: Frigate Cumberland, First Lieutenant Isaac T. Doughty, Second Lieutenant Jacob Read; sloop Albany, Second Lieutenant William Butterfield; steamer Mississippi, First Lieutenant William B. Slack, Second Lieutenant William F Perry.

On the 18th of May a detachment of two hundred Marines and seamen from the Cumberland and Potomac, under the orders of Captain Aulick, sailed up the Rio Grande in the boats of the ships, and effected a junction with a detachment of the army at Barita, on the right bank of the river, and about fifteen miles from its mouth. At this point they established a post without any opposition, and on the same day the army under General Taylor crossed the river and entered Matamoras. The officers of the Corps engaged in this expedition were First Lieutenant D. D. Baker and First Lieutenant Addison Garland.

The first work of importance in which the Marines of the home squadron were engaged was in October, in connection with Commodore M. C. Perry's expedition up the Tobasco River, which resulted in the capture of Frontera and Tobasco. The expedition sailed on the 16th of October, and arrived off the mouth of the Tobasco on the 23d. The town of Frontera was taken without serious resistance, with several sailing vessels and two steamers. On the 25th Tobasco was reached, and the squadron anchored, in order of battle, abreast of the town, within half-musket range. A messenger was sent on shore with a summons to the governor to surrender. An insolent answer was returned, and an invitation sent back to the commander to fire as soon as he pleased. Commodore Perry was extremely reluctant to destroy the town and to subject the inhabitants to the miseries of a bombardment. It was necessary, however, to make some demonstration of his power, and he consequently began a bombardment, which was kept up till the bravado of the official was thoroughly punished, after which Commodore Perry took his departure.

The Mississippi, with a detachment of one hundred seamen and Marines from the Cumberland and the Princeton, put to sea on the 12th of November. At daylight on the morning of the 14th the force reached the bar of Tampico. A detachment of three hundred officers, Marines, and seamen in boats were carried over the bar safely, and passed into the river without opposition, the fort at the entrance having been abandoned. On approaching the town it surrendered. The Marines and seamen were not landed. The officers of the corps engaged in this expedition were First Lieutenant D. D. Baker, First Lieutenant William Lang, First Lieutenant Addison Garland, and Second Lieutenant J. D. Simms.

**1847.**—During the autumn a plan was marked out which contemplated the reduction of Vera Cruz and the march of a large army to the capital. Accordingly, the following year opened with very extensive and active preparations for the attack. On the coast of Mexico was soon assembled what was up to that time the largest force ever under the command of a single American naval officer. The debarkation was appointed to take place on the 8th of March. General orders were therefore issued on the 7th, by the commodore and the commanding-general, prescribing the necessary arrangements. The officers of the Marine Corps who were attached to the battalion in the operations before Vera Cruz were Captain Alvin Edson, First Lieutenant Addison Garland, First Lieutenant Robert C. Caldwell, Second Lieutenant W. B. Slack, Second Lieutenant J. D. Simms, Second Lieutenant George Adams, and Second Lieutenant Felix G. Mayson. During the siege, a detachment of officers and sailors were assigned a place in the trenches, and did noble service. The Marines behaved with their accustomed gallantry, and shared the honors with their gallant brethren of the army and Navy. On the 29th of March the combined forces of the army and Navy took possession of the city and castle, while the American flag floated over both, amid a grand salute from the squadron and the batteries. The following is an extract from Commodore Conner's report, dated on board the Raritan, off Sacrificios, March 10, 1874: "General Scott has now with him upward of eleven thousand men. At his request, I permitted the Marines of the squadron, under Captain Edson, to join him, as a part of the Third Regiment of artillery." On the 28th of March, Brigadier-General Worth issued the following: "The general of brigade avails himself of the occasion on separating from Captain Edson, his officers, and men to express his high appreciation of the energy, zeal, and thorough soldiership which marked their effective co-operation during our association, and also to tender his cordial thanks and respects."

Commodore Perry relieved Commodore Conner on the 21st of March, and turned his attention to a movement against Tuspan. On the morning of the 18th of April, after all due preparations had been made, the whole force was detached from the ship, forming the landing party, and carried in barges, numbering one thousand four hundred and ninety officers, seamen, and Marines, with four pieces of light artillery. The whole was led by Commodore Perry. The Mexicans made but a feeble defence. They rapidly fell back, deserting the batteries before the landing party could get near enough to storm them. The attack proved entirely successful, and the capture was effected, with a loss to the Americans of only three seamen killed and five officers and six men wounded. The following are the names of the officers of the Marine Corps engaged in this expedition: Captain Alvin Edson, First Lieutenant Addison Garland, First Lieutenant Robert C. Caldwell, First Lieutenant W. B. Slack, Second Lieutenant M. R. Kintzing, Second Lieutenant George Adams, and Second Lieutenant Felix G. Mayson.

Some six months had elapsed since the affair at Tobasco, when news reached Commodore Perry that matters at that town had assumed such a phase that the presence of an American force was needed to compel a proper respect for the authority of the United States. Early in June, he started with a formidable force. The officers of the Corps engaged in this expedition were Captain Alvin Edson, Lieutenant W. B. Slack, Lieutenant M. R. Kintzing, Lieutenant Robert C. Caldwell, Second Lieutenant George Adams, and Second Lieutenant Felix G. Mayson. The expedition arrived at the mouth of the Tobasco River, and, after various adventures, reached a point below Tobasco, when a landing was made. The force on shore marched to attack the town by land, while the vessels advanced to lay siege from the river.

The Scorpion pushed on up the stream, and, upon arriving off the town, Captain Bigelow demanded a surrender of the place from the authorities, which was acceded to. The officer sent to hoist the American colors, while in the act of unfurling it to the breeze from the top of the government house, was fired upon several times, one ball passing through the flag, but he escaped untouched. About three hours after the stars and stripes had floated over the town and port, the land expedition, which it had been supposed would have the chief part of the honor and glory of taking the place, came in sight of the fort, when to their surprise they beheld the flag waving over it, which told them they were too late, the work having been done by their more fortunate friends attached to the flotilla. The commodore remained until the 22d inst., and then left to rejoin the squadron outside the bar, leaving as the garrison of the place two officers of Marines and sixty-nine men under Lieutenant Slack, besides the bomb brig Ætna and steamers Spitfire and Scourge, anchored off the town. Captain Van Brunt of the Ætna was left as governor, and Lieutenant Slack as military commandant. On the 24th inst. two field-pieces were landed, which, with the one already ashore, were placed in battery in the main plaza commanding the various entrances to it; the guard, twenty-eight men, were quartered there, and the balance of the sixty-nine Marines, under Lieutenant Adams, were ordered to march directly to the square in the event of an alarm. Two of the pieces were manned by eight sailors, each under their respective officers; the third piece was manned with seven Marines, under charge of a Corporal, they having formerly served in the artillery.

On that very night, between ten and eleven o'clock, an effort was made to drive them in from the plaza by about two hundred Mexicans divided in two parties, the one large, the other small, the latter to be the feint and the former the real attack, led on by Colonel Garcia, an officer of some talent and considerable courage. The sentries, being fired upon by the small body, retired and joined the guard, the large body in the mean time pressing on in another direction to gain an old dilapidated building facing the square, from which they evidently expected to do some telling work.

The guard was under arms in front of the guard-house, and as every man thus stood at his post, a volley of musketry from the old building was fired at them; instantly the three pieces opened fire on the building. The enemy was silenced in a few minutes. From that time forward troops could be seen every day outside the city limits moving from one point to another, and at night the sentinels were sure to be fired upon from various points, thus causing all hands to be continually under arms.

On the night of July 26, Captain Bigelow of the Scorpion arrived from the squadron, intending to return almost immediately, but upon being informed by Captain Van Brunt of the state of affairs ashore, he determined to remain and despatch a boat to the commodore, advising him of the necessity of a reinforcement. On the 29th the Vixen returned, bringing an increase in force of fifty-five Marines and fifty sailors. Captain Bigelow now determined to fit out an expedition and beat up the enemy's quarters wherever he could be found. Accordingly, on the 30th, he left Tobasco with about one hundred and twenty-five sailors as small-arm men, one hundred and fifteen Marines, under Lieutenant Slack, and two field-pieces (6-pounders), and, from information in his possession of the enemy's whereabouts, moved for an Indian village about four miles distant called Tamulté. Just before arriving at the village, and while ascending a gentle acclivity, a fire of musketry was opened by a body of troops occupying the summit, and mostly covered from view by a growth of underwood. This fire was handsomely returned by the Marines, and the two pieces of artillery were soon brought into position. For about twenty minutes a constant and general firing was maintained on both sides, until finally the Mexicans began to give way, at first slowly, but soon rapidly retreating. Throughout the fight Captain Bigelow and Lieutenant Slack acted with a courage and judgment worthy of all praise. During the action the Americans lost two men killed and two wounded, one severely and the other slightly. Lieutenant Kintzing was also slightly wounded by a spent ball, and two men were badly burned by the accidental explosion of powder in a building. The loss of the enemy was not ascertained.

After the above fight, for a day or two they were relieved from all annoyances from the enemy, but the latter soon again began the old system of night attacks, and finally, on the 14th and 15th of July, during the day, went into camp, and from a hill within range of the plaza fired upon the Americans. The latter drove them out, and endeavored to pursue them, but their better knowledge of the country enabled them to escape easily. Among the Americans the sick-list began to increase very rapidly, and soon one-half of the men were down. The commodore, being informed of the extent of sickness prevailing, determined finally to abandon the place. Accordingly, on the morning of the 22d of July, the colors were hauled down, the Marines paying the proper salute upon the occasion. The Americans then left the plaza, marching in column to the air of "Hail Columbia," and embarked on board the vessels bound down the river.

In a communication to the Department, dated on board the Mississippi, July 28, Commodore Perry wrote: "In transmitting the accompanying correspondence, I feel myself called upon to invite the attention of the Department to the zeal and gallantry with which Commander Bigelow, Commander Van Brunt, and all the officers, seamen, and Marines of the detachment left by me at Tobasco performed their arduous duties in defending the city against a most active enemy."

The following is an extract from the report of Commodore Perry, dated United States flag-ship Mississippi, Anton Lizardo, July 4, 1847: "Since the receipt of your despatch of the 21st of May, received on the 1st inst., I have had an interview with Brigadier-General Pierce, whose brigade the Marines ordered to be detached from the squadron are to join. General Pierce has fully agreed with me that the small force of Marines in the squadron, about one hundred and ninety effective men, will be of little advantage to him, and that the consequences of withdrawing them, and the necessity of the immediate evacuation of Tobasco, and of weakening other posts in our possession (measures that will be rendered indispensable by the detachment of the Marines from the squadron), will produce a most pernicious influence with the enemy, who always claim as a triumph any retrograde movement of ours. I beg to be understood that the order of the 21st of May will be obeyed in part immediately, however it may inconvenience the ships of the squadron; and the withdrawal of the Marines from Tobasco, Laguna, Frontera, and Alvarado will be postponed only till I can receive a reply to this communication or hear sooner from the Department on the subject. This arrangement can produce very little delay, as it will require considerable time to collect together the Marines, dispersed as they are along the coast, at points the extremes of which embrace a distance of nearly six hundred miles in extent.

I shall await with much interest your communications, and hope that you will not only soon be able to replace the Marines withdrawn from the squadron, but add to their number."

It is stated that General Pierce urged the commodore to "take the responsibility" of retaining his Marines and holding his captured places, and offered a letter expressive of his conclusion and judgment to this end, which the commodore said was unnecessary.

---

## CHAPTER XV.

1847-48.—War with Mexico, III.—Operations of the Marines attached to the Army under General Scott.

IN June, 1847, a battalion of Marines, under the command of Lieutenant-Colonel Watson, sailed from Fort Hamilton, New York harbor, to join the army under General Scott. The battalion debarked at Vera Cruz and joined the main army at Puebla, August 6. Lieutenant-Colonel Watson's command was assigned as follows:

*Fourth Division.*—Major-General Quitman.

*First Brigade.*—General Shields: New York Regiment, South Carolina Regiment, Steptoe's Battery.

*Second Brigade.*—Lieutenant-Colonel Watson.

*Battalion of Marines.*—Second Pennsylvania Volunteers.

Leaving a competent garrison at Puebla, the army under Scott marched towards the Mexican capital. The brilliant victories of Contreras, Churubusco, and Molino del Rey reflected additional lustre upon the veterans of Vera Cruz. Writing under date of August 27, General Scott said, "I regret having been obliged, on the 20th, to leave Major-General Quitman, an able commander, with a part of his division,—the fine Second Pennsylvania volunteers, and the veteran detachment of United States Marines,—at our important depot, San Augustine. It was there that I had placed our sick and wounded, the siege-, supply-, and baggage-trains. If these had been lost, the army would have been driven almost to despair; and, considering the enemy's very great excess of numbers, and the many approaches to the depot, it might well have become, emphatically, *the post of honor.*"

On the 13th of September the volunteer division, under command of Major Levi Twiggs, of the Marines, accompanied by a pioneer party of seventy men, under Captain J. G. Reynolds, also of the Marines, bearing ladders, crows, and pickaxes, were placed at the head of the column of attack upon the fortress of Chapultepec. These storming parties were supported by the battalion of Marines under command of Lieutenant-Colonel Watson. The brave and lamented Major Twiggs was killed while leading the assault. Immediately after the capture of the fortress, the whole column, under General Quitman, moved directly on the city of Mexico by the Tacubaya Causeway, leading through the Garita Belen into the city. Soiled with dust and smoke, and begrimed with blood, the field officers on foot with the men, they moved on to the charge with banners furled, and no music but the roar of cannon and the rattling roll of small arms. The Garita was taken in a charge at full run at twenty minutes past one o'clock P.M. At break of day of the 14th, a white flag announced the surrender of the enemy's stronghold, the citadel. The division of General Quitman, therefore, was the *first to enter the city*. The honor of first entering the palace, also, and of hoisting upon it the national flag, was accorded to this division, with which the battalion of Marines was connected, thus explaining and justifying the motto afterwards inscribed upon the colors of the Corps: *"From Tripoli to the Halls of the Montezumas."*

The palace was overrun with escaped convicts, and other desperate characters, bent on plunder. Lieutenant-Colonel Watson, with his battalion of Marines, was ordered to clear it of these intruders and protect it from spoliation.

Lieutenant A. S. Nicholson was accordingly detailed, with a detachment of forty men, to perform this duty, which he promptly and successfully accomplished.

The following is a roster of officers of the battalion under the command of Lieutenant-Colonel Watson:

Brevet Lieutenant-Colonel Samuel E. Watson, Major Levi Twiggs, and Major William Dulany. Staff: First Lieutenant and Adjutant D. D. Baker; First Lieutenant and Acting Quartermaster J. S. Develin. Captains: John G. Reynolds, George H. Terrett, and William Lang. First Lieutenants: Jabez C. Rich, Robert C. Caldwell, William L. Young, Thomas A. Brady, John D. Simms, and Daniel J. Sutherland. Second Lieutenants: George Adams, Ed. McD. Reynolds, Thomas Y. Field, Charles G. McCawley, Freeman Norvell, Charles A. Henderson, John S. Nicholson, Augustus S. Nicholson, and Henry Welsh.

The following is a list of the killed and wounded of the Marine battalion under the command of Colonel Watson, incident to the capture of Chapultepec and the City of Mexico, on the 13th and 14th of September: Wounded at Chapultepec—Staff: First Lieutenant and Adjutant D. D. Baker, slightly; First Lieutenant and Acting Quartermaster J. S. Develin, slightly; Sergeant-Major James Montgomery, slightly; Commissary-Sergeant James Orr, severely. Killed: Major Levi Twiggs, at Chapultepec; Corporal Hugh Graham and Privates Anthony A. Egbert, Andrew McLaughlin, John Herbert, Matthew Banks, and Thomas Kelly, all at the gate of the city. Wounded—Company A: Privates Seebeck and Milburn, at the City of Mexico. Company B: Sergeant John Roach and Privates Hugh Rooney, Martin Fogg, John McGihen, P. Phœnix, and S. Williamson, all at Chapultepec. Company C: Privates Briggs and Connor, at Chapultepec. Company D: Sergeants J. Curran, G. Tansill, and W. J. Wilson, and Private Thomas Quinn, at Chapultepec. Company E: Privates Thomas B. Smith and E. Stevens, at Chapultepec. Company F: Second Lieutenant Charles A. Henderson, Musician McDonald, and Privates E. Cooper and James Linns, all at Chapultepec.

The following extract from the report of Major-General Quitman, dated at the National Palace, Mexico, September 29, gives a vivid idea of the work accomplished by the Marines: "During the day I succeeded, under cover of our batteries, in making an important reconnoissance of the grounds and works immediately at the base of the castle. The supporting party on this reconnoissance was commanded by the late Major Twiggs, of the Marines, and sustained during the observation a brisk fire from the batteries and small arms of the enemy, who, when the party were retiring, came out of the works in large numbers; and, although repeatedly checked by the fire of our troops, continued to advance, as the supporting party retired, until they were dispersed, with considerable loss, by several discharges of canister from the guns of Captain Drum's battery, and a well-directed fire from the right of the Second Pennsylvania Regiment, posted on the flank of the battery for its support. During the day my command was reinforced by a select battalion from General Twiggs's division, intended as a storming party, consisting of thirteen officers and two hundred and fifty men and non-commissioned officers and privates, chosen for this service out of the Rifles, First and Fourth Regiments of artillery, Second, Third, and Seventh Regiments of infantry, all under the command of Captain Silas Casey, Second Infantry. At dawn, on the morning of the 13th, the batteries again opened an active and effective fire upon the castle, which was returned by the enemy with spirit and some execution, disabling for a time the 18-pounder in battery No. 1, and killing one of the men at the guns.

"During this cannonade, active preparations were made for the assault on the castle. Ladders, pickaxes, and crows were placed in the hands of a pioneer storming party of select men from the volunteer division, under command of Captain Reynolds of the Marine Corps, to accompany the storming party of one hundred and twenty men, which had been selected from all Corps of the same division, and placed under command of Major Twiggs of the Marines. Perceiving that all the preliminary dispositions were made, Major Gladden, with his regiment, having passed the wall by breaching it, the New York and Pennsylvania regiments having entered over an abandoned battery on their left, and the battalion of Marines being posted to support the storming parties, I ordered the assault at all points. The storming parties, led by the gallant officers who had volunteered for this desperate service, rushed forward like a resistless tide. The Mexicans behind their batteries and breastworks stood with more than usual firmness. For a short time the contest was hand-to-hand; swords and bayonets were crossed, and rifles clubbed. Resistance, however, was vain against the desperate valor of our brave troops. The batteries and strong works were carried and the ascent of Chapultepec on that side was laid open to an easy conquest. In these works were taken seven pieces of artillery, one thousand muskets, and five hundred and fifty prisoners, of whom one hundred were officers, among them one general and ten colonels. . . . The command of the storming party from the volunteer division devolved on Captain James Miller, of the Second Pennsylvania Regiment, by the death of its chief, the brave and lamented Twiggs of the Marine Corps, who fell on the first advance at the head of his command.

"Captain Roberts, of the rifle regiment, who had led the advance company of the storming party at Chapultepec, and had greatly distinguished himself during the preceding day, was detailed by me to plant the star-spangled banner of our country upon the national palace. The flag, the first strange banner which had ever waved over that palace since the conquest of Cortez, was displayed and saluted with enthusiasm by the whole command. The palace, already crowded with Mexican thieves and robbers, was placed in charge of Lieutenant-Colonel Watson, with his battalion of Marines. By his active exertions, it was soon cleared and guarded from further spoliation." Adjutant Baker is mentioned by the general "as conspicuous for his bravery and efficiency."

Lieutenant Robert C. Caldwell was attached to the division of Major-General Pillow as senior officer of the commissary department.

A second battalion of Marines, under the command of Major John Harris, sailed from New York in March, 1848, to co-operate on shore with the squadron, on the Isthmus of Tehuantepec. Upon their arrival, however, at Vera Cruz, the armistice had been concluded, and they were prevented from proceeding there. Major Harris, therefore, was ordered to garrison Alvarado with his battalion. The following is a roster of the battalion under the command of Major John Harris: John Harris, Major, commanding; Algernon S. Taylor, First Lieutenant and Adjutant; W. L. Shuttleworth, First Lieutenant and Assistant Quartermaster.

Company A, ninety men: N. S. Waldron, Captain; J. R. F. Tatnall, Second Lieutenant. Company B, ninety men: Josiah Watson, First Lieutenant, commanding; George R. Graham, Second Lieutenant. Company C, ninety men: James H. Jones, Second Lieutenant, commanding; John H. Broome, Second Lieutenant. Company D, ninety men: George F. Lindsay, Second Lieutenant, commanding; W. S. Boyd, Second Lieutenant. Two drummers and two fifers.

---

## CHAPTER XVI.

**1852-55.**—The Expedition to Japan—Combined attack of English and Americans on the Chinese—Monument in Happy Valley, Hong-Kong.

In the year 1852, Commodore M. C. Perry was ordered to the command of the naval squadron cruising in the East Indies. Besides discharging his ordinary duties he was empowered to make a display of his force in the waters of the empire of Japan, and to endeavor to contract with that government a treaty of amity and commerce. The government of the United States felt itself obliged to make an effort to secure from the Japanese the better treatment of American seamen cast upon their shores, and friendly intercourse with American vessels seeking supplies in their ports. A squadron was therefore placed at the disposal of Commodore Perry, comprising the frigates Mississippi, Susquehanna, Powhatan, the corvette Macedonian, the sloops-of-war Plymouth, Saratoga, Vandalia, and the store-ships Supply, Southhampton, and Lexington. This squadron anchored off the city of Uraga, in the bay of Jeddo, Japan, on the 8th of July, 1853.

Commodore Perry had resolved not to submit to those restrictions and annoyances to which the Japanese government had been accustomed to subject all foreign vessels. He therefore ordered away the guard boats which began to collect around the ships, and intimated that he would employ force if they were not withdrawn. He also refused to confer with any but officers of the highest rank in the empire, keeping himself very secluded, and receiving communications from ordinary officials through his own subordinate officers. All this close of the late sanguinary insurrection in Montevideo, when the government troops charged upon the insurgents who had capitulated, I threw myself between them and thus preserved the latter from destruction.

A sense of justice impels me to say I was not present on the occasion referred to, and that it was Lt. A. S. Nicholson, with the Marine guard of this ship, who very properly interfered to prevent the massacre. I approved his course and thereby became responsible for the consequences, but the generous act was his own, and all the credit is due to him.

Be pleased to make this known in any manner you like, provided the reparation be ample.

Your ob't servant,

(Sg'd) W. F. LYNCH,
*Capt. U. S. Navy.*

**1856.**—The sloop-of-war Decatur, Commander Gansevoort, was stationed off the town of Seattle, Washington Territory, for the purpose of protecting the inhabitants from the savages. On the morning of the 26th of January, 1856, Commander Gansevoort, being satisfied that the Indians were posted in large force in the woods surrounding the town, landed the crew and Marines of the ship in four divisions, with a howitzer to protect the inhabitants. The action was promptly opened by a shot from the howitzer, and a shell from the ship's battery. The Indians returned the fire with spirit from the edge of the woods. A steady fire of small arms, shot, and shell, for half an hour, drove them from their first position to the northward of the town, whence they renewed the attack, and maintained it with bravery for six hours. They finally retired in the night. They numbered several hundreds, and suffered severely.

"One of the most gallant actions in the history of the Navy took place in Chinese waters in the month of November, 1856. It consisted in the storming and destruction of the Canton Barrier Forts by a portion of the squadron under Flag-Officer James Armstrong. Difficulties had arisen in the autumn of that year between the authorities of the city of Canton and the English officials in that region, which caused the British admiral to station a portion of his fleet off that city. Commander A. H. Foote, of the United States sloop-of-war Portsmouth, fearing injury to American interests, took his ship up to the island of Whampoa. Soon afterwards, with several armed boats, he ascended the river to Canton, and established an armed neutrality; the forces from the Levant, Commander William Smith, being added to that of the Portsmouth. In an attack by the British upon the governor's palace several Americans joined, and planted their flag beside the British on the walls of the city. By proclamation Commander Foote disavowed this proceeding, as compromising American neutrality. On consultation with the flag-officer at Whampoa, it was determined to withdraw the American forces from the shore, place them on board the Levant, and anchor her off the city. When Commander Foote was returning from this interview in his unarmed boat, a round shot was fired from the first fort, and, though the American flag was displayed, several more shots were fired, together with grape and canister, from the second fort, compelling the return of the boat to Whampoa. On the following day, the Portsmouth, bearing the flag of Commodore Armstrong for the occasion, and the Levant, their complement of men on duty at Canton being supplied from the officers and men of the flag-steamer San Jacinto, proceeded to the Barrier Forts for the purpose of resenting the insult.

"The Levant grounded before coming within range of the forts, but the Portsmouth, proceeding under a sharp fire till within four hundred and eighty yards of the nearest fort, opened her fire, and maintained it so effectually that, at the end of three hours, during which time she threw over two hundred shells, the enemy ceased firing. The Portsmouth was struck nineteen times; still not a man was wounded.

"Four days of unsatisfactory negotiations with *Yeh*, the governor of Canton, ensued, when Commodore Armstrong, having on account of illness withdrawn from the Portsmouth, gave permission to Commander Foote to renew the attack. The Levant having joined the Portsmouth on the morning of the 20th, both ships commenced a brisk fire, which was returned with great spirit by the forts for an hour, when two hundred and eighty officers, seamen, and Marines were landed in ten boats, four of which had boat-howitzers. The men from the San Jacinto and Levant were under the command of Commanders Smith and Bell, and the whole force was led by Commander Foote. A safe landing was effected opposite the ships, and the nearest fort was stormed in the rear and taken, with the loss, according to their own report, of more than three hundred Chinese. They endeavored with five thousand troops to retake the fort, but were successfully repulsed.

"On the following morning, after another hour's cannonading, the second fort was stormed and carried in a similar manner, but one man of the Americans being killed. One of the launches, however, when advancing towards the fort, was struck by a 68-pound shot, which killed two and wounded six men. In the afternoon of the same day, the third fort, on an island in mid-river, was captured with little opposition. At daylight on the 22d the boats were manned on the lee of the island, whence they were rowed round the point, and crossed the river in the face of a heavy fire from the farther fort and a large the English suffered to about the same extent. Sergeant Hamilton commanded the Marines of the Plymouth on this occasion.

Piracy in the China seas had been practiced by the Chinese for many years previous to 1850, and although condign punishment had more than once been administered by the English and American forces in those waters, yet the pirates were still bold, strong, and indefatigable; vessels were plundered, and the most imaginable scenes connected with the idea of piracy were realities in the career of those lawless bands. Some of the freebooters seemed intent only on plunder, but the majority delighted in the opportunity to gratify more bloody and cruel dispositions, and when rapacity had been satisfied, or hindered in its desires, they gave way to the demands of fiercer passions. Merchant ships were armed to some extent, and in some instances successfully repelled the attacks of the pirates.

In the year 1855 a combined boat attack from the United States ship Powhatan and the English sloop Rattler was made off Khulan against a fleet of piratical junks. The attack was successful, the junks were destroyed, and many prisoners taken.

To commemorate this event a monument was erected in the cemetery at Happy Valley, Hong-Kong, a silent but eloquent tribute to those who fell in the attack.

This monument is built of granite, in the form of a pyramid, about eight feet in height from the base to the apex, and contains on one of the sides the following inscription:

Erected

by the officers and crews of the U. S. Steam Frigate Powhatan and H. B. M. Steam Sloop Rattler in memory of their shipmates who fell in a combined boat attack on a fleet of piratical Junks off Khulan, August 4, 1855.

On the opposite side are the names, as follows:

| | |
|---|---|
| George Mitchell, A. B. | John Pepper, Seaman. |
| James Silvers, Carpenters' Crew. | J. A. Halsey, Landsman. |
| John Maroney, R. M. A. | S. Mullard, Private U. S. M. C. |
| M. Oliff, R. M. A. | B. F. Adamson, Private U. S. M. C. |

## CHAPTER XVII.

**1855-60.**—Insurrection at Montevideo—Engagement with Savages in Washington Territory—The Barrier Forts in China—The Riot in Washington—Savages at Waya—Montevideo—Expedition to Paraguay—Services on Staten Island—John Brown Insurrection—Services at Kisembo—Panama.

DURING the month of November, 1855, an insurrection occurred in Montevideo, which was so sanguinary and disastrous in its character that the foreign residents were in great danger, their lives and property being imperilled.

The United States ship Germantown, Captain W. F. Lynch, together with English, French and Spanish men-of-war, was then lying at anchor in the harbor.

Captain Lynch, who was the senior officer, deemed that this emergency demanded the presence of an armed force on shore. Accordingly, a detachment of Marines, under the command of First Lieutenant A. S. Nicholson, was landed and occupied the custom-house and consulate.

Subsequently the insurrection became so formidable that the commander of the Germantown decided to reinforce Lieutenant Nicholson, and a force of sailors, composed of detachments from the several vessels, amounting to one hundred men, were landed and placed under the command of Lieutenant Nicholson.

The following order will explain the situation, and the subsequent letter to the *National Intelligencer* shows that Marines under their commanding officer prevented a dreadful massacre:

U. S. S. GERMANTOWN, OFF MONTEVIDEO, November 27, 1855.

SIR,—When you see our boats shove off prepare to take a position to cover our landing at the custom-house stairs. After we have landed, our left will be protected by the high wall on the east. You will then advance so as to have in full view the triangular building at the S. E. front of the custom-house, and there await further orders. *Be particular in not permitting an act of hostility unless it becomes necessary in self-defence.* Should the men posted on the roof of the triangular building fire upon you, dislodge them either by musketry or by assault.

Respectfully,

(S'g'd) W. F. LYNCH,
*Commander.*

AUG. S. NICHOLSON, *First Lieut.*

U. S. SHIP GERMANTOWN, AT SEA, May 6, 1856.

TO THE EDITOR OF THE NATIONAL INTELLIGENCER, WASHINGTON, D. C..

GENTLEMEN,—I have recently seen in a New York journal a paragraph copied from your paper of Feb. 21, or thereabouts, wherein it is stated that at the close of the late sanguinary insurrection in Montevideo, when the government troops charged upon the insurgents who had capitulated, I threw myself between them and thus preserved the latter from destruction.

A sense of justice impels me to say I was not present on the occasion referred to, and that it was Lt. A. S. Nicholson, with the Marine guard of this ship, who very properly interfered to prevent the massacre. I approved his course and thereby became responsible for the consequences, but the generous act was his own, and all the credit is due to him.

Be pleased to make this known in any manner you like, provided the reparation be ample.

Your ob't servant,

(Sg'd) W. F. LYNCH,
*Capt. U. S. Navy.*

**1856.**—The sloop-of-war Decatur, Commander Gansevoort, was stationed off the town of Seattle, Washington Territory, for the purpose of protecting the inhabitants from the savages. On the morning of the 26th of January, 1856, Commander Gansevoort, being satisfied that the Indians were posted in large force in the woods surrounding the town, landed the crew and Marines of the ship in four divisions, with a howitzer to protect the inhabitants. The action was promptly opened by a shot from the howitzer, and a shell from the ship's battery. The Indians returned the fire with spirit from the edge of the woods. A steady fire of small arms, shot, and shell, for half an hour, drove them from their first position to the northward of the town, whence they renewed the attack, and maintained it with bravery for six hours. They finally retired in the night. They numbered several hundreds, and suffered severely.

"One of the most gallant actions in the history of the Navy took place in Chinese waters in the month of November, 1856. It consisted in the storming and destruction of the Canton Barrier Forts by a portion of the squadron under Flag-Officer James Armstrong. Difficulties had arisen in the autumn of that year between the authorities of the city of Canton and the English officials in that region, which caused the British admiral to station a portion of his fleet off that city. Commander A. H. Foote, of the United States sloop-of-war Portsmouth, fearing injury to American interests, took his ship up to the island of Whampoa. Soon afterwards, with several armed boats, he ascended the river to Canton, and established an armed neutrality; the forces from the Levant, Commander William Smith, being added to that of the Portsmouth. In an attack by the British upon the governor's palace several Americans joined, and planted their flag beside the British on the walls of the city. By proclamation Commander Foote disavowed this proceeding, as compromising American neutrality. On consultation with the flag-officer at Whampoa, it was determined to withdraw the American forces from the shore, place them on board the Levant, and anchor her off the city. When Commander Foote was returning from this interview in his unarmed boat, a round shot was fired from the first fort, and, though the American flag was displayed, several more shots were fired, together with grape and canister, from the second fort, compelling the return of the boat to Whampoa. On the following day, the Portsmouth, bearing the flag of Commodore Armstrong for the occasion, and the Levant, their complement of men on duty at Canton being supplied from the officers and men of the flag-steamer San Jacinto, proceeded to the Barrier Forts for the purpose of resenting the insult.

"The Levant grounded before coming within range of the forts, but the Portsmouth, proceeding under a sharp fire till within four hundred and eighty yards of the nearest fort, opened her fire, and maintained it so effectually that, at the end of three hours, during which time she threw over two hundred shells, the enemy ceased firing. The Portsmouth was struck nineteen times; still not a man was wounded.

"Four days of unsatisfactory negotiations with *Yeh*, the governor of Canton, ensued, when Commodore Armstrong, having on account of illness withdrawn from the Portsmouth, gave permission to Commander Foote to renew the attack. The Levant having joined the Portsmouth on the morning of the 20th, both ships commenced a brisk fire, which was returned with great spirit by the forts for an hour, when two hundred and eighty officers, seamen, and Marines were landed in ten boats, four of which had boat-howitzers. The men from the San Jacinto and Levant were under the command of Commanders Smith and Bell, and the whole force was led by Commander Foote. A safe landing was effected opposite the ships, and the nearest fort was stormed in the rear and taken, with the loss, according to their own report, of more than three hundred Chinese. They endeavored with five thousand troops to retake the fort, but were successfully repulsed.

"On the following morning, after another hour's cannonading, the second fort was stormed and carried in a similar manner, but one man of the Americans being killed. One of the launches, however, when advancing towards the fort, was struck by a 68-pound shot, which killed two and wounded six men. In the afternoon of the same day, the third fort, on an island in mid-river, was captured with little opposition. At daylight on the 22d the boats were manned on the lee of the island, whence they were rowed round the point, and crossed the river in the face of a heavy fire from the farther fort and a large supporting force of troops; but this fort was captured by a rapid assault, the Chinese retreating beyond the reach of musketry or guns.

"All these forts were constructed by European engineers, with granite walls seven feet in thickness, and mounted one hundred and sixty-eight guns of large and small calibre. Though defended by five thousand men, they were taken by two ships and a force of two hundred an eighty officers, seamen, and Marines; the Portsmouth alone, in the first day's attack, silencing them all, and breaking the walls sufficiently to allow the storming party to enter without difficulty. Their capture settled the difficulty with the Chinese, led to the formation of a treaty of amity and commerce, and caused the American name to be respected by that people. The Secretary of the Navy, in a general order, commended the officers and men in the following words: 'The prompt and decisive course pursued by Commodore Armstrong, his officers and men, caused the flag of the United States to be respected by the Chinese, contributed largely to the security of our citizens in China, and, during the troubles which followed, was probably the means of saving many lives and much property.'"

The following officers of the Corps were in this engagement and in the assault upon the forts: San Jacinto, Brevet Captain John D. Simms; Portsmouth, Second Lieutenant William A. Kirkland; Levant, Second Lieutenant Henry B. Tyler, Jr.

In the latter part of this year, during the efforts to subdue the Indians at Puget Sound, the Marines were engaged. In the fight also at Port Gamble, with Northern Russian Indians, a detachment of Marines in charge of a sergeant behaved with great gallantry.

1857.—The following description of the riot in the city of Washington, June 1, 1857, is taken from the *Washington Star:* "A gang of hired roughs and bullies, 'Plug Uglies,' and other worthies, arrived from Baltimore, having been imported by the 'Know-Nothings' to take full possession of the polls, and, armed with revolvers, billies, and slung-shots, to prevent our citizens from depositing their votes. . . . At the first precinct of the Fourth Ward the officers did all in their power to prevent a riot, placing themselves between the belligerents and voters in the line, with Captain Goddard at the head, who demanded peace, when they rushed on *en masse*, with an impetuosity which nothing short of military force could have withstood; and, although the police fought like heroes, they were forced to abandon the field to the possession of these hired miscreants. Such an exhibition of murderous instruments as the party carried was sufficient to cause the peaceably disposed to keep as far as possible from them. One man was armed with a huge blacksmith sledge, another with a horse-pistol, a third carried a miscellaneous assortment of revolvers, bowie-knives, billies, and an iron bar, while a fourth carried, besides a sack filled with stones, brickbats, etc., a large maul of sufficient strength to fell an ox. The parties brandished them about in a menacing manner, to the terror of all those who were not like themselves participants in these disgraceful scenes.

"Mayor Magruder accordingly sent a dispatch to the President of the United States, asking that a company of Marines should be ordered out for the purpose of preserving the peace. . . . At one o'clock a crowd of men and boys passed up Pennsylvania Avenue from the direction of the Fifth Ward, dragging after them a small brass cannon. Immediately after, two companies of United States Marines, commanded by Captain Tyler, went up to the City Hall and reported their services to the mayor.

"The battalion was under command of Captain Tyler, with Lieutenant Henderson, adjutant. The first company was commanded by Brevet-Major Zeilin, the second by Captain Maddox and Lieutenant Turner. They formed in line, and marched towards the City Hall, the crowd of rioters hooting, yelling, threatening, and insulting them at every step. They were passed in their way up by a crowd of Plug Uglies and others who were hauling the brass cannon, and threatening that none of the Marines should return to the Navy-Yard alive. Arrived at the City Hall, Captain Tyler reported services to the mayor. The Marines then moved to the scene of action, subjected all the while to the most abusive and insulting epithets from the lawless gang who followed. When they arrived oposite the polls, which had been closed for some time, they halted in the middle of the street and ordered arms.

"The cannon was hauled out, and a large crowd of rioters surrounded it, while a delegation was sent to inform the commanding officer that unless the Marines were instantly withdrawn the piece would be discharged into their ranks. Captain Tyler, upon receiving this threatening message, gave the order to move towards the gun, and, when the right of the line was nearly opposite it, General Henderson deliberately went up to the piece and placed his body against the muzzle, thereby preventing it from being aimed at the Marines, just at the moment when it was about to be discharged. The general was dressed in citizens clothes.

"He addressed the rioters, saying, 'Men, you had better think twice before you fire this piece at the Marines.' He informed the crowd that the pieces were loaded with ball cartridge, and warned them repeatedly; warned the citizens to leave the spot. The general, finding that the piece would be fired unless captured instantly, crossed over to Captain Tyler, and gave the order for ten or fifteen of the Marines to take it. During this time a number of pistol-shots were fired at General Henderson, some of the parties standing within a few feet of him. One platoon charged the piece on the run. Those in charge of the piece instantly retreated, and a dozen or more revolvers were discharged at the platoon, which had laid hold of the gun and were taking it away. A sortie was made by the mob to recapture the piece, when Lieutenant Henderson ran up and hauled it across the street to the rear of the detachment.

"A man ran up to the general within two feet of his person, and was about to discharge his pistol, when a private with his musket struck his arm, causing the weapon to fall; at the same time the general seized the villain and marched him off to the mayor, into whose hands he placed him. The pistol-shots now rattled around like hailstones, and the officers had great difficulty in restraining their men from returning the fire. General Henderson and all the officers were constantly admonishing the men not to fire until the order was given, but a shot coming from the crowd struck a private in the cheek, making a dreadful wound, and, several more being hit with stones, the soldiers could stand it no longer; they poured in an answering fire, which was stopped as soon as possible by the officers.

"The two companies now formed and remained standing for some time. Several shots were fired from a distance, by one of which a private was hit in the shoulder.

"The battalion made a movement as if to pour in another fire, which the rioters seeing, took to their heels and fled. The force was finally drawn off to the City Hall, where they remained for some time; and then proceeded to the railroad depot, to intercept, if possible, any more of the murderous gang arriving from Baltimore. It is a fact worthy of notice that, of all the Marines who were called out, sixty were raw recruits, and had not been under drill three days."

1858.—At Waya, one of the Feejee Islands, during the year 1858, two American citizens, engaged in trading with that island, were murdered by the natives. Commander Sinclair, of the Vandalia, made a demand for the perpetrators of the outrage, and was answered with defiance. An expedition was immediately despatched, under Lieutenant Caldwell, to assault and destroy the principal village on the island. Lieutenant Caldwell had with him, among other officers, Lieutenant Ramsay, of the Marines. The force consisted of forty Marines and seamen. A fierce conflict ensued between them and about three hundred native warriors, which resulted in the defeat of the latter. The Hon. Isaac Toucey, Secretary of the Navy, said of the affair: "The gallantry, coolness, and bravery displayed by officers and men was in the highest degree commendable."

In the early part of the year, Flag-Officer Forrest of the Brazil Squadron landed a detachment of Marines at Montevideo. When he arrived there a portion of the people had taken up arms and were advancing to the capital. The government applied to the foreign agents to land an armed force to protect the lives and property of the foreign residents. A sufficient force was therefore landed from the vessels present, and Flag-Officer Forrest selected to command it. A strict neutrality was observed. The detachments returned to their respective ships in about ten days, without coming into collision with the revolutionists, they having capitulated on their approach to the capital. The Marines were commanded by Captain and Brevet Major J. G. Reynolds, assisted by Lieutenants Alexander W Starke and C. D. Hebb, from the flag-ship St. Lawrence and sloop Falmouth.

During the same year, Paraguay having given our government various causes of offence, and refusing to make reparation, a strong naval force, under the command of Captain W. C. Shubrick, was sent out to that country towards the close of the year. A commissioner accompanied the fleet, to settle the difficulty, if possible, without recourse to violence; and he succeeded in so doing. The vessels composing the expedition had the usual complement of Marines. The following officers of the Corps were attached to the squadron: Captain Robert Tansill, Captain and Brevet Major J. G. Reynolds, First Lieutenant C. A. Henderson, Second Lieutenants T. S. Wilson, A. W. Starke, C. D. Hebb, H. L. Ingraham, and J. Schermerhorn.

In the latter part of August a portion of the quarantine buildings on Staten Island were used for the reception of quite a number of yellow-fever patients from merchant vessels. Some of the citizens of the island, believing this to be an improper place to treat patients of this class, and fearing the disease would be disseminated throughout the island if the State persisted in using the buildings for permanent pest-houses, determined to destroy them. On the night of the 1st of September, a large and well-organized mob burned a portion of these buildings. The stores of the Susquehanna, lately arrived from Greytown with fever on board, had been landed and deposited within the buildings, but this part had not been destroyed by the rioters. On the 2d, however, the authorities were informed it was the intention of the malcontents to burn this also. Acting upon this information, a detachment of sixty-five Marines from the barracks and the United States frigate Sabine, under the command of Captain Jabez C. Rich and Second Lieutenant Charles Heywood, were ordered to occupy the government buildings, and to protect them at all hazards. On the second day after the occupation by the Marines, Captain Rich was relieved, the command devolving upon Lieutenant Heywood. The place was held one week, and during that time the rioters made several attempts to enter the enclosure surrounding the government building, but, finding it so well protected, thought it prudent to retire, although they were well supplied with battering-rams and fully armed.

**1859.**—On October 17, 1859, the United States were convulsed by the news of the John Brown insurrection at Harper's Ferry. The President ordered Colonel-Commandant Harris to send forward the available force at head-quarters by special train, and a detachment, consisting of eleven sergeants, thirteen corporals, one bugler, and eighty-one privates, with a battery of Dahlgren howitzers, under the command of Lieutenant Israel Greene, was at the railroad station in one hour. They were met by the Secretaries of War and Navy, who gave Lieutenant Greene his instructions. Upon arrival they found the militia from Shepherdstown, Frederick and Jefferson Counties, and the cities of Baltimore and Frederick held in check by the insurgents, who had been driven into an engine-house. Under the orders of Colonel Robert E. Lee, then in the United States army, the Marines assailed the barricaded engine-house, using a long ladder as a battering-ram, and, after killing two of the band and mortally wounding another, captured the "chief" and the remainder of his followers. Scouting parties sent into the country captured fifteen hundred pikes. One Marine was mortally and another slightly wounded.

**1860.**—In March, Commander Brent, commanding the sloop-of-war Marion, was instrumental in saving from destruction the property of American residents at Kisembo, on the west coast of Africa. On the first of the month they applied to him to send a sufficient force on shore to guard their factories, apprehending they would be destroyed by the natives, who meditated an attack on the Portuguese at that place. Commander Brent promptly responded to the call; proceeded on shore with several officers of the Marion and forty sailors and Marines; established a guard at the factories, maintained it until the next morning, and then returned to the ship. On the 3d a signal was made from the shore for assistance, and Commander Brent immediately landed a force of fifty sailors and Marines with a howitzer. During the day the Portuguese and natives were engaged in conflict, and several attempts were made by the natives to approach the factories to fire them, but were thwarted by the guard under Commander Brent. The next morning the detachment returned to the ships, as no further danger was apprehended. Commander Brent observed a strict neutrality throughout, his only object being the protection of American citizens. The Marines on board the Marion were under the command of a sergeant.

The St. Mary's was stationed on the western coast of Mexico from October, 1859, until ordered to Panama, where she arrived in July, 1860. At the insurrection in the vicinity of Panama, which occurred September 27, a brisk fire was opened from the outskirts of the city, killing six and wounding three of the white inhabitants. Commander Porter immediately went on shore, consulted the American consul, and at the request of the military intendante, and of the agent of the railroad, landed a body of Marines and sailors and a field-piece, and took possession of the railroad station. On the 28th the governor delivered up to the city to the joint occupancy of the forces from the St. Mary's and the British ship Clio. Tranquillity having been restored, the detachment on the 7th of October was withdrawn. The Marines of the St. Mary's were under the command of Lieutenant Edward McDonald Reynolds.

---

## CHAPTER XVIII.

### 1861.—The Civil War.

"In the spring of the year 1861 a civil war was kindled in the United States of America which has neither a pattern in character nor a precedent in causes recorded in the history of mankind. It appears in the annals of the race as a mighty phenomenon, but not an inexplicable one. Gazers upon it at this moment, when its awfully grand and mysterious proportions rather fill the mind with wonder than excite the reason, look for the half-hidden springs of its existence in different directions among the absurdities of theory. There is a general agreement, however, that the terrible war was clearly the fruit of a conspiracy against the nationality of the republic and an attempt, in defiance of the laws of divine equity, to establish an empire upon a basis of injustice and a denial of the dearest rights of man. . . . It was the rebellion of an oligarchy against the people, with whom the sovereign power is rightfully lodged.

"The services of the national Navy during the war, on account of their peculiarity, attracted less attention than those of the army, and were not appreciated by the people. They have an equal claim to the gratitude of the nation, so freely accorded to the other branch of the service. The Confederates having no navy, in a proper sense, and only flotillas here and there, and with some powerful 'rams' on rivers and in harbors, and not a ship on the ocean, excepting roving pirate vessels,—built, armed, furnished, and manned chiefly by the British, and cruising alone,—there were few occasions for purely naval battles. The whole force of the Navy Department was employed in the services of blockade, in assisting the attacks of the armies on fortifications along the rivers and on the borders of the Gulf and the ocean, or in chasing the pirates. In these fields of great usefulness the national vessels performed labors of incalculable value, and officers and men exhibited skill, valor, and fortitude unsurpassed.

"Never in the history of the world were there occasions for such exhausting labors and highest courage in service afloat as the American Navy was subjected to in its operations among the rivers and bayous of the southwestern regions of the republic. Many a victory over which the people have shouted themselves hoarse in giving plaudits to the gallant army might never have been achieved but for the co-operation of the Navy. To the common observer it, in many instances, seemed to be only an auxiliary, or wholly secondary force, when, in truth, *it was an equal, if not chief, power in gaining a victory.* Without it, what might have been the result of military operations at Forts Henry and Donelson, Shiloh and all along the Mississippi River, especially at Vicksburg, Port Hudson, and New Orleans; what at Mobile, Pensacola, Key West, along the Florida seaboard, the sea-coast islands, Charleston, and the borders of North Carolina, and even in holding Fortress Monroe and Norfolk?

"Notwithstanding the weak condition of the naval service, the decree went forth, in the spring of 1861, that all the ports of the States wherein rebellion existed must be closed against commerce by a strict blockade. Foreign nations protested and menaced, *but the work was done.* There were no dock-yards or workmen adequate to construct the vessels needed for the service, yet such was the energy of the department that an unrelaxing blockade was maintained for four years, from the Capes of the Chesapeake to the Rio Grande, while a flotilla of gunboats, protecting and aiding the army in its movements, penetrated and patrolled our rivers, through an internal navigation almost continental, from the Potomac to the Mississippi. Ingenuity and mechanical skill developed amazing inventions. That marine monster, the Monitor, was created, and began a new era in naval warfare; and the world was suddenly enriched by new discoveries in naval service. Vessels of the merchant service were purchased and converted into strong warriors; and men from that service were invited to man them.

Schools were established for nautical instruction; dock-yards were enlarged and filled with workmen; and very soon a large number of vessels were afloat, watching the harbors under the ban. No less than two hundred and eight war-vessels were constructed, and most of them fitted out during the four years; and four hundred and eighteen vessels were purchased and converted into war-ships.

"The blockading service was performed with great vigor and efficiency under the triple stimulus of patriotism, duty, and personal emolument. The British government professed to be neutral, but British merchants and adventurers were allowed to send swarms of swift-winged steamers, laden with arms, ammunition, clothing, and everything needed by the insurgents, to run the blockade. The profits of such operations were enormous, but the risks were equally so; and it is believed that a true balance-sheet would show no profits left, in the aggregate, with the foreign violators of the law. The number of such vessels captured or destroyed during the rebellion by the national Navy was fifteen hundred and four. The gross proceeds of property captured and condemned as lawful prize before the first of November following the close of the war amounted to nearly twenty-two millions of dollars, which sum was subsequently enlarged by new decisions. The value of the vessels captured and destroyed (eleven hundred and forty-nine captured and three hundred and fifty-five destroyed) was not less than seven millions of dollars, making a total loss, chiefly to British owners, of at least thirty millions of dollars." *

---

## CHAPTER XIX.

**1861.**—The Civil War, I.—Resignations—Star of the West—Fort Pickens—Destruction of the Gosport Navy-Yard—Potomac Flotilla—The First Battle of Bull Run—Reorganization of the Corps—Hatteras Inlet.

THE Great Rebellion made the next special demand upon the services of the Marines, and they were rendered "nobly and well" afloat, and "admirably in camp and field." In the beginning, when the defections from the army and Navy occurred, the Marine Corps, being largely composed of Southern men, lost its full share of officers. There were in all six resignations and fourteen dismissals (those who tendered their resignations after May 1, 1861, being summarily dismissed). The first duty to which it was called was on the 7th of January, when the steamship Star of the West left New York with two

*"Civil War in America," Lossing, vol. iii.

hundred and fifty artillerists and Marines during the night to reinforce Sumter. The expedition failed, and on January 12 the Star of the West arrived at New York without landing her troops at Sumter. The captain reported that unexpected obstacles in the removal of the buoys, lights, and ranges, which, though he arrived in the night, compelled him to wait till daybreak outside the harbor, rendered a successful entrance impossible. During the same month a detachment of Marines, under the command of Lieutenant C. D. Hebb, garrisoned Fort Washington, on the Potomac, fourteen miles south of Washington City. On the 22d of January the entire force at the Brooklyn Navy-Yard was put under arms and held in readiness to act immediately, because of some apprehension of an attack by an organized force of persons in sympathy with the Confederates.

On the 12th of April the Marines were engaged in the expedition, under Captain H. A. Adams, for the reinforcing of Fort Pickens. The Marines of the squadron were under Lieutenant Cash. When the reinforcement was complete, the Marines remained in the fort at the request of Captain Vogdes. The whole expedition was under the charge of Commander Charles H. Poor, assisted by Lieutenant Smith, of the Brooklyn, Lieutenants Lewis and Newman, of the Sabine, and Lieutenant Belknap, of the St. Louis; and it is highly creditable to these officers that this service was performed without accident or disorder under unfavorable circumstances. The conduct of Lieutenant Cash and the Marines under his command was thus handsomely acknowledged by Colonel Brown: "The services of the Marines being no longer indispensably necessary, they are relieved from duty at this post. The colonel commanding takes pleasure in publishing his entire approval of the conduct and his appreciation of the services of Lieutenant Cash and his command, which have been of great value, and always cheerfully rendered, and the conduct of the troops uniformly correct and soldierly."

On the night of the 20th of the same month a body of Marines assisted in destroying the Gosport (Virginia) Navy-Yard. About half-past eight o'clock in the evening the Pawnee, with a force of Marines under First Lieutenant A. S. Nicholson, entered Gosport harbor. Her coming was not unexpected, and the men on the Pennsylvania and the Cumberland, several hundred in number, greeted her with a volley of cheers. All Norfolk and Portsmouth were thoroughly aroused by the arrival of the Pawnee. They did not expect her, and were not prepared for her. They were seized with trepidation, thinking, perhaps, she had come down with the Cumberland and Pennsylvania, intending to bombard the town, because they had the night before rifled the United States magazine, just below Norfolk, of about four thousand kegs of powder. Being utterly defenceless, they made no protest against the Pawnee's presence, nor did they venture near the Navy-Yard. The commodore ordered the Marines on the Pennsylvania, the Cumberland, the Pawnee, and in the yard to the work of destruction. All the books and papers, the archives of the establishment, were transferred to the Pawnee. Everything of interest to the government to preserve on the Pennsylvania was transferred to the Cumberland. On this latter it was also said a large amount of gold from the custom-house at Norfolk had been placed. Having made safe everything that was to be carried away, the Marines were next set at work to destroy everything of value on the Pennsylvania and in the yard. Many thousand stands of arms were destroyed. Carbines had their stocks broken by a blow from the barrels, and were thrown overboard. A large lot of revolvers shared the like fate. Shot and shell by thousands went with hurried plunges to the bottom. Most of the cannon had been spiked the day or night before. There were at least fifteen hundred pieces in the yard,—Dahlgren guns and Columbiads. A correspondent of the New York *Times* thus wrote of the work that followed: "It is impossible to describe the scene of destruction that was exhibited. Unwearıedly it was continued from nine o'clock until about twelve, during which time the moon gave light to direct the operations. But when the moon sank behind the western horizon, the barracks near the centre of the yard were set on fire, that by its illumination the work might be continued. The crackling flames and the glare of light inspired with new energies the destroying Marines, and havoc was carried everywhere within the limits of order. But time was not left to complete the work. Four o'clock of Sunday morning came, and the Pawnee was passing down from Gosport harbor with the Cumberland, the coveted prize of the secessionists, in tow, every soul from the other ships and the yard being aboard of them, save two. Just as they left their moorings a rocket was sent up from the deck of the Pawnee. It sped high in the air, paused a second, and burst in shivers of many colored lights. And as it did so, the well-set trains at the ship-houses, and on the decks of the fated vessels left behind, went off as if lit simultaneously by the rocket. One of the ship-houses contained the New York, a ship thirty years on the stocks, and yet unfinished. The other was vacant; but both houses and the old New York burned like tinder." The vessels fired were the Pennsylvania, the Merrimac, the Germantown, the Plymouth, the Raritan, the Columbia, and the Dolphin.

In the early part of 1861 it became necessary to place a flotilla on the lower Potomac. A variety of circumstances combined to render this one of the most arduous duties on the whole insurgent frontier, and it was clearly foreseen that without the active co-operation of the army it would be impossible to prevent the navigation of the river from being obstructed by the batteries on the Virginia side. For several months, however, the Navy succeeded more effectually than could have been expected in keeping the river open for commercial purposes, and restricting, to a great extent, communication between the opposite shores. In the heroic discharge of his duty, Commander J. H. Ward, first commander of the flotilla, lost his life. Commander Ward was killed in an action with a rebel battery, at Matthias Point, on the 27th of June, 1861. He was the first naval officer killed during the rebellion.

The Navy continued to capture every rebel vessel that showed itself on the Potomac, and to give security and protection to the commerce of loyal citizens until the close of October, when the insurgents erected batteries at various points on the Virginia shore, thereby rendering passage on the river dangerous.

Captain Thomas C. Craven succeeded Commander Ward in the command of the Potomac flotilla, and remained on that service until relieved by Lieutenant R. H. Wyman, in the fall of 1861.

Lieutenant Wyman remained in command on the Potomac, constantly engaged in active operations against the enemy until early in 1862, when Commodore A. H. Harwood assumed the command.

In the first battle of Bull Run, or Manassas, a battalion, commanded by Major Reynolds, served in Porter's brigade as the support of Griffin's battery.

The following is Major John G. Reynolds's report of the participation of the battalion in the battle, dated at the Marine Barracks, Headquarters, Washington, July 24, 1861, addressed to Colonel Harris, commandant of the Corps: "I have the honor to report the movements and operations of the battalion of Marines under my command detailed to co-operate with the army. The battalion left the barracks at headquarters in time to reach the Virginia end of the Potomac, 'Long Bridge,' at 3 P.M., July 16, and proceeded up the Columbia turnpike until an officer, purporting to be assistant adjutant-general of Colonel Porter's brigade, came up and assigned us position in the line of march, which placed us immediately in rear of Captain Griffin's battery of flying artillery. This assignment was continued up to the period of the battle at Bull Run. On reaching the field, and for some hours previously, the battery's accelerated march was such as to keep my command, more or less, in double-quick time; consequently the men became fatigued or exhausted in strength. Being obliged at this period to halt, in order to afford those in the rear an opportunity of closing up and taking their proper place in the line, the battery was lost to protection from the force under my command. This I stated to Colonel Porter, who was ever present, watching the events of the day. The position of the battery was pointed out, and I was directed to afford the necessary support. In taking the position the battalion was exposed to a galling fire. While holding it, General McDowell ordered the battalion to cover or support the Fourteenth New York Regiment, which was about to be engaged. The battalion, in consequence, took the position indicated by the general, but was unable to hold it, owing to the heavy fire which was opened upon them. They broke the line several times, but were as frequently formed and urged back to their position, when finally a general rout took place, in which the Marines participated.

"I am constrained to call your attention to the fact that, when taking into consideration that the command was composed entirely of recruits, not one being in service over three weeks, and many had hardly learned their facings, the officers likewise being but a short time in the service, their conduct was such as to elicit the highest commendation. Of the three hundred and fifty officers and enlisted men under my command, there were but two staff officers, two captains, one first lieutenant, nine non-commissioned officers, and two musicians who were experienced from length of service. The remainder were, of course, raw recruits, which being considered, I am happy to report the good conduct of officers and men. The officers, although but little experienced, were zealous in their efforts to carry out my orders. In the death of Lieutenant Hitchcock the Corps has been deprived of a valuable acquisition. On the field he was ever present and zealous. He sought and won the approbation of his commanding and brother officers. Enclosed please find a return of the battalion, showing its present strength, with casualties, etc. The abrupt and hasty retreat from the field of battle presents a deplorable deficiency in both arms and equipments.

"The rout being of such a general character, the men of all arms commingled; the only alternative left was to hasten to the ground occupied by the brigade to which we were attached on the morning of the day of the battle. On my way thither I had the good fortune to fall in with General Meigs, whose consternation at the disastrous retreat was pictured on his countenance. He was of the opinion that the army should hasten to Arlington, fearing otherwise the enemy would follow up their successes and cut us off on the road. My men being weary and much exhausted, without blankets and other necessaries, I determined to strengthen such as should pass the wagons by hot coffee, and move on to head-quarters at Washington, where their wants could be supplied. But few came up; others continued on to the Long Bridge, where, on my arrival, I found some seventy or more, who, at my earnest solicitation, were permitted to accompany me to the barracks. In assuming the responsibility of the return to head-quarters, I trust my course will meet the approbation of authority. Blankets were thrown aside by my order on entering the field, which, from force of circumstances, we were unable afterwards to recover.'

The following are condensed from the report of Colonel Porter, commanding the brigade:

"The Marines were recruits, but through constant exertions of their officers, have been brought to present a fine military appearance.

"The Marines were moving up in fine style in rear of the Fourteenth New York.

"Griffin's battery found its way through the timber to the fields beyond, followed promptly by the Marines.

"Our right was rapidly developed by the Marines.

"Among those who deserve special mention, I beg leave to place the following names,—viz., Major Reynolds of the Marines, whose zealous efforts were well sustained by his subordinates, two of whom Brevet Major Zeilin and Lieutenant Hale, were wounded, and one, Lieutenant Hitchcock, lost his life."

The roster of the battalion was as follows:

Major John G. Reynolds, commanding; Major William B. Slack, quartermaster; Major Augustus S. Nicholson, adjutant and inspector.

Company A: Brevet Major Jacob Zeilin, Second Lieutenants F. Munroe and J. H. Grimes. Company B: Captain James H. Jones, Second Lieutenant Robert W. Huntington. Company C: First Lieutenant Alan Ramsay, Second Lieutenant Hitchcock. Company D: Second Lieutenants W. H. Cartter and W. H. Hale.

Non-commissioned staff, two; first sergeants, four; second sergeants, three; corporals, eight; musicians, four; privates, three hundred and twenty; total, three hundred and fifty-three.

The list of casualties in the action was as follows: Killed: Second Lieutenant Hitchcock, Privates Clegg, Harris, Hughes, Lane, Moore, Perkins, Riley, Ward. Wounded: Brevet Major Zeilin, Lieutenant Hale, Corporal Steiner, Privates Stuart, Bowers, Slemons, Bradford, Dodge, Etchell, Tiger, Lang, McKinney, McCann, Wheelan, McGuigan, Howell, Rannohan, Cook, Potter. Missing: Privates Barrett, Hunt, McChristæ, Clark, McCoy, Lewis, Beans, Dempsey, Kressler, Dermott, Otto, Cannon, Stanley, Duncanson, Foley, and Wood. Recapitulation: Killed, one lieutenant and eight privates; wounded, one brevet major, one lieutenant, one corporal, and sixteen privates; missing, sixteen privates; total, killed, wounded, and missing, forty-four.

The necessity for a better, more thorough, and more efficient organization of the Corps was recognized, and an act was passed by Congress for that purpose July 25, providing that the Corps should consist of the following officers, non-commissioned officers, musicians, and privates,—viz., one colonel-commandant, one colonel, two lieutenant-colonels, four majors, one adjutant and inspector, one paymaster, one quartermaster, two assistant quartermasters, twenty captains, thirty first lieutenants, thirty second lieutenants, one sergeant-major, one quartermaster-sergeant, one drum-major, one principal musician, two hundred sergeants, two hundred and twenty corporals, thirty musicians for band, sixty drummers, sixty fifers, and twenty-five hundred privates. It was further provided that the commissions of the officers then in the Marine Corps should not be vacated by the act; and that the President of the United States might, during the recess of the Senate, first by promotions, and then by selections, appoint the officers thereby authorized, which appointments should be submitted to the Senate, at the next session, for its advice and consent. The appointments of commissioned officers, to be made under the provisions of the act, was to be of persons between the ages of twenty and twenty-five years, and they were to be subjected, under the direction of the Secretary of the Navy, to an examination as to their qualifications for the service to which they were appointed. The act was approved July 25, 1861.

During the following month the services of the Marines were again brought into requisition in the capture of Hatteras Inlet, August 30. Flag-Officer Silas H. Stringham left Hampton Roads, August 26, with the flag-ship Minnesota, Captain G. I. Van Brunt, having in company the United States steamers Wabash, Captain Samuel Mercer; Monticello, Commander John P. Gillis; Pawnee, Commander S. C. Rowan; Harriet Lane, Captain John Faunce; United States chartered steamers Adelaide, Commander Henry S. Stellwagen; George Peabody, Lieutenant R. B. Lowry; and tug Fanny, Lieutenant Pierce Crosby; the transports Adelaide and George Peabody towing schooners with surf-boats on them, and the Monticello and Pawnee, surf-boats only. Major-General Butler took passage in the Minnesota, the transports having parts of two regiments and one company of regulars, under the command of Colonels Max Weber and Hawkins, and Captain Larnard, United States army. On the morning of the 27th they rounded the shoals off Hatteras; at 5 P.M. they anchored at the southward of the cape, hoisted out the surf-boats and made preparations for landing the troops. In the morning, General Butler and the Marines of the Minnesota, the latter under the command of Captain William L. Shuttleworth, United States Marine Corps, were sent to the Harriet Lane. About noon the work of disembarking began, and the Wabash and Cumberland opened fire on Fort Clark. The fire was returned by the fort. The Minnesota, Wabash, and Cumberland continued their fire, passing and repassing the fort until it was abandoned by the enemy. The fire was kept up by the fort, the shot falling short of or passing over the ships. At 2 P.M. the American flag was displayed from Fort Clark by the pickets, who were in possession. A little later the signal to "cease firing" was made, and the squadron hauled off for the night, with the exception of the Monticello, Pawnee, and Harriet Lane, they being ordered to go in shore during the night. The engagement was resumed the next morning, and at 11.10 A.M. a white flag was displayed from Fort Hatteras. The enemy returned the fire throughout the engagement, but with no effect, their shot falling short. Almost at the beginning of the engagement they hauled down their colors, and showed none until the white flag was displayed. Upon the appearance of the white flag the troops marched towards the fort, and, as if by preconcerted signal, but without any order or request, the officers and crews of the squadron gave three hearty cheers for their success. At 11.30 o'clock, Major-General Butler, in the tug Fanny, went into the inlet to the rear of the forts to take possession. No accident to a single officer or man of the Navy, army, or Marines was recorded. Flag-Officer Stringham thus concludes his report: "In conclusion, I beg leave to state to the department and to my government that I have naught but praise to accord to officers, seamen, and Marines, and officers and soldiers of the army who were present, for gallantry and cheerful devotion to duty and to their government, the United States of America, which they all cheerfully and heartily served."

Secretary Welles, in acknowledging the receipt of the report, said, "This brilliant achievement, accomplished without the loss of a man or injury to any one in the Federal service, has carried joy and gladness to the bosom of every friend of the Union."

The following officers of the Marine Corps were in this engagement: Minnesota, Captain W. L. Shuttleworth, Lieutenant Cartter; Wabash, Captain I. T. Doughty; Susquehanna, Lieutenant P. R. Fendall; Cumberland, Lieutenant Charles Heywood. The other vessels, with the exception of the tug Fanny, had their complement of Marines, the guard of each being in charge of a sergeant.

The Marines from the Minnesota, Wabash, and Cumberland, commanded by their own officers, landed with the troops, and entered Fort Clark.

---

## CHAPTER XX.

1861.—The Civil War, II.—The Capture of the Confederate Privateer Judah off Pensacola—Destruction of the Royal Yacht—Battle of Port Royal—Loss of the Governor—The Trent Affair—Occupation of Wassaw Island.

On the night of September 13, 1861, an expedition was fitted out from the frigate Colorado, flag-ship, consisting of the first launch and first, second, and third cutters, under the command of Lieutenants Russell, Sprotson, and Blake, and Midshipman Steece, respectively, assisted by Captain Reynolds of the Marine Corps, Assistant Surgeon Kennedy, Assistant Engineer White, and Midshipmen Forrest and Higginson. The whole force detailed was about one hundred officers, sailors, and Marines. The object of the expedition was the destruction of a schooner which lay off the Pensacola Navy-Yard, supposed to be fitting out as a privateer, and the spiking of a gun in battery in the southeast end of the yard.

The attack was made on the morning of the 14th, at half-past three o'clock. The schooner was moored to the wharf, armed with a pivot and two broadside guns, under the protection of a battery and field-piece. The crew were prepared to receive their assailants, pouring in a volley of musketry as the boats neared the vessel. After a desperate resistance, they were driven from the deck of the schooner on to the wharf, where they rallied and were joined by the guard, a continual fire upon the attacking party being kept up. In the mean time the vessel was set on fire in several places, and while burning was freed from her moorings, and drifted down opposite Fort Barrancas, where she sank.

Of the party assigned to attend to the spiking of the gun, only Lieutenant Sprotson and Gunner Boreton were able to find it, the party becoming separated in the darkness. Fortunately only one man was found in charge of the gun, and he immediately levelled his piece at Lieutenant Sprotson, but was shot down by Gunner Boreton before he could obtain correct aim, both pieces exploding simultaneously. The gun, a 10-inch columbiad, was immediately spiked, and the officers returned to their boat.

The object of the expedition was accomplished in the short space of fifteen minutes, and the whole force of the enemy being aroused, the assailants pulled away, and when a short distance from the shore fired six charges of canister from their howitzers into the yard.

Flag-Officer William Mervine, in his official report, says, "I am grieved to report that this brilliant affair was not unattended with loss on our side. I have to report as killed by shots from the cross-trees of the schooner, while the boats were approaching, boatswain's mate, Charles H. Lamphere, and John R. Herring, seaman and captain of howitzer (two of the best men in our ship), and Marine John Smith (the first man on board the schooner, and who behaved most gallantly), who was, by a sad mistake, having lost his distinguishing mark, killed by one of our own men. . . . The Marines especially seemed to have sustained the reputation borne by their branch of the service, as they receive encomiums from all sides." The following is a full list of the Marines engaged: Captain Edward McDonald Reynolds (wounded), First Sergeant Patrick Fitzsimmons, Sergeant James Gallaher, Corporal John Moore, Corporal George Debbyshire, Privates Martin Barnum, Francis Kelly, James Fuller, Dennis Sullivan, Savillan Coburn (wounded), Anthony Baker, Michael Ryan, Terrence O'Dowd (wounded badly), John Smith (killed), John Smith, 2d, Charles Daily, Edward Eagan, Augustus Hartman, Charles Carberry (wounded), Clarence Barton.

On the night of the 7th of November, 1861, an expedition, consisting of the first and second launches, under command of Lieutenants James E. Jouett and John J. Mitchell, left the frigate Santee, then blockading off Galveston bar, Texas, for the purpose of surprising and burning the man-of-war, steamer General Rusk, lying under Pelican Island Fort.

The expedition entered the harbor at 11.40 P.M., and succeeded in passing the armed schooner guarding the channel, and the Bolivar and Point Forts, without discovery, but unfortunately grounded on the Bolivar spit, and at this juncture was discovered.

Lieutenant Jouett, deeming it imprudent, after this discovery, to encounter a vessel so large and so heavily armed and manned, determined to abandon that part of the expedition. In returning, he boarded, and, after a sharp conflict, captured the armed schooner Royal Yacht. Several stands of arms, thirteen prisoners, and the rebel colors were captured. As the pilot of the expedition had been shot down, and the schooner had received a shell between wind and water, Lieutenant Jouett did not deem it advisable to bring her out. He therefore burned her after spiking the gun, a light 32-pounder. Lieutenant Jouett and Gunner Carter were seriously wounded; also six men, one mortally. A detachment of Marines under the command of a sergeant accompanied this expedition.

In October, 1861, a joint expedition of military and naval forces was organized by the government. Captain Samuel F. Du Pont, an officer of great professional skill and experience, was appointed to its command, the military force being under the control of Brigadier-General W. T. Sherman. The intention of the Department was to seize and occupy one or more points on the southern coast where the blockading squadron might find shelter, possess a depot, and afford protection to loyal citizens. To Flag-Officer Du Pont was intrusted the important and responsible duty of selecting the point of attack. And he, believing that the capabilities of the expedition justified it, determined, after consultation with Brigadier-General Sherman, to make Port Royal, South Carolina, a well-defended harbor, the objective point.

On the 29th of October, 1861, the fleet, consisting of forty-eight vessels, including transports, a larger squadron than ever before assembled under the United States flag, left Hampton Roads. On Friday, November 1, the fleet, being then off Hatteras, encountered one of the severest storms ever known on the coast. The gunboat Isaac Smith was compelled to throw her formidable battery overboard to keep from foundering, and thus relieved, was enabled to go to the assistance of the chartered steamer Governor, then in a very dangerous condition, and on boad of which was the battalion of Marines under Major Reynolds. They were fully rescued by Captain Ringgold, in the Sabine, under difficult circumstances, soon after which the Governor went down. The transport Peerless, in a sinking condition, was met by the Mohican, Commander Godon. All the men on board the transport, twenty-six in number, were saved by the boats of the Mohican under very perilous circumstances.

On Sunday, November 3, the weather moderated, and on the following morning the fleet arrived at Port Royal, and came to anchor off the bar. All aids to the navigation of Port Royal harbor had been removed by the rebels; but, thanks to the skill of Commander Davis, fleet captain, and Mr. Boutelle, of the coast survey, the channel was immediately found, sounded out, and buoyed. By 3 P.M. the transports, with all the gunboats, were sent forward, and before dark they were securely anchored in the roadstead.

On the 8th of November, the squadron, consisting of the Wabash, Commander C. R. P. Rodgers, the leading ship; Susquehanna, Captain J. L. Lardner; Mohican, Commander S. W. Godon; Seminole,

Commander J. P. Gillis; Pawnee, Lieutenant-Commanding R. H. Wyman; Unadilla, Lieutenant-Commanding N. Collins; Ottawa, Lieutenant-Commanding T. H. Stevens; Pembina, Lieutenant-Commanding J. P. Bankhead; Vandalia, Commander F. S. Haggerty, towed by the Isaac Smith, Lieutenant-Commanding J. W. A. Nicholson; Bienville, Commander Charles Steedman; Seneca, Lieutenant-Commanding Daniel Ammen; Curlew, Lieutenant-Commanding P. G Watmough; Penguin, Lieutenant-Commanding T. A. Budd; Augusta, Commander E. G. Parrott, attacked the enemy's batteries on Bay Point and Hilton Head (Forts Beauregard and Walker), and succeeded in silencing them after an engagement of four hours' duration, and after driving away the squadron of rebel steamers under Commodore Tatnall. The defeat of the enemy terminated in utter rout and confusion. Their quarters and encampments were abandoned without an attempt to carry away either public or private property. The ground over which they fled was strewn with the arms of private soldiers, and officers retired in too much haste to submit to the encumbrance of their swords. The Marines and a company of seamen took possession of the deserted ground, and held the forts on Hilton Head till the arrival of General Sherman. The bearer of the despatches giving an account of the victory carried with him the first American ensign raised upon the soil of South Carolina after the rebellion broke out. The following officers of the Corps were attached to the vessels named below during the engagement: Wabash, Captain I. T. Doughty; Susquehanna, First Lieutenant P. R. Fendall; Vandalia, Captain John Schermerhorn. The present colonel, Henry C. Cochrane, who had been appointed to the Marine Corps in the previous August, but was too young to receive his commission, was also present on the gunboat Pembina. The other vessels were supplied with guards, under the charge of sergeants.

Secretary Welles in his reply to the official report of the fight, says, "To you and your associates, under the providence of God, we are indebted for this great achievement by the largest squadron ever fitted out under that flag which you have so gallantly vindicated, and which you will bear onward to continued success."

The following is the full text of Flag-Officer Du Pont's report concerning the Marine battalion, dated on board the flag-ship Wabash, Port Royal Harbor, South Carolina, November 15, 1861, and addressed to Secretary Welles: "I avail myself of the first moment of leisure to transmit to you the report of Major John George Reynolds, commanding the battalion of Marines attached to my squadron, in which he relates all the circumstances attending the loss of the chartered steamer Governor, and the rescue of himself and his command by the frigate Sabine, Captain Ringgold. The Department will find this report exceedingly interesting, and will be gratified to learn that the conduct of the officers and of nearly all the men of the battalion was such as to command Major Reynolds's approval, as it will, I doubt not, receive the favorable notice of the Department. The established reputation and high standing of Major Reynolds might almost dispense with any observation of my own upon the bravery and high sense of honor which he displayed in disputing with Mr. Weidman, though not a seaman, the privilege of being the last to leave the wreck."

The following are extracts from Major Reynolds's report, dated on board the United States ship Sabine, at sea, November 8, 1861, addressed to Flag-Officer Du Pont: "I have the honor to report that the Marine battalion under my command left Hampton Roads on transport steamboat Governor, on the morning of Tuesday, the 29th of October, with the other vessels of the fleet, and continued with them near the flag-ship Wabash until Friday, the 1st of November. On Friday morning, about ten o'clock, the wind began to freshen, and by twelve or one blew so violently that they were obliged to keep her head directly to the wind, and thereby leave the squadron, which apparently stood its course. Throughout the afternoon the gale continued to increase, though the Governor stood it well until about four o'clock." The vessel was much damaged by sea and wind, and was in danger of going down. "At day-break preparations were made for sending boats to our relief, although the sea was running high; and it being exceedingly dangerous for a boat to approach the guards of the steamer, in consequence the boats were laid off, and the men were obliged to jump into the sea, and thence hauled into the boats. All hands were thus providentially rescued from the wreck, with the exception, I am pained to say, of one corporal and six privates, who were drowned or killed by the crush or contact of the vessels. Those drowned were lost through their disobedience of orders in leaving the ranks or abandoning their posts. After the troops were safely reembarked, every exertion was directed to securing the arms, accoutrements, ammunition, and other property which might have been saved after lightening the wreck. I am gratified at being able to say nearly all the arms were saved and about half the accoutrements. Too much praise cannot be bestowed upon the officers and men under my command; all did nobly. The firmness with which they performed their duty is beyond all praise. For forty-eight hours they stood at ropes and passed water to keep the ship afloat. Refreshments in both eating and drinking were passed to them at their posts by non-commissioned officers. It is impossible for troops to have conducted themselves better under such trying circumstances. The transport continued to float some hours after she was abandoned, carrying with her when she sank, I am grieved to say, company books and staff returns. In order to complete the *personnel* of the battalion, I have requested Captain Ringgold to meet a requisition for seven privates, to which he has readily assented. I considered this requisition in order, as I have been informed by Captain Ringgold it is his intention, or orders were given for his ship, to repair to a northern port, in which event he can be easily supplied, and my command, by the accommodation, rendered complete, in order to meet any demand you may make for our services."

The following is a list of the officers of the battalion under the command of Major Reynolds:

Major commanding, John George Reynolds; Adjutant, Second Lieutenant T. L. McElrath; Assistant Quartermaster, Second Lieutenant E. A. Smalley; Assistant Surgeon Ricketts; Captains George R. Graham, James Wiley; First Lieutenants L. M. Goldsborough, Frank Munroe, R. W. Huntington, J. H. Grimes, W. H. Parker, C. H. Nye; Second Lieutenants, H. A. Bartlett, C. A. Stillman, S. H. Mathews, F. H. Corrie, C. H. Bradford, E. B Sturgeon, A. Devereux.

After the capture of Port Royal, Captain Schemerhorn was transferred from the Vandalia to the battalion under Major Reynolds.

On the 15th of March, 1862, Captain C. G. McCawley, Second Lieutenants H. B. Lowry, P. C. Pope, and S. W. Powell joined the battalion, then in camp at Bay Point.

No event of the war attracted more attention from other nations than that which has gone into history as "The Trent Affair." In the capture of Slidell and Mason and their companions no fighting took place, but the presence and active participation of the Marines justifies the introduction of the subject so far as the capture itself was concerned. On the 8th of November, Lieutenant Fairfax, of the United States steamer San Jacinto, received from Captain Wilkes orders to have the second and third cutters of that ship fully manned and armed, and be in all respects prepared to board the steamer Trent, then hove to under the guns of the San Jacinto. On boarding her he was ordered to demand the papers of the steamer, her clearance from Havana, and the list of passengers and crew. The order continued: "Should Mr. Mason, Mr. Slidell, Mr. Eustis, and Mr. McFarland be on board, you will make them prisoners and send them on board the ship immediately, and take possession of her as a prize. All trunks, cases, packages,

ROBERT L MEADE, COLONEL

U. S. M. C. Plate 2

and bags belonging to them you will take possession of and send on board this ship. Any despatches found on the persons of the prisoners, or in possession of those on board the steamer, will be taken possession of also, examined, and retained, if necessary. I have understood that the families of these gentlemen may be with them. If so, I beg you will offer them, in my name, a passage in this ship to the United States, and that all the attention and comforts we can command are tendered them, and will be placed in their service. In the event of their acceptance, should there be anything which the captain of the steamer can spare to increase their comforts in the way of necessaries or stores, of which a war-vessel is deficient, you will please procure them. The amount will be paid for by the paymaster. Lieutenant James A. Greer will take charge of the third cutter, which accompanies you, and assist you in these duties."

At 1.30 o'clock P.M., on the same day, Lieutenant Fairfax repaired alongside the British packet in an armed cutter, accompanied by Mr. Houston, second assistant engineer, and Mr. Grace, the boatswain. He went on board the Trent alone, leaving the two officers in the boat, with orders to wait until it became necessary to show some force. He was shown up by the first officer to the quarter-deck, where he met the captain and informed him who he was, asking to see the passenger-list. The captain declined. Lieutenant Fairfax told him that he had information of Messrs. Mason, Slidell, Eustis, and McFarland taking passage at Havana, in the packet for St. Thomas, and announced his intention to satisfy himself whether they were on board before allowing the steamer to proceed. Mr. Slidell, evidently hearing his name mentioned, stepped forward, and asked if he was wanted. Mr. Mason soon joined them, and then Mr. Eustis and Mr. McFarland, when Lieutenant Fairfax made known the object of his visit. The captain of the Trent opposed anything like the search of his vessel, nor would he consent to show papers or passenger-list. The four gentlemen above mentioned protested also against being arrested. There was considerable noise among the passengers about this time, which led Mr. Houston and Mr. Grace to repair on board with some six or eight men, all armed. After several unsuccessful efforts to persuade Mr. Mason and Mr. Slidell to go peaceably, Lieutenant Fairfax called to Mr. Houston and ordered him to return to the ship with the information that the four gentlemen named in Captain Wilkes's order were on board, and force must be applied to take them out of the packet. A few minutes later there was still greater excitement on the quarter-deck, which brought Mr. Grace with his armed Marines. Lieutenant Fairfax, however, deemed the presence of any armed men unnecessary and only calculated to alarm the ladies present; he directed Mr. Grace to return to the lower deck, where he had been since going on board.

Less than half an hour after they boarded the Trent, the second armed cutter, under Lieutenant Greer, came alongside. He carried in the third cutter eight Marines and four machinists, in addition to a crew of some twelve men. When the Marines and some armed men had been formed just outside the main-deck cabin, where the four gentlemen had gone to pack their baggage, Lieutenant Fairfax renewed his efforts to induce them to accompany him. They still refused to go, unless force was applied. Lieutenant Fairfax called to his assistance four or five officers, and, first taking hold of Mr. Mason's shoulder, with another officer on the opposite side, he went as far as the gangway of the steamer, and delivered him over to Lieutenant Greer, to be placed in the boat. He then returned for Mr. Slidell, who insisted that he must apply considerable force to cause him to go. Lieutenant Fairfax called in three officers, and Mr. Slidell was taken in charge and handed over to Mr. Greer. Mr. McFarland and Mr. Eustis, after protesting, went quietly into the boat. They had been permitted to collect their baggage, but were sent in advance of it, under charge of Lieutenant Greer.

When Mr. Slidell was taken prisoner a great deal of noise was made by some of the passengers, which caused Lieutenant Greer to send the Marines into the cabin. They were immediately ordered to return to their former position outside. Lieutenant Fairfax carried out his purpose without using any force beyond what appears in this report. The mail agent, who was a retired commander in the British navy, had much to say as to the propriety of the seizure, but Lieutenant Fairfax purposely avoided all official intercourse with him. When the American force was finally leaving the steamer, the mail agent made some apology for his rude conduct, and expressed personally his approval of the manner in which Lieutenant Fairfax carried out his orders.

In the report of his participation in the affair, Lieutenant Greer says, "When I first went on board with the Marines, and at intervals during my stay, the officers of the steamer made a great many irritating remarks to each other and to the passengers, which were evidently intended for our benefit. Among other things said were: 'Did you ever hear of such an outrage?' 'Marines on board! why, this looks devilish like mutiny!' 'These Yankees will have to pay well for this.' 'This is the best thing in the world for the South; England will open the blockade.' 'We will have a good chance at them now.' 'Did you ever hear of such a piratical act?' 'Why, this is a perfect Bull's Run!' 'They would not have dared to have done it if an English man-or-war had been in sight!' The mail agent (a man in the uniform of a commander in the Royal navy, I think) was very indignant and talkative, and tried several times to get me into a discussion of the matter. I told him I was not there for that purpose. He was very bitter; he told me that the English squadron would raise the blockade in twenty days after his report of this outrage got home; that the Northerners might as well give up now, etc."

The families of Messrs. Slidell and Eustis having declined Captain Wilkes's offer for their accommodation to accompany their husbands, they proceeded in the Trent. The San Jacinto having been ordered to report at Charlestown, the prisoners were retained on board and conveyed to Fort Warren, where they were committed to the custody of Colonel Dimmick, in command of that fortress.

Captain John Schermerhorn commanded the Marines of the San Jacinto on this occasion.

One of the important expeditions of the war in which the Marines took an active part was the occupation of Wassaw Island, December 5. On that day Commander C. R. P. Rodgers left Tybee Roads before daylight, with the steamers Ottawa, Seneca, and Pembina, and crossed the bar of Wassaw Sound. He approached the fort on Wassaw Island within a mile, and seeing neither guns nor men, did not fire, but sent Lieutenant Barnes to it with a white flag. He found it an enclosed octagonal work, with platforms for eight guns on the water faces. The land faces were protected by abatis. The work was well constructed. The guns had been removed, the platforms cut, and the magazine blown up. From the freshness of the foot-prints and other signs, it appeared to have been abandoned but a short time. Adjoining the fort were huts or sheds for a large garrison. Some lumber and bricks remained; everything else had been carried away.

Another minor success of a similar nature was won by a body of Marines on the 12th of the same month. On that day Lieutenant J. W. Nicholson, commanding the United States steamer Isaac Smith, stationed at St. Helena Sound, South Carolina, went up the Ashepoo, taking the Marines of the Dale, and having Colonel Welsh, of the army, and Lieutenant-Commanding Truxtun, as passengers. He landed the party at Fenwick's Island Fort, to make reconnoissance, and when it was finished, stood up the river as far as Mosquito Creek. Seeing a picket of several mounted men at a house which, on a previous visit, he had discovered to be their head-quarters, he fired at them once with a rifle-shot, and again with an 8-inch shell. This driving them to the woods, he landed the Marines and burned the quarters.

Two weeks later, the Marine guard of the Dale, acting under orders of W. T. Truxtun, Lieutenant-Commanding, had an engagement with a small body of Confederates on the South Edisto River, South Carolina, near a house said to be the property of Governor Aiken. The Confederates took refuge in the house, but were finally driven off, and the Marines returned to the Dale without the loss of a man.

---

## CHAPTER XXI.

**1862.**—The Civil War, III.—Cedar Keys—Appalachicola—Joint Expedition for Operations in the Waters of North Carolina—Cumberland Sound—Fort Clinch—Fernandina.

IN January, 1862, Flag-Officer McKean despatched Commander Emmons, with the steamer Hatteras, to operate against the rebels at Cedar Keys. The expedition was entirely successful, destroying a large amount of public property, including military stores, and capturing a battery of two guns in position at Sea-Horse Key, and several schooners laden and ready to run the blockade. In the latter part of March, Commander Stellwagen, of the Mercedita, arrived off Appalachicola, with that vessel and the Sagamore, Lieutenant-Commanding Drake, and organized a boat expedition, the immediate object of which was the capture of a number of vessels understood to be at or above that city. The place had been evacuated by the soldiers, some six hundred in number, on the first appearance of the naval force. No resistance was offered, and the expedition brought out several vessels and destroyed others, owing to the difficulty of getting them over the bar.

The detachments of Marines in these expeditions were commanded by sergeants.

Early in January a joint expedition of the army and Navy, for operations in the waters of North Carolina, moved from Hampton Roads under command of Flag-Officer L. M. Goldsborough and Brigadier-General A. E. Burnside respectively. The naval force, consisting of seventeen light-draught vessels, with an armament of forty-eight guns, most of them of heavy calibre, arrived at Hatteras

Inlet on the 13th, and in two days succeeded, with great difficulty, in passing over the bulkhead and through the narrow and tortuous channel. It was not, however, until some weeks later that the transports were able to surmount the obstacles and be prepared for active co-operation, which period of delay was employed in obtaining information of the enemy's position and in forming plans for the attack. Roanoke Island lies between the two bodies of water known as Pamlico and Albemarle Sounds, being separated from the main-land by a shallow channel, Croatan Sound. Opposite the southern extremity of the island the main-land juts out in a low marshy point, around which the vessels threaded their way, and on the morning of February 7 moved up Croatan Sound in three columns, commanded respectively by Lieutenants Werden, Murray, and Davenport, the whole under the immediate command of Commander S. C. Rowan. The enemy had formed an extensive obstruction of a double row of piles and sunken vessels, stretching across the sound between the batteries on Pork and Weir Points, and behind this their vessels, eight in number, were drawn up. By half-past ten o'clock, the squadron had approached near enough to begin the attack, directing most of its fire against the fort on Pork Point, but not neglecting the vessels nor the other works, all of which returned the fire, though with but slight effect. By noon the engagement had become general, and was continued so hotly that at two o'clock the battered barracks behind the fort were burning furiously, and at half-past four the batteries for the most part ceased for a while to reply to the firing of the fleet. Five of the enemy's steamers, apparently uninjured, retired behind the point, and the first landing of troops took place. Throughout the sound the depth of water is but slight, and even at a distance of a mile or more from the shore it scarcely exceeds the depth of seven feet.

As none of the vessels, with one or two exceptions, drew less than this amount of water, and some of them drew more than eight feet, the discretion of their commanders was taxed to the utmost in placing them so that their guns would tell effectively. The landing was effected, in light draught steamers and boats, at Ashby's harbor, a large body of the enemy guarding the shore being soon cleared away by some shrapnels from the guns of the Delaware. At five o'clock the batteries again opened, and the vessels of the enemy again came out, but were soon compelled to retire, and at six, the firing being only from Pork Point and at long intervals, the signal to cease firing was made. By midnight some ten thousand troops had been safely landed at Ashby's harbor, where they were joined by six launches from the fleet, with their howitzers, to hold the road during the night, and be ready for active operations in the morning.

It was arranged by General Burnside that his forces should move at a very early hour on the morning of the 8th, and begin their attack upon the enemy; and it was agreed that, as the direction they would have to take would probably soon bring them into the line of fire of the fleet, the vessels should not renew operations until it was known that their fire would not be destructive to friend and foe alike. At nine o'clock, a continuous firing in the interior of the island announced that the army was hotly engaged about midway between the landing and Pork Point, and the vessels at once moved up to re-engage the forts. This they continued until the firing in the interior slackened, when, taking it for granted that General Burnside was carrying everything before him and approaching the rear of the batteries, Flag-Officer Goldsborough gave the order to desist, and proceeded to the task of clearing a passage-way through the obstructions. By four o'clock this was accomplished, and at about the same time that the vessels succeeded in bursting through the barricades, the American flag was unfurled over the battery on Pork Point. A few minutes afterwards the enemy himself fired the works on Redstone Point, together with a steamer which had taken refuge under its guns, and thus ended the eventful struggle of two days, which secured complete possession of the island of Roanoke.

Retreating from Roanoke Island, the rebel naval fleet fled up the sound and into Pasquotank River, towards Elizabeth City, Commander Rowan pursuing them with a flotilla of fourteen vessels, and anchoring for the night a few miles from Fort Cobb. On the morning of the 10th the rebel steamers were discovered drawn up behind the battery, which mounted four heavy guns, and supported by a schooner—the Black Warrior—moored to the opposite bank, and carrying two heavy 32-pounders. When within long range, fire was opened from the battery, the schooner, and the steamers; but the vessels moved on silently and steadily, shot and shell falling thick and fast among them. When within three-quarters of a mile of the battery, Commander Rowan gave the signal for a dash at the enemy; fire was opened with telling effect, and the vessels put at their utmost speed. The enemy was completely demoralized by this bold and wholly unexpected movement; the Black Warrior was set on fire by her officers and destroyed, the fort abandoned, and the entire fleet captured or destroyed.

Passing up the river, the flotilla took possession of Elizabeth City, which the enemy had attempted to fire before hastily leaving it, and

Lieutenant Murray was despatched with a small force to Edenton, of which he quietly took possession on the 12th, and was then sent to obstruct the Chesapeake and Albemarle Canal, a duty which he successfully accomplished. At the same time, Commander Rowan made a reconnoissance of the Chowan River as far as Winston, where a sharp engagement took place on the 19th, but which was, the following morning, occupied by the troops under Colonel Hawkins, who entered the town and destroyed the military stores and quarters found there.

Flag-Officer Goldsborough having been recalled to Hampton Roads, a combined army and Navy expedition, under General Burnside and Commander Rowan, left Hatteras Inlet and arrived at Slocum's Creek, the point selected for the disembarkation of the troops, on the 12th of March. The next morning the landing of troops began, gunboats shelling the woods at the same time. Six naval boat-howitzers with their crews, under Lieutenant R. S. McCook, were landed to assist in the attack on the enemy's works. About 4 o'clock P. M., the first of the enemy's batteries opened on the boats. The firing ceased at sundown. On the 14th, the army having engaged the enemy, Commander Rowan moved up the river with his fleet, and the enemy abandoned the forts in succession under the pressure of the combined columns. On arriving at Newberne, the enemy having fled, Commander Rowan took possession of the place. The approach by river to Newberne was obstructed by piles and torpedoes, from which the vessels sustained some injury, and the passage was disputed by six forts. After the fall of Newberne, Lieutenant A. Murray was despatched with a naval column to take possession of Washington, North Carolina. He was accompanied by a detachment from the army. He arrived on the 21st of March, and the place surrendered to him without resistance.

The batteries on shore having, on the morning of the 25th of April, opened fire on Fort Macon, Commander Samuel Lockwood, senior officer of the blockading fleet off Beaufort, prepared his vessels for action and proceeded within range of the fort. Fire was opened from the steamers Daylight, State of Georgia, Chippewa, and bark Gemsbok, which was continued about an hour and a quarter. In the afternoon a flag of truce was displayed from the fort, which on the next morning surrendered to Major-General Burnside. In these actions the Marines on board the vessels were sergeant's guards.

On the last day of February, 1862, Flag-Officer Du Pont left Port Royal in the Wabash, and on the 2d of March transferred his flag to the sloop-of-war Mohican, and entered Cumberland Sound in that vessel, accompanied by the following vessels, sailing in the order named: Ottawa, Mohican, Ellen, Seminole, Pawnee, Pocahontas, Flag, Florida, James Adger, Bienville, Alabama, Keystone State, Seneca, Huron, Pembina, Isaac Smith, Penguin, Potomska, armed cutter Henrietta, armed transport McClellan (the latter having on board a battalion of Marines under command of Major Reynolds), and the transports Empire City, Marion, Star of the South, Belvidere, Boston, and Georgia, containing a brigade under the command of Brigadier-General Wright.

The first important step in the series of victories was to hoist the flag on Fort Clinch, the first of the captured national forts on which the ensign of the Union resumed its proper place after the first proclamation of the President, Early in the morning of the 4th, the town of Fernandina was occupied by the battalion of Marines under Major Reynolds, and a company of Marines and sailors, under Lieutenant Miller, was sent from the Mohican to hold Fort Clinch. Flag-Officer Du Pont said, "We captured Port Royal, but Fernandina and Fort Clinch have been given to us." One principal and ultimate object of the expedition was, to take and keep under control the whole line of the sea-coast of Georgia, the commander knowing, to use the language of the original paper, "that the naval power that controlled the sea-coast of Georgia controlled the State of Georgia."

On the 7th of March, Commander S. W. Godon was despatched with a division of gunboats, consisting of the Mohican, Pocahontas, and the Potomska, to hold Brunswick; and on the same day another division of the squadron was sent to Jacksonville, both places surrendering without opposition.

On the 12th of March, Commander C. R. P. Rodgers received the surrender of St. Augustine, the citizens raising the United States flag. This expedition gave to Flag-Officer Du Pont possession of Fort Clinch, Fernandina, St. Mary's, Cumberland Island and Sound, Amelia Sound, Jacksonville, St. Augustine, and Brunswick,—in fact, the coast and inland waters from St. Simon's southward.

# CHAPTER XXII.

**1862.**—The Civil War, IV.—Destruction of the Cumberland and Congress—Merrimac and Monitor—Capture of Norfolk—Action at Drury's Bluff—Pass Christian—Expedition up Santee River—Morris Island.

"ONE of the most important contests in the history of modern naval warfare was the fight in Hampton Roads, in which the famous Confederate ram, the Merrimac, was engaged. On the 8th of March one of the lookout vessels of the squadron lying there reported, by signals, that the enemy was coming out from the James River; and soon the iron-plated steam-battery Merrimac, accompanied by several small gunboats, was seen passing Sewell's Point and standing towards Newport News. Passing close by the frigate Congress, to which she delivered a destructive broadside, this formidable monster bore down upon the Cumberland, sloop-of-war, in command, in the temporary absence of Commander Radford, of Lieutenant George U. Morris. The Cumberland at once opened fire, but entirely without effect upon her antagonist, which stood on and struck her under the starboard fore-channels, at the same time delivering her fire. The destruction was terrible. So great, indeed, was the injury inflicted by this crushing blow that, notwithstanding the pumps were kept actively at work, the water rose rapidly in the hold, and in about two hours had drowned the forward magazine. All this time the gallant crew had kept up an active fire, and did not desist until 3.35 P. M., when the water had risen to the main hatchway, the ship canted to port, and, after a parting volley, each man took his chance of life by jumping overboard. All of the wounded who were able to walk had been ordered up, but those who had been carried into the sick-bay were so mangled that it was impossible to save them, and they were left to go down with the vessel they had served so well. Of the gallantry of this action, which has furnished one of the brightest as well as one of the saddest pages to the naval history of the world, it is difficult to speak in fitting terms; and perhaps no better words can be found than the simple sentence in which Lieutenant Morris concluded his report to his commanding officer, who arrived at Newport News only in time to see his vessel go down: 'I will only say, in conclusion, that all did their duty, and we sank with the American flag at the peak.' The Cumberland lost more than a hundred men, nearly one-third of her crew.

"While the Merrimac was engaged with the Cumberland, the smaller vessels accompanying her attacked the Congress, killing and wounding many of her crew; and her commander, seeing the fate of the Cumberland, set sails, and with the assistance of a tug, ran the vessel ashore. At half-past three the Merrimac took position astern of her at a distance of a hundred and fifty yards, and raked her with shells, one of the smaller steamers meanwhile keeping up a fire on her starboard quarter, and two others, approaching from up the James River, opening fire with precision and doing great damage. The two stern guns were now the only means of defence left the Congress, and these were soon disabled, one being dismounted and the other having its muzzle knocked away, and the men swept away from them rapidly with terrible slaughter by the cruel fire of the enemy.

"Meanwhile the steam-frigate Roanoke, the vessel of Captain Marston, the senior officer, and the Minnesota, the most powerful vessel in the Roads, were aground at some miles' distance; and Lieutenant Pendergrast, on whom, at the death of Lieutenant Smith, who fell at his post at half-past four o'clock, devolved the command of the Congress, seeing the rapid slaughter of his men, without any prospect of relief, and being unable to bring his guns to bear upon the enemy, while his ship was on fire in several places, concluded to haul down his colors, and suffer no further loss of life. An officer from the Merrimac boarded the vessel, and soon after a tug came alongside, whose captain demanded of the crew to surrender and leave the ship, as he intended to burn her immediately. A sharp fire from the troops on the shore, however, soon compelled the tug to leave, and the Merrimac again fired several shells, after which she hauled off to engage the Minnesota, and Lieutenant Pendergrast, left to himself, made all haste to get his men ashore, the ship being on fire in several places, and near the magazine.

"The Minnesota, upon the first appearance of the battery, had got under way to engage her, but after proceeding to within a mile and a half of Newport News, had there grounded, and as the tide was ebbing she could not be floated off. Here, then, the great frigate lay at four o'clock, when the Merrimac, with her two consorts, leaving the Congress to her fate, bore down upon her. Fortunately, however, the iron battery drew too much water to come within a mile of her; but the two other steamers, firing rifled guns, did much damage in killing and wounding men, until the heavy guns of the Minnesota drove them off, followed at seven o'clock by the Merrimac, when all three steamed towards Norfolk.

"The firing of her broadside guns had crowded the Minnesota still farther on the sand bank, and although all hands were at work during the night with tugs and hawsers, it was found impossible to move her. The situation of the vessel, hopelessly grounded, with the certainty of the renewal of the attack by her apparently invulnerable antagonist in the morning, was unpleasant in the extreme; but at midnight a new and powerful actor arrived upon the scene. The iron-clad Monitor, Commander John L. Worden, the first of three iron-clad vessels which had been built by the Navy Department, had arrived, most opportunely, at Hampton Roads, at nine o'clock, and immediately received orders from Captain Marston to proceed to Newport News and protect the Minnesota from the attack of the Merrimac; and all untried as the strange little craft was, she was warmly welcomed as she anchored alongside.

"At eight o'clock the following morning the Merrimac was perceived approaching. When she had come within a mile of the Minnesota that vessel opened upon her and signalled the Monitor to attack. Then came the contest which was to exert so important an influence upon naval architecture. Running down the wake of the frigate, the tiny Monitor placed herself alongside of her huge antagonist and fired gun after gun, which were returned by whole broadsides without effect. After a time the little vessel began manœuvring, shooting by her antagonist and sending her shots first into the bow, and again raking her stern, while broadside after broadside was fired from the Merrimac, either passing quite over, or, if they struck, glancing harmless from her bomb-proof turret.

"Finding that she could make no impression on the Monitor, the Merrimac again gave her attention to the Minnesota, returning a tremendous broadside from the frigate with a shot from her rifled bow-gun, which went crashing through the vessel, bursting in the boatswain's room, and setting fire to the ship. The fire was, however, promptly extinguished. Her second shell exploded the boiler of the tug alongside; but an incessant fire from the frigate was now concentrated upon her, some fifty solid shot striking upon her sides, without, however, any apparent effect. The Monitor by this time again came between the contending vessels, forcing the battery to change her position. In doing this she grounded, and the broadsides of the Minnesota were again poured upon her. As soon as she got off she stood down the bay, chased at full speed by the Monitor. Suddenly she turned and made for her antagonist, but a plunging shot through the roof arrested her dash, and for a time the encounter between this seemingly ill-matched pair was again hot and furious. After a time the Merrimac seemed to tire of the fray.

and again headed towards the frigate. It was a trying moment for the Minnesota, fast aground and badly crippled; but the enemy had no mind to renew the experience of the morning, and, it being then shortly after noon, returned to Sewell's Point. During the night, Captain Van Brunt succeeded in getting the ship afloat, and next morning was safely at anchor near Fortress Monroe.

"Towards the close of this terrific engagement a percussion shell exploded against the look-out chink of the pilot-house of the Monitor, where Captain Worden, who so brilliantly fought his little vessel, and who thus made himself in a few hours the hero of the day, was stationed throughout the engagement. The result was a severe injury to the eyes of that officer, which, with the effects of the concussion, so disabled him as to oblige him to place his vessel in command of Lieutenant Greene, executive officer, and to be subsequently removed to Washington."*

Lieutenant Morris, in his supplementary report of the action, says, "Owing to the hurried manner in which my official report to Captain Radford was made, I omitted to mention to you the gallant conduct of Lieutenant Charles Heywood, United States Marine Corps, whose bravery upon the occasion of the fight with the Merrimac won my highest applause. May I respectfully ask that this be appended to my former report." The first shot from the Merrimac killed nine Marines on the Cumberland, who formed part of the Marine division under the command of Lieutenant Heywood. In the account of such a notable fight, it is not out of place to give a list of Marines saved, which is as follows:

Orderly Sergeant Sweeney, Sergeants Brown and Callahan, Corporals Harris, Stevenson, and Craig, Drummer Joshlyn, Fifer Baxter, Privates Lyons, Martin, Bunker, 2d, W. McFaddin, Deady, Lanning, Goetz, Daley, Howard, Murley, Leary, McCarthy, Small, Wilkes, and O'Connor.

It is worthy of record that the last shot at the Merrimac was from the division of Marines.

The St. Lawrence narrowly escaped destruction, while she was hard aground, some distance from the Minnesota, on the first day. The Merrimac, about 7 P.M., planted a 100-pound percussion shell under the starboard counter of the St. Lawrence, which entered the ward-room, passing close to the after magazine hatch without exploding.

Of a portion of the famout battle, Captain Van Brunt, of the Minnesota, said, in his official report, "The Merrimac ran down near

* "Records of Living Officers," Hammersly, First Edition.

to the Rip-raps, and then turned into the channel through which I had come. Again all hands were called to quarters, and when she approached within a mile of us I opened upon her with my stern guns, and made signal to the Monitor to attack the enemy. She immediately ran down in my wake, right within range of the Merrimac, completely covering my ship as far as was possible with her diminutive dimensions, and, much to my astonishment, laid herself right alongside the Merrimac, and the contrast was that of a pigmy to a giant. Gun after gun was fired by the Monitor, which were returned with whole broadsides from the rebels, with no more effect, apparently, than so many pebble-stones thrown by a child. After a while they commenced manœuvring, and we could see the little battery point her bow for the rebels, with the intention, as I thought, of sending a shot through her bow port-hole; then she would shoot by her and rake her through the stern. In the mean time the rebels were pouring in broadside after broadside, but almost all her shots flew over the little submerged propeller, and when they struck the bomb-proof tower the shot glanced off without producing any effect, clearly establishing the fact that wooden vessels cannot contend with iron-clad ones; for never before was anything like it dreamed of by the greatest enthusiast in maritime warfare."

The following is a complete list of the officers of the Corps who were in this engagement: Minnesota, Captain W. L. Shuttleworth and First Lieutenant W. H. Cartter; Roanoke, Captain M. R. Kintzing; Cumberland, First Lieutenant Charles Heywood; Congress, First Lieutenant Joseph F. Baker; St. Lawrence, Second Lieutenant Richard S. Collum.

On the 8th of May the Monitor moved up and shelled Sewell's Point. The Merrimac came out, but though the Monitor kept well up towards her, she refused to engage her pluck adversary, and soon retired under the point. This was her last appearance, for, two days after, Norfolk surrendered to the military forces under General Wool, and the next morning a terrific explosion, in the direction of Craney Island, announced to the Navy in Hampton Roads that the once dreaded battery was no more. Lieutenant Selfridge, of the flag-officer's staff, proceeding in a tug to Sewell's Point, raised the American flag over the abandoned works, and the ships sailed up to Norfolk unmolested.

The following officers of the Corps were attached to the ships which assisted in shelling Sewell's Point on this occasion: Susquehanna, Captain P R. Fendall; San Jacinto, Captain L. L. Dawson; St.

Lawrence, Second Lieutenant R. S. Collum; Dakota, Seminole and Mount Vernon, sergeants' guards, each.

Two days after the fall of Norfolk, three gunboats,—the Galena, Aroostook, and Port Royal,—under Commander John Rodgers, were sent up the James River, in accordance with the orders of the President. On the 11th the Monitor and Naugatuck joined the expedition at Jamestown Island, and the little squadron, after numerous engagements with the enemy's batteries and sharp-shooters, arrived at Drury's Bluff, eight miles from Richmond, where they encountered a heavy battery and two barriers formed of piles and sunken vessels. The Galena and Monitor ran within six hundred yards of the Bluff, but the latter was obliged to drop down again some distance, being unable to elevate her guns sufficiently to make them tell efficiently upon the battery. After an action of three hours, the gunboats, having exhausted their ammunition, returned to City Point, the Naugatuck disabled by the bursting of a gun.

In this engagement the Marine guard of the Galena was commanded by a sergeant. Private Joseph Johnson was killed during the action. Commander Rodgers says, in a report of the action, "The Marines were efficient with their muskets, and they, when ordered to fill vacancies at the guns, did it well."

On the afternoon of the 25th of March two rebel steamers were discovered at Pass Christian. The New London, Lieutenant-Commanding Abner Reed, the blockading vessel, got under way immediately and stood for that place, approaching as near as practicable on account of shoal water. The rebel boats approached within two thousand yards, when the engagement commenced, the New London beginning the action on finding the enemy not disposed to come near her. The fight lasted one hour and fifty minutes, during which time the New London fired over one hundred and sixty shots of all kinds. The steamers of the enemy engaged were the Oregon and Pamlico. After the engagement the enemy left for the lakes. The New London remained on the ground until they were out of sight, and then returned to Ship Island. The detachment of Marines on board the New London was commanded by a sergeant.

On the 25th of the same month the mayor of St. Augustine, Florida, and the members of the City Council joined in the following letter to Flag-Officer Du Pont, commanding the United States naval forces off Florida: "The undersigned, the mayor and City Council and citizens of St. Augustine, beg leave to express their heart-felt gratification and satisfaction at the polite and urban course of Major Isaac T. Doughty and officers of the United States Marines since their arrival in and occupancy of the city, and of the good conduct and discipline of the troops under their command, and also their unfeigned regret at their departure, and respectfully ask that our appreciation of their gentlemanly demeanor be conveyed to Major Doughty and his command. They would also express their desire, and that of the citizens in general, that the battalion of Marines under Major Reynolds, which they understand is now off our harbor, if the same be compatible with the plans of the United States Government, be stationed within our city."

The command of Major Reynolds having been ordered to Washington, Commodore Du Pont detached Lieutenant H. B. Lowry, and ordered him for duty on the Wabash, the flag-ship.

In June, Commander Prentiss, senior officer commanding off Georgetown, South Carolina, informed Flag-Officer Du Pont that if he would send him three small vessels drawing about eight or ten feet of water and a guard of fifty Marines, he would run up the Santee River and destroy the railroad bridge, and thus cut off communication between Charleston and the interior of the State. On receipt of this information, Flag-Officer Du Pont ordered Lieutenant Lowry of the Marines to take passage in the Hope and report to Commander Marchand, and to place on board of the Hall and Henry Andrew the Marine guards of the James Adger, Keystone State, and Albatross. Of these, Lieutenant Lowry assumed command, and with them he proceeded to Georgetown, South Carolina, and reported to Commander Prentiss "for special service in the neighboring waters." The Marines, about sixty in number, were divided up in squads and placed as sharp-shooters on the different vessels. On the 24th the expedition started up the Santee River to destroy the bridge. The Marines were constantly firing at the cavalry, who followed along the banks of the river. Arriving at the plantation of Mr. Arthur Blake, an English subject, Lieutenant Lowry landed the Marines and examined the house, and attempted to capture a battery which had taken possession of the woods near by. They discovered unmistakable evidence that the house had been in use as quarters for the enemy, and found arms secreted there. After a skirmish of about an hour, the force returned to the ship with but two Marines wounded. By order of Commander Prentiss, the house of Mr. Blake was burned. The expedition failed to destroy the bridge, on account of lack of water in the river for the vessels to operate. The Marines remained in the waters about Georgetown, the Santee and Wahamau Rivers until July 1, when they returned to their various ships.

Early in August, Commander Foxhall Parker, commanding the Wabash, was ordered to take one hundred Marines and one hundred sailors to go ashore on Morris Island, and put two 200-pound Parrott and two 200-pound Whitworth guns in battery; the guard of the Wabash, about sixty Marines, and the guard of the New Ironsides, about thirty-five, accompanied by Second Lieutenant James B. Young, all under command of Lieutenant Lowry, landed and reported to Captain Parker. For two weeks the Marines and sailors worked, hauling their guns up the beach and getting them in battery. The Marines remained with the naval battery until the arrival of the Marine battalion on Morris Island, under command of Major Zeilin, when the Wabash men joined them, and the Ironsides men, under Lieutenant Young, returned to their ship.

The officers of the Corps attached to the North Atlantic Squadron during the latter part of the year and on the 1st of January, 1863, were: Flag-ship Minnesota, Captain W. L. Shuttleworth, Second Lieutenant Charles F. Williams; sloop Vandalia, First Lieutenant C. H. Nye.

The officers of the Corps attached to the South Atlantic Squadron during the year 1862 and on the 1st of January, 1863, were as follows: Flag-ship Wabash, Captain James Lewis and First Lieutenant H. B. Lowry; iron-clad steamer New Ironsides, First Lieutenant H. A. Bartlett and Second Lieutenant James B. Young; Powhatan, First Lieutenant Percival C. Pope; store-ship Vermont, Second Lieutenant Alfred Devereux.

A naval station having been created at Cairo, Illinois, one hundred and fifty Marines were ordered there under the command of Captain M. R. Kintzing, with the following officers: First Lieutenants Frank Munroe, S. H. Matthews, and Second Lieutenant F. L. Church.

Lieutenant Church was subsequently ordered to command the guard of the flag-ship Black Hawk.

## CHAPTER XXIII.

**1862.**—The Civil War, V.—The Capture of New Orleans—Passing the Batteries at Vicksburg—The Ariel captured by the Alabama.

On the 3d of February, 1862, Captain D. G. Farragut sailed from Hampton Roads, in the steam-sloop Hartford, to assume the duties of flag-officer of the Western Gulf Blockading Squadron. In addition to the ordinary blockade duties, he was especially charged with the reduction of the defences guarding the approaches to New Orleans, and taking possession of that city. In his confidential instructions he was informed that a fleet of bomb-vessels and armed steamers enough to manage them all, under command of Commander David D. Porter, would be directed to report to him. A large force of vessels, consisting of many of the best frigates and sloops in the service, recently fitting out at the various Navy-Yards, had received orders to report to him at Key West. Eighteen thousand troops, under the command of Major-General Benjamin F. Butler, were to co-operate with the West Gulf Blockading Squadron. Flag-Officer Farragut arrived at Ship Island on the 20th of February, having been detained some time at Key West, and at once commenced active preparation for the attack on the defences of New Orleans. Much difficulty was experienced in getting the larger vessels over the bar, and in the case of the frigate Colorado it was found impossible. On the 16th of March the mortar vessels commenced the bombardment of Fort Jackson, assisted occasionally by the gunboats. On the 1st of April, Flag-Officer Farragut detailed a force to cut and destroy the chain and raft across the river, and this hazardous undertaking was successfully carried through by Captain Bell, assisted by Lieutenant-Commanding Crosby, in the Pinola, and Lieutenant-Commanding Caldwell, in the Itasca. On the 23d of April, 1862, Flag-Officer Farragut made his final preparations for the attack on and passage of Forts Jackson and St. Philip. Every vessel was as well prepared as the ingenuity of her commander and officers could suggest. Chief-Engineer Moore of the Richmond originated the idea of stopping the sheet-cables up and down on the sides of the ships in the line of the engines, which was immediately adopted by all the vessels. Each commander made his own arrangements for protecting the boilers or machinery, by coal, bags of ashes and sand, naval clothes-bags, and in fact every device imaginable. The bulwarks were lined with hammocks by some, by splinter-nettings made with ropes by others. Some rubbed their vessels over with mud to make them less visible. Lieutenant Cummings, the executive officer of the Richmond, made the valuable suggestion that whitewashing the decks would give the men sufficient light in a night attack, and obviate the necessity of using lanterns, which were targets for the enemy to fire at.

At 2 o'clock A.M., April 24, signal was made to get under way, but owing to the great difficulty in purchasing their anchors, the Pensacola and some of the other vessels were not under way until half-past three. The vessels then advanced in two columns, Captain Bailey leading the right and the advance in the gunboat Cayuga, he having been assigned to the first division of gunboats, which consisted of the Cayuga, Lieutenant-Commanding Harrison; the Oneida, Commander Lee; Varuna, Commander Boggs; Katahdin, Lieutenant-Commanding Preble; Kineo, Lieutenant-Commanding Ransom, and Wissahickon, Lieutenant-Commanding Albert Smith, supported by the steam-sloops Pensacola, Captain Morris, and Mississippi, Commander M. Smith. This division was to attack Fort St. Philip. The second division of the column was led by the flag-ship Hartford, followed by the Brooklyn, Captain Craven, the Richmond, Commander Alden, and the second division of gunboats, led by Fleet-Captain Bell, in the Sciota, Lieutenant-Commanding Donaldson, followed by the Iroquois, Commander De Camp; Kennebec, Lieutenant-Commanding Russell; Pinola, Lieutenant-Commanding Crosby; Itasca, Lieutenant-Commanding Caldwell, and Winona, Lieutenant-Commanding Nichols, in the order named.

The enemy's lights, while they discovered the vessels to them, were at the same time guides to the squadron, which soon passed the barrier chains, the right taking Fort St. Philip, and the left Fort Jackson. As the fire became general, the smoke grew so dense that it was very difficult to distinguish friends from foes. Commander Porter had, by previous arrangement, moved up to a certain point on the Fort Jackson side with his gunboats, while his mortar vessels, assisted by the sloop-of-war Portsmouth, engaged the water batteries to the south and eastward of Fort Jackson, and poured a terrific fire of shells into it. A fire-raft was discovered coming down upon the Hartford, and in attempting to avoid it the ship was run on shore, where the rebel ram Manassas, which had not previously been seen, pushed the raft down upon the flag-ship, which was soon on fire half-way up to her tops; she was backed off, and through the good organization of the fire department, and the great exertions of Captain Wainwright and his first lieutenant, officers, and crew, the fire was extinguished.

In the mean time the battery of the Hartford was pouring its missiles of death into Fort St. Philip, which was soon silenced, with the exception of a gun now and then. By this time the enemy's gunboats, thirteen in number, besides two iron-clads, the Manassas and Louisiana, had become visible; they were taken in hand, and in a short time eleven of them were destroyed. The fleet was now fairly past the forts, and the victory was won. Several gunboats were still making resistance. Two of them had attacked the Varuna, which vessel, by her greater speed, was in advance of her consorts; they ran into her and caused her to sink, but not before she had destroyed her adversaries: and when the Hartford passed, the wrecks of the three vessels were lying side by side. Just as the scene appeared to be closing, the ram Manassas was observed coming up at full speed, to attack the Hartford. Flag-Officer Farragut directed Captain Smith, in the Mississippi, to turn and run her down. The order was instantly obeyed by that vessel turning and going at her full speed. But when within fifty yards of each other, the ram put her helm hard aport and ran ashore; the Mississippi poured two broadsides into her, and sent her drifting down the river a total wreck. This closed the morning's fight.

Captain Bailey had preceded the flag-ship up to the quarantine station, and had captured the Chalmette regiment. By order of Flag-Officer Farragut, the officers and men were paroled the same day. Owing to the slowness of some of the vessels, and want of knowledge of the river, the fleet did not reach the English turn until about 10 A.M. on the 25th. The fleet was now formed in two columns as before. Captain Bailey was still far in advance, not having noticed the signal for close order, which was to enable the slow vessels to come up; they opened on him a galling fire from the Chalmette batteries, but the larger vessels soon came to his assistance, and ranged in one after another, delivering their broadsides with such telling effect that the batteries were silenced and the rebel troops driven out.

The fleet then passed up to the city and anchored immediately in front of it. Captain Bailey was sent on shore to demand the surrender of New Orleans from the authorities. The mayor replied that the city was under martial law. General Lovell, who was present, said he would surrender nothing, but, in order to free the city from embarrassment, he would restore the authorities and retire with his troops, which he did. All the steamboats lying at the levee were seized and sent down to quarantine for General Butler's forces. The levee of New Orleans was one scene of desolation. Ships, steamers, cotton, coal, etc., were all in one common blaze, and the ingenuity of the squadron was much taxed to avoid the conflagration.

Flag-Officer Farragut then pushed on to Carrollton, eight miles above, where there were two other forts, which were found deserted.

On the 28th of April, Forts Jackson and St. Philip, after a bombardment of one hundred and forty-four consecutive hours by the mortar flotilla, surrendered to Commander David D. Porter.

Of the resistance offered by the Confederate forts below New Orleans, Admiral Farragut said, "Such a fire, I imagine, the world has rarely seen." In the terrible experience of silencing that fire, reducing the forts, and capturing the city, the Marines bore themselves in such a manner as to elicit the hearty commendation of the officers of the fleet, and to aid materially the success of the engagement. The Varuna sustained the fire of the forts, and at the same time was attacked by two powerful ironclads. After a gallant resistance, in which the two vessels of the enemy were destroyed, the Varuna herself sunk, with colors flying and her guns dealing destruction to the enemy.

Commander Boggs said, "The Marines, although new recruits, more than maintained the reputation of that Corps. Their galling fire cleared the Morgan rifled gun, and prevented a repetition of her murderous fire." Commander John De Camp, commanding the Iroquois, wrote: "The Marines behaved with spirit and gallantry, which we may always expect in well-drilled Americans." Captain T. T. Craven, of the Brooklyn, said, "Lieutenant James Forney, commanding the Marines, had two guns assigned him, and, with his men, fought most gallantly."

But, brave and efficient as were the Marines in the action, a more important work awaited them. On the morning of the 24th of April, immediately after the action with Forts Jackson and St. Philip, and the destruction of the rebel fleet, the Marines, under command of Captain John L. Broome, by order of Admiral Farragut, landed, and took possession of the quarantine, at the same time taking prisoners the rebel troops, with their officers quartered in the quarantine buildings, and hoisting the flag of the United States on the same.

On the morning following the arrival of the squadron opposite New Orleans, a detachment of Marines from the Pensacola received orders to embark in the ship's boats and land.

The detachment consisted of thirty men, under the command of Lieutenant J. C. Harris, U. S. M. C., and was accompanied by Lieutenant Stillwell, U. S. N., and a sailor carrying the flag.

The "levee" at New Orleans was at that time crowded with a multitude of highly-excited men and women, largely composed of the worst element of a notoriously turbulent population, who greeted the Marines with cries of execration and derision; knives and pistols were freely brandished in their faces, but, undaunted by all these hostile demonstrations, that small body of highly-disciplined soldiers calmly but promptly formed, and marched with measured steps and unruffled mien to their destination. Arrived at the United States Mint, a cordon of sentinels was placed around the building, the Confederate flag lowered and replaced by that of the United States. This having been accomplished, the detachment, under Lieutenant Harris, remained in possession until the arrival of the battalion under Captain Broome, when they were consolidated with his command.

Admiral Farragut having determined to take military possession of the city until the arrival of the troops under General Butler's command, a battalion of Marines, under command of Captain Broome, disembarked from the fleet on the 29th of April, and marched to the custom-house. The following is the roster of the battalion: Captain J. L. Broome, commanding; Adjutant, Second Lieutenant J. C. Harris. Company A, fifty men: Captain Alan Ramsay. Company B, fifty men: Captain P. H. W. Fontané. Company C, fifty men: First Lieutenant James Forney. Company D, fifty men: First Lieutenant J. H. Higbee.

Captain Alan Ramsay was detailed with a detachment of Marines to occupy the custom-house and guard the United States flag, then about to be hoisted on the building. At this juncture the Marines were joined by two howitzers, manned by seamen, in charge of Midshipmen J. H. Read and E. C. Hazeltine, from the flag-ship Hartford. After taking possession of the custom-house, Captain Broome received orders from Captain H. H. Bell, senior officer present, to march the Marines to the City Hall, a distance of about half a mile from the vessels of the fleet, and near the centre of the city. The forces under General Lovell had not evacuated the city, and the squares and streets were thronged with an excited mob, brandishing bowie-knives and revolvers, and hailing the Marines with the most abusive language; but the command marched in close order and steady steps to its destination. On arriving at the City Hall, Lieutenant John C. Harris was directed, with a gurd of Marines, to occupy the building and enforce order there while the enemy's flag was hauled down from the flag-staff on the building. After performing this duty, the Marines were marched to the place of embarkation, and returned to the fleet, except the number quartered in the custom-house, who were retained there for the purpose of guarding the United States flag. When the troops of General Butler's command landed at New Orleans, on the 1st of May, the force of Marines still on duty in the city returned to the fleet.

Respectable inhabitants of the city afterwards stated that rarely were men in more imminent danger from assassination than was the small detachment of Marines on that occasion, and that they wondered at the boldness of the act, under the circumstances, and trembled for the safety of the city, fearing, with reason, that any overt act on the part of the excited mob would result in an immediate bombardment, wholly under the command of the ship's broadsides; the calm and steady attitude of the Marines tended to a great extent to the prevention of such a calamity.

And so, three days before the arrival of General Butler, the Marines took possession of and held the public buildings of the city of New Orleans, literally perfecting the conquest of the city by hauling down the Confederate flag and hoisting the stars and stripes. For three days the force under Captain Broome held in subjection the turbulent and rebellious elements of the population of the city, and at the end of that time gave place to the troops under command of General Butler. To rescue this important chapter in the records of the Marines from the oblivion of meagre official documents in the national archives, and to give it a legitimate, and what is trusted will be a lasting prominence, is but an act of justice to the Corps.

The following is a complete list of Marines killed and wounded during the capture of the forts and the city: Killed: Brooklyn, Privates W Lenahan and Henry H. Roff; Iroquois, Private Jacob Schoenfeldt; Mississippi, Corporals George Sanderson, and W. H. Woods. Wounded: Oneida, Private Henry Cooper; Hartford, Second Lieutenant Heisler, Privates Henry King and George White; Brooklyn, Privates Lorin Heath, J. R. Sanders, and Leonard Killion; Pensacola, Lieutenant John C. Harris, Sergeant Stermbergh, Privates George Perkins, Michael O'Bryan, Frederick Davye, Francis Pepper, and John Brogan; Iroquois, Corporal Walter J. White, mortally, Alfred Jackson; Varuna, Privates T. Gordon, D. McLaughlin, J. Logan, and J. McQuinn; Mississippi, Private Richard C. Carman.

The following is a list of the officers of the Corps attached to the Western Gulf Squadron during the year 1862 to January 1, 1863: flagship Hartford, Captain John L. Broome, Second Lieutenant Heisler, who was relieved after the battle of New Orleans by First Lieutenant George W. Collier; steam-sloop Richmond, Captain Alan John C. Harris; steam-sloop Brooklyn, First Lieutenant James Forney; steam-sloop Susquehanna, off Mobile, Captain Philip R. Fendall. Jr.; steam-sloop Mississippi, Captain P. H. W. Fontané; steam-frigate Colorado (this vessel, on account of her heavy draught, was unable to get over the bar), Captain George R. Graham, First Lieutenant Samuel C. Adams; sloop Portsmouth, above the passes, First Lieutenant W. H. Hale; frigate Potomac, off Pensacola, First Lieutenant George W. Collier; steam-sloop Richmond, Captain Alan Ramsay; sloop Vincennes, Ship Island, Second Lieutenant N. L Nokes.

On the 28th of June the fleet under Admiral Farragut passed the batteries at Vicksburg, receiving and answering a terrible fire from the well-mounted guns of the enemy. Captain Broome, commanding the Marines of the squadron, was wounded, as were also Privates

Thomas Nolan and George W. Harris of the Marine guard of the Richmond. Commander R. Wainwright, commanding flag-ship Hartford, says of the conduct of the Marines during the fight: "The Marine guard, under command of Captain John L. Broome, had charge of two broadside guns, and fought them well, thus sustaining the reputation of that distinguished Corps." Officers of the Corps who participated in this engagement were assigned as follows: Hartford, Captain John L. Broome, First Lieutenant John H. Higbee; Brooklyn, First Lieutenant James Forney; Richmond, Captain Alan Ramsay.

On the 15th of July, in the engagement between the Confederate ram Arkansas and Farragut's vessels, above Vicksburg, Captain Broome was again wounded, as was also Private George Rogers of the Marines.

During the early part of December an event occurred which was most deeply regretted by the Marines and by the friends of the Corps. Seldom has the Corps known the unpleasantness of defeat; but in this instance the circumstances of the case forced upon a battalion of Marines the unavoidable necessity of submitting to a superior force without the opportunity to fire a shot. Mortifying as was this fact, however, the firmness and patriotism of the defeated men in refusing to yield to the flattering offers laid before them to tempt them to desert their flag reflected lasting credit upon them. On Saturday, the 1st of December, two companies of Marines, of seventy men each, sailed from New York in the Pacific mail steamship Ariel for Aspinwall. A portion of this command was destined for the garrison at Mare Island, California, and the remainder for the ships composing the Pacific Squadron. On the morning of the 7th, when off the eastern end of Cuba, a steamer hove in sight flying the United States flag. The Ariel continued on her course, followed by the stranger, who rapidly overhauled her. When within a mile, the pursuing vessel fired a blank cartridge, at the same time hauling down the stars and stripes, and hoisting in their stead the flag of the Confederacy. Captain Jones, commanding the Ariel, paid no attention to this unmistakable summons, but continued on his course, saying, "I will not heave to." A solid shot was then fired, which passed through the rigging of the Ariel, followed by a shell, which caused much damage, cutting away the foremast. At this, Captain Jones hove to. In a few moments a boat filled with armed men came alongside, in charge of Lieutenant Armstrong, of the Confederate Navy, who announced to the captain of the Ariel that the latter had surrendered to the famous Alabama. The Marines were ordered to surrender their arms, and the officers their swords. Any resistance on their part would have endangered the lives of the women and children. The order having been complied with, Lieutenant Armstrong addressed the command, calling for twenty volunteers to fill up the Marine guard of the Alabama, offering the best pay and plenty of prize money if they would join the Pride of the Ocean, also saying he knew they thought more of Jeff. Davis than of Lincoln. Finding that not a man would prove recreant to his trust, or a traitor to his country, the lieutenant withdrew. The Ariel being in possession of a prize crew of sailors, Major Garland, at the request of Captain Semmes, detailed sentinels, though prisoners of war, to guard the spirit rooms to prevent the rebel crew from making use of the liquor they contained. The officers and men of the Ariel were paroled, and, after considerable discussion and delay, the vessel was ransomed for the handsome sum of two hundred and sixty-one thousand dollars, and permitted to proceed on her journey. The battalion arrived in due course of time at Mare Island, where they were shortly afterwards exchanged.

During this year a detachment of Marines, under the command of First Lieutenant McLane Tilton, garrisoned Pilot Town, Louisiana.

## CHAPTER XXIV.

**1863.**—The Civil War, VI.—Raid of Rebel Ironclads among the Vessels of Charleston, South Carolina—Attack on Port Hudson—The Draft Riots in New York City—Morris Island—Attack on Fort Sumter—Potomac Flotilla—East Gulf Squadron.

About four o'clock on the morning of January 31, 1863, during the obscurity of a thick haze, two iron-clad gunboats ran out of Charleston by the main ship-channel, unperceived by the squadron, and began a raid upon the blockading fleet. Most of the latter were of the light class of purchased vessels, two of the heaviest men-of-war, the Powhatan and Canandaigua, being then at port coaling and repairing. The Mercedita, Captain Stellwagen, was the first vessel attacked, and was rendered powerless without having an opportunity to fire a shot. Unable to use his guns, and being at the mercy of the enemy, who was lying alongside, all resistance was deemed hopeless by Captain Stellwagen, and he surrendered. The crew and officers were paroled. The ironclad, leaving the Mercedita to her fate, to sink or float, next engaged the Keystone State, Commander Le Roy, who was also attacked by the enemy. The fire was gallantly returned, but the superior power of the enemy soon disabled the Keystone State. In the meantime the Augusta, Commander Parrott, the Quaker City, Commander Frailey, and the Memphis, Acting Lieutenant Watmough, kept up a fire upon the enemy, diverting his attention from the Keystone State, which was soon after taken in tow by the Memphis and drawn away from the fire. The Augusta and Quaker City were both struck in their hulls, the Memphis only in her rigging. The Housatonic, Captain Taylor, gave chase, and a shot from her struck the pilothouse of one of the ironclads, carrying away one of her flags. The enemy's vessels passed to the northward, receiving the fire of the ships, and took refuge behind the shoals. The only casualties were on the Mercedita and the Keystone State. On the Keystone State they were very large; about one-fourth of her crew were killed and wounded, and among the former was the medical officer of the ship, Assistant-Surgeon Jacob H. Gotwald, who was scalded to death while rendering surgical aid to one of the wounded men. The following is a list of casualties in the Marine guard of the Keystone State: Killed, Orderly Sergeant Edward Livermore, Corporal William A. Graw, Privates Thomas Riley, John W. Armstrong, William Deitz, John P. Conway, William Peyton, and Patrick Herrick. Wounded, Private Michael Scott.

In the attack on Port Hudson, by Admiral Farragut, on the 13th of March, the following officers of the Marine Corps were engaged: Flagship Hartford, Captain John L. Broome and First Lieutenant J. H. Higbee; Richmond, Captain Alan Ramsay; Mississippi, Captain P. H. W. Fontané. The following is a list of casualties in the Marine guards: Hartford, killed, Private Thomas F. Butler. Mississippi, missing, Privates Patrick Bannon, eter Doyle, Patrick Flaherty, James McMullin, John Redding, William Talant, and John Kelly; killed, Private David Kelly. Richmond, killed, Privates Charles Catherwood and John Thompson; wounded, Privates Thomas Nolan, Joseph P. Mullin, George W. Harris, Michael O'Neil, Robert Staples, and Edward Conover; Corporals John S. Gross and Robert H. Neely. Monongahela, Corporal Francis Marr, wounded. In his report of the participation of the steamer Richmond in the engagement, Commodore James Alden says, "Captain Ramsey, who deserves special mention, in charge of the Marine division of great guns, *had nearly a whole gun's crew swept away by a single cannon shot.*" While passing the batteries on Red River, on the 19th, Privates John Brown, Michael Corcoran, and Patrick Cook were wounded.

The enemy's steamer Virginia, having been chased by the Wachusett, was captured when near a shoal close to the island of Majores.

Lieutenant George P. Houston, of the Marines, volunteered to bring the vessel out from its proximity to the dangerous shoals. Accordingly, with a detachment of his guard, he took possession of the vessel and ran her out into the open sea, fifteen miles off the coast. A Marine was placed at the wheel, another as oiler in the engine-room, others in the fire-room, while the gallant lieutenant ran the engines himself. So efficiently and promptly was this duty performed, that Rear-Admiral Wilkes expressed his approbation in a letter to the Department.

On the 13th of July a brigade of seamen and Marines, under the command of Lieutenant-Commander R. W. Meade, United States Navy, left the Navy-Yard in Brooklyn, in response to a request from the authorities of the city of New York. A strong resistance having been made to the draft by evil-disposed and lawless persons, and many lives having been lost in consequence, it became necessary to quell the disturbance by force of arms. The brigade marched to the City Hall, and from that point the battalion of Marines, under Captain J. C. Grayson, was sent in different directions, clearing the streets in some localities and assisting the police in making arrests. In addition to this, the district in which the rioting had taken place was thoroughly patrolled, and sentinels posted over public buildings and property in danger of destruction by the disaffected. This duty was performed from the 13th to the 20th of July, inclusive, to the satisfaction of the city authorities, and won their marked approbation. Captain Grayson's battalion consisted of two companies, the first consisting of ninety men, under First Lieutenant C. A. Stillman, and the second of ninety men, under Second Lieutenant R. L. Meade.

In the following August a battalion, under the command of Major Jacob Zeilin, sailed from New York to co-operate with the South Atlantic Squadron, under the command of Rear-Admiral J. A. Dahlgren. The battalion was debarked on Morris Island, where the camp was established. It participated in all of the subsequent engagements which resulted in the capture of the outer defences of Charleston. The roster of the officers of the battalion was as follows: Major Jacob Zeilin, commanding; Lieutenant J. C. Harris, adjutant; Lieutenant C. H. Bradford, quartermaster; Company A: Captain C. G. McCawley, Lieutenants R. L. Meade and F. T. Peet; Company B: Captain C. D. Hebb, Lieutenants George C. Stoddard and L. E. Fagan; Company C: Captain L. L. Dawson, Lieutenants William Wallace and E. P. Meeker. Major Zeilin having been allowed to return home on sick leave shortly after the arrival of the battalion, he was succeeded by Captain E. McD. Reynolds, from the Wabash. Lieutenant-

Colonel John George Reynolds subsequently was ordered to command the battalion. Captain Dawson and Lieutenant Harris returned home on sick leave soon after the arrival of Colonel Reynolds, and Lieutenant H. B. Lowry was appointed adjutant, and Lieutenant William Wallace succeeded to the command of Company C. The first important work in which the battalion engaged was the attack on Fort Sumter, under Rear-Admiral Dahlgren, on the 8th of September. On the night of the 6th the enemy evacuated Morris Island. This offered an opportunity for assaulting Fort Sumter, which, if successful, would enable Admiral Dahlgren to pass the obstructions in the channel. He therefore directed the commanding officers of the different vessels off Charleston to send detachments of sailors and Marines, with boats, to the flag-ship, prepared for the proposed movement. Captain Reynolds, commanding the Marine battalion on Morris Island, was ordered to send one hundred men and six officers to the flag-ship to participate in the assault. Having called for volunteers, the following officers responded: Captain C. G. McCawley, First Lieutenants H. B. Lowry, P. C. Pope, Charles H. Bradford, and John C. Harris; Second Lieutenants R. L. Meade, William Wallace, and L. E. Fagan. The naval battery was also ordered to send its quota, consisting of four officers and fifty-three men. At dark all the detachments had arrived, and at eleven o'clock the expedition, consisting of over twenty boats, and with twenty-four officers and four hundred and thirteen men, was placed under the command of Commander T. H. Stevens, of the Patapsco, with Lieutenant Commander Bunce and Lieutenant Moreau Forrest as aids. Lieutenant Commander Williams, of the Wissahickon, was placed in charge of the first division; Lieutenant Remey, of the naval battery, of the second; Flag Lieutenant Preston, of the third; Lieutenant Higginson, of the Powhatan, of the fourth; and Captain McCawley, commanding the Marines, of the fifth division. Ensign Craven, of the Housatonic, was also attached to this division. Commander Stevens and his aids led the flotilla in the admiral's barge.

The plan of attack was to assail the fort on three sides,—one party landing on the gorge wall and attempting to ascend the *debris* and gain the parapet; a second was to attempt to gain entrance through the lower embrasures, and a third was to act as a reserve.

The Daffodil took the boats in tow and steamed up to a short distance from Sumter, where they cast off and formed in line of attack.

The boats pulled cautiously along and made slow progress necessarily, as the proper line of attack had to be observed.

At half-past one the first line of boats approached closely the fort, and were discovered by the sentry on the walls of the work and sharply challenged. No reply was made to the question of "What boat is that?" A second challenge of the same nature failing to elicit a reply, the sentry discharged his musket, and called to the officer below to "Turn out the guard." The boats on being hailed pulled quickly to the fort; but before they could reach it several shots had been fired at them. The boats had dashed rapidly up, the formation of the line of advance being broken, and each boat striving to effect the first landing. Seven boats succeeded in getting alongside the *debris* on the gorge wall; the others while pushing up were met with a sharp fire of musketry.

Signal lights were burned from Sumter, and in a moment all the rebel batteries bearing on the fort opened a fire of shell and shrapnel on the fort itself, and, of course, on any party that might be about its base. About one hundred and fifty sailors and Marines got ashore, and instead of finding a slope of *debris* up to the parapet of the gorge wall they found a perpendicular range of masonry, which the rebels had constructed, meeting them full in the face. All their efforts to find a place of ascent were fruitless. Not a soul could ascend the wall, and the party found themselves in a critical position. The rebels had manned the parapet with infantry, and were also firing through loop-holes, formed of sand-bags, in the upper slope of the *debris*. In addition to this, hand grenades were hurled upon the assaulting party's heads, and bricks were detached and tumbled down upon them. Three of the boats were torn to pieces by hand grenades or shells from the distant rebel batteries and retreat was rapidly being cut off. At this juncture a rebel ram came down and opened fire with grape and canister upon the boats, the rebels on the fort throwing flashes of light upon the dark waters about them from a large locomotive lamp. As each boat was brought to light, volleys of musketry, canister and grape, now poured in from the fort and gunboat, and many were killed and wounded. The only mark for our men to fire at was this light, and a volley or two was thrown at it, but to little effect. Finally, a continuance of their effort to carry out the plan being evidently of no avail, and fearing a heavy loss of men without any gain, the order to retire was given. Four boats came off from the landing at the gorge, and three, being destroyed, were left there. Only a small portion of the storming party succeeded in regaining their boats. Many were killed and the balance taken prisoners.

About ten officers and one hundred and four men were missing,—eighty known to be killed.

The officers taken prisoners were Lieutenant-Commander Williams,

Lieutenant Remey, Flag-Lieutenant Preston, Lieutenant Brower, Ensign Porter, Acting Master's Mates Henry and McCarthy, Lieutenants Bradford and Meade of the Marines. Lieutenant Bradford was mortally wounded, and died in the hands of his captors. He received every attention and kindness from Dr. Mackey, a prominent citizen of Charleston, and an old friend of the gallant officer's father. His remains were interred in the doctor's family lot; but the feeling among the populace was intense against this act of humanity and kindness, and, by order of the authorities, the body was removed to Potter's field. One of the first acts of Admiral Dahlgren, after the fall of Charleston, was to remove and bury, with all the honors of war, the remains more befitting the career of the brave and lamented officer. Lieutenant Meade suffered all the privations of prison-life, and remained in Columbia, South Carolina, until regularly exchanged; not, however, until more than a year had elapsed.

Lieutenant Higginson, who was to have gained access to the fort through the embrasures of the lower casemates, in the northeast face, found a landing impracticable, and was compelled to give up the attempt, as his boats were thumping heavily on sharp rocks at the base of the fort. Lieutenant Lowry, of the Marines, who had charge of two boats, discovered the rebel steamer bearing down upon his boats before he had reached the fort, and as she opened fire, and seemed determined to run the boats down, he ordered his division to pull for shoal water. The steamer soon turned and steamed towards Fort Moultrie, and Lieutenant Lowry ordered his boats to pull up again to the fort. Other boats were driven off by this steamer and prevented from co-operating as effectually as they might have done. It may, perhaps, be regarded as a fortunate circumstance that all the boats did not effect a landing, as it would only have resulted in an increase of the rebel gain in prisoners.

The want of success of the expedition was due to causes over which neither the admiral nor Commander Stevens had any control. The appearance of the *debris* on the gorge face was deceptive. Instead of sloping gradually from the parapet to the water's edge, it was found to slope only from the parapet to the top of the sandbag barricade or traverse, which the rebels piled up to protect the wall against our breaching batteries. This wall of sand was, as least, twelve feet high, and, without the aid of scaling ladders, no one could possibly reach its top, from which, perhaps, it might have been possible to mount the parapet or the mass of the *debris*. This fact was not discovered until the attempt to scale the wall had been made, and then it was under a heavy fire. Retreat was cut off so quickly by the destruction of a portion of the boats and the appearance of the rebel steamer, at a short distance only from the gorge, that the only sensible alternative between death and surrender was made by the party remaining on the landing, and they gave themselves up as prisoners, and were speedily inside Fort Sumter, although not as captors.

The casualties among the Marines were as follows:

*Powhatan.*—Wounded, Corporal Hennis, Privates Hariland and Ball; taken prisoner, Corporal Cully, Privates Scanlan, Foy, Kelly, White, Allen, McConigley, and Hall.

*Housatonic.*—Taken prisoner, Private Bannon.

*Marine Battalion.*—Taken prisoner and mortally wounded, First Lieutenant C. H. Bradford, quartermaster.

Company A: Private Rodgers, wounded and sent to the Memphis. Company B: Taken prisoner, First Lieutenant R. L. Meade, First Sergeant Chisholm, Corporal Piggott, Privates McKenna, Mullin, Van Zant, McIntyre, Hurshfield, Johnson, Keiffer, Himes, Gordon, Long, Stansbury, Reynolds, Murphy, and Mallady.

Company C: Taken prisoner, Sergeant Mulhall, Corporal Black, Privates Haynes, Hurley, Preston, Gettings, McKinley, McNeal, Bradford, Walsh, Kassman, Siddell, and Martin.

On the 28th of December, Lieutenant-Commander R. W. Meade, Jr., of the gunboat Marblehead, was sent, at the head of a sufficient force of Marines and sailors, to complete the victory of the 25th at Stono, South Carolina, to capture and destroy the property in the earthworks of the enemy. The expedition was entirely successful, the work of capture and destruction being prompt and effectual. In his report of the affair the commander said, "The conduct of the blue-jackets, and especially the Marines, was excellent."

The vessels of the Potomac flotilla found constant employment in keeping a close watch and guard to intercept and prevent, as far as possible, communication with the rebels, and many captures were made. To provide against possible contingencies at the time of the invasion of Maryland and Pennsylvania, which terminated in the battle of Gettysburg, vessels were stationed at what were deemed available points along the upper waters of the Chesapeake, to co-operate with the military authorities. A gunboat was sent up the Susquehanna to Havre de Grace, another up the Gunpowder, a third up the Buckwater, while one was posted at Annapolis, and another at Wilmington. Detachments of Marines, under the command of sergeants, served on the larger vessels of the flotilla.

In the East Gulf Squadron many important boat expeditions for cutting out vessels and destroying salt-works were projected and executed with success. More than one hundred blockade-runners were captured or destroyed by the squadron during the year, and violating the blockade became so precarious a business that few were desperate enough to attempt it. In the latter part of the year the limits of this squadron were extended so as to embrace within its cruising-grounds the waters of the Bahamas in the vicinity of Cuba.

In all these expeditions the Marines participated, a number of the non-commissioned officers receiving honorable mention. Until May of this year the Marines of the flag-ship St. Lawrence were under the command of First Lieutenant R. S. Cullum.

The following officers of the Corps were attached to vessels in the South Atlantic Squadron during the year: Flag-ship Wabash, Captain E. McD. Reynolds, First Lieutenant H. B. Lowry, subsequently Second Lieutenant L. E. Fagan; iron-clad steamer New Ironsides, First Lieutenant Henry A. Bartlett and Second Lieutenant James B. Young (this vessel, in which these officers served, participated in twenty-six engagements with the forts commanding the defences of Charleston); Vermont, at Port Royal, Second Lieutenant Henry J. Bishop. The following officers of the Corps were attached to the North Atlantic Squadron during the year: Minnesota, flag-ship, Captain John Schermerhorn, Second Lieutenant Charles F. Williams; ironclad steamer Roanoke, First Lieutenant Frank Munroe.

---

## CHAPTER XXV.

**1864.**—The Civil War, VII.—Condition of the Corps—Sounds of North Carolina—The Ram Albemarle—Murrill's Inlet—The Defeat of the Alabama—Engagement at Simonosaki—Havre de Grace—Admiral Farragut's Attack on the Defences of Mobile Bay—Safety of the Wabash due to the Action of the Marines.

THE following is an extract from the report for 1864 of the Hon. Gideon Welles, Secretary of the Navy: "The colonel commandant of the Marine Corps reports his command in a good state of discipline. Although its number is now fully equal to the quota authorized by law, he is unable to comply with all the requisitions for guards for sea-going vessels. The reports from the several squadrons and vessels of the service show that, in the gallant deeds of the Navy, the Marines have borne an honorable part."

In the sounds of North Carolina important events were in preparation. The value of the possession of these waters was evidenced by the frequent and persistent efforts made by the enemy to repossess himself of them and of the important points upon their shores held by the army and Navy. In the spring of 1864 the possession of the sounds seemed to be very insecure; the land force was small and scattered; most of the gunboats were slightly built, the iron-clads then at the disposal of the government being unsuited for operations in shallow waters; and far up in the almost inaccessible waters of the Roanoke and the Neuse it was known that the construction of armored vessels, as well as of others of light draught, was in progress.

On April 17 the enemy besieged Plymouth, and two days after the ram Albemarle, a formidable iron-plated battery, descended the Roanoke and attacked the wooden gunboats lying off that town. Lieutenant-Commander Flusser, in anticipation of the attack, had chained together his vessels, the Miami and the Southfield, intending to fight the ram in that way, but at three o'clock in the morning she made her appearance, and in half an hour had sunk the Southfield, disabled the Miami, whose gallant commander was killed, and obtained possession of the river. The next day the defences of the town were carried, the garrison taken prisoners, and thus the entire command of the upper sound passed into the hands of the enemy.

To prevent further disaster, vigorous measures were at once adopted. Captain Melanchton Smith was ordered to assume command in the sounds, with instructions to attack the ram, at all hazards, in the best manner to insure its destruction. On the 5th of May the enemy besieged Newberne, and on the same day the ram again came out. Captain Smith promptly engaged her, with four vessels,—the Metabessett, Wyalusing, Sassacus, and Whitehead. The engagement began about half-past four, and continued furiously for three hours, the gunboats firing rapidly and repeatedly, and ramming the battery with such effect that at dark she retired up the river, and did not again make her appearance until the 24th, when she was seen near the mouth of the river, but, on a shell being thrown from the Whitehead, immediately returned. It was not, however, until the destruction of the ram by Lieutenant Cushing, that Commander Macomb was enabled to drive the rebels from their works at Plymouth, and re-establish the supremacy of the government in the waters of North Carolina.

The following-named officers of the Corps were attached to the North Atlantic Squadron on the 1st of January, 1864: Flag-ship Minnesota, Captain John Schermerhorn and Second Lieutenant Charles F. Williams; iron-clad steamer Roanoke, First Lieutenant Frank Munroe, subsequently First Lieutenant L. P. French. The remainder of the vessels of this squadron, with some few exceptions, were supplied with sergeant's guards, and in all the operations on James River, in the sounds of North Carolina, and off Cape Fear River and adjacent inlets, the Marines shared in the labor and success of their brethren of the Navy, and won the hearty approbation of their commanding officers, of the Department, and of the country.

On the 1st of January, Lieutenant Louis E. Fagan of the Marines, in command of thirty men, participated in the expedition which, under orders from Admiral Dahlgren, destroyed a Confederate vessel and valuable cargo in Murrill's Inlet. For his conduct in this affair, Lieutenant Fagan received honorable mention in the official report of Admiral Dahlgren to the Secretary of the Navy.

On the night of April 18 the Marines on the Wabash, off Charleston, ably assisted in repelling the attack of a torpedo-boat. They were stationed by Ensign Charles H. Crane, who happened to be officer of the deck at the time, in such a position that their continued and well-directed musketry fire had an excellent effect in defeating the purposes of the enemy.

On the 12th of May, Colonel-Commandant John Harris, who "had served his country faithfully and well" for fifty years, died at headquarters, Washington, District of Columbia, after a short illness. Major Jacob Zeilin was commissioned on the 10th of June, to succeed him.

While the attention of the whole nation was directed to the contests going on at home, news was received of an engagement off the coast of France, the result of which served to encourage the North, and to create a corresponding feeling of regret, if not despondency, in the South.

At 10.20 A. M. on Sunday, June 10, 1864, the Kearsarge, while off the port of Cherbourg, France, discovered the piratical steamer Alabama, accompanied by the English yacht Deerhound, standing out. Captain Winslow at once cleared his ship for action, and when the Alabama had reached the distance of seven miles from shore, and was about nine hundred yards from the Kearsarge, the engagement commenced. Captain Winslow, fearing that his opponent would, in case of injury, steam in within the line of jurisdiction for protection, determined to run under her stern and rake. To avoid this, Semmes, the commander of the Alabama, sheered, and keeping broadside on to the Kearsarge, was forced into a circular track; at the seventh rotation, the Alabama was disabled, and headed for shore; another shot brought down the rebel flag, and a white one was run up; at 12.10 an officer from the Alabama came alongside the Kearsage and surrendered his vessel, which was reported in a sinking condition, and at 12.30 P. M. the Alabama went down. Captain Semmes escaped to the shore in the English yacht Deerhound, as did many of his officers and men. The remainder were picked up by the boats of the Kearsarge and taken on board that vessel. Three of the crew of the Kearsarge were wounded. The total number of killed and wounded on the Alabama has never been given. Seventeen of her wounded men were picked up by the boats of the victorious vessel. The battery of the Kearsarge consisted of seven guns,—two eleven-inch Dahlgren, one 30-pounder rifle, and four light 32-pounders. That of the Alabama consisted of eight guns, one heavy 68-pounder, of nine thousand pounds weight; one 100-pounder rifle, and six heavy 32-pounders. For this gallant action, the only sea fight of the war, Captain Winslow was promoted to the grade of Commodore, his commission dating July 19, 1864.

In his report of the action, Lieutenant-Commander Thornton says, "The Marines fought the rifle gun upon the top-gallant forecastle, under the charge of Acting Master's Mate Charles H. Danforth. The action on our part was commenced by this gun, and its fire was rapid and effective throughout. The high reputation of their service was nobly sustained by the Marine guard of this ship. The orderly sergeant, C. T. Young, and the master-at-arms, Jason R. Watrous, also deserve special mention for admirable performance of their duty." The following is a complete list of the Marines who were on board the Kearsarge: Orderly Sergeant Charles T. Young; Corporals Austin Quimley and Henry Hobson; Privates R. G. Dolley, Patrick Flood, James Kerrigan, John McAlun, George A. Raymond, James Tucker, Isaac Thorton, and John G. Batchelder.

During the early part of July, in an engagement between the United States steamship Wyoming, Commander D. McDougal, and the batteries at Simonosaki, Japan, Private Alexander Furlong, of the Marines, was killed, and Private Michael Doyle was wounded.

Later in the same month, when the city of Washington was threatened by the Southern army, a battalion of Marines and a battery of howitzers, under the command of Captain James Forney, assisted by Second Lieutenant George B. Haycock, were ordered from Philadelphia to Havre de Grace, Maryland, to open the railroad to Baltimore, then in possession of the enemy. Major General French says, "The battalion commanded by Captain Forney attracted my attention by its fine military appearance, its discipline, and the admirable manner in which it was handled. The arrangements made by Captain Forney for the artillery to repel the attack threatened upon the station had a great influence in preventing one. The rapid manner in which the order concentrating the troops at Havre de Grace was obeyed by him, and the valuable and effective services performed by the battalion under his direction entitle him and them to the recognition of the government."

The bay of Mobile, guarded at its entrance by two formidable fortifications, constructed by the government in former years, was difficult to blockade, and was one of the principal ports for trade with the rebels. It had been the intention of the Navy Department to get possession of that bay as soon as operations on the Mississippi would permit the detachment of a sufficient co-operative military force for the expedition. In this there was delay, caused by the army being fully occupied in other quarters. In the mean time the rebels, availing themselves of the advantages of their protection, proceeded to the construction and collection of a formidable Navy, with the view of raising the blockade. The information received was of such a character that the Department deemed it important that Rear-Admiral Farragut should resume his command, which he did, and on the 18th of January, 1864, arrived off Mobile.

Knowing the disadvantages of attacking iron-clad vessels with wooden ones, and that, too, in the face and under the guns of heavy fortresses, without a co-operative land force, he deferred the movement until the necessary elements of success could reach him. But in the mean time he stood ever ready to meet and measure his strength with the iron-clad fleet of Buchanan, should it venture to come out. Thus he constantly threatened an attack on Mobile, thereby aiding the army in its general movements elsewhere.

Military co-operation was secured early in July, and two ironclads from the James River and two from the Mississippi having reached him, Rear-Admiral Farragut made his final preparations for his attack on the rebel defences of Mobile Bay.

On the 8th of July, Rear-Admiral Farragut held a consultation with Generals Canby and Granger, on board the Hartford, on the subject of an attack upon Forts Morgan and Gaines, at which it was agreed that General Canby would send all the troops he could spare to co-operate with the fleet. Circumstances soon obliged General Canby to inform Rear-Admiral Farragut that he could not spare a sufficient number of troops to invest both forts; and in reply Farragut suggested that Fort Gaines should be the first invested, engaging to have a force on the sound ready to cover the landing of the army on Dauphin Island, in the rear of that fort. Lieutenant-Commander De Kraft, of the Conemaugh, was detailed to that duty.

A second consultation between Rear-Admiral Farragut and General Granger was held on board the Hartford on the 1st of August, and the 4th of the month was fixed upon as the day for the landing of the troops, and the entrance of the fleet into the bay. But owing to the unavoidable delay of the iron-clad Tecumseh and Pensacola, the fleet was not ready to move. General Granger, however, was up to time, and the troops actually landed on Dauphin Island. In the light of subsequent events the delay proved an advantage, as the rebels were busily engaged during the 4th in throwing troops and supplies into Fort Gaines, all of which were captured a few days afterwards.

The Tecumseh arrived in the evening of the 4th, and, everything being propitious, the attack was commenced on the following morning. The fleet was under way by 5.40 A.M., in the following order, two abreast and lashed together: Brooklyn, Captain James Alden, with the Octorara, Lieutenant-Commander C. H. Green, on the port side; flag-ship Hartford, Captain Percival Drayton, with the Metacomet, Lieutenant-Commander J. E. Jouett; Richmond, Captain T. A. Jenkins, with the Port Royal, Lieutenant-Commander B. Gherardi; Lackawanna, Captain J. B. Marchand, with the Seminole, Commander E. Donaldson; Monongahela, Commander J. H. Strong, with the Kennebec, Lieutenant-Commander W. P. McCann; Ossipee, Commander W. E. Le Roy, with the Itasca, Lieutenant-Commander George Brown; Oneida, Commander J. R. Mullany, with the Galena, Lieutenant-Commander C. H. Welles. The ironclad Tecumseh, Commander T. H. M. Craven; the Winnebago, Commander T. H. Stevens; the Manhattan, Commander J. W. A. Nicholson; and the Chickasaw, Lieutenant-Commander G. H. Perkins, were already inside the bar, and had been ordered to take up their position on the starboard side of the wooden vessels, or between them and Fort Morgan, for the double purpose of keeping down the fire of the water battery and the parapet guns of the fort, as well as to attack the ram Tennessee as soon as the fort was passed.

At the urgent request of the captains and commanding officers of the fleet, Rear-Admiral Farragut yielded to the Brooklyn being the leading ship-of-the-line, as she had four chase guns, and an ingenious arrangement for picking up torpedoes; and because, in their judgment, the flag-ship should not be too much exposed. The attacking ships steamed steadily up the main ship-channel, the Tecumseh firing the first shot at forty-seven minutes past six o'clock.

At six minutes past seven o'clock, the fort opened upon the fleet, and was replied to by a gun from the Brooklyn, and immediately afterwards the action became general.

It was soon apparent that there was some difficulty ahead. The Brooklyn, having got into shoal water, stopped, and by so doing arrested the advance of the fleet, while at the same time the guns of the fort were playing with great effect upon that vessel and the Hartford. A moment after, the ironclad Tecumseh was struck by a torpedo, and disappeared almost instantaneously beneath the waves, carrying down her gallant commander and nearly all her crew. At this juncture, Rear-Admiral Farragut, after ordering the Metacomet to send a boat to save, if possible, any of the perishing crew of the Tecumseh, dashed ahead with the Hartford, closely followed by the Brooklyn and the other ships. The Hartford steamed through the buoys where the torpedoes were to have been sunk, Farragut believing that from their having been some time in the water they were partially innocuous, and determined to take the chance of their explosion. From the moment the vessels turned to the north-westward to clear the middle ground, they were enabled to keep such broadside fire upon the batteries of Fort Morgan that the rebel guns did comparatively little injury.

Just as Farragut passed the fort, about ten minutes before eight o'clock, the ram dashed at the flag-ship, as had been expected, and in anticipation of which the Monitors had been ordered on the starboard side. He took no further notice of the ram than to return her fire. The rebel gunboats Morgan, Gaines, and Selma ran ahead, and the latter particularly annoyed the flag-ship with a raking fire which her guns could not return. At 8 A.M. Farragut ordered the Metacomet to cast off and go in pursuit of the Selma. Captain Jouett was after her in a moment, and in an hour's time he had her as a prize. The Morgan and Gaines succeeded in escaping under the guns of Fort Morgan. The Gaines was so injured that she had to be run ashore, where she was subsequently destroyed; but the Morgan escaped to Mobile during the night, although she was chased and fired upon.

Having passed the forts and dispersed the enemy's gunboats, most of the vessels were ordered to anchor, when the ram Tennessee was seen standing for the flag-ship. This was at forty-five minutes past eight. The monitors and such of the wooden vessels as were best adapted for the purpose were immediately ordered to attack the ram, not only with their guns, but with bows on at full speed.

The Monongahela, Commander Strong, was the first vessel that struck her, and in doing so carried away her own iron prow, together with the cutwater, without apparently doing her adversary much injury. The Lackawanna, Captain Marchand, was the next vessel to strike her, which he did at full speed; but though her stern was cut and crushed to her plank-ends for the distance of three feet above the water's edge to five feet below, the only perceptible effect on the ram was to give her a heavy list. The Hartford was the third vessel to strike her, but, as the Tennessee quickly shifted her helm, the blow was a glancing one, and as she rasped along the side the flag-ship poured a whole port broadside of 9-inch solid shot within ten feet of her casement. The monitors, working slowly, delivered their fire as opportunity offered. The Chickasaw succeeded in getting under her stern, and a 15-inch shot from the Manhattan broke through her iron plating and heavy wooden backing, though the missile itself did not enter the vessel.

Immediately after the collision with the flag-ship, Captain Drayton was directed to bear down for the ram again. He was doing so at full speed, when unfortunately the Lackawanna ran into the Hartford just forward of the mizzen-mast, cutting her down to within two feet of the water's edge. The flag-ship was soon got clear again, and was rapidly approaching the Tennessee, when she struck her colors and ran up the white flag. Just at this time she was sorely beset: the Chickasaw was pounding away at her stern, the Ossipee was approaching her at full speed, and the Monongahela, Lackawanna, and the Hartford were bearing down upon her, determined upon her destruction. Her smoke-stack had been shot away, her steering chains were gone, compelling a resort to her relieving tackles, and several of her port shutters were jammed. From the time the Hartford struck her until her surrender she never fired a gun.

During this contest with the rebel gunboats and the ram Tennessee, and which terminated by her surrender at ten o'clock, many more men were lost than from the fire of the batteries of Fort Morgan. Admiral Buchanan was wounded in the leg, two or three of his men were killed, and five or six wounded. Commander Johnson, formerly of the United States Navy, was in command of the Tennessee, and came on board the flag-ship to surrender his sword and that of Admiral Buchanan. Thus terminated the famous naval battle of August 5, 1864.

The wounded of both sides were sent to Pensacola for medical treatment.

On the following day one of the ironclads shelled Fort Gaines with such effect that Colonel Anderson, the commander, sent a communication to Rear-Admiral Farragut offering to surrender. General Granger, commanding the military forces, was sent for, and the terms of capitulation were signed by the respective parties on board of the Hartford.

From this time active movements were in progress for the reduction of Fort Morgan, and on the 22d of August, at daylight, a bombardment was opened from the shore batteries, the monitors and ships inside, and the vessels outside the bay. At 6 A.M. of the 23d a white flag was displayed by the rebels, and at 2 P.M. the fort was unconditionally surrendered to the Navy and army of the United States. Fort Powell had been attacked on the night of the 5th and blown up.

The capture of Forts Powell, Gaines, and Morgan and the destruction of the rebel fleet gave the government possession of the bay and closed the port against all illicit trade with the rebels. As late as September 13, Rear-Admiral Farragut informed the Department that he was engaged in removing torpedoes which had been strewn in the bay to obstruct naval operations.

The Marines in this engagement were among the foremost at every point where bravery and discipline could aid in winning victory. Captain Percival Drayton, commanding the Hartford, said, in reference to his vessel, "The two after-guns were entirely manned by Marines, who, under the direction of Captain Charles Heywood, performed most efficient service." Captain James Alden, commanding the Brooklyn, said that Captain Houston of the Marines fought his guns "nobly and well." Lieutenant-Commander George Brown, commanding the Itasca, reported that "The Marines conducted themselves with the usual distinguished gallantry of their Corps. Sergeant James S. Roantree is particularly deserving of notice." Similar testimony was given by others of the naval officers. The following is a list of the officers of the Corps attached to vessels in this action: Flag-ship Hartford, Captain Charles Heywood; Brooklyn, Captain G. P. Houston; Richmond, First Lieutenant C. L. Sherman. The remaining vessels, with the exception of the monitors, carried sergeant's guards, as follows: Octorara, Metacomet, Port Royal, Seminole, Kennebec, Itasca, Galena, Lackawanna, Monongahela, Ossipee, and Oneida. The list of Marines killed and wounded was as follows: Monongahela, wounded, Private William Feeney. Brooklyn, killed. Privates Michael Murphy and W. Smith; wounded, Private George R. Leland. Lackawanna, wounded, Privates James Keefe and Frederick Hines. Oneida, wounded, Corporal David Johnston and Private John Kilroy.

Below is a list of Marines who received especial honorable mention from their commanding officers, and who received "medals of honor" from the Department: On board the Brooklyn, Sergeants J. Henry Dennig and Michael Hudson, and William M. Smith and Miles M. Oviatt, corporals, "for conspicuous good conduct at their guns." On board the Richmond, Orderly Sergeant David Sprowls, "recommended for coolness and for setting a good example to the Marine guard, working a division of great-guns; joined the Richmond September 27, 1860; was in the actions with Fort McRea, the head of the passes of the Mississippi, Forts Jackson and St. Philip, the Chalmettes, the rebel ironclads and gunboats below New Orleans, Vicksburg, Port Hudson, and present at the surrender of New Orleans." On the same vessel, Sergeant Andrew Miller, "recommended for coolness and good conduct as captain of a gun in action; was on board the Brooklyn in the actions with Forts Jackson and St. Philip, the Chalmettes, the rebel ironclads and gunboats below New Orleans, batteries below Vicksburg, and present at the surrender of New Orleans." Same vessel, Sergeant James Martin, "recommended for coolness and good conduct as captain of a gun in the action; was in the actions with Forts Jackson and St. Philip, the Chalmettes, the ironclads and gunboats below New Orleans, Vicksburg, Port Hudson, and present at the surrender of New Orleans, on board of the Richmond."

Fort Powell, situated at Grant's Pass, the entrance to Mobile Bay on the west, was rendered useless to the enemy after a heavy bombardment by the Chickasaw, Lieutenant-Commander G. H. Perkins. The next morning a detachment of twenty-five Marines, commanded by Captain Charles Heywood and Lieutenant Sherman, was sent to take possession of the fort and to prevent the retreat of the enemy at Fort Gaines to the mainland. The detachment remained in Fort Powell thirty days.

On the 1st of October the Wabash went ashore on the outer edge of Frying Pan Shoals. The "crew" consisted of a motley assemblage of negroes, landsmen and boys, who had been temporarily shipped to do duty until the new crew could be received at Norfolk. Either through ignorance or unwillingness, the men failed to obey orders, and the frigate was in danger of serious injury, if not of becoming a total loss. The launch was ordered to be manned and a kedge anchor carried out. None obeyed the summons except a few old sailors, and it was found impossible to get the boat from the side of the frigate. In this emergency the Marines, under Lieutenant L. E. Fagan, were called upon by Captain De Camp, and with alacrity they obeyed the order. In a few moments they were away from the ship, pulling like "old salts" at the oars. The anchor was placed, and when the tide came in the ship floated and was safe. The efficiency of the Marines in that hour of emergency, won the praises of the commander of the frigate, who frequently alluded to their action in the affair in words of commendation.

In February, Lieutenants R. S. Collum and H. C. Cochrane were ordered to Mound City, Illinois, with a detachment of forty men, to guard the ordnance stores which supplied the Mississippi squadron. Subsequently Lieutenant Cochrane was ordered to the command of the Marines of the flag-ship Black Hawk.

---

## CHAPTER XXVI.

1864.—The Civil War, VIII.—The Expedition under Admiral Porter up Red River—The First Attack on Fort Fisher.

On the 7th of March, Rear-Admiral Porter had assembled at the mouth of Red River a formidable fleet of ironclads composed of the following vessels: Essex, Commander Townsend; Eastport, Commander Phelps; Benton, Lieutenant-Commander Greer; Lafayette, Lieutenant-Commander Foster; Choctaw, Lieutenant-Commander Ramsay; Louisville, Lieutenant-Commander Owens; Carondelet, Lieutenant-Commander Mitchell; Osage, Lieutenant-Commander Selfridge; Ouchita, Lieutenant-Commander Byron Wilson; Fort Hindman, Lieutenant Pierce; Pittsburg, A. V. Lieutenant Hoel; Chillicothe, A. V. Lieutenant Conthony, who was killed on the way up the river. On the death of Lieutenant Conthony Lieutenant-Commander Watson Smith was ordered to the command. Mound City, A. V. Lieutenant Langthorne; Neosho, A. V. Lieutenart Howard; and a force of light-draught vessels, consisting of Black Hawk, Lieutenant-Commander K. R. Breese; Lexington, Lieutenant Bache; Cricket, Acting Master Gorringe; Gazelle, Acting Master Thatcher. The fleet was joined at the mouth of Red River by a portion of General Sherman's forces in transports, under the command of General A. J. Smith. The joint forces moved up the river on the 10th of March to form a junction with Major-General Banks at Alexandria In their progress some of the vessels branched off into the Atchafalaya, while the main portion continued up Red River. The rebels were driven in turn from Simmsport and Fort de Russy, the latter being again captured with its guns and munitions of war and a few prisoners. Some of the fleetest vessels were despatched to Alexandria with the hope of cutting off the rebels in their retreat, but without success. The place was occupied by the combined forces, and about the 1st of April both Army and Navy commenced to move up the river towards Shrevesport. A part only of the naval force could proceed farther up the river than Alexandria, and it was with difficulty that they reached that point. But the assistance of the gunboats was so essential to success that some risks had to be taken, and extraordinary exertions were made to pass the vessels over the falls, so as to secure the required co-operation. Main force had to be used to haul the gunboats. Grand Ecore was reached without accident and occupied without opposition. There were at this time indications of the usual rise of the season in the river, and everything promised success. Twenty-three heavy guns had been captured from the enemy since their entrance into the river.

Springfield landing was designated as the point for the next juncture of the co-operating forces, and it was reached at the appointed time, three days after leaving Grand Ecore, by six gunboats and twenty heavy transports. Here they learned that the army under General Banks had met a reverse, and was falling back to Pleasant Hill, some distance below. Rear-Admiral Porter was therefore compelled to turn back, with the full knowledge that in retracing his steps he would be interrupted at every assailable point. The rebels, flushed with their success against the army, availed themselves of every opportunity which offered for harassing the gunboats and transports. The cavalry and artillery, taking advantage of the winding stream, moved rapidly from point to point, attacking on every available occasion. But the gunboats successfully fought their way, and from time to time repelled their assailants with terrible slaughter.

On the 14th of April, Rear-Admiral Porter got back to Grand Ecore, where he found the vessels which he had left at that point still detained above the bar. The river, instead of rising as usual at this season, had fallen during his absence. The army was preparing to move back upon Alexandria; the water having so receded, there was little hope of getting the vessels out, and destruction apparently awaited the best portion of the squadron. But, in the words of the admiral, "Providence provided a man for the occasion." Lieutenant-Colonel Joseph Bailey, acting engineer of the Nineteenth Army Corps, an intelligent and efficient officer, devised a plan for the construction of a series of dams across the rocks at the falls, thus, by artificial means, supplying that which nature withheld,—a sufficient depth of water for the passage of the vessels.

Extraordinary as was the project, and received with incredulity, the mind that conceived it was enabled to carry it into successful operation. Men were set to work, wood-cutters collected, quarries opened, and after some weeks the undertaking was accomplished. The dams were built; the vessels passed safely over the falls, to the delight of the assembled army and Navy, who had mutually participated in this work, and on the 16th of May, Rear-Admiral Porter had the satisfaction of announcing that the fleet was relieved from danger.

There is probably, in naval history, no other instance of such peril and difficulty so successfully and skillfully surmounted. Congress appropriately acknowledged the services of Lieutenant-Colonel Bailey on this occasion, and they were still further recognized by his promotion.

A division of the Red River expedition that was unable to get above the falls was sent, under command of Lieutenant-Commander James P. Foster, up the Washita River as far as Monroe. The force captured three thousand bales of Confederate cotton, brought out eight hundred negroes, and destroyed much rebel property.

While above the falls, Rear-Admiral Porter received intelligence of the capture of Fort Pillow; he despatched a force of ironclads to that point to prevent its permanent occupation by the rebels, and to keep the river open to commerce. On the 25th of March the rebels made an attack upon Paducah and demanded its immediate surrender, saying they would give no quarter if refused. The gunboats Pesota, Paw Paw, and Fort Hindman, at once opened fire upon the rebels and with such effect as to drive them off with great loss.

Captain Pennock, naval commandant at Cairo, upon receiving intelligence of the attack upon Paducah, at once sent up reinforcements of gunboats to the assistance of those already there, and the same officer saved Columbus, Kentucky, and recovered Fort Pillow, thus preventing a blockade of the river by his zeal and personal exertions in hastening reinforcements to those points.

The attacks upon Fort Pillow, Columbus, and Paducah were made at the time when the greater portion of the Mississippi squadron was operating on the Red River and its tributaries. Anticipating that the enemy would avail themselves of the absence of his more formidable vessels to attack certain points within the limits of the upper portion of the squadron, Rear-Admiral Porter had intrusted to Fleet-Captain Pennock the entire control of the vessels stationed on the Ohio, Tennessee, and Cumberland Rivers, and on the Mississippi as far down as Memphis, thus enabling him to act promptly when the time arrived. In his detailed report of the naval operations before Vicksburg, Rear-Admiral Porter acknowledged, in the handsomest terms, his indebtedness to Fleet-Captain Pennock, who was also commandant of the naval station at Cairo, for the promptness with which he kept the squadron supplied with all that was required or could be procured.

Rear-Admiral Porter, having been for nearly two years on arduous and exhausting duty on the Mississippi, received leave to return East in the summer, and was subsequently detached in order to take command of the North Atlantic Squadron. Captain A. W. Pennock, senior officer on the station, was left in charge. On the 1st of November, Acting Rear-Admiral S. P. Lee assumed command of the Mississippi Squadron and entered on the discharge of his duties.

For a long time, since early in 1862, the Navy Department had been fully aware of the importance of closing the port of Wilmington. It was the last port remaining to the rebels, and it was through it that supplies and munitions were now obtained; through it cotton was sent to Europe and the rebel credit abroad chiefly sustained. Owing, however, to the many claims upon its attention, the co-operation of the War Department, though often asked for, could not be obtained until, in the fall of 1864, Lieutenant-General Grant, having given the subject the closest consideration, was of the opinion that, under cover of the guns of the Navy, a landing might be effected on the beach. A part of the naval vessels might, he thought, force a passage beyond the batteries, and thus, the works on Cape Fear being isolated, a combined attack upon them might be followed by their reduction.

This plan of operations the heads of departments adopted, and it became necessary to select, for the command of the very large fleet which would be required, that officer whose past career would be the best guarantee of future services. The choice fell upon Admiral Farragut. But the health of this distinguished officer had been so impaired by long and arduous service in the gulf that he was obliged to decline the command, and Rear-Admiral David D. Porter, who had become distinguished in connection with the series of brilliant operations of the Mississippi Squadron, was selected in his place, and ordered to take command of the North Atlantic Blockading Squadron. This placed the port of Wilmington and its defences within the sphere of his operations. Major-General Butler was to have command of the land forces.

The 1st of October was first named for the sailing of the expedition, but so many delays occurred that it was not until the month of December that the expedition was in motion. This delay seems to have been unavoidable on the part of the army, owing to the many operations in which it was engaged, but the naval preparations were complete long before. In the meantime every squadron was suffering a depletion to furnish the vessels which lay idle at Hampton Roads and Beaufort. At length, however, all was in readiness, and, on the 16th of December, the troops were embarked in transports, only to be once more delayed, however, by a severe gale, which lasted three days, and caused them severe sufferings in their crowded quarters.

On the 18th, Admiral Porter sailed from Beaufort with all the monitors and smaller vessels, and was joined at the rendezvous, twenty miles east of New Inlet, on the North Carolina coast, by the larger war vessels and by the transports which had there assembled. The next day another heavy gale set in, which lasted two days. This somewhat scattered the fleet, but a calm succeeded, with weather so favorable that at length the propitious time seemed to have arrived.

A novel experiment was, however, to be first made: a powder magazine was to be exploded, so close to the fort that it was thought by many that the fort itself would be levelled to the ground, or that the magazines of the fort would be ignited, and that thus the rebels and all their works would be swept from the earth. A vessel called the Louisiana, which had been brought from Norfolk, loaded with an immense charge of powder, and carefully fitted with long fuses and machinery, was to do this deadly work. On the night of the 23d she was towed by another steamer close in shore, so near that the guns in the casements of Fort Fisher could be distinguished from her deck. The rebels, mistaking her for a blockade runner, welcomed her with the usual signals, and the brave little party having her in charge lighted their fuses and fires in the cabins almost in sight of the garrison. They then took to their small boat, and escaped to the vessel which had towed the powder-boat to the shore. An hour after, the explosion followed, but the result was by no means what was hoped for. The enemy was somewhat stunned for a time, but little real damage was done.

In the meantime the fleet remained at a distance of twelve miles from the bar, and the transports at an equal distance down the coast, but with orders to stand in shore as soon as the noise of the explosion should be heard. A carefully-prepared plan had been distributed to each commander of a vessel, and, early in the morning, they were to take their places accordingly, and to open fire as each got its assigned position. The whole fleet, consisting of about fifty vessels, was formed in three divisions, each with its reserve close at hand, and all placed upon the radius of a circle about a mile from the fort. At 11.30 the New Ironsides took its position, and at once opened fire, followed by the Monadnock, Canonicus, and Mahopac. At first the enemy replied briskly, but as the larger vessels were followed by the others, each adding to the deadliness of the fire, his guns were served with less vigor, and, in an hour and a quarter after the first shot was fired, the fort relapsed into silence. The effect within the fort of such a concentrated fire must have been fearful. Two of the magazines were blown up, and shot and shell so rained upon the garrison that no living thing could stand it. The bravest would have been forced to take refuge in the bomb-proofs. A moderate fire was continued for three hours after the enemy had ceased to reply, but the army transports not having arrived, the fleet retired for the night to a safe anchorage.

During the day's operations, a few of the vessels were struck, and one or two were severely injured, but they were of the smaller and less serviceable class. Much damage, however, was done by the bursting of large guns in the fleet. No less than six of the 100-pound Parrotts thus exploded, killing and wounding a large number of officers and men. The next day was Christmas, and all the transports had come in. After consultation with General Weitzel, on the part of General Butler, a plan of operations was decided upon. The forts were to be attacked again by the Navy, while the army should be landed and an assault made. A hundred small boats were to assist in landing the troops. At seven o'clock, on the 25th, signal was made to get under way, and the vessels proceeded to take position as before, which they did promptly and in the best order. Again the terrible rain fell around and over the fort, but this day little reply was made from the fort. The landing of the troops was begun about five miles farther up the beach, and was quite unopposed. Three thousand men were put ashore, and a reconnoisance made of the neighborhood of the works. One soldier even went inside the fort; an orderly with despatches was captured, and the flag-staff, which had been shot down, was brought away; but the enemy kept himself so close that scarce a human being was to be seen.

Two light batteries and a few men were captured. But General Butler decided that the fort was substantially uninjured as a place of defence; that so soon as the fire of the fleet should be withdrawn the enemy would be as strong as ever, and that only a regular siege, for which he was not prepared, could reduce the works. He therefore decided to withdraw the troops to the transports again. This he did, and the next morning the army returned to Fortress Monroe. The war vessels in the meantime remained, keeping up a slow but constant fire, hoping to tire out the enemy or to dismount their guns.

Unsuccessful as this affair proved, there was not lacking an exhibition of bravery and coolness on the part of all, which tended to insure victory in the next assault on the fort. Commodore Lanman says that the conduct of the Marines on the Minnesota was entirely satisfactory. Every one performed his duty to the utmost of his ability. The working and practice of the guns could not have been better. Many excellent shots were made, and in this respect the two guns worked by the Marines were equal to any other division. Captain Butler of the Marines, commanding his battery, being upon the sheer-pole of the mizzen rigging, pronounced the practice excellent. Commodore S. W. Godon, commanding the Susquehanna, says, "First Lieutenant William Wallace, with his fine company of Marines, handled most effectively two extra 9-inch guns." Similar testimony was borne by other naval officers to the bravery and efficiency of the Marines on the various vessels connected with the fleet.

The following is a complete list of the officers of the Corps who were engaged in this affair: On board the Colorado, Captain L. L. Dawson and First Lieutenant E. P. Meeker; Minnesota, Captain George Butler and Second Lieutenant George M. Welles; Powhatan, First Lieutenant F. H. Corrie; New Ironsides, First Lieutenant Richard S. Collum and Second Lieutenant Albert B. Young; Susquehanna, First Lieutenant William Wallace; Wabash, First Lieutenant L. E. Fagan; Vanderbilt, First Lieutenant W. H. Parker; Juniata, Second Lieutenant Jones Pile; Brooklyn, Captain G. P. Houston; Ticonderoga, First Lieutenant C. F. Williams. The following vessels carried sergeant's guards: Santiago de Cuba, Fort Jackson, Shenandoah, Tuscarora, Rhode Island, Mohican, Keystone State, Malvern, Onondaga, Shamrock, Chicopee, Nereus, Mendota, Iosco, Osceola, Pawtuxet, Mackinaw, R. R. Cuyler, Mattabessett, Massasoit, Agawam, Quaker City, Pontoosuc, Eutaw, Mercedita, and Miami. The following Marines were killed and wounded: On the Juniata, Second Lieutenant Jones Pile, killed, and Private William Kennedy, wounded; Ticonderoga, Private Cornelius Collins, wounded.

The following is a report from Admiral Porter announcing the result of the engagement:

FLAG-SHIP MALVERN,
OFF NEW INLET, NORTH CAROLINA, December 26, 1864.

SIR,—I have the honor to forward with this a somewhat detailed report of the two engagements with Fort Fisher and the surrounding works.

We attacked with the whole fleet on the 24th instant, and silenced every gun in a very short time.

On the 25th instant we again took up our position, within a mile of the fort (the iron vessels within twelve hundred (1200) yards), without a shot being fired at us; shelled it all day, with now and then a shot from the rebels, and stopped firing after sunset.

The army landed and re-embarked, considering it impracticable to assault the place.

I shall remain here and keep shelling the enemy's works on every occasion whenever the weather permits.

I have the honor to be, very respectfully, your obedient servant,

DAVID D. PORTER,
*Rear-Admiral.*

HON. GIDEON WELLES,
*Secretary of the Navy, Washington, D. C.*

The following are the vessels engaged in the fight, with the names of their commanders:

Minnesota, Commander Joseph Lanman; Mohican, Commander D. Ammen; Colorado, Commander H. K. Thatcher; Tuscarora, Commander J. M. Frailey; Wabash, Captain M. Smith; Susquehanna, Commodore S. W. Godon; Brooklyn, Captain James Alden; Powhatan, Commodore J. F. Schenck; Juniata, Captain W R. Taylor; Kansas, Lieutenant-Commander P. G. Watmough; Yantic, Lieutenant-Commander T. C. Harris; Maumee, Lieutenant-Commander R Chandler; Mackinaw, Commander J. C. Beaumont; Ticonderoga, Captain C. Steedman; Pawtuxet, Commander J. H. Spotts; Shenandoah, Captain D. B. Ridgley; Seneca, Lieutenant-Commander M. Sicard; New Ironsides, Commodore William Radford; Monadnock, Commander E. G. Parrott; Canonicus, Lieutenant-Commander George E. Belknap; Mahopac, Lieutenant-Commander E. E. Potter; Saugus, Commander E. R. Colhoun; Nyack, Lieutenant-Commander L. H. Newman; Unadilla, Lieutenant-Commander F. M. Ramsay; Huron, Lieutenant Commander T. O. Selfridge; Pequot, Lieutenant-Commander D. L. Braine; Pontoosuc, Lieutenant-Commander Wm. G. Temple; Neresus, Commander J. C. Howell; Vanderbilt, Captain C. W. Pickering; Fort Jackson, Captain B. F. Sands; Santiago de Cuba, Captain O. S. Glisson; Tacony, Lieutenant-Commander W. T. Truxtun; Osceola, Commander J. M. B. Clitz; Chippewa, Lieutenant-Commander A. W. Weaver; Sassacus, Lieutenant-Commander G. W. Young; Rhode Island, Commander S. D. Trenchard; Mount Vernon, Acting Volunteer Lieutenant James Trathen; Britannia, Acting Volunteer Lieutenant Samuel Huse; Quaker City, Commander W. F. Spicer; Iosco, Commander John Guest; Howquah, Acting Volunteer Lieutenant J. W. Balch; Wilderness, Acting Master H. Arey; Cherokee, Acting Volunteer Lieutenant W. E. Dennison; A. D. Vance. Lieutenant-Commander J. H. Upshur; Moccasin, Acting Ensign James Brown; Gettysburg, Lieutenant R. H. Lamson; Alabama, Acting Volunteer Lieutenant Frank Smith; Keystone State, Commander H. Rolando; Nansemond, Acting Master John H. Porter; Emma, Acting Volunteer Lieutenant T. C. Dunn; Tristram Shandy, Acting Ensign Ben. Wood; Governor Buckingham, Acting Volunteer Lieutenant J. McDiarmid; Little Ada, Acting Master S. P. Crafts.

## CHAPTER XXVII.

**1864-65.**—The Civil War, IX.—Services in the Pacific Squadron—Marines at Boyd's Neck and Tulifinney Cross-Roads—Second Attack on Fort Fisher—Its Capture by Porter and Terry—Occupation of Georgetown—Report of Admiral Dahlgren.

On the 9th of November, 1864, Admiral Pearson, commanding the Pacific Squadron, wrote to Commander H. K. Davenport, saying that the commander of the American steamer Salvador had stated that among his passengers for the next trip from Panama were a number of desperate men, bent on taking his vessel after leaving her moorings. Commander Davenport was directed to go on board this vessel with a sufficient force, and, in the event of a difficulty, to protect the passengers from harm. For the further protection of the vessel and passengers against these supposed pirates, Admiral Pearson stated that he should accompany the Salvador to sea in his ship, and, should any pirates be found on board, the commander was directed, when beyond the legal jurisdiction of the government of New Grenada, to capture them and convey them to the admiral's ship. On the morning of the 10th the commander left his ship with four armed boats, and went alongside the Pacific Mail Steamship's steamer Sacramento, as had been previously agreed upon with the master of the steamer Salvador. He subsequently overhauled and and boarded the Salvador. By shrewd management his plan was entirely successful, and he captured the whole party without bloodshed. In his report of the affair, Admiral Pearson spoke in the highest terms of the good conduct of Captain D. M. Cohen of the Marines.

On the 26th of November, Admiral Dahlgren began arrangements for a joint army and Navy movement, co-operating with General Foster to establish connection with General Sherman. A brigade from the fleet was organized under the command of Commander George H. Preble, consisting of a battalion of naval artillery, under command of Lieutenant-Commander E. O. Mathews, a battalion of sailor infantry, under command of Lieutenant James O'Kane, and a battalion of Marines, two hundred and fifty strong, under command of First Lieutenant G. G. Stoddard.

On the evening of the 28th the brigade was embarked on board the Pontiac, Sonoma, and Mingo.

At 4.30 A. M., the 29th, all the vessels of the expedition got under way and proceeded up the Broad River, under cover of a dense fog.

At 8.30 A. M. the brigade landed at Boyd's Landing, and was immediately pushed forward on the road, in advance of the military forces, about two miles.

The artillery was advanced up the road, with the battalion of Marines and sailor infantry deployed to the right and left, and in advance as skirmishers. After skirmishing through the swamps and woods all day, the command went into camp at 11 P. M. at the first fork.

At 7 A. M., the brigade was again in motion, and at 8 A. M joined the army at the church, where Brigadier-General Hatch had established his headquarters.

At 8.15 A. M. the army moved on up the Grahamsville road, leading to the right of the church, and at 8.30 A. M. the enemy opened fire upon the head of our column from a movable piece of rifled artillery placed in the road. This was replied to by one 12-pounder Napoleon; A skirmish fire of musketry ensued, and the enemy was driven back, firing as he retreated. As the head of our column reached the turn of the roads at Honey Hill, it was opened on by a battery placed in position behind a strong earthwork. A sharp and stubbornly-contested fight ensued, which lasted until sundown, when it was found impossible to drive the enemy from his position; our troops were withdrawn in good order. Our total loss in killed, wounded, and missing amounted, as officially reported, to seven hundred and forty.

By direction of Brigadier-General Hatch the brigade brought up the rear column on the march, and was mostly in reserve at the commencement of the action. At Meridian, however, the Marine battalion was ordered into the line of battle on the right, in the rear of the Twenty-fifth Ohio Volunteers, and was continued in line and under fire until called out, by order from the general, at 6 P.M.

During the night following the battle our artillery took up, by order, a defensive position at the forks of the road, supported by the infantry of the brigade and the Thirty-fourth and Thirty-fifth regiments of colored troops.

The next morning, December 1, the brigade was ordered into a position on the Grahamsville road, where it subsequently intrenched and remained until the evening of the 5th, when it was withdrawn, by order of the general, to Boyd's landing, and embarked for an expedition up the Tulifinney.

On the 6th a successful attack was made on the enemy near the mouth of Tulifinney Creek, and that night our forces camped on the battle-field. During the night a skirmishing fire from the picket-line drew the whole force out under arms. The next morning, the 7th, the enemy attacked our skirmish line from the woods in front and to the right of our line. Our howitzers firing to the right prevented his getting around on that flank, while they assisted the musketry in front to disperse them. In this attack the enemy fired a few shots on the right that had not before been heard from. In this affair we had two wounded.

On the morning of the 9th a forward movement was made, and the command of the reserves was given to Commander Preble, consisting of the Thirty-second, Thirty-fourth, and Thirty-fifth United States Colored Infantry. The forces being in position at 9 A. M., the artillery, in all ten pieces, army and Navy, opened a rapid fire in front to clear the woods, which was continued ten minutes. On its ceasing, the skirmish-line, commanded by Colonel Silliman, and consisting of the One Hundred and Twenty-seventh and One Hundred and Fifty-seventh New York and the Marine Battalion, in all about four hundred strong, entered the woods in front, and was followed by the main line of battle, consisting of the One Hundred and Forty-fourth and Fifty-sixth New York, and One Hundred and Second colored regiments, and the sailor infantry. This line was followed by and designed to protect the Twenty-fifth Ohio, with axes, who had orders to cut a road one hundred feet wide through the woods, to open the railroad to the fire of our batteries on the right. As the troops advanced all our artillery commenced and continued to shell on the right and left flanks, and kept up the fire during the whole advance, the signal to recall being the cessation of the artillery fire. At 9.55 A. M. the skirmish firing commenced, and the musketry was continued until near sundown. At 11.30 A. M. the Thirty-fourth colored regiment of the reserves was ordered to the front to report to General Potter. At 3 P. M., the Twenty-fifth Ohio having done its work, the artillery fire ceased. At 4 P. M. the One Hundred and Forty-fourth New York, having expended all its ammunition, returned, and at 4.15 the Thirty-second colored was ordered to the front, and the One Hundred and Forty-fourth New York reported to Preble and took its place in the intrenchment. The enemy throughout the day continued a fire of grape, shot, and shell from a battery on the right. About sundown the firing on both sides ceased, and our forces were withdrawn in good order, the Twenty-fifth Ohio having made the cutting which was the intent of the attack. Our loss in the fleet-brigade was one killed, eighteen wounded, and three missing. It is needless to add that officers and men behaved with gallantry and spirit. After the actions on Derang's neck of the 6th, 7th, and 9th of December no forward move was attempted, and the brigade remained in its intrenchments near Talbird's house until permanently withdrawn. Two of the howitzers, however, were posted in a battery at the front, and two at the 30-pounder Parrott battery on the extreme left of the lines for its support. On the 18th and 19th two rifled howitzers, which had been brought up in place of the two light howitzers sent to the rear, were placed in the swamp battery, and opened fire upon the railroad at a distance of eight hundred yards, doing it some damage; the enemy did not return the fire. On the morning of the 26th it was planned for Lieutenant O'Kane, with one hundred and twenty sailors and Marines, to take the two-gun battery at Dawson's Point, assaulting it in flank; the morning proved very rainy, and on his making the attempt the intervening marsh was found to be impassable, and the enterprise was reluctantly abandoned. On the 28th the brigade embarked, the battalion of Marines returning to Bay Point.

*Casualties* (in action at Boyd's Neck, November 30, 1864).—Killed: First Sergeant Philip Napp, Company B. Wounded: Company A, Private James Walker (mortally), Patrick Neilly, and Michael Maloney; Company B, Sergeant Philip Clancy, Corporal John Young, and Private William Wilson. Missing: Company B, Private Robert Joyce. At Tulifinney Cross-Roads, December 6, 7, and 9. Wounded: Company C, Sergeant Edgar Kermerer; Company B, Private Thomas Mallen. Missing: Company C, Private Charles Brandt. Wounded: Company A, Sergeant R. C. Bates, Private James Gorman; Company B, Sergeant Christopher Lutz, acting first lieutenant, mortally, left on the field, Privates Robert Kellon, Thomas Quinn, and William Gordon; Company C, Corporal William McMurray. Steamer Donegal, Private P. W. Wood, drowned December 10. Company C, Privates Bernard Dyer and Manuel de Jesus.

In his official report, Admiral Dahlgren said, "I desire to bring First Lieutenant Stoddard to the notice of the department. He did good service in the field with the Marines of the fleet brigade at Boyd's Neck and the Tulifinney, and now has the command of the largest force of Marines that has been collected for some time. He has always acquitted himself with credit. I would respectfully suggest a brevet."

On the 29th of December the Secretary of the Navy again addressed General Grant, at the suggestion of the President, as follows:

NAVY DEPARTMENT, December 29, 1864.

LIEUTENANT-GENERAL GRANT, CITY POINT, VIRGINIA:

The substance of despatches and reports from Rear-Admiral Porter, off Wilmington, is briefly this: The ships can approach nearer to the enemy's works than was anticipated. Their fire can keep the enemy away from their guns. A landing can easily be effected upon the beach north of Fort Fisher, not only of troops, but all their supplies and artillery. This force can have its flanks protected by gunboats. The Navy can assist in the siege of Fort Fisher precisely as it covered the operations which resulted in the capture of Wagner. The winter, also, is the most favorable for operations against Fort Fisher. The largest naval force ever assembled is ready to lend its co-operation.

Rear-Admiral Porter will remain off Fort Fisher, continuing a moderate fire to prevent new works from being erected, and the ironclads have proved that they can maintain themselves in spite of bad weather. Under all these circumstances, I invite you to such a military co-operation as will insure the fall of Fort Fisher, the importance of which has already received your careful consideration.

This telegram is made at the suggestion of the President, and in hopes that you will be able at this time to give the troops which heretofore were required elsewhere. If it cannot be done the fleet will have to disperse, whence it cannot again be brought to this coast.

GIDEON WELLES, *Secretary of the Navy.*

Admiral Porter was accordingly informed of General Grant's intention in the following despatch:

NAVY DEPARTMENT, Saturday, December 31, 1864.

SIR,—Lieutenant-General Grant will send immediately a competent force, properly commanded, to co-operate in the capture of the defences on Federal Point. It is expected that the troops will leave Hampton Roads next Monday or Tuesday. This is all the information the Department has to give you, but relies upon your skill and judgment to give full effect to any move that may be arranged.

The Department is perfectly satisfied with your efforts thus far, and you will convey to all hands the satisfaction the Department feels.

I am, sir, etc., GIDEON WELLES.

REAR-ADMIRAL D. D. PORTER,
*Commanding N. A. Blockading Squadron, off Wilmington.*

Major-General Terry arrived at Beaufort on the 8th of January, and a new plan of operations was arranged. Two days of stormy weather succeeded, but on the 12th every preparation had been made, and the vessels got under way once more, ranged in three lines, with the transports in company. This time the order of attack was different. At 3 A. M., on the 13th, the New Ironsides, Commodore Radford, leading the monitors Saugus, Monadnock, Canonicus, and the Mahopac, moved towards the fort, and received its fire unnoticed until they reached a position within a thousand yards of it, when they opened their batteries and a sharp fight ensued.

At 8.30 A. M. a signal was made to the fleet to transport to land the troops. This was carried on so rapidly that by 2 P. M. eight thousand men were all ashore, with twelve days' provisions and their intrenching tools. Then Porter ordered his wooden vessels to engage in the conflict. The bombardment was continuous until dark, to the severe damage of the armament of the fort, when the wooden vessels fell back to their anchorage. But the ironclads fired slowly throughout the night, by which the garrison was much demoralized. "By sunset," says Admiral Porter, in his report, "the fort was reduced to a pulp; every gun was silenced by being injured or covered with earth, so that they could not work." The bombardment continued with unremitting vigor throughout the 14th, and a fire, magnificent alike for its power and accuracy, demonstrated the efficiency of the fleet.

The 15th was the day designated for the grand attack by sea and land. The troops by this time were rested, after their long confinement on shipboard, and were eager for the attack. At 9 A. M. the squadron was signalled by Admiral Porter to attack in three lines, and about eleven o'clock they were all in position. Each had opened fire as they took position in line, and the bombardment was kept up furiously all day. At first the rebels replied with some effect, from the same batteries as before, but they were soon driven away into the bomb-proofs. Sixteen hundred sailors and four hundred Marines had been detailed to assist in the assault; they were to attack the sea face of the fort, while the soldiers assaulted from the land side. By three o'clock the troops were all in position, and the signal to change the direction of the fire was given, when the guns were turned on the upper batteries, away from the point where the assault was to be made.

As soon as the army was found to be in motion, the men from the fleet were ordered to advance also. This they did on the run, along the beach. The sailors, being armed with cutlasses and pistols, were expected to treat the fort as a vessel, and board it with a dash. Before the assault made on the fort by the Marines and seamen, a strong force with shovels and picks threw up within six hundred yards of the fort a well-protected breastwork, and from that gradually advanced to within two hundred yards a succession of rifle-pits, which were most promptly occupied by a line of skirmishers composed of Marines, under Second Lieutenant L. E. Fagan. One sergeant and six privates, who became detached from Lieutenant Fagan's guard while in the intrenchments at the front, accompanied the army in the storming of the fort, and behaved with gallantry and coolness during the hand-to-hand conflict that ensued. In his report, Lieutenant Fagan called especial attention to the conduct of Corporal Tomlin, who, under a heavy fire from the enemy's sharp-shooters, advanced into an open plane close to the fort, and assisted a wounded comrade to a place of safety. The assaulting party was divided into four lines, the first line being composed of Marines under Captain L. L. Dawson. The Marines were divided into four companies, under command of Captain Butler, First Lieutenant Wallace, First Lieutenant Corrie, and First Lieutenant Parker, giving First Lieutenant Williams charge of some twenty-five skirmishers, and First Lieutenant Meeker acting as adjutant. The first line advanced promptly to the rifle-pits, not without some loss to officers and men. The second, third, and fourth lines advanced rapidly up to the sea face of Fort Fisher, and nearly gained the parapet, supported by the fire of the Marines. The parapet was, however, lined with one dense mass of the enemy who had mistaken this for the main assault, and the advance was finally brought to a halt. The men evinced great determination and valor, but they failed to get any further, though a few gained the parapet, only to be instantly swept away. The attack on the north-east bastion, however, had the effect of diverting the attention of the enemy, and the troops assaulting met with less opposition until a lodgment was made on the parapet not far from the river. Now commenced the terrible struggle. Up to this time the national loss had been comparatively small, for the Navy had kept the garrison quiet. Now it was compelled to cease firing at that part of the fort, for its shells would be as hurtful to friends as to foes. Instantly the garrison sprang to its guns, and musketeers swarmed upon the parapet. But General Curtis, who had effected a lodgment on the parapet, held his own until General Pennypacker came to his support. The latter advanced rapidly to General Curtis's right, drove the Confederates from the palisades, extending from the west end of the land face of the fort to the river, and captured a number of prisoners. The brigade broke through the palisades and joined General Curtis.

The Confederates used the huge traverses of the land front for breast-works, and over the tops of these the combatants fired in each other's faces, while the fleet kept up a continuous fire farther to the southward, to prevent reinforcements reaching the fort from the Mound Battery, or Battery Buchanan. The struggle was desperate. The Confederates were steadily pushed back, until, at dusk, they had lost nine of these traverses. At six o'clock Abbot entered the fort with his little brigade, and at 9.30 o'clock, when two more traverses had been carried, the contest ceased. Two thousand and eighty-three prisoners were captured.

Admiral Porter immediately sent the following despatch:

FROM OFF FORT FISHER, January 15, 1865.

SIR,—Fort Fisher is ours. I send you a bearer of despatches with a brief account of the affair.

General Terry is entitled to the highest praise and gratitude of his country for the manner in which he has conducted his part of the operations. He is my *beau ideal* of a soldier and a general. Our co-operation has been most cordial. The result is victory, which will always be ours when the army and Navy go hand in hand. The Navy loss in the assault was heavy. The army loss is also heavy.

D. D. PORTER, *Rear-Admiral.*

HON. GIDEON WELLES, *Secretary of the Navy.*

The following vessels composed the fleet in the second attack:

Ironclad division: New Ironsides, Commodore William Radford; Saugus, Commander E. R. Calhoun; Canonicus, Lieutenant-Commander George E. Belknap; Mahopac, Lieutenant-Commander A. W. Weaver; Monadnock, Commander E. G. Parrott.

Line No. 1, Brooklyn, Captain James Alden, leading, consisted of the Mohican, Commander Daniel Ammen; Tacony, Lieutenant-Commander W. T. Truxtun; Kansas, Lieutenant-Commander P. G. Watmough; Yantic, Lieutenant-Commander T. C. Harris; Unadilla, Lieutenant-Commander F. M. Ramsay; Huron, Lieutenant-Commander T. O. Selfridge; Maumee, Lieutenant-Commander Ralph Chandler; Pequot, Lieutenant-Commander D. L. Braine; Pawtuxet, Commander J. H. Spotts; Seneca, Lieutenant-Commander M. Sicard; Pontoosuc, Lieutenant-Commander W. G. Temple; Nereus, Commander J. C. Howell.

Line No. 2, Minnesota, Commodore Joseph Lanman, leading, consisted of the Colorado, Commodore H. K. Thatcher; Wabash, Captain M. Smith; Susquehanna, Commodore S. W. Gordon; Powhatan, Commodore J. F. Schenck; Juniata, Lieutenant-Commander T. S. Phelps; Shenandoah, Captain D. B. Ridgley; Ticonderoga, Captain Charles Steedman; Vanderbilt, Captain C. W. Pickering; Mackinaw, Commander J. C. Beaumont; Tuscarora, Commander J. M. Frailey.

Line No. 3, Santiago de Cuba, Captain O. S. Glisson, leading, consisted of the Fort Jackson, Captain B. F. Sands; Osceola, Commander J. M. B. Clitz; Sassacus, Lieutenant-Commander J. L. Davis; Chippewa, Lieutenant-Commander E. E. Potter; R. R. Cuyler, Commander C. H. B. Caldwell; Maratanza, Lieutenant-Commander George W Young; Rhode Island, Commander S. D. Trenchard; Monticello, Lieutenant W. B. Cushing; Alabama, Acting Volunteer Lieutenant A. R. Langthorne; Montgomery, Acting Volunteer Lieutenant T. C. Dunn; Iosco, Commander John Guest.

The reserve division, under Lieutenant-Commander J. H. Upshur, in the A. D. Vance, consisted of the Britannia, Acting Volunteer Lieutenant W. A. Sheldon; Tristam Shandy, Acting Volunteer Lieutenant F. M. Green; Lillian, Acting Volunteer Lieutenant T. A. Harris; Fort Donelson, Acting Master G. W. Frost; Wilderness, Acting Master H. Arey; Aries, Acting Volunteer Lieutenant F. S. Wells; Governor Buckingham, Acting Volunteer Lieutenant J. McDiarmid; the Nansemonde, Acting Master J. H. Porter; Little Ada, Acting Master S. P. Crafts; Eolus, Acting Master E. S. Keyser, and Republic, Acting Ensign J. W. Bennett, being used as despatch vessels.

The fall of Fort Fisher rendered all the other works at the mouth of the Cape Fear River untenable, and during the nights of the 16th and 17th the Confederates blew up Fort Caswell, on the right bank of the river. They also abandoned Battery Holmes, on Smith's Island, and their extensive works at Smithville and Reeves' Point, and fled towards Wilmington.

The triumph of the army and Navy was now complete.

The army lost in the attack six hundred and eighty-one men, of whom eighty-eight were killed, five hundred wounded, and twenty-two missing. Among the wounded were General Bell, mortally, and Generals Curtis and Pennypacker, severely. On the morning after the victory, while the exultant soldiers, sailors, and Marines were swarming into the fort, its principal magazine, buried deep in the earth, at the centre of the parade, exploded, killing two hundred and wounding one hundred men.

The fleet lost about three hundred men during the action, among whom were the gallant Lieutenants Preston and Porter. During the bombardment about fifty thousand shells were exploded, and during the first attack about eighteen thousand.

Major-General Terry, in his official report of the engagement, says, "I should signally fail to do my duty were I to omit to speak in terms of the highest admiration of the part borne by the Navy in our operations. In all ranks, from Admiral Porter to his seamen, there was the utmost desire not only to do their proper work, but to facilitate in every possible manner the operations of the land forces. To him, and to the untiring efforts of his officers and men, we are indebted that our men, stores, tools, and ammunition were safely and expeditiously landed, and that our wounded and prisoners were embarked for transportation to the North; to the great accuracy and power of their fire it is owing that we had not to confront a formidable artillery in the assault, and that we were able, with but little loss, to push forward the men, preparatory to it, to a point nearly as favorable for it as the one they would have occupied had siege operations been undertaken and the work systematically approached. The assault of the sailors and Marines, although it failed, undoubtedly contributed somewhat to our success, and certainly nothing could surpass the perfect skill with which the fleet was handled by its commander."

After the engagement, five officers of the Corps were recommended for gallantry in the face of the enemy, by Admiral Porter,--viz., Captains L. L. Dawson and George Butler, and First Lieutenants William Wallace, Charles F. Williams, and Louis E. Fagan.

Lieutenant-Commander Parker, in his report to Admiral Porter, mentions Captain George Butler and Lieutenant William Wallace as being at the front. Commodore S. W. Godon, commanding the Susquehanna, said, "Lieutenant Wallace was wounded severely in the charge on the fort. I particularly regret the injury to this officer, as he cannot well be replaced by his Corps in the affections of this ship." The following is an extract from the report of Fleet Captain K. R. Breese: "I wish to bear witness to the handsome manner in which Lieutenant Fagan of the Marine Corps did his duty with his sharp-shooters, and to the gallantry he exhibited in advancing his men so close to the enemy's works." Lieutenant-Commander F. B. Blake, of the Susquehanna, says, "We had fourteen wounded; among them, I regret to state, Lieutenant Wallace, whose gallantry was conspicuous, especially in his efforts to rally his men." The bravery and efficient service of the present Captain Porter, son of the admiral, who at the time held a clerkship, won for him a commission in the Marines. The reports of the officers of the various vessels bear testimony to the nature of the services rendered by those Marines who did not land. Commodore J. F. Schenck, commanding the Powhatan, said, "It is due to the Marines of this ship, and to their commander, First Lieutenant F H. Corrie, that the utmost alacrity and cheerfulness was shown in the performance of their duty."

The list of officers of the Corps attached to vessels in this engagement was as follows: Colorado, Captain L. L. Dawson and First Lieutenant E. P. Meeker; Minnesota, Captain George Butler; Brooklyn, Captain G. P. Houston; Vanderbilt, First Lieutenant W. H. Parker; Powhatan, First Lieutenant F. H. Corrie; New Ironsides, First Lieutenant R. S. Collum and Second Lieutenant A. B. Young; Susquehanna, First Lieutenant William Wallace; Ticonderoga. First Lieutenant C. F. Williams; Wabash, First Lieutenant L. E. Fagan. The following is the list of the killed and wounded and missing: From the Colorado, killed in the assault, Privates Andrew Munn, William Thomas, and Patrick Caine; wounded in the assault, Sergeant Daniel Kelley. Minnesota, killed in the assault, Privates Daniel O'Neil and William Burke; wounded in the assault, Corporal Thomas O'Leary, Privates Joseph Simmerman, John Calhoun, Edward O'Brien, William Duggan, and Nathaniel O'Donnell. Powhatan, missing, Corporal Rulon, Musician Bourke; Privates Benson, Wasmuth, Hicks, and O'Brien, 2d; wounded in the assault, Corporal Abijah Ball, Privates Owen Daly, James Conan, James Casgriff, Patrick O'Donnel, Will Kegsl, and Frank Alick. Susquehanna, killed in the assault, Privates Thomas Brennan, N. S. Minkoff; wounded in the assault, First Lieutenant William Wallace, Corporal White Henlett, and Private John Costello. Santiago de Cuba, wounded in the assault, Privates James Sherry, Charles McCarty, and John Mahan. Wabash, wounded in the assault, Privates William Daly (since dead), Esek Bailey, and —— Moran; Privates George Brown and —— Scanlan, by explosion of magazine in Fort Fisher. Shenandoah, killed in the assault, Private Charles Jones. Ticonderoga, killed by explosion in Fort Fisher, Private Charles Colcord. Brooklyn, wounded in the assault, Private James Jackson. Tuscarora, wounded in the assault, Corporal A. Collum and Private J. B. Jenkins. Rhode Island, wounded in the assault, Private John Daisy. Mohican, wounded in the assault, Sergeant Oscar Smith, Corporal John Graham, Privates John L. Russell, Stephen Moore, and by the explosion in Fort Fisher, Privates William F. Ryan and Richard Broderick. Osceola, wounded in the assault, Private William Williams. Juniata, killed in the assault, Private Edward Brennan; wounded, Private John Redden. Maratanza, wounded in the assault, Private Charles Sullivan.

The following is a list of Marines who received honorable mention: On the Ticonderoga, Orderly Sergeant Isaac N. Fry, "recommended for coolness, good conduct, and skill as captain of a gun;" Shenandoah, Privates Charles Smith, Adolph Burton, Wilmer D. Lyne, James Whyte, and Thomas S. Brown, mentioned "for gallant conduct in the assault;" Minnesota, Corporal John Rannahan and Privates John Shivers and Henry Thompson "behaved bravely."

The capture of Fort Fisher by the combined efforts of the army and Navy, under those gallant and distinguished officers, Porter and Terry, gave the liveliest satisfaction to the loyal people, for it seemed like a sure prophecy of peace nigh at hand. All the participants in the contest were regarded with gratitude and honored everywhere. When the Ticonderoga, Captain Steedman, and the Shenandoah, Captain Ridgley, of Admiral Porter's fleet, arrived at Philadelphia, a pleasing incident in illustration of the public feeling occurred.

The officers, seamen, and Marines of both vessels were invited to dine at the Soldiers' Reading-Room, which was under the immediate management of Miss McHenry, a lady well and widely known by her acts of benevolence and patriotism. The invitation was accepted. An elegantly-arranged and sumptuous dinner was prepared, and a military band was in attendance. After dinner one of the men of the Shenandoah presented to the ladies two flags, one of which was shot from the mast-head of his ship during the bombardment of Fort Fisher.

The following despatch from Admiral Dahlgren, dated on board the flag-ship Harvest Moon, Rebellion Roads, Harbor of Charleston, February 18, shows the progress made up to that date: "Charleston was abandoned this morning by the rebels. I am now on my way to the city." On the 26th of February Admiral Dahlgren reported to the Department an important result of the work in which he was engaged, —the occupation of Georgetown, South Carolina; that town and its defences having been occupied by the forces under his command. The town was held by six companies of Marines, under Lieutenant Stoddard, and the battery, fifteen guns, by one company of Marines, under Lieutenant Breese. First Lieutenant Charles F. Williams, of the Marines, was ordered to the battalion under the command of Lieutenant Stoddard, and remained on duty with it until the occupation of Georgetown.

The following officers of the Corps were attached to this squadron during the year 1864, and part of 1865: First Lieutenant George G. Stoddard, First Lieutenant Charles F. Williams, and Second Lieutenant James B. Breese; in the Gulf Squadron, during the same time, First Lieutenant Eugene A. Smalley; West Gulf Squadron, Second Lieutenant C. L. Sherman.

The following is the report of the final disposition of the remains of the gallant Bradford:

REMOVAL OF THE REMAINS OF LIEUTENANT BRADFORD, UNITED STATES MARINES, FROM POTTER'S FIELD.

REPORT OF REAR-ADMIRAL J. A. DAHLGREN.

FLAG-STEAMER PHILADELPHIA,

CHARLESTON HARBOR, March 17, 1865.

SIR,—I have just caused the remains of Lieutenant Bradford (United States Marines) to be restored to the cemetery wherein they were first deposited. He was mortally wounded in the assault on Sumter, September 8, 1863, and taken prisoner. After death, Dr. Mackey, of Charleston, had the body interred in the Magnolia Cemetery, but when it became known, so much excitement was manifested that it was found necessary to remove it and place it in the Potter's field.

As soon as the public service would permit, I took the necessary measures to reverse the savage and disgraceful act of the rebel community. The body was taken to St. Paul's, where service and a sermon were gone through by Chaplain Blake, United States Navy; then taken back to Magnolia Cemetery, and there reinterred with all the honor due to the memory of a gallant officer who had fallen in the cause of his country. I was present, also General Hatch, and many officers of the army and Navy. A detachment of United States Infantry attended, a detachment of seamen from the ships, the escort being Marines. I have to thank General Hatch for the promptness and courtesy with which means were supplied for the occasion.

I have the honor to be, very respectfully, your obedient servant,

J. A. DAHLGREN,

*Rear-Admiral, Com'dg S. A. B. Squadron.*

HON. GIDEON WELLES, *Secretary of the Navy.*

---

## CHAPTER XXVIII.

1865.—Assassination of the President—The Arrest of Semmes.

ON the 15th of April, the day of the death of President Lincoln, and for several days afterwards, the wildest rumors were circulated in the city of Washington; the sentinels at the various posts were doubled, but the indefatigable and untiring energy of General Augur brought order out of chaos and allayed the fears of the timid citizens.

The guard at the Navy-Yard gate, under the temporary command of First Lieutenant R. S. Collum, was strengthened, sentinels were doubled, and two monitors, the Saugus and the Montauk, were moored at the Navy-Yard wharf, head and stern, upon which a strong detachment of Marines were stationed under the command of Captain F. Munroe, with Lieutenants J. M. T. Young, D. P. Mannix, E. R. Miller, and H. Bigelow as his subordinates, to provide for contingencies. The authorities had determined to confine those who might be arrested as suspects, in connection with the assassination and great conspiracy, on board the monitors named.

The first man arrested was Paine, who was brought at midnight to the Navy-Yard in a closed carriage. He was received at the gate from the detectives, and transferred to the monitor, where he was confined in double irons, in a cell closely guarded. The next one received was Atzerodt; then came Spangler, Arnold, O'Laughlin, Dr. Mudd, and a Portuguese, whose name cannot now be recalled. Finally Harold was brought in with Booth's body, and all of the prisoners were guarded by the Marines with the exception of Mrs. Surrat.

Paine, in the extremity of his despair, attempted suicide by beating his head against the iron walls of his cell. To prevent any future attempt, a padded hood was constructed which completely covered his head, leaving nothing but his mouth and nostrils exposed, and secured under his chin by a throat-latch securely locked. The other conspirators were secured in the same manner. These men were kept on the monitors for a period of a week after the autopsy on Booth's body, and were then transferred to the arsenal.

After their arrival at the arsenal the Portuguese was released.

Booth's body was brought to the Navy-Yard, and placed upon a carpenter's bench on the monitor Montauk, securely guarded. Visitors were excluded, no one being permitted to enter except with a pass signed jointly by the Secretaries of War and Navy. An autopsy was performed upon the body of Booth by distinguished medical officers of the army and Navy, and Dr. May, an eminent civilian, in the presence of other officials; also in the presence of Colonel Baker.

It now became a question with the authorities how to dispose of the body. It was at first determined to place the remains in the hands of Colonel Baker, with orders to sink it at sea; this idea was abandoned at the eleventh hour, and at midnight, after the autopsy was performed, the body was delivered to Baker. He conveyed it to the old penitentiary in the arsenal grounds, where it was buried in a cell. Subsequently it was removed and buried at the foot of the gallows with the other malefactors who were executed. Some years afterwards, Edwin Booth, who was always known as a loyal man, and who went into dignified retirement after the tragedy, requested permission from the government to remove his brother's remains. This request was granted, and the dust of John Wilkes Booth reposes in the family lot in a cemetery of Baltimore.

About the 1st of December, 1865, on returning to his station at Portsmouth, New Hampshire, from which he had been absent a few days, Lieutenant L. P. French of the Marines found a telegram awaiting him, directing him "to report in person, without a moment's delay, to the Secretary of the Navy." He lost no time in complying with this mysterious summons, which had created no little excitement among the officers at the yard, and on arrival at Washington he found much curiosity aroused among the officers there in regard to the matter. He was soon ordered, under injunctions of the strictest secrecy, by the Secretary of the Navy, to prepare at once to find and arrest Raphael Semmes, the commander of the Alabama. He was directed to select a sufficient force from the non-commissioned officers of the Corps to accompany him, to give all necessary aid and protection. After making a suitable selection, he reported to the Secretary of the Navy as ready for final instructions. The Secretary gave him an order to proceed to Vicksburg or elsewhere, wherever he might find Semmes, and return with him to Washington. On receiving this order, the question arose as to whether it was not in excess of the authority of the Secretary of the Navy, as Semmes was no longer a naval officer, and if upon land in the locality where he was supposed to be, he was under military jurisdiction; especially as he had been paroled as a brigadier-general of the Confederate service. The order was, therefore, taken to Mr. Stanton, the Secretary of War, and by him endorsed as follows: "All officers subject to the orders of the War Department will render Lieutenant French whatever aid he may deem necessary to execute the within order. (Signed) Edwin M. Stanton, Secretary of War."

Equipped with this order, and with charges and specifications duly setting forth the offences for which it was proposed to try Semmes, Lieutenant French proceeded as directed. On reaching Louisville, Kentucky, he was led to believe that Semmes was not at Vicksburg. He, however, proceeded down the Mississippi, stopping at Memphis, where his views were confirmed, and upon going to Vicksburg, he found that Semmes had not been there for several months, nor could be gain any tidings of him. He continued on down the river, but learned nothing of Semme's whereabouts until he reached New Orleans. There he met Major Lovell, the efficient provost marshal general on General Canby's staff. It was learned from him that Semmes had been in New Orleans a few weeks before, and had gone, as was thought, to Mobile, where it was understood his family resided. French proceeded directly to that place, and a few moments after leaving the steamer met the object of his search on the street. Although he had never seen Semmes before, no one who had ever seen his photograph could possibly mistake him. As he had not announced his arrival or the object of his mission to the military authority of the place, he deferred making the arrest until he had conferred with the officer in command. He soon had reason to regret his delay, for although he made all haste to report himself to the commanding officer of the visit, during the time thus occupied Semmes left the city, as he was doubtless about to do when the lieutenant met him. Lieutenant French had been thrown off his guard in this particular by being informed, very positively, that Semmes's home was in the city. But it was found that he resided some miles out of town. Lieutenant French was not a little surprised to find that few even of the army officers seemed at all familiar with the history of Semmes, or the fact that he had played a very conspicuous part in the great war drama just at its close.

After availing himself of every means of information that could be obtained in the city of Mobile, he became satisfied that the Spring Hill road, where it was said Semmes lived, was not honored by the residence of that distinguished individual. Fearing that to delay and await his return might put further off his arrest, French determined to find him that night, if possible. It was 2.30 P. M. when he arrived in Mobile, but, although he did not allow himself a moment for refreshment, darkness had already set in. The force to be taken with him was left to his discretion, and he took a guard of about thirty men, and ordered three carriages, in one of which were two sergeants, who had accompanied him from Washington, and a detective. He was not long in discovering that the information which the detective furnished "sprang from an over-vivid imagination." He therefore reduced the otherwise useless individual to the rank of driver, and substituted an intelligent contraband in his place. After driving out on the road referred to, and making many inquiries, a man was finally found who had heard of this noted privateersman. He informed Lieutenant French that Semmes resided some six or seven miles out of the city, on the Durham road. To reach this road they had to retrace their steps some two miles to a cross-road, and by a circuitous route the Durham road was finally reached. The night was cloudy and the darkness intense, and, as the military authorities at Mobile were under the impression that the country round about the city was still infested by guerrillas, the guard that accompanied the expedition felt a good deal of trepidation. Following the direction of the new-found guide, Lieutenant French proceeded along the road some two or three miles, and finally reached a gate at the foot of an avenue leading to the house where Semmes lived.

Taking the contraband and the two sergeants of Marines with him, French groped his way up the dark avenue, some sixty rods, to the house. He rapped at the door, and a little girl, some twelve or fourteen years of age, opened it. He inquired if Captain Semmes was at home. She answered that he was, when French turned to the colored man and told him to go down to the carriages and send up the guard. The little girl was alarmed at seeing the men; she hastily shut the door, and ran through the hall. Lieutenant French quickly opened the door, and saw her enter a door at the end of the hall. Presuming that she ran directly to her father, he approached the door and knocked, and was greeted by a gruff voice, telling him to come in.

On entering the room he recognized Semmes. Introducing himself,

RICHARD S. COLLUM, MAJOR
U. S. M. C. **Plate 3**

he said, "I am Lieutenant French of the Marine Corps, and I ame here, by orders of the Navy Department, to present you with this order," handing him the order for his arrest.

Semmes took it with evident astonishment, and, seating himself by a table, scanned the document with the greatest interest. After deliberating for a few minutes, he turned to French, and, in a contemptuous manner, said, "Well, sir, what are your orders in the premises?" French replied, "For you to get ready to start for Mobile with me in five minutes." Semmes rose from his seat and demanded to know if General Woods, the general in command of the Department, knew of this "outrage" being perpetrated. French asked, "What outrage?" He said, "Tearing me from the bosom of my family in this summary and violent manner." French replied, "I know of no one more familiar with summary proceedings than yourself, and, although General Woods has been informed that you are to be arrested, it could make no difference; my orders, as you have seen, are from a higher authority."

At this moment Semmes's family, consisting of his wife, three grown daughters, a son twenty-five years of age, and the little girl, rushed into the room, the ladies much agitated. The lieutenant began to reassure them, when Semmes ordered his family all out of the room, and commanded them to be quiet. Not knowing but that they might appeal to their neighbors, if they had any, or some power that would interfere with the object of his mission, French inquired of Semmes if his friends would attempt to interfere with his arrest. He replied, very emphatically, "No, sir! it is enough for my family and my friends to know that the United States government orders this." He then requested to be allowed until ten o'clock the next morning to make preparations for his journey. This was finally conceded, but when informed that he must select his room, and remain in it under guard, his temper again forsook him, and he protested violently against what he termed "a gross outrage" to perpetrate on an officer of his rank, claiming that his parole should be accepted. Lieutenant French replied that he did not concede to his prisoner any military or naval rank whatever, and should treat him accordingly, and that his treatment would be measured only by his good behavior. Semmes then went to his room, and French proceeded to station a guard in and about his house.

Nothing occurred worthy of note until the appointed time on the following morning, when Semmes took his departure with his captor for Mobile, where they took the steamer for New Orleans. Lieutenant French was urged by the military authorities to take a large guard as far as New Orleans, as they anticipated that the fact of the arrest would be telegraphed there, and that the friends of Semmes might rally and attempt to rescue him. But French's interview, while at New Orleans, with Major Lovell convinced him that the discipline in that city was too efficient to permit any thing of the kind. Nor was he in this regard mistaken, for, although when they reached the city the depot was surrounded by a crowd, no effort was made to rescue the prisoner.

At Semmes's request, French drove with him at once to General Canby's head-quarters, as Semmes wished to see that officer and entreat that he should not be locked up in a prison, of which he seemed to be in greater dread than of anything else. On reaching Canby's head-quarters, and being shown into a reception-room, French entered the parlor and informed General Canby of the object of his visit. General Canby returned with him to the reception-room, when French said, "General Canby, this is Captain Semmes." Semmes, on rising and bowing to the general, said, very imperiously, "Admiral Semmes, if you please, sir!" The general returned his salute with only a dignified bow. Either from the failure to have his rank and title recognized in accordance with his demand, or moved by the dignified manner of General Canby, or possibly by a realization of his position, the prisoner was much embarrassed and ill at ease. General Canby remarked that he would not interfere with any disposition that French desired to make of the prisoner, and so the interview terminated. French took Semmes to the St. Charles Hotel, to await the sailing of the steamer for New York, three days after.

The trip to New York was without incident, and upon arrival at Washington, Semmes was delivered to the commandant of the Navy-Yard in that city. He remained a close prisoner at the Marine Barracks under the special care of Lieutenant Cochrane, until April, when he was released without trial and restored to his parole.

## CHAPTER XXIX.

1866-71.—The Great Fire in Portland, Maine—Grade of Brigadier-General created—Services in Aid of Civil Power—Fight at Formosa—Capture of Filibusters—Aiding the Civil Authorities—Capture of the Forward—Darien Expedition—Valuable Services in aiding the Revenue Officers.

1866.—A fire broke out in the city of Portland, Maine, late on the afternoon of July 4, 1866, and spread with such fearful rapidity that in a very brief time a valuable portion of the city was burned. As is usually the case in times of public distress and excitement, thieves and lawless persons from other cities gathered in Portland, anticipating a rich harvest of plunder from the effects of the suffering inhabitants. The city became so infested with these birds of prey that, on the evening of the 6th, it was determined by the city authorities to ask for military aid. Early on the morning of the 7th a telegram was sent to the commandant of the Navy-Yard at Kittery, Maine, asking for as many Marines as could be spared. Promptly on receipt of the despatch, with scarcely an hour's warning or preparation, Lieutenant-Colonel James H. Jones left Portsmouth in command of two companies of Marines, and arrived in Portland at 12.30 P. M. Upon arrival, a strong line of sentinels was posted throughout the burned district and in other avenues of the city; many persons were arrested and quiet soon established. The Marines remained on duty until the 11th of July, when they returned to the station. Colonel Jones, his officers and men, received the thanks of the city government, in a series of resolutions, for the very valuable services rendered on that occasion. The full strength of the battalion was as follows: Lieutenant-Colonel James H. Jones, commanding; first company, Captain John Schermerhorn, three sergeants, three corporals, two musicians, and fifty privates; second company, First Lieutenant Henry J. Bishop, two sergeants, two corporals, two musicians, and fifty privates.

1867.—In the latter part of February, 1867, by act of Congress, the grade of colonel-commandant was abolished, and that of brigadier-general and commandant substituted. Accordingly, on the 2d of March, Colonel-Commandant J. Zeilin was commissioned brigadier-general.

At the time of the raid on the illegal whiskey establishments in Philadelphia, a force of Marines, consisting of about one hundred men, under command of Brevet Major L. L. Dawson, Brevet Captain Fagan, and Lieutenant Ford, aided the civil authorities in preserving peace, and in protecting the officers from being assaulted by the mob.

During the excitement over the efforts of Brooklyn, New York, whiskey men to defraud the government, the Federal officers called on the commandant at the Navy-Yard for aid in enforcing the law. In accordance with the call, Brevet Lieutenant-Colonel Broome of the Marines, with four companies, was ordered to the scene of the troubles at "Irishtown." The two companies which first arrived at the spot were commanded by Brevet Captain William Wallace and Second Lieutenant James B. Breese. The other companies, which left the barracks at a later hour, were officered by Brevet Major G. P. Houston, Captain William J. Squires, Brevet Captain E. P. Meeker, and First Lieutenant Sherman. The duty which the Marines were called to execute at this time was of a peculiarly delicate nature; but their courageous bearing, fine discipline, and gentlemanly deportment, in the face of much peril, as well as of the most exciting provocation, won the highest commendation alike of the city authorities and the Federal officials.

On the 7th of June, Admiral Bell, in accordance with instructions, left Shanghai in the Hartford, accompanied by the Wyoming, Lieutenant-Commander Carpenter commanding, for the south end of the island of Formosa, to destroy, if possible, the lurking-places of the band of savages inhabiting the southeast end of that island, who murdered, the preceeding March, the shipwrecked officers and crew of the American bark Rover. The landing of one hundred and eighty-one officers, sailors, and Marines, provided with four days' rations and water, was made on the 13th, under the command of Commander G. E. Belknap, of the Hartford, accompanied by Lieutenant-Commander Alexander S. McKenzie, fleet lieutenant, as second in command; the latter earnestly sought to go on the expedition; but his valor was rewarded only by death, as he was killed in the fight. In the terrible heat, and surrounded by savages in ambush, the Marines and sailors fought desperately, burning a number of native huts, and chasing the warriors until they could chase them no longer, though at a grievous cost of life.

The Marines were under command of Captain James Forney, who submitted the following report to Commander Belknap, dated on board the flag-ship Hartford, at sea, June 17: "I have the honor herewith to submit a brief report of the part taken by the Marines on the 13th inst., on the island of Formosa. On first landing, by your order, I took charge of twenty Marines, deploying them forward as skirmishers. A dense and almost impenetrable thicket of bush prevented the men from advancing very rapidly. I penetrated with them to a creek about half a mile from the beach without meeting with any of the enemy, and was then recalled for further orders. You then instructed me to leave a sergeant and five men on the beach, and to advance with the main body, headed by yourself. In consequence of all further operations coming under your own observation, I have nothing further to report, except that the men behaved gallantly, and deserve credit for the manner in which they marched over such a rough and hilly country, and under such intense, scorching heat. Orderly-Sergeant Gottermeyer was of great assistance to me during the entire day, and deserves favorable mention. A few of the men were struck down by the oppressive heat of the sun, but were not seriously injured, and are now doing their usual duty. The entire number of Marines on shore was forty-three, thirty-one of whom were from this ship, and twelve from the Wyoming." Having accomplished all that was possible, Admiral Bell returned to Takao on the 14th, and there buried the remains of the brave McKenzie in the grounds of the British consulate.

1869.—On the 13th of July, Lieutenant Breese was ordered from the Marine Barracks, Brooklyn, to report to General Barlow, United States marshal, to aid in preventing a violation of the neutrality laws. He proceeded, in the revenue cutter Mahoning to Gardiner's Island, at the east end of Long Island, where, on the 16th, with a company of fifty Marines, he surrounded a camp of Cuban filibusters, and captured one hundred and twenty-five of them. The prisoners were sent on board of the Mahoning and carried to New York, where the expedition arrived the next day.

1870.—On the 25th of March, Brevet Lieutenant-Colonel J. L. Broome, commanding the Marine barracks at the Brooklyn Navy-Yord, received a communication from Rear-Admiral S. W. Godon, to the effect that the latter had been applied to by the assistant assessor of the Third New York District for a force of Marines, to assist the United States Internal Revenue officers in seizing and destroying a number of illicit distilleries. Colonel Broome was ordered to prepare a force of at least one hundred men, and as many more as he could conveniently spare for the purpose, and make the necessary arrangements with Assistant Assessor J. B. Wass to carry out the law for the suppression of the illegal operations, affording every protection to the assessors. In obedience to the order, he arranged with the assistant assessor to accompany him to "Irishtown," on the 28th, for the purpose of affording protection to the United States revenue officers. The Marines, when joined by a detachment from the United States receiving-ship Vermont, numbered one hundred and twenty-nine men. This force was, with loaded muskets and forty rounds of ball cartridges, in readiness to leave the barracks at the appointed time. Colonel Broome and his command left the barracks, accompanied by the Internal Revenue officers. They proceeded to "Irishtown," where a large crowd of rioters had assembled. The mob greeted the Marines with yells and shouts of defiance, crying out, "Irishtown is ready for you!" "The rangers will warm you!" etc. Some stones were thrown at this time, but no one was injured. On arriving at Plymouth Street, the immediate neighborhood was occupied by the Marines. The mob was violent, and stones, brick-bats, etc., began to fly rapidly. One of these missiles, evidently thrown at the Marines, struck a police officer on the head, injuring him severely. At this, Colonel Broome made a slight demonstration with the command, which seemed to remind the mob that a continuance of their proceedings might be attended with some danger to themselves, for they became in a measure quiet, and showed evident signs of timidity. Bricks and stones, however, after this, were thrown at intervals from the house-tops. In the mean time, while the Marines were subjected to these annoyances in the streets, the Internal Revenue officers were engaged in destroying a large illicit distillery. When this was accomplished, the command moved to another distillery, which was soon put out of working order by the Internal Revenue officers. The assessors, having accomplished all they intended for the day, were escorted to a locality half a mile from "Irishtown," followed by a crowd of ruffians, who were cursing and abusing them all the way. The revenue officers, supposing that they no longer required protection, wheeled out of column, and had proceeded up Myrtle Avenue a short distance, when the mob rushed after them. Colonel Broome made a movement to assist the revenue officers, seeing which the mob scattered rapidly. First Lieutenant Henry J. Bishop was on duty at the place where most of the disturbances occurred, and was very energetic and efficient in the performance of his duties. The thanks of the numerous officers were duly presented to Colonel Broome, and the conduct of the Marines was mentioned in the most complimentary terms by General McDowell.

On the 6th of June information was received by the United States consul at Mazatlan, from the consul at Guaymas, Mexico, that the steamer Forward, bearing the San Salvador flag, had landed between one hundred and two hundred men of different nationalities, who had taken possession of and robbed the custom-house, forced the foreign merchants to contribute funds and goods to a large amount, compelled the United States consul, under protest, to supply coal for the steamer, and committed other outrages. On learning these facts, Commander

W. W. Low of the Mohican, at Mazatlan, immediately made preparations for getting under way, it being reported that the Forward was still in the Gulf of California. A few days later the Forward was captured and burned in the harbor of Boca Teacapan, a place about half-way between Mazatlan and San Blas. The destruction of the piratical craft was accomplished by the Marines and sailors of the Mohican in six boats, under command of Lieutenant Brownson, who in his report mentions, among others, the conduct of First Sergeant Philip Moore and Corporal F. Moulton of the Marines as deserving special notice.

Accompanying the important Darien exploring expedition of the year, under Commander Thomas O. Selfridge of the Navy, was a force of sixty Marines, under Brevet Major George P. Houston, Lieutenants Goodrell and Allen. Although the mission was one of a peaceful nature, the detachment had the honor to serve in an expedition whose results were of great importance to the commercial interests of the world, under a naval officer who had done more to solve the great problem of interoceanic communication than has been accomplished by any one else.

On the 2d and 3d of November, of the same year, the Marines were again called on to assist the revenue officers. The force numbered two hundred and forty-five officers, non-commissioned officers, and privates, and was employed in protecting the officers of the Internal Revenue Department, while engaged in "Irishtown," in seizing and breaking up illicit stills. They were employed in these localities where the people were most unruly, and it became necessary, in consequence of the mob stoning the men, to charge upon the rioters on two occasions, occupy the roofs and upper stories of many houses, tear down fences, etc., all of which was promptly done by the Marines. Colonel Broome was ably assisted by Captain and Brevet Major G. P. Houston, First Lieutenant and Brevet Captain E. B. Meeker, First Lieutenant F. D. Webster, First Lieutenant E. R. Robinson, and Second Lieutenant G. C. Goodloe, who did all they could to make the duty they were engaged in a success, and all deserved great credit. It became necessary in some instances to place detachments in charge of Orderly-Sergeant James W. Buckley, Sergeant Thomas B. Isham, and Sergeants Ethelbert Reese, William Stanley, and Rudolph Bamberg, who performed their duty well.

In a communication from General Zeilin, commandant of the Corps, to Colonel Broome, dated at Head-quarters Marine Corps, Washington, District of Columbia, November 27, the general said, "I acknowledge the receipt of your communication of the 14th inst., enclosing copies of letters of thanks and commendations of Supervisor L. B. Dutcher, and the colonel commanding troops during the recent operations at Brooklyn, New York, etc. I have read with peculiar pleasure and satisfaction the testimony borne by these authorities as to the valuable services rendered by the Marines on the occasion referred to. The services which the officers and men were called upon to perform on this trying occasion were of a delicate nature, requiring firmness and discretion. To the display of these qualities is due the successful and bloodless result of these operations. I highly appreciate your conduct throughout as commanding officer of the Marines. The ready and able assistance afforded you by Captain G. P. Houston, First Lieutenants E. P. Meeker, Frank D. Webster, E. R. Robinson, and Second Lieutenant G. C. Goodloe, as well as by the non-commissioned officers and privates, is deserving of the warmest commendation, and reflects credit upon all concerned."

1871.—On the 12th and 13th of January, 1871, the services of the Marines at the Brooklyn Yard were again called into requisition in aiding the revenue officers in that portion of Brooklyn known as "Irishtown." A detachment from the Eighth Infantry was also employed. On the 12th two large distilleries were broken up, and in accomplishing this the whole day was consumed. On the next day, Brevet Major-General Vogdes informed Colonel Broome that, in consequence of the irregularity of some of the soldiers, he desired that the Marines should be exclusively employed in protecting the workmen at all the different distilleries in "Irishtown," and also to guard all public property and material seized. In consequence of this arrangement, the Marines were formed into detachments, occupied all distilleries seized in "Irishtown," guarded all the material seized and all the workmen employed. The different guards were in charge of Captain and Brevet Major G. Porter Houston, First Lieutenant and Brevet Captain E. P. Meeker, First Lieutenant Erastus R. Robinson, Second Lieutenant S. K. Allen, Second Lieutenant E. T. Bradford, Second Lieutenant G. C. Goodloe, Second Lieutenant J. D. Smyser, Orderly Sergeant James W. Buckley, Sergeant William Ring, Sergeant Thomas B. Isham, Sergeant Morris Clarrison, Sergeant Howard Fisher, Sergeant William H. Gardner, and Corporal James Cassidy. These officers and non-commissioned officers performed their duties to the great satisfaction of the revenue officers, and in consequence of the good conduct of the Marines on this occasion, Colonel Broome was directed by the admiral commanding the station to express to them his high appreciation of their good conduct. During the 12th, First Lieutenant F. D. Webster, attached to the United States steamer Tennessee, was with the battalion of

Marines, and, with the officers referred to, did all that was necessary to assist in carrying out the wishes of the United States Internal Revenue Department. General Zeilin, in acknowledging the receipt of Colonel Broome's report of the affair, wrote: "The manner in which these delicate duties were performed reflects credit upon all concerned, and meets with my entire approbation."

During the following July the Marines were once more called on to render aid to the revenue officers, who were endeavoring to break up the illicit business in "Irishtown." At about three o'clock on the morning of the 14th the command proceeded to "Irishtown," for the purpose of protecting a large force of United States officials, under General James Jordan, while engaged in arresting certain desperadoes and proprietors of the illicit stills. General Jordan requested Colonel Broome to halt the Marines in the Navy-Yard near the York Street gate until he should require their assistance, which was done. The colonel then proceeded with the revenue officers out of the Navy Yard to Dixon's Alley, each party of officers proceeding to the posts assigned. The party at the entrance of or near Dixon's Alley had been there but a few minutes, when, to the surprise of all, a rapid fire from different points was opened on them from men who were lying flat in the street or hidden in areas of houses and in wagons. The firing had not been going on more than a few seconds when Captain Clinton Gilbert fell, mortally wounded. Major Weeks was wounded in the hand and another officer in the leg. As the party were hard pushed, Colonel Broome ordered the Marines up immediately, and opened on the gang with his revolver. In a few minutes the Marines, under Lieutenant Welles, made their appearance, on the double-quick. The affair momentarily grew exciting. As soon as the ruffians heard the Marines coming, they broke at a run and hid themselves in the alleys and dens of the locality, where it was impossible to find them. Colonel Broome then placed guards at all important points, and ordered all the residents of houses to leave their windows, which they did. Guards of Marines proceeded to different parts of "Irishtown," and protected the revenue officers and United States marshals while engaged in forcing open houses to make arrests. The conduct of the enlisted men on the occasion of this raid was admirable; and First Lieutenant George M. Welles, First Lieutenant E. R. Robinson, and Second Lieutenant Samuel K. Allen, who composed all the commissioned officers under Colonel Broome, performed their duty exceedingly well, and received the warmest thanks of the civil authorities. The conduct of the officers and men was subsequently complimented by General Zeilin, as "deserving of all praise."

In the following September, the revenue officials were again in need of armed assistance, and, for the fourth time, the Marines at the Brooklyn Navy-Yard were ordered to the work. At about one o'clock in the morning, Colonel Broome with his command proceeded on board the tug Catalpa to aid General J. Jordan, assessor, in making seizure of vessels with contraband whiskey on board. The tug proceeded first to Williamsburg, where a search was made for vessels and lighters containing whiskey, but none was found. Other points were also searched without success. At the first appearance of daylight, the tug proceeded to "Irishtown," where the wharves were searched for a vessel said to be there loading with whiskey in barrels, but the vessel was not found. The information that a vessel was there loading with whiskey was positive, but the law-breakers managed to get her away.

The fifth "whiskey raid" in which the Marines were engaged was on the 17th of October. The Marines were occupied during the day in guarding the property of illicit distillers in "Irishtown," and in protecting and escorting working parties and revenue officers from one locality to another. No trouble occurred until the latter part of the day, at which time a gang of ruffians attacked a small party of Marines, under Lieutenant Bradford, with stones, wounding Sergeant Casey. This gang was speedily dispersed, and one of their number was arrested and turned over to the civil authorities. The seizures made on this occasion were very extensive. The officers under Colonel Broome, throughout this raid, were Captain H. A. Bartlett, First Lieutenant Erastus R. Robinson, Second Lieutenant Samuel K. Allen, and Second Lieutenant Edward T. Bradford. Lieutenant Smyser was officer of the guard for the security of prisoners.

---

## CHAPTER XXX.

1871.—The Corean Expedition—A surveying party fired upon by the Corean Forts—Punishment inflicted on the Coreans—The Marines in Advance—The Final Triumph—Report of Captain Tilton.

THE American Minister to China, Mr. Low, having been instructed, in 1871, to arrange and conclude, if possible, a convention with Corea for the protection of sailors and others shipwrecked on the coast of that country, it became necessary that he should put himself in direct communication with the authorities. With this view, and with the assent of the authorities of China, to which country the government of

Corea was tributary, the minister sailed, in May, 1871, from Nagasaki to Boisée anchorage, on the Salée River in Corea, in the Colorado, the flagship of Rear-Admiral Rodgers commanding the Asiatic fleet. After the arrival of the fleet at the anchorage, and after communication with and visits from the local authorities, the Coreans appeared satisfied that the expedition was not only peaceful but friendly in its objects. A party engaged in the peaceful object of making surveys and soundings, in the interest of science, and for the safety of commerce, with the understood consent of the authorities, without protest or challenge by any one, passed up the channel to a point above the Corean forts, beyond the reach of assistance from the main body of the fleet. This party was suddenly and treacherously attacked while in the difficult navigation of an unknown passage swept by strong tides and filled with hidden rocks. The surveying boats were obliged to repass the forts, under a fierce fire of cannon suddenly opened upon them and maintained with spirit and energy. The small vessels which had accompanied the party, hurrying into action, answered the fire of the forts, and, driving the Coreans from their works, rescued the surveying boats from danger, with only two of our men wounded.

It was determined, by the concurrent judgment of the Minister and Admiral Rodgers, that an explanation should be at once demanded, and that ample time should be given the Coreans to understand the situation and make proper reparation. This was done, and ten days were allowed to pass, during which no explanation was offered by the Coreans. An attack was then planned and carried out upon the forts and citadel from which the outrage had been committed.

The work which followed proved the most important engaged in by any portion of the Marine Corps since the close of the war. On the 10th of June the Monocacy and Palos, with four steam-launches, left Boisée Island, and proceeded up the river Salée. The landing-party comprised a brigade of seamen infantry, under Lieutenant Commander Casey, with seven field-pieces under Lieutenant-Commander Cassell, five hundred and seventy-five strong, and a battalion of Marines under the command of Captain McLane Tilton, as follows: Second Lieutenant W. J. Macdonald, Jr., adjutant; First Company, fifty-three men, First Lieutenant J. B. Breese; Second Company, fifty-two men, First Lieutenant F. A. Mullany. This battalion was in ship's boats in tow of the Palos and Monocacy.

Captain Blake, of the Alaska, was detailed as commander-in-chief of the forces, and Commander Kimberly was assigned to the command of the division at his own request. To Lieutenant-Commander Picking was assigned the command of the steam-launches.

The general programme was that the Monocacy, Palos, and launches should advance up the river, and, by attacking the forts on that side create a diversion in favor of the landing-party, who should advance by land and capture and demolish the forts as they advanced.

The expedition advanced promptly, in obedience to a signal from the flag-ship, and the vessels soon received a hot fire from one of the forts, which was quickly silenced by the Monocacy and Palos.

The launches containing the Marines were promptly cast off and pulled for the shore, immediately followed by the naval brigade under Lieutenant-Commander Casey.

The battalion under the command of Captain Tilton landed on a gently-sloping beach, two hundred yards from high-water mark. The mud proved a serious obstruction, coming up over the knees of the tallest of the party, a fact which would have given a deadly advantage to a better-appointed enemy. On landing, a line of skirmishers was thrown out, facing a tongue of land jutting out into the river, covered with scrub and strengthened by a square redoubt on the right. As the Marines advanced, the garrison of the fort were seen running through the brush and long-grown crops, turning a few times, and firing shots which took no effect. The Marines then reconnoitred the grounds and entered the fort, where they remained till the main body came up. The main body proceeded to dismantle the fort, spiking and dismounting the guns and throwing down the works. Meanwhile, the Marines advanced with the river on their right, extending as far to the left as possible, so as to cover a wide surface of country. At length they took position on a wooded knoll, which appeared to be used as a cemetery, being covered with hemispherical mounds. The main body was three-quarters of a mile behind. The evening was now far advanced, and a general order was issued to the force to bivouac for the night where they were. Accordingly, for the first time, a Western force spent the night on the soil of Corea. On the following day the little army advanced on the second line of fortification. One-third of the Marines were ordered to march on the face of the works looking landward, two-thirds being held in reserve. But the assault proved a bloodless one; the enemy had decamped, probably in consequence of the shot and shell of the squadron; and they entered the works without opposition. With despatch they set to work, and in a remarkably short space of time the battlements were dismantled, fifty or sixty insignificant pieces of cannon being spiked and thrown over the cliff into the river. The ramparts were then demolished on the front and right face of the works. But the main object the party had in view had still to be dealt with, being no other than the destruction of the horseshoe-shaped citadel, which first opened fire on the surveying party. For it, therefore, our forces now started, the Marines in advance. For a time only a few unarmed natives were dislodged, who were left unmolested. At last, as the crest of a ridge was reached, the enemy were seen occupying a parallel ridge at no great distance, and discharging their wretched matchlocks at the foreigners as they came in view, their great black heads popping up and down all the time. But, short as the distance was, it was beyond their range. The enemy were grouped on a knoll some distance off; but a few shells judiciously planted among them exerted a wonderful influence, and they fled in all directions. A skirmishing party of Marines was thrown out, and advanced along a narrow ridge leading directly to the horseshoe citadel, the grand stronghold of the Coreans. Of the six thousand troops said to have been employed to oppose the Americans, one thousand formed the garrison of the citadel. The main body followed the skirmishers close up, in column of fours. A line of Marines was thrown out, to advance parallel to the right flank of the redoubt, which was selected as the point of attack, and where the advance was concealed from the enemy. This advance was successfully accomplished, till the party took position along the crest of the hill, about one hundred and fifty yards from the enemy, with their right resting on a path leading to the redoubt, along which path were planted in line about twenty-five banners a few feet apart. Parallel to this ridge was another, about thirty yards in advance, but to reach it the whole line must expose themselves to view.

The banners being regarded as a decoy, Captain Tilton detailed only four men to advance towards them. They had secured fifteen, when a tremendous fire was opened on them, a perfect hail of bullets, lasting, happily, only half a minute. As soon as it slackened, a rush was made for the ridge in advance, which movement was accomplished with the loss of only one Marine, Private Dennis Hemahan, although another tremendous volley was opened on the them as soon as the rush was made. They were now only one hundred and twenty yards from the redoubt, but the rank vegetation and the shoulder of the hill on which it stood screened them from its fire. Firing now became general and rapid on both sides, the deadly effect of the American practice being witnessed afterwards by the number of Coreans killed and wounded. High above the sharp rattle of the firing rose a melancholy chant of the Coreans as they fought. As little assaulting parties of the Americans advanced to close quarters, the Coreans, their ammunition apparently expended, assailed them with stones. The citadel was built upon the apex of a conical hill, about one hundred and fifty feet high from the bottom of the ravine, through which our men had to pass to reach it. The hill-side was very steep, and the walls of the fort joined the acclivity with scarcely a break in the line. Had not the face of the walls been somewhat shattered by the shells from the Monocacy and the howitzers on shore, the ascent would have been most difficult. Nothing could check the Marines and sailors; on they rushed. In his report, Admiral Rodgers says, " The heroic McKee was first to mount the parapet, and the first to leap into a hand-to-hand conflict. There he fell, as his father fell in Mexico, at the head of his men, first inside the enemy's stormed works." Other officers and men were quickly over the parapet. The fighting inside the fort was desperate. The resolution of the Coreans was unyielding; they apparently expected no quarter, and probably would have given none. They fought to the death, and only when the last man fell did the conflict cease. The enemy made no organized resistance in the forts lower down, on the point towards the river. These were opened to a rear attack by the capture of the citadel, and the garrison fled. Many of them, however, fell under the fire of our musketry and howitzers, which had nearly cut them off from retreat. The yellow cotton flag, about twelve feet square, with a large Chinese cabalistic character in black in the centre, which flew over the fort, was captured by the Marines. It was torn down by Corporal Brown, of the Colorado's guard, and Private Purvis, of the Alaska's guard, by Captain Tilton's orders. Private Purvis, of the Alaska's guard, had his hand on the halliards a second or two before any one else, and deserves the credit of the capture. The command, to a man, acted in a very creditable manner. Fifty flags were taken, including that of the generalissimo; four hundred and eighty-one pieces of ordnance fell into our hands, besides very many match-locks and gingals. The guns comprised eleven 32-pounders, fourteen 24-pounders, two 20-pounders, and the remainder, four hundred and fifty-four, were 2- and 4-pounders. Two hundred and forty-three dead Coreans were counted in the works. Few prisoners were taken, not above twenty, and some of these were wounded. These last were treated with all the attention possible, and finally released. Thus, in the language of Admiral Rodgers, "was a treacherous assault upon our people, and an insult to our flag, redressed."

On the morning of the 12th, at daylight, after having occupied the field of battle eighteen hours, the entire force re-embarked in one hour, and returned to the anchorage in the Monocacy and Palos. Commander Kimberly, in his report of the affair, says, " To Captain Tilton and his Marines belongs the honor of first landing and last leaving the shore, in leading the advance on the march, in entering the forts, and in acting as skirmishers. Chosen as the advanced guard, on account of their steadiness and discipline, and looked to with confidence in case of difficulty, their whole behaviour on the march and in the assault proved that it was not misplaced." Lieutenant-Commander Casey, reporting the action to his superiors, bore similar testimony, as follows: "The Marines were always in the advance, and how well they performed their part I leave you to judge. Their conduct excited the admiration of all. I cannot express in too high terms my admiration for the gallant conduct, under fire, of the officers attached to my command, and their cheerfulness in executing orders under trying circumstances."

The following Marines were honorably mentioned: Steamer Alaska, Marines who captured flags, Privates Hugh Purvis, commanding-general's flag, John Kelly, H. M. Tolman, and J. B. Butler; steamer Benicia, Marines who captured flags, Corporal Thomas H. Baker and Privates Daniel Barry, John Bourke, Charles C. Collins, William Dervees, George McIntyre, and Michael McNamara. Commander Kimberly mentions the following-named Marines who distinguished themselves in the attack: "Private Dougherty, of the guard of the Benicia, for seeking out and killing the commanding officer of the Corean forces; Private McNamara, of same guard, for gallantry."

Commander Kimberly says, "To Lieutenant-Commander Picking, who had charge of the steam-launches and boats, we are under great obligations for the manner in which he was at all times near us to assist with his guns and boats, and it was to his exertions in getting the boats up that our very successful embarkation from the upper forts took place, and which was really a serious matter, taking everything into consideration."

The following is the report of Captain McLane Tilton, commanding United States Marines on this occasion:

United States Flag-Ship Colorado,

at anchor off Isle Boisee, Corea, June 16, 1871.

Sir,—In conformity with your direction, I have the honor to make the following report of the part taken by the Marines of the Asiatic Fleet in the late expedition against the Coreans:

On Saturday, the 10th instant, the guards of the Colorado, Alaska, and Benicia, numbering one hundred and five, rank and file, and four officers, equipped in light marching order, with one hundred rounds of ammunition and two days' cooked rations, were embarked from their respective ships and towed up the Salée River by the United States ship Palos. Upon nearing the first line of fortifications, extending up the river on the Kang Hoa Island side, the Palos anchored, and by order of the commanding officer all the boats cast off and pulled away for the shore, where we landed on a wide sloping beach, two hundred yards from high-water mark, with the mud over the knees of the tallest man, and crossed by deeper sluices filled with softer and still deeper mud. After getting out of the boats, a line of skirmishes was extended across the muddy beach, and paralled to a tongue of land jutting through it to the river, fortified on the point by a square redoubt on the right, and a crenellated wall extending a hundred yards to the left, along the river, with fields of grain and a small village immediately in its rear.

The fortification had been silenced by the cannonade from the United States ship Monocacy, and the steam-launch and the garrison fled through the brush and fields, firing a few shots as they retired at a distance. The Marines, by order, then advanced on the place, sweeping through the grain-fields and village, meeting no opposition, and remained in possession until the main body came up, when we were again ordered to push forward, which we did, scouring the fields as far as practicable from the left of the line of march, the river being on our right, and took a position on a knoll, covered with hemispherical mounds, and commanding a fine view of the beautiful hills and inundated rice-fields immediately around us, and distant about half a mile from the main body. A reconnoissance was then made towards the next fort, a square work of hewn granite foundation, with a split rock, mud, and mortar rampart, crenellated on each face, with a front of about thirty paces, and a messenger despatched to headquarters with the information that the road was clear and passable for artillery. Pickets were posted on the flanks of our little position, five hundied yarde to the right and left, a rice-field inundated being in front, and a Dahlgren 12-pounder planted so as to command the junction of the only two approaches, which the commanding officer had ordered up to us as a support. An order having been sent to hold our position till morning, we bivouacked with our arms by our sides, dividing our force in three reliefs, one of which was continually on the alert. No incident occurred during the night except rapid firing of small-arms and howling from a hill inland from us, about a third of a mile distant. Two or three shots from the artillery with the main body were fired across the left of our picket, in the direction of the noise, which presently ceased.

Sunday morning, the 11th of June, the main body came up, and we received orders to push forward, which we did, and after reaching the fields in rear of the next of the line of fortifications, we threw a line of skirmishers across the peninsula of hills on which the fort stood, and after the main body came up we advanced towards the rear face, with two-thirds of our guards in reserve. We entered this second place after reconnoitring it without opposition, and dismantled the battlements by throwing over the fifty and sixty insignificant breech-loading brass cannon, all being loaded, and tore down the ramparts on the front and right face of the work to the level of the tread of the *banquette*. The ramparts consisted of a pierced wall of chipped granite, with a filling of earth in the interstices, and coated over with mortar, giving it the appearance of being more solid than it really was. The cannon were rolled over the cliff into the water by Bugler English without much trouble, who climbed down for this purpose. I cannot give the weight, but the bore was not over two inches in diameter. A photographer came on shore from the Monocacy and succeeded in taking a negative picture of the place. We were then ordered by the commanding officer to push forward and find the road leading to our objective point, and to cover the flanks of the main body, which we did with two-thirds of the

Marines deployed, the remainder in reserve. We scoured the scrubby woods and fields of grain, stirring up two or three unarmed natives, refugees from the village we had just passed, who were not, however, molested; and after progressing half a mile, down deep ravines and the steepest sort of hills, were fired upon from a high ridge a little to the left of us, up which our skirmish-line cautiously wheeled, and upon reaching the summit saw the enemy on a parallel ridge opposite, who blazed away at us with their gingals or matchlocks, their black heads popping up and down the while from the grass, but only one spent ball struck us without any injury. A piece of artillery was here brought up from the valley beneath us, by direction of Lieutenant-Commander Cassel, by superhuman exertion on the part of his men, and several shells landed among the enemy grouped on a knoll, scattering the party, when our skirmish-line pushed on down the narrow range leading to the circular redoubt, our objective point, and known to us as the citadel, being the third work of the line of fortifications, the main body following in column of fours.

Upon reaching a point a third of a mile from this work, a general halt was ordered to rest the men, who were greatly fatigued after their comparatively short, although extremely deep, march; the topography of the country being indescribable, resembling a sort of chopped sea, of immense hills and deep ravines lying in every conceivable position. We then advanced cautiously, with our line of skirmishers parallel to the right face of the redoubt, which was our point of attack, concealed from view from the enemy, and took position along the crest of a hill one hundred and fifty yards from him, closing intervals to one pace from the right skirmishers, the line extending along the ridge, our right resting in a path leading to the redoubt, upon which were placed about twenty-five banners in single file, a few feet apart, and at right angles to our line, the first banner being only four paces from our right skirmishers. Thirty paces in front of us was another ridge, parallel to the one we now occupied, but in order to reach it, the whole line would be exposed to view. The main body came up and formed close behind us. The banners seemed to be a decoy, and several of us went from our right, took about fifteen of them, which drew a tremendous hail of bullets from the redoubt, which relaxed in half a minute, when away we pushed, availing ourselves of the opportunity to get to the next ridge, accomplishing the move with the loss of only one man, a Marine from the United States ship Alaska, although for several seconds exposed to a galling fire, which recommenced immediately after the rush began.

Our lines were now only one hundred and twenty yards from the redoubt, but the abrupt slope of the hill and the weeds covered us very well.

The firing now commenced rapidly from both sides; ours increasing as the men settled comfortably, and their fire was effective, as the forty or fifty killed and wounded inside the redoubts show. The firing continued for only a few minutes, say four, amidst the melancholy songs of the enemy, their bearing being courageous in the extreme, and they exposed themselves as far as the waist above the parapet fearlessly; and as little parties of our forces advanced closer and closer down the steep ravine between us, some of them mounted the parapet and threw stones, etc., at us, uttering the while exclamations seemingly of defiance. One of these little parties of our forces, the very first to enter the redoubt, was led by our beloved messmate, the heroic, the brave McKee, who fell pierced with a bullet in a hand-to-hand struggle on the ramparts.

The yellow cotton flag, about twelve feet square, with a large Chinese character in black on the center, thus "B. T.," which flew over the fort, was captured by the Marines. It was torn down by Corporal Brown, of the Colorado's guard, by my direction, while Private Purvis, of the Alaska's guard, has his hands on the halliards a second or two before any one else, and deserves the credit of the capture.

Corporal Brown deserves equally with him to be honorably mentioned for his coolness and courage. The command, to a man, acted in a very creditable manner; all deserve equal mention. The officers of Marines were Lieutenant Breese, Mullany, and McDonald, who were always to be found in the front.

The wounded were soon attended to by the surgeon's corps, who removed them to the Monocacy, lying in the stream. The place was occupied all Sunday night, the artillery being posted on the heights, and commanding the rear approaches, the men bivouacking with their companies on the hills. Early Monday morning the entire force re-embarked on board the Monocacy, the Marines being the last to leave.

The re-embarkation was accomplished in a masterly manner, in the space of an hour, no confusion whatever occurring, although the current was very strong, the rise of the tide being nearly twenty feet. The Monocacy then steamed to the fleet, some ten miles below, where we all joined our respective ships.

Of the Marines, there was one killed and one severely wounded, the first being Private Dennis Harrigan, of the Benicia, and the wounded man Private Michael Owen, of the Colorado, shot through the groin as he was charging towards the redoubt, falling about forty paces from the parapet. The accoutrements and arms of the guard of this ship were returned, and no loss of property occurred. The expenditure of ammunition was sixteen hundred cartridges, about forty rounds to each man. . . .

---

## CHAPTER XXXI.

1871-74.—Services in Boston—Panama—The Delaide—Review at Key West—The Grade of Brigadier-General Abolished.

During the great fire in Boston, which broke out on the night of the 9th of November, 1872, the city authorities called on the commandant at the Navy-Yard for a force of Marines to aid the police in maintaining the peace, and preventing the wholesale robbery that was threatened by gangs of thieves who swarmed to the city. Lieutenant-Colonel Jones, commanding the post, placed the command of the force detailed, in answer to the request, in the hands of Captain Richard S. Collum. Between one and two o'clock on the morning of the 10th, Captain Collum started for the scene of duty with his detachment, consisting of First Lieutenants William Wallace, George M. Welles, three sergeants, two corporals, and nineteen privates. He marched directly to the City Hall, and, upon arrival, reported to the Hon. William Gaston, mayor, for such duty as he might require. Captain Collum was directed to march to the scene of the conflagration, with discretionary power to act as he might think best for the preservation of order and the public safety. At 2 o'clock A. M. the force arrived at the point where their services were required. At 6 A. M. First Lieutenant J. H. Sherburne reported with a detachment of fifteen men from the United States ship Ohio. The Marines remained on duty until noon, when, the State militia having arrived, they returned to the barracks. In his report of the affair, Captain Collum said, "I would respectfully call your attention to the noble and gallant conduct of the men, who were placed in situations of extreme peril, and, by their steadiness and firmness in executing the orders given them, prevented confusion among the crowd of by-standers, and saved many lives. To First Lieutenants William Wallace, George M. Welles, and John H. Sherburne great praise is due for their untiring exertions in the performance of their duty. Always ready, willing, and prompt, their conduct merits my hearty commendation."

At 2 o'clock P. M., on the 11th, Captain Collum received orders to take command of two companies of thirty men each, commanded by First Lieutenant William Wallace and Orderly Sergeant Francis Groll, and proceed to the subtreasury for further orders. Upon arrival, he was ordered to guard the removal of government funds from that building to the custom-house. After performing that duty, he returned with the command to the barracks. At 5 o'clock P. M., of the same date, he was ordered to take command of a detachment, consisting of one sergeant, one corporal, and twenty-two privates, and report to Postmaster William L. Burt for duty. First Lieutenant William Wallace was detailed to accompany him. As soon as they arrived, quarters were assigned them in the post-office building, the rear of which was much damaged by the fire, thereby exposing a large amount of public property to the depredations of the lawless persons then in the city. Sentinels were posted in the most exposed positions with loaded muskets, and this duty was continued for several days. In closing his report, Captain Collum says, "I cannot mention too highly the valuable aid and important services rendered by First Lieutenant Wallace throughout the whole arduous duties we have performed, and the uniformly good conduct of the men under very trying circumstances."

Postmaster Burt addressed a letter to Colonel Jones, after the fire, in which he said, "I desire to convey to Captain Collum and Brevet

Captain Wallace of your command, and the men under them, who were assigned at my request as a guard to the subtreasury and post-office during the week of the great fire, my thanks for the service they rendered us. They remained without relief from Monday to Saturday night. Their discipline and soldierly bearing were marked, and they performed their duties thoroughly, gaining great credit from all our citizens." General Zeilin, in acknowledging the receipt of Colonel Jones's report, says, "I have to say that the conduct of those comprising the detachment is a matter of pride and gratification to me. I had learned from other sources of the discipline and soldierly conduct, as well as of the efficient aid rendered by the detachment, on this occasion, and I desire you will extend to Captain Collum, Lieutenants Wallace and Welles, as also to the non-commissioned officers and privates of the detachment, my thanks and most complimentary recognition of their services. And I further desire to say, that the promptitude displayed in responding to the call for the services of the Marines, and their efficiency and discipline in rendering such service, speak most highly for the energy and zeal of their commanding officer."

1873.—At the time of the destructive fire in Boston, May 30, when the famous Globe Theatre and other valuable buildings were burned, the city authorities again called on the Marines for aid; and in obedience to an order received by Colonel Jones from Commodore Parrott, commandant of the station, Captain Collum assumed command of a detachment, consisting of First Lieutenant William Wallace, Second Lieutenants J. C. Shailer and J. T. Brodhead, four sergeants, three corporals, and fifty-nine privates, and proceeded to the City Hall, where orders were received from the Hon. H. L. Pierce, mayor. At 2 o'clock P. M. the detachment was reinforced by fifteen men from the guard of the United States ship Ohio, under the command of First Lieutenant J. H. Sherburne; also twenty-one men from the guard of the United States ship Powhatan, in charge of a sergeant. At 3 o'clock P. M. they were relieved by a detachment of the Fifth Artillery until 5 o'clock P. M., when they again went on duty and remained until the next morning. The streets within their lines were thoroughly patrolled during the night, and the utmost vigilance was observed. Captain Collum thus closes his report to Colonel Jones: "With great gratification, I respectfully present for your favorable consideration the men of the command, who won the respect of all by their uniformly good conduct, vigilance, and soldierly bearing. To First Lieutenants William Wallace and J. H. Sherburne and Second Lieutenants J. C. Shailer and J. T. Brodhead my hearty thanks are due for their promptitude and the able manner with which they conducted the arduous duties assigned them."

General Zeilin expressed great satisfaction on the receipt of Colonel Jones's report, and wrote to that officer as follows: "It affords me great satisfaction to express to you my thanks for the uniform good conduct and promptitude of the men under your command. On all occasions, when called upon, have they maintained the past renown of the Corps. You will please express to Captain Collum, the officers and men in his charge, my thanks for the good conduct, vigilance, and soldierly bearing displayed by them on all occasions." The city government of Boston officially recognized the services of the Marines, and Mayor Pierce sent the following to Commodore Parrott: "I have great pleasure in complying with the request of the City Council to communicate to you the thanks of the city government and citizens of Boston for the very valuable services rendered by Captain Richard S. Collum, Lieutenants William Wallace, J. H. Sherburne, J. C. Shailer, and J. T. Brodhead, and men under their command attached to the Marine Corps, in preserving order and protecting property during and after the fire."

Rear-Admiral Steadman, on arriving at Panama, May 7, 1873, found hostilities in progress between the opposing parties contending for possession of the government of the state of Panama, and, at the request of the United States consul and a number of influential American and other foreign citizens, sent on shore a force of two hundred Marines and sailors, with four pieces of artillery. A portion of this force was withdrawn on the 11th, and the remainder on the 22d, all differences having been settled. A second landing was made September 24, under the orders of Rear-Admiral Almy, the revolutionary movements having been renewed. A force of one hundred and thirty Marines and sailors, well armed and equipped, was landed, and was afterwards increased to one hundred and ninety men. Detachments were posted to protect the American consulate, and other American houses and American property. The landing parties were from the Pensacola and Benicia, the latter vessel belonging to the North Pacific Station. Hostilities ceased October 8, and the force was withdrawn, excepting a detachment of thirty men, which was left a few days longer, to guard the depot and the railroad, should the troubles revive. The landing of these detachments during the two emergencies, while quieting the fears of foreign residents, secured the safe transit of the passengers and their effects, and of the freight and specie of four lines of steamers, two of which were not of our nationality, depending on this road for prompt transportation. The officers and men who composed the landing parties received the commendation of their respective commanders-in-chief for the creditable and admirable manner in which they discharged their duty. The detachment of Marines, numbering twenty-two, was under the command of Second Lieutenant James D'Hervilly.

On the 25th of June, the Delaide, an Italian merchant vessel, lying in the harbor of Callao, Peru, was discovered to be on fire. The flames gained such headway that her crew were powerless to overcome them. As the vessel was loaded with explosive and combustible materials, it was feared that the shipping in the crowded harbor would be greatly damaged, and possibly many lives lost, unless the progress of the flames could be checked. At this juncture several of the officers and men of the United States ship St. Mary's volunteered their assistance. Among these was Captain P. C. Pope, fleet-marine officer of the squadron. The volunteers were entirely successful in extinguishing the flames; not, however, without severe exertion and great danger. Captain Pope, with the other officers, was the recipient of a letter of thanks from the Minister of foreign affairs of Italy, by order of the king, for his distinguished services on that occasion. This was transmitted to him in a complimentary letter from the Navy Department, accompanied by the following letter from Admiral Almy, addressed to Captain Pope personally: "It affords me much gratification to transmit you a copy of a letter from the Italian Minister at Washington to the Department of State, conveying the thanks of the Italian government to yourself and other officers for gallant and skilful services rendered on the occasion of a fire on board of the Italian bark Delaide, in the port of Callao, on the 25th of June, 1873. I will take this occasion to add, that no more agreeable duty can ever fall to the lot of a true naval or Marine officer than that of relieving those who may be in imminent danger of shipwreck and destruction, to whatsoever nation they may belong. Under such circumstances the genuine man, of whatever profession he may be, knows no nationality. To him, all belong to the same family. These sentiments seem to have actuated you in your spirited behavior upon the occasion of the fire on board of the Italian bark Delaide."

1874.—On the 24th of January, 1874, the harbor of Key West, Florida, presented an unusual scene. There was assembled one of the largest and most imposing fleets since the fleet of Admiral Porter in 1864. Vessels had been recalled from the different squadrons in anticipation of war with Spain, growing out of the Virginius affair. This difficulty having been settled, it was determined, before ordering the vessels to their stations, to exercise the fleet in squadron tactics, apply the teachings of the torpedo system, landing in surf-boats, target practice, and exercise the Marines of the fleet in battalion movements on the island, under the command of the fleet-marine officer, Brevet Lieutenant-Colonel Charles Heywood.

The Marines were landed in conjunction with the sailors of the fleet, and on several occasions by themselves. A large number of citizens from different parts of the country assembled at Key West, as well as many officers of the army and Navy, to witness these drills on sea and land. The sailors exhibited remarkable proficiency, and the battalion under the command of Colonel Heywood won the applause, not only of civilians, but of the officers of the two services. The precision in marching, the physique of the men, the evolutions, and the perfection in the manual delighted the distinguished veterans who were there as spectators. The following is the roster of the battalion:

Captain and Brevet Lieutenant-Colonel Charles Heywood, colonel; Captain George W. Collier, lieutenant-colonel; Captain Frank Monroe, major; Lieutenant William F. Zeilin, adjutant; Dr. Robert A. Marmion, surgeon; George A. Deering, paymaster. Company A: Captain W. R. Brown, four sergeants, one drummer, one fifer, forty-eight corporals and privates. Company B: Captain W. B. Remey, four sergeants, one drummer, one fifer, forty-eight corporals and privates. Company C: First Lieutenant and Brevet Captain G. B. Haycock, five sergeants, one drummer, one fifer, forty-eight corporals and privates. Company D: First Lieutenant E. R. Robinson, four sergeants, one drummer, one fifer, forty-eight corporals and privates. Company E: First Lieutenant F. H. Harrington, four sergeants, one drummer, one fifer, forty-eight corporals and privates. Company F: Second Lieutenant G. C. Goodloe, four sergeants, one drummer, one fifer, forty-eight corporals and privates. Company G: Second Lieutenant B. R. Russell, four sergeants, one drummer, one fifer, forty-eight corporals and privates. Company H: Second Lieutenant R. D. Wainwright, four sergeants, one drummer, one fifer, forty-eight corporals and privates. Company I: Second Lieutenant F. Scott, four sergeants, one drummer, one fifer, forty-eight corporals and privates. Company K: Second Lieutenant D. Whipple, four sergeants, one drummer, one fifer, forty-eight corporals and privates.

On the 6th of June, 1874, the grade of brigadier-general was abolished, and it was provided that "thereafter the commandant shall have the rank and pay of a colonel, and shall be appointed by selection by the President from the officers of said corps."

## CHPTER XXXII.

1877.—The Labor Riots—Services of the First Battalion.*

In 1877, the terrible and unprecedented labor riots, which paralyzed business in nine States, and led to wholesale arson and murder, culminating in Pittsburg, called all of the available forces into the field.

Colonel Charles G. McCawley, commandant of the corps, received orders from the Secretary of the Navy, at 3 A. M. of Saturday, the 21st of July, to have all the marines available in Washington ready to move to Baltimore, Maryland, at a movement's notice. The following was the strength of the battalion at this time:

Brevet Lieutenant-Colonel Charles Heywood, commanding; First Lieutenant B. R. Russell, adjutant.

Passed Assistant-Surgeon A. M. Owen, United States Navy; Passed Assistant-Surgeon T. D. Myers, United States Navy; Captain G. W. Collier, First Lieutenant Aulick Palmer, Second Lieutenant W. S. Schenck, Second Lieutenant J. T. Brodhead, and one hundred and twenty-seven non-commissioned officers, musicians, and privates.

At 6 P. M. orders came to proceed to the Baltimore and Ohio Depot, and take a special train for Camden Station, Baltimore. With great alacrity the battalion marched to the depot, to learn that no engineer could be obtained to run the locomotive. After a delay of an hour, a man was found who said that he would undertake it if promised protection. Threats were made that the troops would not be permitted to leave the city. Lieutenant Russell and ten men were placed upon the engine and tender with loaded rifles, and instructions to shoot any one who attempted to interfere with the train in the slightest degree. In five minutes the force was embarked and the cars moving. An instant later, three torpedoes exploded under the locomotive and caused a momentary sensation; but the mob fell back and contented itself with shouting. Meantime, the train proceeded very cautiously, stopping at way-stations to receive advices from the authorities in Baltimore, who were fearful of surprise or accident. At the bridges it was halted, and the structures examined before attempting to cross; scouts were sent out to reconnoitre suspected localities, and upon entering the city all the lights were extinguished, and the train stole

*Captain Cochrane in *United Service Magazine.*

Colonel Commandant CHARLES G. McCAWLEY,
U. S. M. C. **Plate 4**

quietly up to a point within a square of its destination. Before starting, the following despatch was sent:

WASHINGTON DEPOT, July 21, 1877.

GOVERNOR CARROLL, CITY HALL, BALTIMORE, MARYLAND:

Will leave immediately for Baltimore with Marines.

COLONEL HEYWOOD, *Commanding Battalion.*

And upon reaching the Relay House the governor was notified of their safe arrival. There Colonel Heywood found three despatches, as follows:

(1) CAMDEN STATION, July 21, 1877.

COMMANDER TROOPS ON TROOP TRAIN:

Better be prepared for action between Relay House and Camden Station.

T. FITZGERALD.

(2)

COMMANDER TROOP TRAIN:

Hold train till F. O. T. FITZGERALD.

(3)

COLONEL HEYWOOD, COMMANDING MARINES:

Come to Camden Station as soon as possible. Report to General Herbert.

JOHN LEE CARROLL, *Governor.*

Upon receipt of the last the trains continued, and arrived without acciden at 11.30 P. M. An aid of Governor Carroll's met them with the following letter:

BALTIMORE, July 21, 1877.

COLONEL HEYWOOD, COMMANDING MARINES:

SIR,—I am in receipt of your despatch announcing the departure of the Marines, and have to request, on arrival, that they proceed to Fort McHenry, to support General Barry's Artillery in case it should be called to the city. Colonel Gilmore will deliver this note, and will furnish you any further information you may require.

I remain your obedient servant,

JOHN LEE CARROLL, *Governor.*

With this document came verbal orders to disperse the mob around the depot, and in less than ten minutes the battalion was advancing in double time, and the turbulent rabble, which for hours had been cheering the strikers, hissing the police, and groaning the railroad company, was fleeing before it. The police, who behaved gallantly, charged almost simultaneously and captured over a hundred of the principals, which made their prisoners for the day aggregate upward of three hundred. The Marines then marched around the station, clearing all of its approaches simply by their advent, and after a short halt again started for the enemy, guided by a detail of policemen and detectives. For several hours the battalion marched through the worst portions of the city without molestation. Not a stone or brick was thrown, and the crowds broke upon its approach, which was an agreeable surprise. At last it halted in the vicinity of Fort McHenry, where they found Brevet Major-General W. F. Barry, United States army, with Ramsay's battery of the second artillery awaiting them. The battalion faced about, and after being so disposed, with skirmishers ahead and on the flanks, as to have the two pieces of artillery guarded on all sides, marched back to the Camden depot, which was reached at 3.30 A. M. The fire-bells had been ringing nearly all night, and the sky was lighted by the blaze of burning buildings, cars, and lumber-yards. When the rioters were driven away from the depot it was not believed that their deviltry had been entirely checked, and so it proved. They had resorted to the torch.

At four o'clock Company A, Captain Collier and Lieutenant Schenck, was detailed to go to Mount Clair Station, the principal oil and cattle depot of the road, and disperse a mob which was burning oil-trains. A locomotive with two passenger cars was placed at its disposal; and Colonel Harry Gilmor, police commissioner, accompanied as a guide and representative of the municipality. A dozen men were stationed in the cab and on the engine, with the usual orders about obstruction of the road or interference with the train. The crowd at the depot threatened to shoot the engineer, and swore that the train should never reach Mount Clair; but it did. At Camden Junction, one mile from Mount Clair, several thousand noisy and excited men and women were found collected about some burning oil-cars, but they gave way to the Marines, who, deploying as skirmishers, drove the rioters back to the Frederick road, and then pulled apart the consuming trains. They were insolent and defiant, but made no assault. After performing this duty Captain Collier and a part of the company returned and found a guard regularly mounted, and thirteen sentry posts established in the streets leading to the depot. The men having been on duty all night, with muskets in their hands and blankets over their shoulders, were marched to the Eutaw House, at 9.30 A. M., for breakfast, and then were allowed to sleep by companies on the platforms of the depot. The Fifth Maryland Regiment, which had behaved so handsomely on the 20th, was also quartered at the depot, with their wounded in a hospital-car inside of the yard.

Lieutenant Schenck was left with thirty men and six of the Metropolitan police to protect the bridge and workshops at Camden Junction, where another mob was threatening arson. He succeeded in saving the property, and remained until 7.30 P. M., when he was relieved by a battalion of one hundred and fifty artillerymen from New York, under Major H. G. Gibson, of the Third Artillery. Owing to the threatening demeanor of this crowd, Lieutenant Schenck was obliged to keep up a cordon of eighteen sentinels and the utmost vigilance, and several times had to use force to press the aggressive rioters back.

At 3 A. M., the following despatch was received:

WASHINGTON, D. C., July 22, 1877.

COLONEL CHARLES HEYWOOD, U.S.M.C., BALTIMORE:

SIR,—You will proceed, by command of General Hancock, with the Marines under your command to Philadelphia.

R. W. THOMPSON, *Secretary of the Navy.*

General W. S. Hancock, commanding the Division of the Atlantic, accompanied by his staff, had arrived in Baltimore on Sunday morning, and made Barnum's Hotel his head-quarters. Finding that peace was partially restored, and the police, militia, and regulars, now represented by portions of the Second and Third Artillery, some general service troops under Lieutenant Sage, Eleventh Infantry, and a detachment of engineers commanded by Major Abbott, were sufficient to control the mobs, it was decided to send the Marines to Philadelphia, which was now having its turn and seriously threatened. It was desired to get off at daylight, but, for some cause, the necessary transportation was not furnished until about nine o'clock. This gave the men an opportunity to breakfast and better prepare them for the journey. A special train received them, and at noon they were opposite the United States Arsenal, in Philadelphia, where Officer Taggart, of the Philadelphia, Wilmington and Baltimore Railroad Company, met them, and reported to Colonel Heywood that the track was torn up and blockaded on the Pennsylvania Railroad to prevent their passage.

The battalion was then disembarked and was soon *en route* for West Philadelphia. It was met by large crowds of roughs; but, as in Baltimore, they made way for the troops, and contented themselves with threats. Upon crossing the Schuylkill, an oil-train was discovered to be on fire upon a branch road near the Blockley Almshouse.

The rioters, getting word of the coming of the Marines, moved off towards the buildings of the Pennsylvania Railroad, which caused a change to be made in the direction of the column, which then drove the miscreants before it as far as the new passenger depot, through which it passed, and continued up the tracks to the round-house. At this point they climbed the bluff on the west side, and, having stationed a guard, again drove back the crowd, now numbering thousands, to the first line of houses beyond the Callowhill Street bridge.

A line of pickets was established, and the hill-top permanently occupied, as it is a commanding position, overlooking the round-houses, tracks, and depots. About 2 P.M., General Hancock and staff and Colonel Thomas A. Scott, president of the Pennsylvania Railroad, visited the bivouac, and inspected the posts of the sentinels. The day was a very warm one, and the march from the arsenal so dusty that the command presented a sorry sight. Neither officers nor men had their clothes off since leaving Washington (three days and two nights before), and the former were very much gratified when some of the good people living within the lines invited them to their houses and tables, and extended bathing facilities. The mayor of Philadelphia, Hon. W. S. Stokley, and his chief of police also visited the camp, and presented a letter of introduction to Colonel Heywood from General Hancock, in which he was informed that in the event of further riotous proceedings the Marines were to act as a *posse comitatus,* under the orders of the mayor, for the preservation of the peace. Mr. Stokley cautioned him to look out for trouble soon, possibly that night. Here seemed to be a fair prospect of repeating the Baltimore experiences, but the police managed several rushes by the crowd to cross the bridge without asking for help. The men slept upon the ground, and the officers upon a neighboring porch, and next morning all hands were wet from dew, but there was no complaint. On the 24th, Brevet Major-General Brannan, United States army, lieutenant-colonel First Artillery, arrived, and assumed command of all the united forces in Philadelphia, by order of General Hancock, who had his headquarters, temporarily, at the St. George Hotel, and three companies of the Third Artillery joined the Marines. Passenger trains ran as usual, but no freight was moved. That night was passed as the previous one, but next day tents began to arrive. Captain H. B. Lowry, assistant quartermaster, United States Marine Corps, reported for duty, and Mr. Geo. W. Childs, with great kindness, sent each man a present of a pipe and a supply of tobacco, and the officers some liquid nourishment. Mr. James Paul also sent for the medical department some superior wines, whiskeys, and brandy, and Colonel Heywood received the following very encouraging despatch:

HEAD-QUARTERS, PHILADELPHIA, July 25, 1877.

COLONEL HEYWOOD, COMMANDING UNITED STATES MARINES, MARINE CAMP, WEST PHILADELPHIA:

Is there anything I can do for you in the way of commissary supplies or other matters to make your command comfortable? You have performed important services, which I shall recognize in a proper manner when I find more time.

HANCOCK, *Major-General.*

On the 26th fifty-five "A" tents were received, and were at once pitched in regular order, giving much needed shelter from the scorching sun, and the camp was christened "Camp Scott."

On the 27th, Lieutenants C. P. Porter and R. D. Wainwright, with a company of fifty-four men, joined the battalion through a misapprehension in issuing orders, as they were intended for the Washington Arsenal. General Hancock, therefore, ordered them back as far as Baltimore to report to General Barry. The admirable behavior of the men had now become a subject of remark, particularly as up to this time all hands were confined most stringently to the limits of the camp with arduous patrol and guard duties. The city police guarded the front of the depot, with orders to call on the Marines if they needed assistance, and their determined conduct, backed by the moral effect of the presence of the artillery and Marines, no doubt prevented any further outbreak. A mass-meeting of workingmen, advertised to be held in Kensington, on Saturday night, the 28th of July, was such a failure that it virtually marked the end of the strikes in Philadelphia.

On the 1st of August the only disorder remaining in Pennsylvania was in the coal regions, and the troops stationed at Reading, fifty-eight miles from Philadelphia, being needed there, it was determined to relieve them with Marines.

Colonel Heywood was thereupon ordered to provide his command with ten days' rations and two hundred rounds of ammunition, and proceed to that point by special train.

A ride of two hours up the valley of the Schuylkill brought them to Reading, where they were warmly welcomed. The men were quartered in some sheds attached to the railroad buildings, and the officers were allowed to retain the Pullman car in which they came, and that night obtained the first really good sound sleep since leaving Washington.

The following day, the principal men of the community, including the mayor, the Hon. Heister Clymer, member of Congress, and Mr. G. A. Nichols, vice-president of the Reading Railroad, called upon the officers and bade them welcome. At sundown the dress-parade drew thousands of visitors to the camp at the depot, and public curiosity seemed unable to satisfy itself. Every evening witnessed the same ceremony, and the interest and attendance increased until the local papers estimated the numbers present "as high as five thousand." The routine as to fatigue duties, formations, inspections, roll-calls, etc., was almost as minutely conformed to as in permanent garrison, and it was, without question, this strict discipline which gained the battalion such substantial and gratifying recognition from the people and from the military authorities.

At the head-quarters of the Military Division of the Atlantic it was never lost sight of. Before getting fairly settled the following telegram came:

PHILADELPHIA, August 3, 1877.

COLONEL HEYWOOD, COMMANDING MARINES, READING, PENNSYLVANIA:

If I can assist you in procuring supplies of clothing, medicine, ammunition, or any other supplies you may stand in need of, let me know, and I will be glad to do so.
HANCOCK, *Major-General.*

Colonel Heywood replied, "I am not in want of supplies of any kind at present. If I should need anything I shall accept your kind offer."

A few days after this, Colonel and Surgeon John M. Cuyler, of General Hancock's staff, and medical director of the division, visited the battalion to inquire into its comfort and wants, by order of the general, whose distinguished kindness and solicitude for the Marines at all times while under his command, as well as his subsequent most generous appreciation of their services in the official papers which are to follow, will long be remembered. Dr. Cuyler examined the temporary hospital, its conveniences, stores, appliances, etc., very carefully, but found no patients. He also inspected the men and their quarters, including an improvised bath-room in a tent, the cooking utensils, and arrangements, and said that everything was scrupulously clean and neat, and that he had nothing to suggest.

On the 12th, General Hancock again manifested his deep interest by sending General Easton, the quartermaster-general of the division, to Reading, to see if he could do anything for the Marines; owing, however, to the admirable arrangements in the quartermaster's department, by Captain Lowry, he found them in need of nothing,—contented and happy. On the 13th, Colonel Heywood was officially informed that "the division commander intends to express in formal orders, before you are relieved at Reading, his high appreciation of the valuable services and good conduct of the officers and men of your command. Should his intention through accident not be acted on before your departure, it will only be delayed for a brief period."

The organization of this battalion was as follows: Major and Brevet Lieutenant-Colonel Charles Heywood, commanding; First Lieutenant B. R. Russell, adjutant; Captain and Assistant-Quartermaster H. B. Lowry, quartermaster; Passsed Assistant-Surgeon A. M. Owen, surgeon; Passed Assistant Surgeon T. D. Myers, assistant surgeon; J. W. Herbert, first-class apothecary; First Sergeant Charles Lombardy, surgeon-major; Sergeant Henry Ulean, quartermaster-sergeant.

Company A: Captain George W. Collier, commanding; Second Lieutenant W. S. Schenck, three sergeants, five corporals, one drummer, one fifer, and twenty-nine privates. Total, thirty-nine.

Company B: First Lieutenant Aulick Palmer, commanding; Second Lieutenant J. T. Brodhead, three sergeants, four corporals, and thirty privates. Total, thirty-seven.

Company C: First Lieutenant A. C. Kelton, commanding; Second Lieutenant Jesup Nicholson, two sergeants, four corporals, and thirty-one privates. Total, thirty-seven.

Upon leaving Reading it was difficult to discover upon which side the regret was greater. A short and sudden campaign, begun in danger, privation, and fatigue, had ended in security and rest. The citizens of Pennsylvania had treated the strangers so well that in their behalf each man would have fought as three. The approbation of all superior officers had been won, the railroad authorities had been hospitable and attentive, in every quarter they had met with appreciation and commendation, and, although in triumph returning, they were yet sorry, as soldiers, that "the strikers' war" was over.

Before the departure of the First Battalion from Reading, Colonel Heywood and his command received the following complimentary order from head-quarters:

HEAD-QUARTERS MILITARY DIVISION OF THE ATLANTIC,
NEW YORK CITY, August 13, 1877.

GENERAL ORDERS No. 46.

The Marines now at Reading, Pennsylvania, under command of Colonel Heywood, when relieved by the detachment of United States artillery, ordered there for that purpose, will proceed to the Marine Barracks, Washington, District of Columbia, reporting to the proper naval authorities. The major-general commanding desires to express his high appreciation of the excellent conduct and soldierly qualities of the Marines during the entire period of the recent disorders. Citizens and soldiers are united in admiration of the soldierly bearing, excellent discipline, and devotion to duty displayed by them while at Baltimore, Philadelphia, and Reading. A copy of this order will be forwarded to the Honorable Secretary of the Navy, that he may be informed of the valuable services and arduous duties rendered by the Marines while serving in the Military Division of the Atlantic.

By command of Major-General Hancock.

(Signed) JOHN S. WHARTON,
*Captain Nineteenth Infantry, Act'g Ass't Adj't-General.*

Official.

(Signed) JOHN S. WHARTON,
*Captain Nineteenth Infantry, Aide-de-Camp.*

COLONEL HEYWOOD,
*United States Marines.*

Colonel Heywood was again the recipient, two months later, of a complimentary extract from the report of General Hancock to the adjutant-general of the army, as follows:

HEAD-QUARTERS MILITARY DIVISION OF THE ATLANTIC,
NEW YORK, October 26, 1877.

TO MAJOR CHARLES HEYWOOD, UNITED STATES MARINES, MARINE BARRACKS, WASHINGTON, D. C.:

SIR,—I am directed by Major-General Hancock to furnish you with the following extract from a special report made by him on the 24th inst. to the adjutant-general of the army:

HEAD-QUARTERS MILITARY DIVISION OF THE ATLANTIC,
NEW YORK CITY, October 24, 1877.

TO THE ADJUTANT-GENERAL, U. S. ARMY, WASHINGTON, D.C.:

SIR,—

* * * * * * * * * *

While all the officers are entitled to commendation in their respective spheres, I deem it incumbent upon me to mention the names of the following, who held commands at important points during the disturbances, and certain staff-officers as well, whose services were especially valuable.

MAJOR CHARLES HEYWOOD,
*United States Marines.*

* * * * * * * * * *

Very respectfully, your obedient servant,
(Signed) WINFIELD S. HANCOCK,
*Major-General Commanding Division.*

I am, sir, very respectfully, your obedient servant,
(Signed) JAMES B. FRY,
*Assistant Adjutant-General.*

## CHAPTER XXXIII.

The Labor Riots—Services of the Second Battalion—Separate Detachments.*

THE Second Battalion of Marines was commanded by Captain and Brevet Lieutenant-Colonel James Forney, and composed of men belonging to the Norfolk Barracks, to the monitor fleet, and to the several ships of the North Atlantic Squadron. During the night of July 26 it had marched from head-quarters of the Corps to the Washington Arsenal to replace the artillery force of Major McMillan, United States army, which had gone north by rail in the afternoon. On the

*Captain Cochrane in *United Service Magazine.*

morning of that day reports came of a strike among the employés of the Baltimore and Potomac Railroad, who up to that time had stood firm, thus preserving a line of communication with Baltimore. A heavy special train of regular troops was being made up to pass north over the road, and their commander, fearing interference at the tunnels and the long bridge over the Anacostia River, or eastern branch of the Potomac, points easily obstructed, requested that guards of Marines be stationed there. Accordingly, First Lieutenant H. C. Cochrane, with a detachment of thirty-five men, was sent to the bridge near the naval magazine, and First Lieutenant F. A. Mullany and Second Lieutenant G. F. Elliott, with similar detachments, to the tunnels. The bridge was examined, and, being found secure, sentinels were posted and possession held until the train had safely passed. The battalion under the command of Colonel Forney, after being joined by the companies of Captain H. J. Bishop and First Lieutenant H. H. Coston, numbered one hundred and sixty officers and men. Sentinels were promptly posted at the main entrance of the arsenal, along the water front, and over the principal buildings, and the Marines, with their marvellous aptitude for any condition of service, entered at once upon garrison duties. Next day the battalion was reported to Major-General John M. Schofield, United States army, who had been assigned to the command of the District of Columbia during the emergency, and remained undisturbed until the morning of the 28th, when it was ordered to the Baltimore Depot by the general, to be in readiness to proceed to Baltimore, which city was still restless. A small company, under First Lieutenant F. A. Mullany, was detached and sent back for the protection of the arsenal, which had been left unguarded, and at night, owing to the receipt of more pacific advices, Colone Forney was instructed to return with the others and await developments.

Early on the morning of the 29th the following telegram was received:

WAR DEPARTMENT, A. G. O.,
WASHINGTON, D. C., July 29, 5.30 A. M.

COLONEL FORNEY, COMMANDING DETACHMENT OF MARINES:

Hold your detachment in readiness to proceed to the depot, as it is possible it may be needed on the line of the Baltimore and Ohio Railroad.

(Signed) T. M. VINCENT, A. A. G.

At noon orders to move came, and at 3 P. M. the battalion was in Baltimore, when it was reinforced by First Lieutenant C. P. Porter's company, which had been sent back from Philadelphia by order of General Hancock. Colonel Forney reported to Brevet Major-General

W. F. Barry, United States army, who directed him to join Lieutenant-Colonel and Brevet Major-General R. B. Ayres, Third Artillery, the commanding officer of a post established at Camden Station, who assigned the companies to quarters on the platform of the depot, and the officers to a sleeping-car. Later in the day General Ayres notified Colonel Forney to detail ten non-commissioned officers and seventy-five privates for the purpose of guarding twenty-five freight trains, which were to be started the following day with the hope of getting them as far as Martinsburg, West Virginia.

General Barry's order to General Ayres, prescribing the details of this important duty, is as follows:

HEAD-QUARTERS UNITED STATES TROOPS,
CAMDEN STATION, BALTIMORE, MD., July 29, 1877.

LIEUTENANT-COLONEL R. B. AYRES, THIRD ARTILLERY, COMMANDING POST:

COLONEL,—The colonel commanding directs, inasmuch as the Baltimore and Ohio Railroad Company proposes to start twenty-five trains westward to-morrow morning, commencing at 10 A. M. that you have in readiness at 9.30 A. M. a guard of seventy-five men and a sufficient number of officers to be distributed among these trains as guards. The men will be provided with one day's cooked rations. A locomotive engine and car will be at this station at 9.30 A. M. to-morrow, to take the whole party to Riverside Station, from which place the freight trains will start, and where the guards will be distributed among the various trains by the senior officer of the detail. This guard will return from either Martinsburg or Cumberland on the first eastward-bound freight trains, acting as guards to them. The senior officer will be instructed that no train is to start without a guard.

Very Respectfully,
(Signed) CLARENCE O. HOWARD,
*First Lieutenant and Adjutant, Second Artillery.*

This order was referred to Colonel Forney to execute, with an endorsement by General Ayres, requiring him to have three commissioned officers ready to accompany the detail, which was to be in light marching order, with forty rounds of ammunition per man. The companies of Captain Bishop and First Lieutenant Porter were selected, and Lieutenant W. P. Biddle, the third officer.

Captain Bishop was instructed by General Ayres as follows: "Send two Marines on each locomotive, and scatter your command in detachments through the twenty-five trains. Fire on any man who attempts to desert his engine, and on anybody who interferes with the trains." Extra ammunition was issued to each man, and what could not be carried in the cartridge-boxes was stowed in the knapsacks. They took the train designated in General Barry's order and proceeded to Riverside Station, where was assembled a large force of the strikers and their sympathizers.

The twenty-five trains were made up without interruption, and the Marines distributed as prescribed. Lieutenant Porter was sent in the first train with twelve men, Lieutenant Biddle in a later one, and Captain Bishop, with the last detachment, embarked on the 25th. The engineers and firemen were looked upon with more or less suspicion by the railroad company, and many of them were new men.

Only a few days before, the strikers in the towns of Keyser and Grafton had declared that the road should not be opened, and Governor Mathews had in consequence called on the President for more troops. Companies E, D, and I, Fifth Artillery, were sent from Baltimore in response, and it was this disposition which caused the battalion of Marines to be sent from the Washington Arsenal to replace them. Knowledge of their declaration caused it to be feared that some of the trains would be fired into, or be thrown from the track, by the co-strikers, who were seen in groups along the road, and had many opportunities for doing serious mischief afforded them by the numerous bridges, cuts, curves, and tunnels for which the road is famous. The well-disposed people along the route seemed glad of the presence of the Marines, and cheered lustily, and after a twelve hours' run all the trains reached Martinsburg safely. The last got in at midnight, and the Marines, who less than ten days before had been pacing their monotonous beats on ship-board, hundreds of miles away, thus found themselves in the mountains of Virginia, and in the streets of the little city where the greatest labor strike in history had its origin. The opening of the road to this point having been confirmed, the eastward-bound trains were made ready, and at 2 A.M. next day Lieutenant Porter started back with a convoy. At daylight Lieutenant Biddle followed, and at 9 A.M. Captain Bishop again brought up the rear. Baltimore was reached without accident or incident, and the captain had the pleasure of being complimented by General Ayres upon the success of the expedition, with request to convey the general's thanks to his command.

It was then learned that at about the same hour of the night that they reached Martinsburg a train on the next western division of the railroad had been thrown from the track by rioters while transporting troops between Keyser and Piedmont, severely wounding a soldier named Hamilton, of Battery E, Second Artillery, and slightly wounding several others.

Colonel Forney was again ordered, August 1, to make a detail for service as train-escort, and on the 2d, Lieutenant G. F. Elliott, one non-commissioned officer, and five privates were sent to guard the paymaster of the road from Baltimore to Martinsburg, with funds to pay off the employés. This duty was successfully performed, and upon his return Lieutenant Elliott rejoined the battalion at Fort McHenry, to which point it had proceeded at noon of the 2d, by order of the Secretary of War. While at the fort the battalion performed the usual garrison duties, with daily drills, parade, guard-mounting, etc. Each night, until the 6th of August, it was considered necessary to send a special detail, consisting of one officer, two non-commissioned officers, and eighteen privates, to guard the grain-elevators and wharves of the Baltimore and Ohio Railroad at Locust Point from retreat until reveille, which duty was performed by Lieutenants Coston, Elliott, L. C. Webster, and Biddle.

The roster of the battalion at this time was as follows: Captain and Brevet Lieutenant-Colonel James Forney, commanding; First Lieutenant G. F. Elliott, adjutant; Assistant-Surgeon E. H. Green, surgeon; First Sergeant H. Carlisle, sergeant-major; Sergeant J. E. Dudley, quartermaster-sergeant. Company A: Captain H. J. Bishop, commanding; Second Lieutenant L. C. Webster, three sergeants, five corporals, one drummer, one fifer, thirty-four privates: total, forty-four. Company B: First Lieutenant H. H. Coston, commanding; Second Lieutenant W. P. Biddle, two sergeants, five corporals, one drummer, one fifer, thirty-four privates: total, forty-three. Company C: First Lieutenant C. P. Porter, commanding; two sergeants, four corporals, one drummer, one fifer, thirty-five privates: total, forty-three. Company D: First Lieutenant R. J. Wainwright, commanding; two sergeants, four corporals, one drummer, one fifer, thirty-seven privates: total, forty-five.

General Order No. 45, August 15, announced that the "Marine battalion would be relieved and proceed to Washington by rail at 4 P.M." In this order General Barry said, "The colonel commanding the United States troops in and near Baltimore avails himself of the present occasion to express in this public manner the recognition of Major-General Hancock, and of himself, of the valuable services rendered by the various detachments of United States Marines while serving with the army in this command."

At four o'clock the battalion marched out of the fort and returned to head-quarters, where it was preceded only a short time by the battalion of Colonel Heywood, from Pennsylvania.

The company which garrisoned the Washington Arsenal was composed of two detachments, one of twenty-five men, under Lieutenant Mullany, and the other of two sergeants, two corporals, and twenty-five privates, under Lieutenant H. C. Cochrane.

The composition of the detachment of Marines was highly illustrative of the mobility of the service. Fifty-four enlisted men represented the head-quarters of the Corps, the Norfolk Barracks, the frigate Franklin, the steam-sloops Hartford, Powhatan, Plymouth, Ossipee, and Swatara, and the ironclads Ajax, Saugus, Lehigh, Mahopac, Manhattan, and Catskill,—fourteen different commands,—and yet, in three days, no person could have told that they had not served together for years. On the afternoon of August 9, the companies of seamen were embarked on their respective ships, and Lieutenant Cochrane detached from the command of the Marines, and ordered back to the Plymouth, being relieved by Captain Hamilton's battery of the Second Artillery.

The guard at the United States Arsenal, at Watervliet, New York, being very small and insufficient for its protection, Captain F. H. Corrie was ordered at 5 P.M., July 24, to report to Captain D. L. Braine, United States Navy, commanding the Colorado. At midnight, Captain Corrie left the New York Navy-Yard, on the steam-tug Fathomer, with Lieutenant Henry Whiting and fifty men, detailed from the guard of the Colorado. Captain Corrie's orders were marked "confidential," and enjoined upon him to use all the means in his power to protect the government property at the place for which he was destined. To conceal all knowledge of this movement, the tug was ordered to steam up the Hudson River under cover of night, and at daylight land at the nearest town, to transfer the Marines to the morning express bound north. This was done, and the company was landed at Dobb's Ferry, a small station twenty-two miles from New York. Discovering that the express train did not usually stop, authority was asked, which was promptly granted, and an extra car attached for their use. At Poughkeepsie trouble was expected. A large mob was found at the depot, some of whom got on the platform of the car containing the Marines, and looked in to get their measure. The spokesman turned and said, "Well, them fellers look like reg'lars; guess we don't want nothin' to do with them," and got off. Another said to Captain Corrie, "Where are you going?" He replied, "Up the road." Next, "And what are you going to do?" "Obey orders." This was altogether too indefinite, and he retired.

No attack was attempted, and at 2 P.M. the company reached West Troy safely, and marched to the arsenal. The company remained at the arsenal until the 20th of August, when, in obedience to a telegraphic order, Captain Corrie returned with his command to New York. General Hagner, the commandant of the arsenal, in post orders, acknowledged "the satisfactory manner in which the duties assigned to the Marine detachment had been performed," and in his order to Captain Corrie, said, "It gives me pleasure to add, in wishing yourself and Lieutenant Whiting a safe return to your regular station, my personal acknowledgments for our very agreeable social intercourse." General Hancock, in special order No. 189, remarked, "The division commander is gratified to announce that Colonel Peter V. Hagner, commanding Watervliet Arsenal, has commended the officers and men of this detachment for their cheerful assistance and soldier-like conduct in discharge of the duties assigned them."

The important arsenal at Frankford, in Philadelphia, was almost as defenceless as that at Watervliet, and in a much more exposed condition.

On July 22, Colonel Thomas A. Scott, president of the Pennsylvania Railroad Company, informed Captain C. H. Welles, commandant of the League Island yard, that he had evidence that an attempt would be made by the rioters to secure arms and ammunition from the Frankford Arsenal.

Captain Welles acted promptly, and detailed Captain W. R. Brown, of the Corps, to proceed with fifty Marines, forty rounds of ammunition, and one field-piece, on the steam-tug Pilgrim to the arsenal. He also ordered Captain Adams, commanding the training-ship Constitution, to send a party of Marines, under Lieutenant J. B. Breese, to the naval magazine at Fort Mifflin. An extra guard of Marines, under First Lieutenant S. K. Allen, was stationed at the main gate of the yard with a field-howitzer. Captain Brown, Lieutenant George T. Bates, and the detachment arrived, and were quartered in the grounds. The Marines remained one month at Frankford, and returned to League Island August 21.

Major Whittemore, the commandant of the arsenal, in his order for the detachment to return, expressed his "thanks to the commander, and the men composing it, for the services rendered by them when most needed." Captain Brown and his detachment were specially mentioned in General Hancock's order No. 189.

Before disbandment, the First and Second Battalions were consolidated, under the command of Colonel Heywood, for a street parade and review by the Hon. R. W. Thompson, Secretary of the Navy. Eight companies, in fatigue uniform, numbering over three hundred and fifty men, marched from the head-quarters barracks up Pennsylvania Avenue to the Navy Department, on Seventeenth Street, and gave a merching salute to Secretary Thompson, the several chiefs of bureau, and Colonel Charles G. McCawley, commandant of the corps, who were assembled on the west steps of the building. The Marine Band, fifty-two pieces, and a drum corps of twenty-two pieces accompanied them, and the papers of the day spoke highly of the precision of the marching, as well as of the fine order of the arms and other equipments of the entire command.

The Secretary issued the following congratulatory letter:

NAVY DEPARTMENT, August 17, 1877.

SIR,—The Department has received from Major-General Hancock an official order, wherein he bestows the highest praise upon the Marines who have just returned from Maryland and Pennsylvania, under command of Major and Lieutenant-Colonel by Brevet Heywood and Captain and Lieutenant-Colonel by Brevet Forney. The major-general expresses his high appreciation of the excellent conduct and soldierly qualities of both officers and men, and adds that citizens and soldiers are united in admiration of the soldierly bearing, excellent discipline, and devotion to duty displayed by them while at Baltimore, Philadelphia, and Reading.

The Secretary of War also forwarded to the Department a copy of a report from Colonel and Surgeon J. M. Cuyler, medical director of the Division of the Atlantic, who, in accordance with instructions, inspected the sanitary conditions of the Marines at Reading, Pennsylvania, in which Medical-Director Cuyler says,—

"The sanitary state of the command is excellent, and the officers evidently take great pride in looking after the health and comfort of the men. I do not recollect of ever having seen a more soldierly set of men, or a more orderly set, proving that they are under most excellent discipline. These men seem to be so well taken care of by their officers that I really cannot see what the major-general commanding can do to add to their health and comfort. It is quite remarkable that men performing such service are able to keep themselves and their arms, etc., so very clean and neat."

This was to be expected from the well-established character of the Marines, but is none the less gratifying to the Department, and it desires to express through you to both officers and men its high appreciation of them, and to assure them of its readiness to co-operate in the adoption of any measure necessary to their comfort and an increase of their efficiency. They are a most important arm of the national defence, and the readiness with which they have responded to the recent summons of the Department to aid in the suppression of unlawful combinations of men, no less than their proficiency in discipline, prove that they may confidently be relied upon whenever the public exigency shall call them into active service.

Very respectfully, R. W. THOMPSON,
*Secretary of the Navy.*

COLONEL CHARLES G. MCCAWLEY,
*Commandant of the Marine Corps, Washington, D. C.*

## CHAPTER XXXIV.

**1878-82.**—Universal Exposition—Marines at Alexandria.

In 1878, during the Universal Exposition at Paris, a company of Marines under Lieutenants B. R. Russell and W. F. Zeilin, were sent out in the Constitution for duty in that city. They had charge of the American Department, and the Hon. R. C. McCormick, commissioner-general for the United States, reported to the Navy Department that "their excellent conduct, both on and off duty, was spoken of in the most complimentary terms by the French authorities, by the citizens of Paris, and the visitors of the Exposition," and at the ceremonies attending the distribution of prizes, in October, their military bearing elicited "a grand outburst of enthusiasm, and they throughout reflected honor upon our flag by their admirable performance of an important duty in a foreign land."

**1882.**—After a lapse of seventy-seven years a detachment of Marines again demonstrated their efficiency and gallantry amid scenes of riot and bloodshed in the Land of the Pyramids.

Much surprise had been excited by the apparent supineness of the French and English naval forces in the harbor of Alexandria on the occasion of a serious outbreak in that city on the 11th of June, 1882, and which resulted in the destruction of the greater part of the European quarter and the death or wounding of more than two hundred European residents. The outbreak was in no sense an ordinary disturbance of the peace. Its promoters and agents were Egyptian soldiers, its fury was directed almost exclusively against European residents, and included in its havoc civil and military functionaries of the very powers whose guns frowned voicelessly upon the scene. Whatever the reason, whether military or diplomatic, for the inaction of the forces, it was none the less a painful ordeal for the gallant men who lay that Sunday afternoon in the harbor of Alexandria undergoing a harder battle with their knightly impulses than any that Egyptian fellaheen could have offered them.

The flag-ship Lancaster, Captain B. Gherardi, bearing the flag of Rear-Admiral J. W. A. Nicholson, arrived on the 27th of June at Alexandria. It was found that the English admiral was making preparations to administer condign punishment to Arabi Pacha and his troops. The Lancaster became a refuge for men, women, and children, and her decks were crowded with those unfortunate and innocent foreigners residing in the city. On the 11th were enacted those scenes of horror so graphically described by the newspapers of the day,—pillaging, conflagrations, rapine, etc. It became necessary, therefore, that armed forces should occupy the city as soon as possible, and the naval representatives of the different nationalities responded promptly. On the 14th a detachment of naval artillery, under the command of Lieutenant-Commander Goodrich, consisting of Lieutenants Hutchins and Graham, Midshipmen Dent and Capeau, sixty-nine sailors and two Gatlings; also a detachment of Marines, under the command of Captain H. C. Cochrane, consisting of Lieutenants F. L. Denny and L. W. T. Waller, and sixty non-commissioned officers, musicians, and privates, landed and marched through hot and smoky streets, filled with *débris* from fallen walls and merchandise from plundered shops, and lined with burning stores and dwellings. They finally reached the Grand Square of Mehemet Ali, and occupied the building containing the American consulate. Their presence was immediately felt and recognized. *They were the first troops in the centre of the city,* which they thoroughly patrolled, having sentinels posted at the bourse, telegraph offices, and banks that had not been pillaged.

On the morning of the 15th the majority of the force was withdrawn, leaving on shore Lieutenant Hutchins, Master Burnett, Midshipman Dent, two sailors, with one Gatling, Lieutenants Denny and Waller, in command of twenty-five Marines. This detachment was withdrawn on the 18th, with the exception of Lieutenant Denny and six Marines, who remained as a guard to the consulate until the 24th.

Information was received during our occupation that Arabi Pacha was returning with a large force to attack the city. The French and Italians hastened back to their vessels, leaving the English and American Marines and sailors to receive the rumored night attack.

The correspondent of the London *Times,* at Alexandria, telegraphed as follows: "Lord Charles Beresford states that without the assistance of the American Marines he would have been unable to discharge the numerous duties of suppressing fires, preventing looting, burying the dead, and clearing the streets."

Colburn's *United Service Magazine* for September, of that year, has the following note: "But the Americans did not limit their friendly actions to expressions of sympathy. After Arabi's retreat to Kafir Dowar, it was rumored that he intended to attack Alexandria in force; on learning which the other nationalities prudently withdrew to their ships. The American Marines, however, sixty in number, expressed their determination to 'stick by the British and take their chances,' and, small as were their number, and great as was the risk, they did stick as long as was necessary."

Under date of October 24, 1882, the following letter was addressed to Lieutenant Denny, from Eaton Square, London:

It gives me great pleasure to testify to your very valuable services and true courage during those irksome, terrible days after the bombardment of Alexandria. I can confidently say without the services of American troops order could not have been restored, and the great fires in the city subdued. To Lieutenant Hutchins, and you, and your smart, faithful force great credit is due. I have represented these facts to my government. . .

(Signed) CHARLES BERESFORD.

## CHAPTER XXXV.

### 1885.—The Expedition to Panama.

In accordance with the terms of the treaty of 1848 with New Grenada, now one of the Colombian states, the United States of America guaranteed the protection of traffic across the Isthmus of Panama.

The political complication which resulted in the revolution of 1885 may be briefly stated as follows:

The president of each of the eight states forming the United States of Colombia, as well as the national executive, are elected for terms of two years. Each state is independent in the management of its local affairs. The elections in Panama occurred in the summer of 1884, and two candidates were balloted for,—viz., Señors Arosemena and Lambert. The former is said to have had the support of the government at Bogota. Although the results of the balloting were disputed, it was claimed that Señor Lambert had been elected.

The Constitutional Assembly, composed of thirty-two members, met on the 6th of January, 1885, decided that there had been no election, and proceeded to elect a president, as provided for by law.

General Santo Domingo Vila, who came to the Isthmus in November, 1884, as a fiscal agent of the general government to examine into the contracts between the Panama Railroad Company and the state, was chosen president of the state of Panama January 7, 1885, by the Constitutional Assembly. Señors Arosemena and Vivas Leon were chosen or appointed first designado and second designado.

The office and duties of the designado are similar to those of vice-president, and in the event of death or absence of the president they succeed successively to the presidency.

The other officers of the state government, such as secretary of state, governor of Panama, prefects, etc., are appointed by the president. The commander-in-chief of the Colombian (national) forces is appointed from Bogota, and that office was filled by General Gonima.

The steamer Boyaca, which later on performed valuable service for the national government, was lying off Panama, formed part of Gonima's force, and was commanded by Colonel Ulloa.

In February a portion of the national forces stationed on the Isthmus was sent to Buenaventura, the seaport of the state of Cauca, about three hundred miles south of Panama, to aid in suppressing the revolution which had broken out in that state. About the 1st of March, General Vila, obtaining two months' leave of absence, sailed himself with more national troops for Carthagena, to help suppress the rebellion in the state of Bolivar. The strength of the national forces having been thus reduced in Panama, those who were discontented with the result of the election took opportunity to attempt a revolution.

First Designado Arosemena had succeeded to the presidency of the state in the absence of Santo Domingo Vila, and General Gonima, the commander-in-chief, was in Colon.

General Aizpuru, who had been president of the state of Panama for one term, in 1874, and who had served in the national legislature, was recognized as the leader of the liberal party in the state. On the 16th of March he made a demonstration against the government, which caused Señor Arosemena to take refuge on her British Majesty's ship Heroine, then lying off Panama; broke open cars, opened and blocked switches, obstructed the road, prevented repairs to the same, cut telegraph wires, seized and held railroad employés, claimed the right to exercise a censorship over the telegrams, and made it necessary to close the transit. General Gonima, hearing of this demonstration on the part of Aizpuru, proceeded by train to Panama with the national force then stationed at Colon. When Gonima arrived, on the 17th of March, Aizpuru retired from the city, and Señor Arosemena returned from the Heroine.

Colon was now without troops. Prestan, a Haytien negro, with a trace of white blood, the leader of a faction of the radical wing of the liberal party, took advantage of the opportunity and seized that city. Aizpuru and Prestan were pronounced in their dislike to all foreigners on the Isthmus, especially to the Americans living there, and this feeling was made use of to incite their adherents and hold them together.

About the 20th of March, Arosemena resigned. Vivas Leon should then, in the absence of Santos Domingo Vila, have succeeded to the presidency of Panama, but was prevented by General Gonima, who declared himself the "Military and Civil Chief of Panama." This title and office are established by law; the national government may, under certain circumstances, make such an appointment. It is very questionable, however, whether General Gonima had a right to seize the reins of government, more particularly since the second designado, Señor Vivas Leon, was in Panama, and should, according to precedent and law, have succeeded to the presidency.

On the night of the 30th of March, General Gonima sent Colonel Ulloa (then commanding the Boyaca) by train to Colon, with a portion of the troops in Panama, to put down Prestan's revolution. In order to prevent a conflict within the limits of Colon, Mr. Burt, the superintendent of the Panama Railroad Company, notified Colonel Ulloa that he must disembark at Monkey Hill, about two miles outside of Colon. This was also made necessary by Prestan, who, hearing of the departure of national troops, went out with his force to meet Ulloa, tearing up the railroad tracks between Colon and Monkey Hill.

The two forces, numbering about one hundred and fifty men each, met on the Panama side of Monkey Hill. After a short conflict Prestan was driven from his position into Colon and behind his barricades in that city. After fighting for several hours on the morning of April 1, the insurgents were dislodged and put to flight by the national forces, led with great gallantry by Colonel Ulloa and his second in command, Colonel Broun, who was chief of police at Panama. Both officers were severely wounded in the engagement. During the conflict the city was fired by the insurgents and destroyed.

The departure of Colonel Ulloa and his command from Panama still further reduced the force of that city; and on the day of the fight in Colon, April 1, Aizpuru, with the ammunition taken on the 16th of the preceding month from a sealed car in transit to Central America, captured Panama. General Gonima, with less than one hundred soldiers in the cuartel, made a gallant defence, and held out for some time against a much larger force. He finally surrendered to Aizpuru to prevent the destruction of the cuartel by dynamite taken from contractors employed by the canal company and used for blasting.

When General Gonima surrendered, the Boyaca was included in the stipulations, but Señor José Obaldia, who was in command of her refused to be governed by the terms of the surrender, and sailed for Buenaventura, then held by the national government. General Aizpuru, having seized the government, declared himself president of the state of Panama, and filled the offices with his friends.

Colonel Ulloa, with less than one hundred national troops, held Colon; the rest of the Isthmus included in the zone of transit was held by the insurgents under Aizpuru.

This was the condition of affairs upon the arrival of the First Battalion of Marines at Colon, in the Pacific Mail Steamship Company's steamer City of Para, on the 11th of April, 1885.

On Thursday, April 2, at 12 M., an order was received by Colonel-Commandant C. G. McCawley from the Navy Department to detail a battalion of Marines to sail the next day on the City of Para for Colon. Within twenty-four hours after the receipt of the telegraphic order, this battalion, organized from officers and men detached from Portsmouth, New Hampshire, New York, Philadelphia, and Washington, sailed in the City of Para from New York, fully equipped, commanded by Brevet Lieutenant-Colonel Charles Heywood, United States Marine Corps, a veteran of the last war, who came out of that war brevetted *twice* for distinguished gallantry in the presence of the enemy.

The battalion arrived at Colon on the night of the 11th, and disembarked the next morning at six o'clock, in obedience to the following order:

UNITED STATES FLAG-SHIP TENNESSEE (first rate),
ASPINWALL, U.S.C., April 11, 1885.

SIR,—Proceed to Panama with the battalion of Marines under your command, for the protection of American lives and property in that vicinity. The details of this service are left to your discretion.

Panama is now in the hands of the revolutionary forces, and it is feared that if the place is attacked by the regular Colombian troops, these revolutionists will attempt to destroy the city, or portions of it, by burning. As the burning of Panama would involve the destruction of much American and other foreign property, you will prevent it if possible.

Please advise Captain Norton, commanding United States ship Shenandoah, now at Panama, of your arrival there.

I enclose herewith a copy of my telegraphic instructions from the department, and also some extracts of a sworn protest by Mr. George A. Burt, general superintendent of the Panama Railroad Company, for your information.

I have detailed a Gatling gun and a 12-pounder S. B. howitzer, with officers and crew, for service with your command.

The officers are directed to report to you.

Very respectfully, JAS. E. JOUETT,
*Rear-Admiral, Commanding U. S. Naval Force on N. A. Station.*

MAJOR CHARLES HEYWOOD, U.S.M.C.,
*Commanding Marine Battalion, Isthmus of Panama.*

At six o'clock the same night, Colonel Heywood, with the First Battalion, occupied the railroad company's buildings at the Panama end of the line, forty-seven and a half miles distant, situated outside the walls of the city, *and the transit was opened.*

When the battalion arrived at Panama, the city was convulsed, and great excitement ensued. The Marines were confronted by a desperate and turbulent mob, but, owing to the firm bearing and formidable appearance of these troops, an attack was not made by the insurgents as was apprehended.

The transit was kept open by a detachment on every train, and, although attempts were made by the insurgents to attack the trains at various points, a few well-directed shots, together with the admirable discipline of the men, prevented any disaster.

Although the Marines were scattered along the coast from Maine to Florida, a second battalion was quickly detailed and equipped, and sailed in the steamer Acapulco, on Tuesday, April 7, under command of Captain J. H. Higbee, United States Marine Corps. With this battalion also sailed a detachment of one hundred and fifty sailors, the whole under command of Commander B. H. McCalla, United States Navy, in accordance with the following order:

NAVY DEPARTMENT, WASHINGTON, April 5, 1885.

SIR,—You will proceed with a detachment of Marines and blue-jackets, per Pacific Mail steamer Acapulco to Aspinwall and report to the senior naval officer present for duty in connection with opening the transit and protecting and caring for the lives and property of American citizens upon the Isthmus of Panama. You will have command of the entire force sent from New York for the above purpose, subject to the orders of the admiral commanding the North Atlantic Squadron, or of the senior officer present, the force sent by the City of Para and by the Acapulco being considered as your particular command.

You will take every precaution to protect the health and lives of the officers and men placed under you, and while taking all the necessary steps to have your command in the highest state of efficiency, you will expose it to the trying climate of the Isthmus no more than is absolutely necessary.

Very respectfully, W. C. WHITNEY,
*Secretary of the Navy.*

COMMANDER BOWMAN H. MCCALLA, U.S.N.,
*Navy-Yard, New York.*

The Acapulco came to anchor off Colon at 9 P.M. of the 15th of April. During that day, in order to prepare for landing at once in case of necessity, two days' rations were served out, and the ammunition-boxes of the 3-inch rifles and Gatlings were supplied with shell, shrapnel, and small-arm ammunition. Each man carried forty cartridges, his blanket rolled with a change of clothing, canteens, and haversacks.

Early the following morning the Acapulco hauled alongside Wharf No. 1 of the Panama Railroad Company, which had been partially rebuilt, and at nine o'clock Rear-Admiral Jouett inspected the command, expressing himself as well pleased with its appearance.

Rear-Admiral Jouett had arrived on the afternoon of the 10th of April, in the flag-ship Tennessee, accompanied by the Swatara. The

Marine garrison stationed at Pensacola, Captain R. L. Meade, Second Lieutenant Arthur H. Clark, seven non-commissioned officers, twenty-eight privates, and two musicians came down in the flag-ship.

The following order was issued by Rear-Admiral Jouett:

North Atlantic Station,
United States Flag-Ship Tennessee (first rate),
Aspinwall, U.S.C., April 15, 1885.

Sir,—By direction of the Navy Department you will please assume command of the forces which have been sent from New York by steamer for the protection of American interests on the Isthmus of Panama. You will also include in your command all the forces which have been landed from the North Atlantic Squadron.

You will land the men now on board the Acapulco at your own discretion, as soon as they can be properly accommodated on shore.

The transit is open and trains have been running regularly since the 11th instant. Aspinwall is held by the Colombian troops, and Panama is in the hands of the insurgents.

I enclose for your information a memorandum showing the number and disposition of the forces on shore; also copies of a squadron general order relative to sanitary precautions to be observed, and of my orders to Lieutenant Colahan and Colonel Heywood.

Very respectfully, Jas. E. Jouett,
*Rear-Admiral, Commanding U. S. Naval Force on N. A. Station.*

Commander B. H. McCalla, U.S.N.,
*Pacific Mail Steamer Acapulco, Aspinwall, U.S.C.*

On the 16th the following communication was addressed to Commander McCalla by Rear-Admiral Jouett:

North Atlantic Station,
United States Flag-Ship Tennessee (first rate),
Aspinwall, U.S.C., April 16, 1885.

Sir,—I send enclosed for your information a copy of my instructions from the honorable Secretary of the Navy relative to the purpose for which our government has sent so extensive an expedition to the Isthmus of Panama, together with a copy of a letter which I addressed, upon my arrival here, to Colonel Ramon Ulloa, of the Colombian Guard, the only representative of the constitutional government of Colombia who was in this vicinity at that time.

You will please require your command to carefully comply with the Department's instructions relative to the conduct of our forces while they remain on the Isthmus.

As I have already informed you, the city of Panama is now in the hands of the revolutionary forces, commanded by one Aizpuru, who assumes the title of president of the state. I do not wish to recognize the authority of the said Aizpuru in any way.

The regular Colombian forces have recently been very successful in the interior of Colombia, and I trust they will soon come to this Isthmus, re-establish the authority of the constitutional government, afford adequate security to life and property, and enforce order.

It is rumored that Aizpuru has threatened to burn Panama in case the government forces attack him there. As the burning of Panama would cause the destruction of much property belonging to Americans, interrupt the transit, besides being an act of vandalism, you will prevent it, if possible, with the force under your command.

I gave orders to this effect to Colonel Heywood before you assumed command of the force on the Isthmus.

I also learn, by rumor, that there are a number of evil-disposed persons at Matachin. You will, therefore, please direct the officer commanding that post to be vigilant, in order that excesses may be prevented in that vicinity.

I do not intend to allow any arms or ammunition to be introduced into this country during the present disturbance. It is reported to me that one hundred and forty-five boxes of cartridges have recently been landed here for parties on the Isthmus by the steamer City of Para. I shall have them sent back to New York by the same steamer. You will please see that all freight arriving at the Isthmus is examined by some officer of your command, and will not allow arms or ammunition to be landed without my authority.

In order to preserve the strict neutrality of the Isthmus of Panama, and to avoid interruption to the transit, you will please prevent any insurgent force from landing or operating in this vicinity.

Please keep me fully informed in regard to all matters affecting your command and our interests on the Isthmus.

Send daily sick report. Very respectfully, JAS. E. JOUETT,

*Rear-Admiral, Commanding U. S. Naval Force on N. A. Station.*

COMMANDER B. H. MCCALLA, U.S.N.,

*Commanding Naval Force, Isthmus of Panama.*

In obedience to the general instructions received from Rear-Admiral Jouett, the following orders were issued:

HEAD-QUARTERS U. S. NAVAL FORCE ON THE ISTHMUS OF PANAMA,
COLON, April 16, 1885.

SIR,—Proceed to Matachin, in a special train that will be provided by the Panama Railroad Company to follow the 3 P.M., with the company now under your command, and relieve Lieutenant Impey and the garrison now at that place.

I have directed the commanding officers of a section of artillery and of a Gatling gun to report to you as part of your command.

Your duty will be to keep the transit open, to protect the lives and property of American citizens, and to have your command in the highest state of efficiency. Take every precaution to preserve the health and lives of your command.

Prohibit the use of wines, spirituous and malt liquors. Dispense with all unnecessary ceremony and parades. Instruct your men in the early morning and late in the afternoon, avoiding the sun in the middle of the day. Take with you three thousand rounds of rifle ammunition in excess of what has been served out. I enclose for your guidance a copy of "Notes on Sanitary Precautions" and a copy of General Order No. 15, North Atlantic Station.

Very respectfully, B. H. MCCALLA,

*Commanding U. S. Naval Force on the the Isthmus of Panama.*

CAPTAIN R. W. HUNTINGTON, U.S.M.C.,

*Commanding Company A, Second Battalion U. S. Marines, Colon.*

HEAD-QUARTERS U. S. NAVAL FORCE ON THE ISTHMUS OF PANAMA,
COLON, April 16, 1885.

SIR,—You will relieve the command at Garrison No. 1 at 5.30 to-day. You will dispose of your force so as to protect the water approach to Colon. A 3-inch rifle with thirty rounds shrapnel, twenty rounds shell, and the necessary primers and fuses will be issued to you.

Place a commisioned officer in charge of this gun, and direct him to instruct in its use the men necessary to work it. Issue forty rounds of ammunition to each man, and take one thousand rounds additional. Protect the lives and property of American citizens. Do not expose your men to the sun in the middle of the day. Instruct them early in the morning and late in the afternoon, and take every precaution to protect the health and lives of your command. Prohibit the use of wines, spirituous and malt liquors. Enclosed you will find for your guidance General Order No. 15, North Atlantic Station, and "Notes on Sanitary Precautions."

Continue to carry out the instructions now in force at Garrison No. 1.

Very respectfully, B. H. MCCALLA,
*Commanding U. S. Naval Force on the the Isthmus of Panama.*

CAPTAIN R. S. COLLUM, U.S.M.C.,
*Commanding Company B, Second Battalion, Colon.*

HEAD-QUARTERS U. S. NAVAL FORCE ON THE ISTHMUS OF PANAMA,
COLON, April 16, 1885.

SIR,—Order Captain George C. Reid, with his company, to report to Lieutenant Mason for duty in connection with the protection of trains between Colon and Panama; with the remaining two companies of Second Battalion relieve Nos. 2 and 3 garrisons at 5.30 P.M. to-day. Acquaint yourself with the disposition of the naval forces in the vicinity of Colon, as well as the approaches to that place.

Garrison No. 3 is head-quarters. In my absence you will have command of the entire naval force at Colon. Your duty is to protect the lives and property of American citizens. There is a force of one hundred men of the Colombian government in Aspinwall. Carry out the orders now in force in the garrison. Prohibit the use of all wines, spirituous and malt liquors in your command, and take every precaution to protect the lives and health of your officers and men. Avoid unnecessary parades and ceremonies. Keep your men out of the sun in the middle of the day as much as possible. Instruct them in the early morning and late in the afternoon. Issue forty pounds of ammunition to each man and take one thousand rounds extra with you for each company. I enclose for your guidance copies of General Order No. 15 and "Notes on Sanitary Precautions."

One Gatling gun with one thousand rounds and caisson will be assigned to you. Be pleased to place it in charge of a commissioned officer and direct him to instruct the men necessary to its use.

Very respectfully, B. H. MCCALLA,
*Commander, Commanding U. S. Naval Force.*

CAPTAIN J. H. HIGBEE, U.S.M.C.,
*Commanding Second Battalion Marines.*

HEAD-QUARTERS U. S. NAVAL FORCE ON THE ISTHMUS OF PANAMA,
COLON, April 17, 1885.

SIR,—You will relieve Lieutenant Hosley, in command of No. 4 garrison, at 5.30 to-day, taking with you the remainder of your battalion.

Protect the lives and property of American citizens. Take every precaution to protect the health and lives of your command. Keep your men out of the sun in the middle of the day. Instruct them early in the morning or late in the evening, and avoid all unnecessary parades and ceremonies. Prohibit the use of wines, spirituous and malt liquors. Enclosed you will find for your guidance General Order No. 15, North Atlantic Station, and "Notes on Sanitary Precautions." Acquaint yourself with the regulations in force at the garrison and carry them out until further orders.

With regard to this garrison, the importance of guarding the causeway by which an insurgent force would be apt to approach is called to your attention.

Very respectfully, B. H. MCCALLA,
*Commanding U. S. Naval Force on the the Isthmus of Panama.*

LIEUTENANT C. O. ALLIBONE, U.S.N.,
*Commanding Gatling Battalion, Colon.*

HEAD-QUARTERS U. S. NAVAL FORCE ON THE ISTHMUS OF PANAMA,
COLON, April 16, 1885.

SIR,—You will relieve Lieutenant Kimball in command of the train guards in time to place guards on the early train to-morrow morning. Captain George C. Reid, United States Marine Corps, has been directed to report to you with the company under his command. In addition, you will have the two remaining sections of your battalion.

Acquaint yourself with the duties now performed by Lieutenant Kimball's command. Keep the transit open. Protect the lives and property of Americans. Enforce the regulations of the railroad company. Consult with Mr. Burt, the superintendent of the Panama Railroad, and do anything in your power to aid him and the officials of the trains in guaranteeing safe conduct to passenger and freight. Take every precaution to preserve the health and lives of your command. Omit unnecessary ceremonies and parades. Prohibit the use of wines, spirituous and malt liquors. Protect the men from the sun as much as possible in the middle of the day, instructing them early in the morning and late in the afternoon. Your attention is called to the necessity of exercising tact in the important duty assigned to you.

Enclosed you will find for your guidance General Order No. 15, North Atlantic Station, and "Notes on Sanitary Precautions."

Very respectfully, B. H. MCCALLA,
*Commanding U. S. Naval Force on the the Isthmus of Panama.*

LIEUTENANT T. B. M. MASON, U.S.N.,
*Commanding Battalion of Rifled Guns, Colon.*

At 3 P.M., Captain Huntington departed with his company, by special train, for Matachin. At 4 P.M. the remainder of the Second Battalion and the detachment of sailors disembarked, and at 5.30 P.M. were distributed as follows:

COLON GARRISON.

Captain J. H. Higbee, United States Marine Corps, commanding; First Lieutenant W. F. Spicer, United States Marine Corps, adjutant.

*Post 1.*—Captain R. S. Collum, United States Marine Corps, commanding; First Lieutenant S. H. Gibson, United States Marine Corps; and Second Lieutenant H. K. Gilman, United States Marine Corps; Company B, Second Battalion, fifty men, and one 3-inch B. L. R.

*Post 2.*—First Lieutenant F. H. Harrington, United States Marine Corps, commanding; First Lieutenant Henry Whiting, United States Marine Corps; Company E, Second Battalion, fifty men.

*Post 3.*—Lieutenant T. B. M. Mason, United States Navy, commanding; Lieutenant Wainwright Kellogg, United States Navy; Ensign H. M. Witzel, United States Navy, and thirty-one blue-jackets and one Gatling. Captain W. S. Muse, United States Marine Corps; First Lieutenant R. Wallach, United States Marine Corps; Second Lieutenant F. E. Sutton, United States Marine Corps; Company C, Second Battalion, fifty men. Captain G. C. Reid, United States Marine Corps; Second Lieutenant S. L. Jackson, United States Marine Corps; Company D, Second Battalion, fifty men.

*Post 4.*—Lieutenant C. O. Allibone, United States Navy, commanding; Junior Lieutenant C. J. Badger, United States Navy; Gunner J. J. Walsh, United States Navy; fifty blue-jackets, one Gatling, one 3-inch B. L. R.

Armored car, Hotchkiss Rev. Cannon, one Gatling, one S. B. howitzer.

Paymaster Stephen Rand, United States Navy, clerk, and yeoman.

Passed Assistant-Surgeon Robert Whiting, United States Navy, and apothecary.

GARRISONS OF MATACHIN DISTRICT.

Captain R. W. Huntington, United States Marine Corps, commanding.

*Matachin Post.*—Second Lieutenant Carroll Mercer, United States Marine Corps, thirty-nine Marines; Junior Lieutenant J. C. Colwell, thirty-eight blue-jackets, one Gatling, and one 3-inch B. L. R.

*San Pablo Post.*—First Lieutenant George F. Elliott, United States Marine Corps, commanding, thirteen Marines; Junior Lieutenant Alexander Sharp, twelve blue-jackets, one 12-pounder S. B.

The forces were now disposed on the Isthmus as follows:

Commander B. H. McCalla, commanding United States naval force.

*Personal Staff.*—Lieutenant W. S. Cowles, United States Navy; Lieutenant W. H. Reeder, United States Navy.

*General Staff.*—Paymaster Stephen Rand, Jr., United States Navy; Passed Assistant-Surgeon Robert Whiting, United States Navy.

NAVAL BRIGADE.

*Headquarters, Panama.*

Major and Brevet Lieutenant-Colonel Charles Heywood, United States Marine Corps, commanding; First Lieutenant H. G. Ellsworth, United States Marine Corps, adjutant-general; First Lieutenant A. C. Kelton, United States Marine Corps, brigade commissary and quartermaster; Passed Assistant-Surgeon F. N. Ogden, United States Navy, surgeon; Assistant-Surgeon M. H. Crawford, United States Navy.

PANAMA GARRISON.

*First Battalion.*—Captain R. L. Meade, United States Marine Corps, commanding; Second Lieutenant F. E. Sutton, adjutant; Sergeant M. Harrington, sergeant-major. Company A: Captain E. P. Meeker, commanding; Second Lieutenant T. G. Fillette; three sergeants, five corporals, one drummer, one fifer, forty-two privates. Company B: Captain L. E. Fagan, commanding; three sergeants, five corporals, one drummer, one fifer, forty-one privates. Company C: Captain H. C. Cochrane, commanding; First Lieutenant F. L. Denny; four sergeants, five corporals, forty-one privates. Company D: First Lieutenant O. C. Berryman, commanding; First Lieutenant W. C. Turner; three sergeants, four corporals, forty-one privates. Company E: First Lieutenant Jesup Nicholson, commanding; Second Lieutenant A. H. Clark; three sergeants, four corporals, one drummer, one fifer, forty-one privates.

COLON GARRISON.

*Second Battalion.*—Captain J. H. Higbie, United States Marine Corps, commanding; First Lieutenant William F. Spicer, adjutant; First Sergeant E. G. Arnot, sergeant-major. Company A: Captain R. W. Huntington, commanding (Matachin District); First Lieutenant G. F. Elliott; Second Lieutenant Carroll Mercer; four sergeants, four corporals, one drummer, one fifer, forty privates. Company B: Captain R. S. Collum, commanding; First Lieutenant S. H. Gibson; Second Lieutenant H. K. Gilman; four sergeants, four corporals, one drummer, one fifer, forty privates. Company C: Captain William S. Muse, commanding; First Lieutenant Richard Wallach; four sergeants, four corporals, one drummer, one fifer, forty privates. Company D: Captain G. C. Reid, commanding; Second Lieutenant S. L. Jackson; five sergeants, four corporals, one drummer, one fifer, thirty-nine privates. Company E: First Lieutenant F. H. Harrington, commanding; First Lieutenant Henry Whiting; four sergeants, four corporals, one fifer, forty privates.

THIRD BATTALION.

*From Vessels on the North Atlantic Station, Colon.*—Captain J. M. T. Young, United States Marine Corps, commanding; First Lieutenant C. M. Perkins, adjutant. First Sergeant R. McClure, sergeant-major.

Company A: First Lieutenant M. C. Goodrell, commanding; three sergeants, three corporals, one drummer, one fifer, twenty-six privates. Company B: First Lieutenant L. J. Gulick, commanding; two sergeants, three corporals, one drummer, one fifer, twenty-six privates. Company C: Second Lieutenant C. A. Doyen, commanding; two sergeants, three corporals, one drummer, one fifer, twenty-six privates.

*Independent Company from the United States Ship Shenandoah, Pacific Station, Panama.*—First Lieutenant T. N. Wood, commanding; two sergeants, two corporals, one drummer, one fifer, eighteen privates.

*Independent Company from the United States Ship Iroquois, Pacific Station.*—First Lieutenant H. C. Fisher, United States Marine Corps, commanding; two sergeants, two corporals, one drummer, one fifer, eighteen privates.

*Battalion of Gatling Guns.*—Lieutenant C. O. Allibone, commanding. First Section: Lieutenant C. J. Badger, United States Navy; twenty-three seamen. Second Section: Gunner J. J. Walsh, United States Navy; twenty-four seamen. Third Section: Lieutenant Alexander Sharp, United States Navy; twenty-six seamen.

*Gatling Detachment from United States Ship Alliance.*—Naval Cadet C. P. Plunkett, United States Navy, commanding; twenty seamen.

*Detachment from United States Ship Shenandoah.*—Lieutenant E. M. Hughes, commanding; Ensign J. H. Oliver, United States Navy; 3-inch breech-loading rifle, twenty seamen; 12-pounder howitzer, twenty seamen; Gatling gun, twenty seamen.

*Battallion of 3-inch Rifles.*—Lieutenant T. B. M. Mason, United States Navy, commanding. First Section: Lieutenant Wainwright Kellog, United States Navy; twenty-five seamen. Second Section: Lieutenant J. C. Colwell, United States Navy; twenty-six seamen. Third Section: Ensign H. M. Witzel, United States Navy; twenty-six seamen.

*Artillery Detachment from United States Ship Swatara.*—Lieutenant F. E. Sawyer, United States Navy, commanding; Naval Cadet J. K. Seymour, United States Navy; twenty-two seamen.

*Force at Camp Jouett* (during occupation of the city).—Lieutenant L. Young, United States Navy, commanding; thirty seamen.

On the morning of the 21st of April, Commander McCalla transferred his head-quarters from Colon to Panama. On the same day Admiral Jouett inspected the several garrisons, and returned to Colon on the same afternoon.

At this time it was reported that the national troops at Buenaventura were being embarked in the iron steamer Guayaquil. On that day Captain Higbee was directed to order Companies B and D, Second Battalion, under the command of Captain Collum, to proceed to Panama by the three o'clock train. After the arrival of these companies, they, with a Gatling from the Alliance, Lieutenant Sawyer, United States Navy, and a howitzer from the Swatara, Ensign Plunkett, United States Navy, all under the command of Captain Collum, were quartered at night in cars at the new passenger station, extending our lines to the bridge crossing the railroad.

On the 23d information was received that seven hundred national troops had embarked in the Boyaca, the hulk Guayaquil, and a schooner, with a tow of three whale-boats to land the men at Panama.

Commander McCalla says that, "On the 24th it was reported that barricades were being erected in the streets of Panama, an indication that General Aizpuru would resist the national forces, and that fighting in the streets would follow." At 11.55 A. M. our consul-general called on Commander McCalla, and notified him that two barricades were being erected at certain points, and that in consequence communication would be cut off with the Central and South American cable office.

Commander McCalla considered it necessary, therefore, to occupy the city, and he notified Rear-Admiral Jouett of his intention. Captain Higbee was directed to send the entire force at Colon by special train. At one o'clock the disposition of the force for an entry into Panama was completed, and as soon as the boats from the Shenandoah were in sight, Colonel Heywood was ordered to advance with his whole force, and occupy the positions in the city as indicated by previous orders.

Tanks of powder and a Farmer's dynamo-machine, with wire and fuzes, had been obtained from the Shenandoah, to be used, if necessary, to blow up the buildings.

The advance was made in three columns. The right column, under the command of Captain Collum, composed of Companies B, First Lieutenant Gibson, Second Lieutenant Gilman, D, Captain Reid, Second Lieutenant Jackson, Second Battalion, one field-piece and one Gatling, Lieutenant Sawyer, United States Navy, and Cadet Plunkett,

United States Navy, advanced from the new passenger station at the bridge, along the chemin de la Savanne, and the Carrera del Isthme, to the Plaza Santa Anna. The centre column, under the command of Captain Meade, composed of Companies C, Captain Cochrane, First Lieutenant Denny, E, First Lieutenant Nicholson, Second Lieutenant Clark, First Battalion, one field-piece and one Gatling gun, Ensign Witzel, United States Navy, and Cadet Seymour, United States Navy, advanced from its position at the Panama Railroad Company's buildings to the cable office by the Carrera de Colon. The left column, under the immediate command of the brigade-commander, Colonel Heywood, composed of Companies A, Captain Meeker, Second Lieutenant Fillette, B, Captain Fagan, First Battalion, one 3-inch rifle and one Gatling gun, Lieutenant Hughes, United States Navy, and Ensign Oliver, United States Navy, advanced, by a street running parallel with the beach, to the Plaza Alfaro; thence by the Carrera de Cordova to the American consulate and the Pacific Mail office. The advance was made without music, by the right flank, deployed for street firing, a distance of one yard between each file, the front and rear ranks marching as close as possible to the houses on either side and the artillery in the centre. Each man carried eighty rounds; the artillery was ready for action, and four men, on the right of each company, carried axes slung. The Plaza Santa Anna, a part of the city occupied by the worst classes, was reached by the right column, and the cathedral occupied without meeting resistance.

A barricade in the Carrera de Caldas was captured by the column under Captain Meade and destroyed. It was made of heavy wooden posts set in the ground three feet apart, connected by boarding to be backed by earth. A barricade in the Carrera de Ricuarté, constructed of sand-bags, was overcome by Colonel Heywood, occupied, and a Gatling mounted behind it trained on a cuartel garrisoned by three hundred insurgents. In the Cathedral de Santa Anna a howitzer was placed in position, loaded with shrapnel, at the front entrance, and a Gatling at the rear trained on a cuartel of insurgents. The insurgent president, Rafael Aizpuru, and his secretaries of war and state were arrested by Colonel Heywood near the consulate of the United States, and confined in the Grand Central Hotel, in the custody of Lieutenant Reeder, United States Navy, with a guard from Company A, First Battalion.

While the military necessity for the occupation of the city has been much criticised, yet the plan was boldly conceived, and the execution of this conception was brilliantly and successfully carried out by Colonel Heywood. In thirty minutes from the time the signal was made the United States forces had complete possession of the city.

The force was now disposed as follows:

Head-quarters, Commander McCalla, with his staff, occupied the Grand Central Hotel.

Head-quarters of the brigade, Colonel Heywood, with his staff, occupied the American consulate.

Company C, First Battalion, Captain Cochrane, was at the Cable Company's office; Company E, of the First Battalion, Lieutenant Nicholson, a short Gatling, and twenty sailors, under Ensign H. M. Witzel, were at the sand-bag barricade in the Carrera de Ricuarté; Company B, First Battalion, Captain Fagan, was stationed at the consulate; Company A, First Battalion, Captain Meeker, was stationed at the Pacific Mail Company's office, as also the commanding officer of the First Battalion, Captain Meade; Lieutenant Hughes, United States Navy, with twenty sailors and a short Gatling, was stationed at the junction of the Carrera de Sucre and the Cathedral Plaza. Ensign Oliver, United States Navy, with twenty sailors and a 3-inch breech-loading rifle, was stationed at the junction of the Carrera de Sucre and the Carrera de Cordova. Captain Collum occupied the Plaza Santa Anna, the most important position in the city, with Company B, Second Battalion, Lieutenants Gibson and Gilman; Company D, First Battalion, Lieutenant O. C. Berryman, forty sailors with two pieces of artillery, Junior Lieutenant F. E. Sawyer, United States Navy, and Naval Cadet Seymour, United States Navy. A line of sentries was posted from the Plaza Santa Anna along the Carrera Chiriqui to the bay, and another along the Carrera de Bocas Toro to the sea, the former connecting with a line running parallel with the bay to Camp Jouett. At Camp Jouett there were the detachments of Marines from the Shenandoah and Iroquois, under Lieutenants Wood and Fisher, and a detachment of sailors under Lieutenant Lucien Young, United States Navy. Lieutenant Mason, United States Navy, was placed in command at Camp Jouett.

The garrison from Colon, consisting of Campanies C, Captain Muse, and E, Lieutenant Harrington, Second Battalion, two sections of Gatling and two sections of rifled guns, under command of Lieutenant Allibone, United States Navy, arrived about 5 P. M. Before sunset a strong picket-line was thrown across the peninsula on which the city was built, extending from Camp Jouett to the shore on the south. This line was occupied by Companies D, Captain Reid, and E, Lieutenant Harrington, three Gatlings under Lieutenant Badger, Naval Cadet Plunkett, and Gunner Walsh, the whole under the command of Lieutenant Allibone.

Company C, Captain Muse, was sent to the office of the Pacific Mail Steamship Company.

About 7 P.M. there was a quarrel between some citizens in the Cathedral Square, during which several shots were exchanged. The Gatling gun in the Calle Sucre, sufficiently elevated to clear the tops of the houses, was fired across the plaza, clearing it in a few seconds. In accordance with strict orders from Colonel Heywood, the city was thoroughly patrolled during the night, and no disturbance occurred.

The Third Battalion, under the command of Captain Young, United States Marine Corps, arrived at Panama at 10 P. M., and was quartered for the night in the railroad-station.

On the following day the commanding officer of the insurgent troops asked Commander McCalla if the force would be withdrawn to Camp Jouett, provided that a guarantee were given that no barricades should be erected, and that no street fighting should be permitted. An affirmative reply was made, and, at the request of the officer, Commander McCalla saw General Aizpuru. He offered to give the guarantee before mentioned. An agreement was therefore signed by both parties. At 5.30 P. M. Company D, Captain Reid, and the Gatling from the Alliance, reported to Captain Collum, at the Plaza Santa Anna. The naval force was withdrawn from the city at 8 P. M., except the force in the Plaza Santa Anna, which remained until 9 P.M.

The position at the railroad depot, occupied by Colonel Heywood when he opened the transit, was the best one stragetically that could have been selected. From the new passenger station all roads leading from the city could be occupied in a very short time.

On the 26th there was a fight between Jamaicans and Colombians, at Paraiso Station, seven and a half miles from Panama. The railroad officials reported that the station-house had been fired into the previous night, and that the workmen were in fear of their lives. Captain Reid, with his company, was sent to Paraiso to quell the disturbance there, which duty he performed successfully with great credit to himself and his command.

Early on the morning of the 28th the Boyaca, with the canal tug, bulk Guayaquil, schooner, and three whale-boats were seen in the bay. Lieutenant Reeder was sent to call on the commander-in-chief to present the compliments of the commander of the forces, with a communication requesting the commander of the national forces not to land within our lines.

Colonel Montoya, who had been appointed military and civil chief of the state of Panama, was on board the Boyaca, as well as Colonel Reyes, the commander-in-chief of the military force, which consisted of about eight hundred men.

Rear-Admiral Jouett having come to Panama on the afternoon of

Tuesday, the 28th, on Wednesday Colonels Reyes, Montoya, and Aizpuru, met in the railroad office, Rear-Admiral Jouett presiding at the conference. The conference resulted in an agreement being signed between the representatives of the Colombian government and Aizpuru, by which the latter was to surrender.

At daylight on the morning of the 30th the government troops landed at the railroad wharf and encamped near the bridge occupied by two companies of the Second Battalion. By the terms of this agreement the national troops marched into the city from their camp in the savannah, at 1 P. M., on Thursday, the 30th; a battalion under Captain Collum, consisting of Companies B, Lieutenant Gibson, and E, Lieutenant Harrington; Second Battalion, one Gatling section and one B. L. R. section, Lieutenant Sawyer and Ensign Seymour, were drawn up on the road in front of the new passenger station, and presented arms as the Colombian force under Colonel Reyes marched past. The latter halted his command and returned the salute. On the arrival of Colonel Reyes at the palace, the Colombian flag was hoisted in Camp Jouett, and a national salute fired from head-quarters. On Friday, May 1, the four companies of the Second Battalion and the Naval Artillery returned to Colon, and on the 7th of May this expeditionary force sailed in the Pacific Mail steamship Colon for New York, where it arrived on the 16th.

GENERAL ORDER NO. 19.

NORTH ATLANTIC STATION,
U. S. FLAG-SHIP TENNESSEE (first rate),
ASPINWALL, U. S. C., May 7, 1885.

The commander-in-chief takes great pleasure in expressing to all the officers and men of his command his high appreciation of their services on the Isthmus of Panama. Their strict attention to duty, their high state of discipline and efficiency, their promptness in emergencies, and, above all, their forbearance in delicate and critical situations, have won the respect and admiration of all.

He will not fail to represent fully to the Navy Department how well all have done, and with what alacrity and devotion officers and men have hastened at every call of duty.

He extends his sincere congratulations to all his command, that the object of their mission has been accomplished peacefully, and with honor to themselves and to their country.

To those who are about to return to their homes he wishes a joyous meeting with their friends, and to each and all long life and happiness.

This order will be read at muster on board each vessel of the squadron, and to the battalions on shore.

JAS. E. JOUETT,
*Rear-Admiral, Commanding U. S. Naval Force on N. A. Station.*

The First Battalion, the first to arrive and the last to leave the Isthmus, which was only when entire tranquillity had been restored, sailed in the Acapulco for New York, on the 22d of May.

Before sailing, the following letter was addressed to Colonel Heywood by Admiral Jouett:

ASPINWALL, U. S. C., May 22, 1885.

COLONEL CHARLES HEYWOOD, U. S. M. C., COMMANDING MARINE BATTALION, ISTHMUS OF PANAMA:

SIR,—

* * * * * * * * * *

Your departure from the Isthmus with your command gives me occasion to express my high estimation of the Marine battalion. You and your battalion came from home at the first sound of alarm, and you have done hard and honest work. The Marine battalion has been constantly at the front, where danger and disease were sure to come, first and always. When a conflict has seemed imminent, I have relied with most implicit confidence on that body of tried soldiers. No conflict has come, but I was well aware how nobly and steadily, through weary and anxious nights, exposed to a deadly climate, the Marines have guarded our country's interest.

Please communicate to your command my grateful acknowledgment of their faithful service on the Isthmus of Panama, and accept my sincere thanks for your earnest and valuable assistance. (Signed) JAS. E. JOUETT,

*Rear-Admiral Commanding.*

The following table gives the total force of Marines under the command of Colonel Heywood during the occupation of the Isthmus.

| | Major and Brevet Lieutenant-Colonel. | Captains. | First Lieutenants. | Second Lieutenants. | Sergeants. | Corporals. | Drummers. | Fifers. | Privates. | Total. |
|---|---|---|---|---|---|---|---|---|---|---|
| First Battalion . . . | 1 | 4 | 6 | 3 | 17 | 23 | 3 | 3 | 206 | 266 |
| Second Battalion . . | . . | 5 | 6 | 2 | 22 | 20 | 4 | 5 | 199 | 263 |
| Third Battalion . . . | . | 1 | 2 | 2 | 8 | 9 | 3 | 3 | 78 | 106 |
| First Independent Co. . | . | . . | 1 | . . | 2 | 2 | 1 | 1 | 18 | 25 |
| Second Independent Co. | . . | . . | 1 | . . | 2 | 2 | 1 | 1 | 18 | 25 |
| | 1 | 10 | 16 | 7 | 51 | 56 | 12 | 13 | 519 | 685 |

Extract from official report of Commander McCalla:

While all the officers did well, I feel it my duty to call the attention of the Department to those who came under my immediate observation.

Lieutenants Cowles and Reeder, of whose professional attainments I had formed the highest opinion, were invaluable. My association with them on the Isthmus only served to increase my respect for their ability and qualifications as naval officers. Paymaster Rand proved himself thoroughly familiar with the duties required of him, and was most energetic in providing for the wants of the detachments stationed along the line. Passed Assistant-Surgeons Whiting and Ogden had not only their own legitimate duties to perform, but the work of apothecaries as well. Passed Assistant-Surgeons Percy, Drake, and Ashbridge, temporarily assigned to duty from the North Atlantic Squadron, were most watchful and attentive to the sick men under their care. Lieutenant-Colonel Heywood, whose reputation as a soldier is well known in the Navy, was constantly on the alert and was indefatigable in his efforts to prevent the garrison at Panama from being surprised. Captain Huntington, of whom I had formed a high opinion, commanded the district of Matachin in the most satisfactory manner. Lieutenant Allibone is one of the best officers I have ever known. It was only necessary to outline to him instructions to feel sure that the details would be thoroughly and promptly executed. Constantly looking after the welfare and comfort of his men, they were always in a high state of efficiency. Captain Collum proved himself in every way deserving of the high reputation he enjoys in the service. He was at all times to be depended upon, and when directed to hold the Plaza Santa Anna I knew it would be done. Captains Reid and Harrington proved themselves thoroughly familiar with the details of their profession. Lieutenant Elliott, who commanded the garrison of San Pablo under the general direction of Captain Huntington, is a remarkably good officer. Thrown upon his own resources, he at once established an excellent set of regulations for the government of the post.

It was afterwards found necessary to increase the force under Lieutenant Elliott, and to order Second Lieutenant Gilman, from Colon to San Pablo. Lieutenant Gilman is a valuable and accomplished officer, always ready at a moment's notice to go where his services were required.

First Lieutenant A. C. Kelton, the brigade quartermaster, was always equal to the many and sometimes unusual demands made upon him. During the occupation of Panama he displayed great energy in supplying the various detachments with food. First Lieutenant F. C. Denny performed his duty thoroughly while in command of a platoon of his company in the hotel in which General Aizpuru was detained.

My attention was called to the coolness of Ensign Witzell, who had charge of the Gatling gun at the barricade in front of the cuartel. Ensign James M. Oliver, attached to the Shenandoah, had been ill from the effects of fever for some days, and was directed to return to his ship for treatment on the morning of the 24th of April. To my surprise I found him at his gun in Panama during the night of the 24th, although such exposure was likely to cause serious effects. Such behavior is very commendable.

The conduct and efficiency of the non-commissioned and petty officers are worthy of the highest praise. The behavior of the enlisted men was admirable; less than three per cent. were found to be unreliable on duty or in camp. When the peculiar temptations to which they were subjected in the trying climate of the Isthmus are considered, this fact is highly creditable to them.

In conclusion, I desire to state that I consider it a great honor to have commanded so fine a body of officers and men.

Very respectfully your obedient servant,

B. H. McCalla,
*Commander U. S. Navy.*

Hon. W. C. Whitney,
*Secretary of the Navy.*

Consul-General Adamson has officially stated that the firm bearing, strict discipline, and splendid conduct of the Marines reflected great credit upon Colonel Heywood and his officers, and the presence of these fine troops allayed anxiety and distrust, enabling the foreign residents to feel that security which could not be obtained under any other circumstances. The commander of an English gunboat lying off Panama, when told that a brigade of Marines was in possession of the Isthmus, remarked that "tranquillity was then assured," that "he knew what they were, having seen a detachment of the Corps at Alexandria."

---

## CHAPTER XXXVI.

**1889.**—Disaster at Apia, Samoa.

On the 15th and 16th of March, a hurricane of great force swept the waters of the harbor, in which were lying at anchor the following men-of-war: United States ships Trenton, Vandalia, and Nipsic; H. B. M. ship Calliope, and H. I. G. M. ships Adler, Olga, and Eber. There were also a few merchant vessels and small craft. The Nipsic had the inner berth, and the Trenton (last to arrive) had the outer berth.

The following are the official reports of the commanding officer of the Trenton and the executive officer of the Vandalia, the two American vessels sunk on that occasion:

### REPORT OF CAPTAIN N. H. FARQUHAR, COMMANDING UNITED STATES SHIP TRENTON.

APIA, SAMOA, March 19, 1889.

SIR,—I have the honor to inform you that the United States flag-ship Trenton, under my command, was driven ashore in a hurricane on Saturday, March 16, about 8 P. M., alongside the wreck of the United States ship Vandalia in this harbor.

The ship has since filled with water and, in my opinion, has broken in two places,—abreast the mizzen mast and near the smoke-stack.

It is possible the ship might be floated to dock with the assistance of powerful pumps used by wrecking companies, but, as there are no facilities of this kind, she will be a total loss.

On Thursday, March 14, the wind came out approximately from the southward, with much rain, the barometer slowly falling, being 29.70 at noon, 29.60 at midnight; the force of the wind from 2 to 4. During the mid-watch, March 15, got up steam, wind remaining in the same direction and varying from 0 to 4, the barometer falling from 29.60 to 29.56. At 4 A. M. the wind, and until 8 A. M., was from 2 to 6 in force, the barometer at 8 A. M. being 29.42.

By noon, though the barometer fell to 29.30, the wind had not increased nor practically changed in direction.

During this watch prepared to send down lower yards and house topmasts. From meridian to 4 P. M. sent down lower yards, housed topmasts, and made preparations for bad weather. At 1 P. M. the wind was east, force 1 to 2, barometer 29.24; at 2 P. M. wind variable, force 1 to 2, barometer 29.20; at 3 P. M. wind northeast, force 2 to 4, barometer 29.20; at 4 P. M. wind north, force 4 to 7, barometer 29.29.

This seemed to indicate that the gale had broken and that the wind would haul to the west. Instead, however, it backed to the northeast, barometer rising at 8 P. M., being 29.36, wind northeast, force 4 to 8.

Up to this time the ship was moored with fifty-two fathoms on port bower, forty-five fathoms on starboard bower, with starboard sheet under foot. At 7.57 parted port bower chain, let go port sheet, steamed ahead, veering to sixty fathoms on it and starboard bower.

The barometer during this watch, 8 P.M. to midnight, was steady about 29.40, wind northeast by north to north-northeast, force 7 to 8. Midwatch of March 16, barometer 29.36 to 29.38, wind north-northeast, force 7 to 8, steaming ahead slowly to relieve strain upon anchors. At 6 A.M. barometer fell to 29.23, wind north by east, force 6 to 9, tremendous sea. About 7 A.M. wheel-ropes carried away, rudder broken in two pieces, so as to be useless. About 9 A.M. wind came out from north, force 8 to 10, barometer rising to 29.30 at 11 A.M. At 9.30 fires were extinguished by water in fire-room, which came from hawse-pipes, notwithstanding every precaution in the way of jackasses, hammocks, etc., having been used to keep the water out, but being on the berth-deck, low down, and a full-bowed ship, it was impossible to keep the water out. The hand as well as steam-pumps were going, with men bailing besides, and the hatches battened down.

I attribute the loss of the ship primarily to the location of the hawse-pipes. I have several times reported officially against their location to the Navy Department.

Up to noon the ship had dragged very little, if any; at noon, barometer 29.29, wind blowing hurricane from the north.

At 1 P.M. the barometer fell to 29.19, the lowest reading, the hurricane continuing with the same force until about midnight, the barometer rising slowly, reaching 29.52 at that hour; wind north-northwest.

From 4 to 8 P.M., dragging slowly at times, set storm, mizzen, and mizzen-stay sail, with sheets amidships to prevent any sheering of the ship. About 3 P.M. parted port sheet chain, and shortly after starboard bower veered to ninety fathoms on remaining anchor.

Shortly after 8 P.M. dropped alongside the Vandalia, took off her crew from the tops and rigging, and made fast to her. Pounding terribly all night, but the wreck of the Vandalia kept us off the reef. Notwithstanding every effort, could not keep the water down in the holds. About midnight the wind had somewhat abated. At daylight, the morning of the 17th, a line was sent to us from the shore.

As much of the provisions as could be handled were gotten on the spar-deck In the afternoon of the 17th, the water still gaining, it was deemed advisable to land the officers and crew.

Permit me to express my most heart-felt thanks to you for your most valuable counsel and for keeping us in good cheer, particularly when in our greatest danger, by your good example.

During these trying days the officers sustained the reputation that our Navy is proud of. The crew generally worked well.

Lieutenant R. M. G. Brown, the navigator, was by my side the whole time, and to his excellent judgment, one time at least, the ship was cleared of a reef. Had we struck it I fear few of the four hundred and fifty people on board of the Trenton would be alive to-day.

Several officers and men were injured, but only one man killed,—J. Hewlett, landsman, whose skull was broken by the sea.

Very respectfully, N. H. FARQUHAR,
*Captain U. S. Navy, Commanding U. S. Flag-Ship Trenton.*

REAR-ADMIRAL L. A. KIMBERLY, U.S.N.,
*Commanding U. S. Naval Force on Pacific Station.*

## REPORT OF LIEUTENANT J. W. CARLIN, EXECUTIVE OFFICER OF THE VANDALIA.

APIA, SAMOA, March 21, 1889.

SIR,—On the 15th and 16th of March a violent gale swept over Apia, wrecking and beaching every craft in the harbor except H. B. M. ship Calliope, a powerful steamer, which saved herself by going to sea.

The gale was long foretold, and the Vandalia was prepared for heavy weather, having sent down lower yards, housed topmasts, and got up steam in obedience to signals from the flag-ship. We were moored with our anchors northeast by east, and southwest by west, forty-five fathoms on each chain, starboard anchor well over to the eastern reef. The shifts of wind previous to the final blow put an elbow in the hawse.

When the strength of the gale came upon us we were directly in the hawse of the Calliope, rendering it injudicious to veer.

All the vessels in the harbor were to leeward of us except the Trenton. About midnight of the 15th we began to drag, and commenced to steam up to our anchors, and continued to steam, with but few intermissions, until we struck, the speed varying from eighteen to forty-two revolutions. The engines worked well, with the exception of a short time when the steam ran down to twenty-five pounds, caused by an accident to one of the boilers. We had on seven boilers of the eight, and they furnished all the steam the engines could use.

We had but one sheet-anchor, and the stock of that was broken in the last gale. This anchor was held in reserve as long as possible, in expectation of the wind coming out from the northward and westward, but was let go about 2 A.M. of the 16th, as near the weather reef as possible.

At daylight we saw that the vessels in the inner harbor were in trouble. The Eber soon disappeared entirely, and the Adler struck the reef, the seas heaving her on top of it and capsizing her. The Calliope was our nearest neighbor, and from daylight until 9 A. M. we were in imminent danger of getting foul of each other, and also of striking the reef, being frequently within fifty feet of it, and sometimes within twenty feet. The current was so strong that spanker, helm, and two anchors on weather bow were not sufficient to bring the ship's head to wind. The seas were of immense force, and the steering-gear was carried away about 8 A.M. The ship was steered thereafter by the relieving tackles.

The Calliope put to sea about 9 A.M., and one obstacle was removed from our path. Feeling that we must go on the reef if we remained where we were, we made every exertion to get into the inner harbor, hoping to escape the extraordinary current we had heretofore encountered. In doing so, we passed between the Olga and the lee reef with but a few feet to spare on either hand.

We slipped the sheet-chain to avoid fouling the Olga's ground-tackle, and veered on both bowers to clear the ship herself. After passing the Olga we made strenuous exertions to bring her head to the wind, but they were of no avail, and the stern took the inner point of the reef at 10.45 A.M. The engines were kept going until we were convinced that the ship was hard and fast. They were then stopped, safety-valves opened, and the firemen called on deck. The ship's head swung slowly to starboard, she began to fill and settle, and the rail was soon awash, the seas sweeping over her at a height of fifteen feet above the rail. We were within two hundred yards of the shore, but the current was so strong and the seas so high that swimming was a reckless undertaking.

We found the Nipsic beached and abandoned just inside of us. A line was got on board her, but more were drowned than saved. It was imposible to get a line to shore. There is no apparatus for throwing a line, and our buoys floated to seaward. E. M. Hammar, seaman, attempted to carry a line, but was swept back against the ship and killed. Many men attempted to swim, but so many were drowned that the remainder were deterred.

The commanding officer was washed overboard from the poop about half an hour after striking; his strength had been exhausted by constant work for so long a time, and he was unable to sustain the shock of the heavy seas. He was surrounded by officers and men and had been rescued several times, when a sea of unusual violence swept him, and all in his immediate vicinity, overboard. The crew then deserted the poop and forecastle, and took refuge in the tops and rigging, where they remained until about 8 P.M., eight hours. The Olga was driven on shore about 5 P.M.

About dark the Trenton dragged down on us, and we expected her to carry away our masts and throw us into the sea, but she came down so gently, and was so expert in throwing us lines, that nearly all our men escaped from their perilous position in the rigging to the comparative safety of the Trenton's decks. The mizzenmast and the mainmast soon went by the board. In the morning the seas had gone down considerably and the crew was sent on shore in boats.

The following is the list of the lost: C. M. Schoonmaker, captain; Frank H. Arms, paymaster; F. E. Sutton, United States Marine Corps, first lieutenant; John Roche, pay-clerk; George Murrage, boatswain's mate; E. M. Hammar, seaman; George Gorman, carpenter; M. Craigin, captain after-guard; William Brown, first quartermaster; T. G. Downey, paymaster's yeoman; B. F. Davis, engineer's yeoman; Thomas Riley, landsman; John Kelly, ordinary seaman; Henry Baker, landsman; John Hanchett, sergeant; Frank Lissman, sergeant; Michael Cashen, corporal; Nicholas Kinsella, corporal; H. C. Gehring, private marine; Adolph Goldner, private marine; Frank Jones, private marine; N. B. Green, bayman; H. P. Stalman, bayman; C. H. Hawkins, steerage steward; G. H. Wells, private marine; C. E. G. Stanford, landsman; W. Bransbane, cabin steward; Joseph Griffin, first-class fireman; M. Erickson, ordinary seaman; Thomas Kelley, second-class fireman; W. Howat, coal-heaver; C. P. Kratzer, ordinary seaman; M. H. Joseph, equipment yeoman; George Jordan, private marine; John Wilford, private seaman; Henry Wixted, private marine; Aylmer Montgomerie, private marine; John Sims, private marine; Charles Kraus, private marine; Ah Keau, cabin cook; Ah Pack, seamen's cook; Ten Dang, landsman; Yee Hor, wardroom cook.

Colonel GEORGE C. REID, Adjutant and

Inspector U. S. M. C. Plate 5

The Vandalia is a total loss. Her rail is awash, and she is filling with sand. There is nothing standing except the foremast. The safe has been brought on shore, but aside from that the articles recovered will be insignificant. The crew is at work doing everything possible in the way of wrecking.

All records having been lost, this report is devoid of data concerning wind, weather, barometer, etc.

The loss of the Vandalia was due to the extreme violence of the gale, the great height of the seas, the extraordinary strength of the current, poor holding ground, and the unprotected condition of a small harbor fringed with coral reefs, and crowded with vessels. Very respectfully, J. W. CARLIN,

*Lieutenant U. S. Navy,*
*Executive Officer U. S. S. Vandalia, Commanding Survivors.*

REAR-ADMIRAL L. A. KIMBERLY, U.S.N.,
*Commanding U. S. Naval Force on Pacific Station.*

In a supplementary report Lieutenant Carlin commends Sergeant John Coleman and Private Carroll of the Marines for gallant conduct.

These men and several others of the detachment volunteered for duty in the fire-room, and remained in the coal-bunkers, "*where danger appalled the stoutest heart.*" Sergeant Coleman had previously received a medal for "*personal valor*" at Corea, under Captain Tilton.

By reference to the list of the lost it will be observed that *one-half* of the Marines of the Vandalia were drowned, including their promising young officer, Lieutenant Sutton, one of the star members in his class at the Naval Academy.

The survivors of the two vessels were comfortably quartered on shore. The Marines, with their marvellous adaptability for service of any nature, were immediately placed in charge of the town, under the command of the senior officer of Marines, Captain R. W. Huntington, assisted by First Lieutenant T. G. Fillette.

It is on record that Captain Huntington, Lieutenant Fillette, and the Marines under their command were commended for the faithful performance of their duty under very trying circumstances.

Without the steady and reliable discipline maintained by the Marines, trouble would have ensued, and it was to this small detachment of faithful soldiers that the naval authorities were indebted for the maintenance of law and order.

The following graphic description of the disaster is condensed from the sketch of John P. Dunning, Esq., correspondent at Samoa for the Associated Press, and published in the *St. Nicholas* for February, 1890.

"The harbor in which the disaster occurred is a small semicircular bay, around the inner side of which lies the town of Apia. A coral reef, which is visible at low water, extends in front of the harbor from the eastern to the western extremity, a distance of nearly two miles. A break in this reef, probably a quarter of a mile wide, forms a gateway to the harbor. The space within the bay where ships can lie at anchor is very small, as a shoal extends some distance out from the eastern shore, and on the other side another coral reef runs well out into the bay.

"The American consulate is situated near the centre of the line of houses composing the town, and directly in front of it is a long strip of sandy beach. The war-vessels were anchored in the deep water in front of the American consulate. The Eber and Nipsic were nearest the shore. There were ten or twelve sailing vessels, principally schooners, lying in the shallow water west of the men-of-war.

"The storm was preceded by several weeks of bad weather, and on Friday, March 15, the wind increased and there was every indication of a hard blow. The war-ships made preparation for it by lowering top-masts and making all spars secure, and steam was also raised to guard against the possibility of the anchors not holding.

"By eleven o'clock at night, the wind had increased to a gale. The crews on most of the sailing vessels put out extra anchors and went ashore. Rain began to fall at midnight, and the wind increased in fury. Great waves were rolling in from the open ocean, and the pitching of the vessels was fearful. The Eber commenced to drag her anchors at midnight, and an hour later the Vandalia was also dragging. However, by using steam they succeeded in keeping well off the reef and away from the other vessels. The wind blew more and more strongly, and rain fell in torrents. By three o'clock the situation had become alarming. Nearly every vessel in the harbor was dragging, and there was imminent danger of collisions. There was no thought of sleep on any of the ships, for every man was needed at his post.

"On shore the howling of the wind among the trees and houses, and the crash of falling roofs, had aroused many persons from their beds, and figures were soon seen groping about the street looking for some spot sheltered from the tempest. The tide was coming in rapidly, and the surf was breaking all over the street, a hundred feet above the usual high-water mark. The spray was thrown high in the air and beat into the windows of houses nearest the shore. Rain fell like sleet, and men and women who were wandering about in the storm shielded their faces with small pieces of board or with any other article that could be used as a protection against the wind and sand.

"I had spent the evening in-doors and had retired about eleven o'clock. The house which I occupied was some distance from the shore and was surrounded by a thick growth of trees. Several of these had fallen with a crashing sound, and I found it impossible to sleep. I arose and determined to go down to the beach, for I felt that the vessels in the harbor must be in great danger. I reached the street with the greatest difficulty, for I had two treacherous little foot-bridges to cross, and the night was so dark and the force of the wind so great that I felt I was wandering about like a blind man. When I had walked down to the beach, I looked across the angry waters at the lights of the vessels and realized far more clearly than before that the storm was something terrible. I wandered along the beach for a distance of half a mile, thinking it possible that I might find some one, but the whole place seemed to be deserted. The only light visible on shore was at the American consulate. I *found a solitary Marine on duty as sentry there.* I exchanged a few words with him and then retired to a temporary shelter for several hours, until a number of natives and a few white persons commenced to collect on the street. The natives seemed to know better than the rest that the storm would result in awful destruction. People soon gathered in little groups and peered out into the darkness across the sea of foaming waters. Fear was depicted upon every face. Men stood close together and shouted to make themselves heard above the roar of the tempest.

"Through the blackness of the night could be seen the lights of the men-of-war, and even above the rushing and roaring of the wind and waves, the shouting of the officers and men on board came faintly across the water. It could be seen that the vessels were dragging, as the lights were moving slowly in different directions and apparently crossing and recrossing each other. Every moment it seemed as if two or more of the great war-ships were about to come together, and the watchers on the beach waited in breathless anxiety to hear the crash of collision.

"A little after five o'clock, the first rays of dawn broke upon the scene and revealed a spectacle not often witnessed. The position of the vessels was entirely changed. They had been swept from their former moorings and were all bearing down in the direction of the inner reef. Black smoke was pouring from their funnels, showing that desperate efforts were being made to keep them up against the storm. The decks swarmed with men clinging to masts or to anything affording a hold. The hulls of the ships were tossing about like corks, and the decks were being deluged with water as every wave swept in from the open ocean. Several sailing vessls had gone ashore in the western part of the bay. The Trenton and Vandalia, being farther out from the shore than the other ships, were almost obscured by the blinding mist. The vessels most plainly visible were the Eber, Adler, and Nipsic. They were very close together and only a few yards from the reef.

"The little gunboat Eber was making a desperate struggle, but every moment she was being driven nearer and nearer the reef. Her doom was certain. Suddenly she shot forward as if making a last effort to escape destruction. The current, however, bore her off to the right, and her bow struck the port quarter of the Nipsic, carrying away several feet of the Nipsic's rail and one boat. The Eber then fell back and fouled with the Olga, and after that she seemed unable to make any effort to save herself. Awful seas broke over the little vessel as she swung around broadside to the wind. Presently she was lifted high on the crest of a great wave and hurled with awful force upon the reef. In an instant there was not a vestige of her to be seen. She struck upon her bottom, rolled over towards the sea, and disappeared from view. Every timber must have been shattered, and half the poor creatures aboard of her crushed to death before they felt the waters closing above their heads. Hundreds of people were on the beach by this time, and the work of destruction had occurred within full view of them all. They stood for a moment appalled by the awful scene, and a cry of horror arose from the lips of every man who had seen nearly a hundred of his fellow-creatures perish in an instant. Then with one accord they all rushed to the water's edge nearest the point where the Eber had foundered. The natives ran into the surf far beyond the point where a white man could have lived, and stood waiting to save any one who might rise from the water. There was no thought of the war between Germany and Samoa; there was no sign of enmity against the people who had banished their king and carried him off to a lonely isle thousands of miles from his native land; the savage forgot the oppression which a civilized people had placed upon him, and now held out his hand to save a human life, caring little whether it was that of friend or foe.

"At first it seemed as if every man on the ill-fated steamer had gone to his death. Not even a hand appeared from the depths where the Eber sank. But the breakers on the reef had hidden a few struggling men who had come to the surface and struck out feebly for shore.

"Presently a man was seen clinging to the piling of a small wharf nearby. Willing hands soon grasped him and drew him up on shore. He was a young man with a handsome, boyish face, and wore the uniform of an officer. He proved to be Lieutenant Gædeke, and was the only officer of the Eber who was saved. Lieutenant T. G. Fillette, the marine-officer of the Nipsic, who for several months had been stationed on shore in charge of the guard of Marines at the American consulate, took the German officer under his care.

"Four sailors from the Eber were found struggling in the water near shore about the same time. They were quickly rescued by the natives and also taken to the American consulate. There were six officers and seventy men on the Eber when she struck the reef, and of these, five officers and sixty-six men were lost.

"Lieutenant Gædeke, the survivor, was almost heart-broken over the sad fate of his fellow-officers and men. He was the officer of the watch and was on the bridge when the Eber went down. He said that all the other officers were below, and he supposed they were crushed to death. It was about six o'clock in the morning when the Eber foundered.

"During the excitement attending that calamity the other vessels had been for the time forgotten; but we soon noticed that the positions of several of them had become more alarming. The Adler had been swept across the bay, being for a moment in collision with the Olga.

"She was now close to the reef, about two hundred yards west of the point where the Eber struck, and, broadside on, like the Eber, she was approaching her doom.

"In half an hour she was lifted on top of the reef and turned completely over on her side. Nearly every man was thrown into the water. They had but a few feet to swim, however, to reach the deck, as almost the entire hull was out of water. Only twenty men were drowned when the steamer capsized.

"The others clung to the guns and masts in safety, and as the bottom of the vessel was towards the storm, the men on the deck were well protected. Natives stretched a rope from the shore to the Adler during the day and a number of sailors escaped by that means. But the rope parted before all had left the vessel, and the others were not taken off until next day. They clung to the wreck during the long weary hours of the day and night, and were greatly exhausted when they finally reached the shore.

"Just after the Adler struck, the attention of every one was directed towards the Nipsic. She was standing off the reef with her head to the wind, but the three anchors which she had out at the time were not holding, and the steamer was being beaten towards the point where the Eber went down. It was only by the most skilful management that her officers and crew were saved from the same fate that befell the Eber. The Nipsic also narrowly escaped destruction by being run into by the Olga, and it was the blow she received from that vessel which finally sent her ashore. As she was trying to avoid a collision with the Olga, the little schooner Lily got in her track and was cut down. There were three men on the Lily, two of whom were drowned, but the third swam to the Olga and reached her deck in safety.

"As the Nipsic's anchors were not holding, orders were given to attach a hawser to a heavy 8-inch rifle on the forecastle and throw the gun overboard. As the men were in the act of doing this, the Olga bore down on the Nipsic and struck her amidships with awful force. Her bowsprit passed over the side of the Nipsic and, after carrying away one boat and splintering the rail, came in contact with the smokestack, which was struck fairly in the centre and fell to the deck with a crash like thunder. For a moment it was difficult to realize what had happened, and great confusion followed. The crew believed the ship was going down, and men ran up in the rigging for safety. The iron smokestack rolled from side to side with every movement of the vessel until finally heavy blocks were placed under it. By that time the Nipsic had swung around and was approaching the reef. It was an anxious moment for all on board. They had seen the Eber strike a few yards from where they now were, and it seemed certain that they would go down in the same way. Having lost her smokestack, the vessel was unable to keep up her steam power. Captain Mullan was upon the bridge at the time with Ensign H. P. Jones, the latter being the officer of the watch. The captain remained cool and collected during the dangerous moment. He saw that in another moment the Nipsic would be on the reef, and probably every man on board be lost. Any further attempt to save the vessel would be useless, so he gave the orders to beach her. The limited amount of steam which could still be carried was brought into use and her head was put around towards the shore. She had a straight course of about two hundred yards to the sandy beach in front of the American consulate.

"There were then several hundred natives and about fifty white persons, principally Americans and Englishmen, standing near the water's edge watching the critical manœuvres of the Nipsic, and I remember the feeling of dread which came over me as I saw the vessel running alongside the dangerous reef, liable at any moment to be dashed to pieces upon it. As she came nearer the shore I could easily distinguish the faces of officers who were my personal friends, and I did not know but that I might be looking upon them for the last time. Near me were standing United States Vice-Consul Blacklock, and Ensign J. L. Purcell, an officer of the Nipsic, who had been on shore during the night. I could judge from their faces that their fears were the same as mine. But the Nipsic escaped the reef and her bow stuck fast in the sand about twenty yards from the water's edge. She then swung around, forming an acute angle with the line of the shore.

"Just as the vessel struck, five sailors jumped into a boat and commenced to lower it, but the falls did not work properly and one end of the boat dropped, throwing the men into the water, and drowning all of them. Another boat, containing Dr. E. Z. Derr, the ship's surgeon, and a half-dozen sick men, was lowered in safety, but it capsized before it reached shore.

"The men were within a few feet, however, of the natives who were standing waist-deep in the surf, and they were pulled up on the beach and taken to the consulate. Several men on the Nipsic ran to the rail and jumped overboard. All these reached the shore except two sailors, who were unable to swim through the current and were swept out into the bay and drowned.

"By this time every man aboard had crowded to the forecastle. A line was thrown to the natives, and double hawsers were soon made fast from the vessel to the shore, and the natives and others gathered around the line to assist the men off. Seumanu Tafa, chief of Apia district, and Salu Anae, another chief, directed the natives in their work. The scene was one of intense excitement. The seas broke upon the stern of the Nipsic with awful force, and it seemed as if the vessel would be battered to pieces before the men on her decks could be saved. The waves were rolling high on the beach, and the undertow was so strong that the natives narrowly escaped being washed out into the bay. The rain continued to pour, and the clouds of flying sand grew thicker every moment. The voices of officers shouting to the men on deck were mingled with the loud cries and singing of the Samoans as they stood battling against the surf, risking their lives to save the American sailors.

"To one who saw the noble work of these men during the storm it is a cause of wonder that they should be called savages by more enlightened races. There seemed to be no instinct of the savage in a man who could rush into that boiling torrent of water that broke upon the reef and place his own life in peril to save the helpless drowning men of a foreign country.

"While Americans and Germans were treated alike, it was plain that their sympathies were with the Americans, and they redoubled their efforts when they saw an opportunity to aid the men who represented a country which had insisted that their native government should not be interfered with by a foreign power. During the trying hours of that day they never faltered in their heroic efforts when it was possible to save a life.

"As the Nipsic lay helpless on the beach, they gathered about the vessel and showed a determination to risk everything to save the officers and men aboard. Nearly all the American and English residents of Apia were on the shore in front of the consulate, and there seemed to be a willingness on the part of every man to render whatever assistance was in his power. Ensign Purcell of the Nipsic and several other Americans were up to their waists in the water ready to lend a hand to the men as they left the ship. The position was a most dangerous one, as the waves were washing far up on the beach and great pieces of floating wreckage were being swept back and forth. The force of the water was so strong that it was necessary to hold on firmly to the life-line which was stretched from the Nipsic's bow, and I remember once that my grasp upon the line was broken by an immense wave which completely enveloped me. I was thrown violently across the rope, and then as the water receded I was carried out with it. Fortunately, two natives caught me before I had gone too far, and with their assistance I grasped the line again. There was no attempt to leave the Nipsic in disorder. Captain Mullan and several other officers stood by the rail where the hawsers were made fast, and directed the movements of the men. They came down the ropes quickly, but the seas were rolling so high under the bows of the ship that the men were often entirely submerged and their hold upon the lines broken. Nothing but the noble efforts of the natives saved them from being swept out into the current and drowned. As soon as each man would come within reach, he would be grasped in the strong arms of half a dozen Samoans and carried out of the water. Captain Mullan insisted upon being the last man to leave the ship, and he finally found himself on the deck with Lieutenant John A. Shearman by his side. The captain, being unable to swim, did not care to descend the rope by means of his hands and legs, as all the others had done, so he procured an empty water-cask and attached it to the hawser. When he was seated in the cask, Lieutenant Shearman stood alone upon the deck and started his brave commander down the line. The young officer then climbed down the rope, and the Nipsic was left alone to battle with the waves.

"The Nipsic, Adler, and Eber were the smallest war-ships in the harbor. The four large men-of-war, the Trenton, Calliope, Vandalia, and Olga, were still afloat and well off the reef. They remained in a comparatively safe position for two hours after the Nipsic was beached, but persons on the shore were watching them intently all the time. About ten o'clock in the morning the excitement on shore began to grow more intense as the Trenton was seen to be in a helpless condition. The great vessel was lying well out in the bay, and, with every wave that rolled in, her stern would be lifted out of the water, and it was seen that her rudder and propeller were both gone, and there was nothing but her anchors to hold her up against the unabated force of the storm.

"The Vandalia and Calliope also were in dangerous positions, bearing back towards the reef near the point where lay the wreck of the Adler. Great waves were tossing the two vessels about, and they were coming closer together every minute. The Vandalia attempted to steam away, but in so doing a collision occurred. The iron prow of the Englishman was lifted high in the air and came down with full force upon the port quarter of the Vandalia. The jib-boom of the Calliope was carried away, and the heavy timbers of the Vandalia were shivered. Every man who stood upon the deck of the Vandalia near the point of collision was thrown from his feet by the shock.

"A hole was torn below the rail, and water rushed into the cabin. It was impossible to ascertain the extent of the damage at the moment, but it seemed as if the Vandalia had received her death-blow. Men rushed up the hatches in the belief that the steamer was sinking, but they afterwards returned to their posts. Just after this collision, Captain Kane of the Calliope determined to make an effort to steam out of the harbor, as he saw that to remain in his present position would lead to another collision with the Vandalia or throw his vessel upon the reef. He accordingly gave orders to let go all anchors. The Calliope's head swung around to the wind and her engines were worked to their utmost power. The steamer seemed to stand still for a moment, and then the rapidly revolving propeller had its effect, for the vessel moved up slowly against the great waves which broke over her bows and flooded her decks from stem to stern. Clouds of black smoke poured from her funnel as more coal was thrown into the furnaces. Every tension was strained in her heroic struggle against the storm. She seemed to make her headway at first inch by inch, but her speed gradually increased until it became evident that she could clear the harbor.

"This manœuvre of the gallant British ship is regarded as one of the most daring in naval annals. It was the one desperate chance offered her commander to save his vessel and the three hundred lives aboard. An accident to the machinery at this critical moment would have meant certain death to all. Every pound of steam which the Calliope could possibly carry was crowded on, and down into the fire-rooms the men worked as they never had worked before. To clear the harbor, the Calliope had to pass between the Trenton and the reef, and it required the most skilful seamanship to avoid a collision with the Trenton, on the one hand, or total destruction upon the reef, on the other. The Trenton's fires had gone out by that time, and she lay helpless almost in the path of the Calliope. The doom of the American flag-ship seemed but a question of a few hours. Nearly every man aboard felt that his vessel must soon be dashed to pieces, and that he would find a grave under the coral reef. The decks of the flag-ship were swarming with men, but, facing death as they were, they recognized the heroic struggle of the British ship, and as the latter passed within a few yards of them a great shout went up from over four hundred men aboard the Trenton. 'Three cheers for the Calliope!' was the sound that reached the ears of the British tars as they passed out of the harbor in the teeth of the storm; and the heart of every Englishman went out to the brave American sailors who gave that parting tribute to the Queen's ship.

"A well-known London journal afterwards remarked: 'The cheer of the Trenton's men was the expression of an immortal courage. It was distressed manhood greeting triumphant manhood, the doomed saluting the saved.' The English sailors returned the Trenton's cheer, and the Calliope passed safely out to sea, returning when the storm had abated. Captain Kane, her commander, in speaking of the incident afterwards, said, 'Those ringing cheers of the American flag-ship pierced deep into my heart, and I will ever remember that mighty outburst of fellow-feeling which, I felt, came from the bottom of the hearts of the gallant admiral and his men. Every man on board the Calliope felt as I did; it made us work to win. I can only say, "God bless America and her noble sailors!"'

"When the excitement on the Vandalia which followed the collision with the Calliope had subsided, it was determined to beach the vessel.

"Lieutenant J. W Carlin, the executive officer, was practically in command, as Captain Schoonmaker had been thrown across the cabin the night before and severely injured. The captain was in a dazed and weakened condition, able to do little towards directing the movements of the ship; but, notwithstanding his injuries, he faced the storm like a hero and stood by the side of his first officer until the sea finally swept him off to his death. Of all the officers who did their duty nobly in the face of danger, none received more commendation than Lieutenant Carlin.

"Officers and men alike spoke of his conduct in the highest terms of praise, and said that his cool and calm demeanor kept the men at work when panic was almost breaking out among them. He had been on duty since the morning before, and had not tasted food in all that time.

"The Vandalia was obliged to move along the edge of the reef, a distance of several hundred yards, in order to reach a point in front of the American consulate where it was thought safe to run her aground.

"Every one on shore stood near the consulate and watched the vessel steam across the harbor. Her draught was greater than the Nipsic's, and it was known that she would not be able to get very close to the shore. She came on until her bow stuck in the soft sand, about two hundred yards off shore and probably eighty yards from the stern of the Nipsic.

"Her engines were stopped and the men in the engine-room and fire-room below were ordered on deck. The ship swung around broadside to the shore, and it was thought at first that her position was comparatively safe, as it was believed that the storm would abate in a few hours and that the two hundred and forty men aboard could be rescued then.

"It was nearly eleven o'clock when the Vandalia struck, and, notwithstanding her easy position, it soon became apparent that her officers and crew were in great danger. Nearly all the officers were on the poop-deck, but their faces could not be distinguished from the shore on account of the blinding mist. The men were scattered about on the gun-deck and on the forecastle, holding on to the masts and sides of the ship. In half an hour it was noticed that the vessel was settling down. Lying as she did, almost broadside to the wind, the seas broke over her furiously and poured down the hatches.

"One by one the boats were torn from their davits and swept away. Efforts were made to fire lines ashore, but it was impossible to do so, as the ammunition was saturated with water.

"One brave sailor, named F. M. Hammar, volunteered to swim through the surf with a line, in the hope that his comrades might be rescued in that way. It was an undertaking which seemed to mean certain death, for a boiling torrent of water, covered with floating wreckage, was rushing between the Vandalia and the shore.

"Notwithstanding the peril of such an act, the man fastened a small cord to his body, stood by the rail a moment, and then plunged into the sea. He had hardly touched the water when he was thrown violently against the side of the ship and knocked into insensibility.

"There was no possibility of saving him, and he drowned in sight of all who had witnessed his heroic action.

"By noon the entire gun-deck of the Vandalia was under water, and from that time on the condition of those aboard was the most pitiable that can be imagined.

"The torrents of water that swept over the ship knocked the men from their feet and threw them against the sides. Several were badly injured. Most of the men sought refuge in the rigging. A few officers still remained upon the poop-deck, but a number had gone aloft. The wind seemed to increase in fury, and as the hull of the steamer sank lower the force of the waves grew more violent. Men on shore were willing to render aid, but were powerless.

"No boat could have lived a moment in the surf, and it was impossible to get a line to the vessel, as there was no firing-apparatus on shore.

"The remembrance of those hours when the sea was washing over the Vandalia has come to me many times since then, and the scene is as vivid as it was when I stood on the beach in that blinding storm and watched the awful spectacle. I recalled then that a few days previous Captain Schoonmaker had been ashore and had given me an invitation to go aboard the Vandalia and spend some time with him. Circumstances had prevented me from accepting his invitation at the time, but I had intended to avail myself of the opportunity of passing a few days at least on a man-of-war, and in fact had made arrangements to go aboard on the day before the hurricane, but the early stage of the storm had already set in then, and the bay was so rough that the ship could not be reached in a small boat with safety, so I had been obliged to postpone my visit. I confess that, as I watched the vessel that day and saw the waves sweeping men into the sea, I felt that I had had a fortunate escape.

"When the distressed condition of the Vandalia became apparent, three officers of the Nipsic—Lieutenant Shearman, and Ensigns Purcell and Jones—made every effort to rescue the men; and during the whole day and night, with the assistance of several other Americans and the natives, they labored incessantly to reach the doomed vessel and used every means to save the lives of the men.

"A long hawser was procured, and three natives were found who were willing to venture out in the surf with a cord and attempt to reach the Vandalia. The men entered the water a quarter of a mile above the spot where the steamer lay, and struck out into the surf with the cord tied to their bodies.

"Shouts of encouragement went up from the shore, and the Samoans struggled bravely to reach the sunken ship. But, expert swimmers as they were, they were unable to overcome the force of the current, which rushed down like a cataract between the Vandalia and the shore, and the men were thrown upon the beach without being able to get within a hundred yards of the vessel. Seumanu Tafa, their chief, urged the men to try again, and several other attempts were made, but without success.

"It was now evident that many of those on the Vandalia would not be able to withstand the force of the waves much longer, and would be swept into the sea. Natives waded into the water, and stood just on the edge of the current ready to grasp any one who should float near them. The seas continued to break over the vessel, and it was not long before several men were washed over the side. As soon as they touched the water they swam for the Nipsic, where they grasped ropes hanging over the side and attempted to draw themselves on deck. A number succeeded in doing this, but others were so weak that, after hanging to the ropes a moment, their grasp was broken by the awful seas which crashed against the side of the vessel, and they would fall back into the current.

"The first man who came ashore was Chief-Engineer A. S. Greene. When he was washed from the deck of the Vandalia, he was fortunately able to catch a piece of floating wreckage.

"He soon drifted into the current, and was swept down along the shore. The natives saw his head above the water, and they clasped each other's hands and formed a long line stretching out into the current. As the chief engineer swept by, the native farthest out grasped him by the arm and brought him to shore.

"Just before he was rescued, another man, who had been washed from the Vandalia, was seen clinging to a rope by the side of the Nipsic. The waves had torn away all his clothing.

"There were several Vandalia sailors aboard the Nipsic by this time, and he shouted to them to draw him up, but his voice was lost in the roar of the wind, and, after clinging to the rope a while longer, he let go and grasped a piece of board which was floating past him. He also was drawn into the current, and commenced to sweep along the shore. He was farther out than Chief-Engineer Greene had been, but the Samoans were making every effort to reach him, and had advanced so far into the current that they were almost carried away themselves. Just as the drowning man was within a few feet of the mouth of a small river, where another current would have swept him far out into the bay, the natives caught him and drew him ashore. He proved to be H. A. Wiley, a young naval cadet. He was carried to the consulate insensible, and it was only after great exertions that he was resuscitated.

"It was not long after Greene and Wiley were washed overboard that the four officers who were drowned were swept from the deck. Captain Schoonmaker was clinging to the rail on the poop-deck. Lieutenant Carlin was standing by him trying to hold the captain on, as the latter was becoming weaker every minute. Every one on the deck saw that he could not stand against the rush of water much longer. No one knew it better than himself, and he several times remarked to those about him that he would have to go soon. Lieutenant Carlin tried to get him up in the rigging, but the captain said he was too weak to climb up, and would have to remain where he was as long as possible. He had no life-preserver; one had been offered him several times, but he had refused it. At last a great wave submerged the poop-deck. Captain Schoonmaker held on to the rail with all the strength he had left, but the torrent of water wrenched a machine-gun from its fastenings, and sent it whirling across the deck. The captain was bending down at the time, and the gun struck him on the head, and either killed him outright or knocked him insensible, for the wave swept him from the deck. He sank without a struggle, and was seen no more.

"Paymaster Frank H. Arms and Pay-Clerk John Roche were lying upon the deck exhausted, but clinging with all the strength they possessed to anything which came within their grasp. They were washed off together. The paymaster sank in a moment, but Roche drifted over to the stern of the Nipsic, where he grasped a rope. He was a large, fleshy man, and, being greatly exhausted, could not possibly draw himself up on the deck. His hold upon the rope was soon broken, but he continued to float under the stern of the Nipsic several minutes, wildly throwing out his arms in a vain attempt to clutch something. He finally sank under the vessel. Lieutenant Frank E. Sutton, the marine-officer, died in nearly the same way. Weakened by long exposure and the terrible strain to which he was subjected, he was unable to retain his hold any longer, and was washed overboard and drowned.

"During the remainder of the afternoon there followed a succession of awful scenes of death and suffering not soon to be forgotten. The storm had not abated in the least. The wind continued to blow with terrible force; waves that seemed like mountains of water rolled in from the ocean and broke upon the reef and over the ill-fated Vandalia. The sheets of water which fell from the clouds, and the sand which was beaten up from the shore, struck like hail against the houses. White men who stood out in the storm were often obliged to seek shelter to escape the deluge of rain and sand, which cut the flesh like a knife, and even the natives would occasionally run for safety behind an upturned boat or a pile of wreckage.

"The Vandalia continued to settle, and the few men who had not already taken to the rigging stood on the poop-deck or on the forecastle, as the vessel amidships was entirely under water.

"Almost twenty-four hours had elapsed since any one aboard had tasted food, and all were weak and faint from hunger and exposure. Men were now washed from the decks and rigging a half-dozen at a time, and a few, who felt that they were growing too weak to hold on, jumped into the water, determined to make one last effort to save themselves.

"Nearly every man who jumped or was washed overboard succeeded in reaching the Nipsic, and a number of them climbed upon the deck by the aid of ropes. Those who reached the deck assisted others who were struggling in the water, and several lives were saved in that way. But many a poor fellow who reached the Nipsic's side was unable to hold on to a rope long enough to be drawn up, and the seas would wash him away and sweep him into the current. None of them came near enough to the shore to be reached by the natives, and those who once got into the current were carried out into the bay and drowned.

"As I stood on the beach that afternoon, I saw a dozen men go down before my eyes. I was with Lieutenant Shearman and Ensigns Purcell and Jones, Nipsic officers, and Consul Blacklock nearly all the time. We had been dragging heavy hawsers up and down the beach all day, and had adopted every means in our power to render some assistance to the drowning men. As we watched them struggling in the water, far beyond any human aid, I remember how we felt, that we must do something to reach them; but we were powerless. We had seen a hundred German sailors go down early in the morning, and while we had recognized the horror of that calamity we were not impressed with the same feeling which came upon us as we saw men of our own country suffering the same fate. Here there was a bond of sympathy which appealed to us as Americans, and one who, in a foreign land, has ever seen such death and suffering befall his fellow countrymen can appreciate the feeling with which we watched those scenes in Apia harbor.

"By three o'clock the Vandalia was resting her whole length on the bottom, and the only part of her hull which stood out of water was the after part of the poop-deck and the forward part of the forecastle. Every man was in the rigging. As many as could be accommodated there had climbed into the tops and sunk down exhausted upon the small platforms. Others clung to the ratlines and yards with the desperation of dying men, expecting every moment to be their last. Their arms and limbs were bruised and swollen by holding on to the rough ropes. A number had been greatly injured by falling about the decks, and many a poor creature was so benumbed with cold and exposure to the biting rain, and so weak from want of food, that he sank almost into insensibility, and cared not whether he lived or died.

"More than one man who was clinging to the ratlines gave way under the terrible strain and fell to the deck, only to be washed over the submerged side of the ship and drowned.

"A hawser had been made fast from the deck of the Nipsic to the shore, and the Vandalia men who had escaped to the Nipsic reached shore in that way.

"The Nipsic had by that time swung out straight from the shore, so that the distance between the two vessels was not more than fifty yards. A small rope was made fast from the foremast of the Vandalia to the stern of the Nipsic, and a few men escaped by it, but before all in the fore-rigging were rescued, the line parted, and could not afterwards be replaced.

"The terrible scenes attending the wreck of the Vandalia had detracted attention from the other two men-of-war which still remained afloat; but about four o'clock in the afternoon the positions of the Trenton and Olga became most alarming. The flag-ship had been in a helpless condition for hours.

"At ten o'clock in the morning her rudder and propeller had been hawse-pipes (the large openings in the bow through which the anchor-to her discomfiture, great volumes of water poured in through the chains pass). From ten o'clock in the morning until six in the evening, when she grounded, the Trenton held out against the storm without steam or rudder, and her escape from total destruction was miraculous. Admiral Kimberly, Captain Farquhar, and Lieutenant Brown, the navigating officer, stood upon the bridge the whole day and directed the movements of the ship. For two hours before the fires were extinguished the water was rushing in through the hawse-pipes and pouring down the hatches into the fire-room and engine-room. The men at work there were in a most perilous position, as they were so far down below the deck that if the vessel had gone upon the reefs suddenly and sunk, they never could have escaped. Engineers Galt and Matthews were in charge of the engine-room during the time that the water was pouring down the hatches. All the men there stood at their posts until they were waist-deep in the water and the fires were extinguished. The berth-deck also was flooded, and efforts were made to close the hawse-pipes. Lieutenant W. H. Allen remained below all day superintending this work, but, though he was partially successful, the force of the water was so great that everything placed in the pipes was torn out. It was a most dangerous post, as the men stationed there had two decks above them, and in case the vessel should go down their escape was shut off. Allen and his men were deluged with the torrents of water which rushed in through the openings with every pitch of the vessel. It was necessary to work the pumps early in the day, and this was kept up constantly. Men never fought against adverse circumstances with more desperation than the officers and men of the Trenton displayed during those hours when the flag-ship was beaten about by the gale. There was not an idle man on the ship. The entire supervision of affairs outside of the manœuvring of the vessel fell upon Lieutenant-Commander H. W. Lyon, who afterwards received the commendation of his superior officers for the efficient services which he rendered during the storm. Among the officers who rendered most valuable assistance were Lieutenants Graham, Scott, and Allen, and Ensign Blanden.

"By the skilful use of a storm-sail, the Trenton kept well out in the harbor until the middle of the afternoon, and then she was forced over towards the eastern reef. Destruction seemed imminent, as the great vessel was pitching heavily, and her stern was but a few feet from the reef. This point was a quarter of a mile from shore, and if the Trenton had struck the reef there, it is probable that not a life would have been saved. A skilful manœuvre, which was suggested by Lieutenant Brown, saved the ship from destruction. Every man was ordered into the port rigging, and the compact mass of bodies was used as a sail. The wind struck against the men in the rigging and forced the vessel out into the bay again. She soon commenced to drift back against the Olga, which was still standing off the reef and holding up against the storm more successfully than any other vessel in the harbor had done. The Trenton came slowly down on the Olga, and this time it seemed as if both vessels would be swept on the reef by the collision and crushed to pieces. People on the shore rushed to the water's edge and waited to hear the crash which would send to the bottom both men-of-war and their loads of human lives. Notwithstanding the dangerous situation of the ships, a patriotic incident occurred at this time which stirred the hearts of all who witnessed it. The storm had been raging so furiously all day that not a flag had been raised on any of the vessels. As the Trenton approached the Olga, an officer standing near Admiral Kimberly suggested that the flag be raised. The Admiral, whose whole attention had been absorbed in directing the movements of the ship, turned for a moment to the group of officers near him, and said, 'Yes, let the flag go up!'

"In an instant the stars and stripes floated from the gaff of the Trenton, and to those on shore it seemed as if the gallant ship knew she was doomed, and had determined to go down with the flag of her country floating above the storm. The Olga, seeing the approach of the Trenton, attempted to steam away, but just as she had commenced to move up against the wind, her bow came in contact with the starboard quarter of the flag-ship. The heavy timbers of the Trenton's quarter were shivered, several boats were torn from the davits, and the American flag, which had just been raised, was carried away and fell to the deck of the Olga. Fortunately, the vessels drifted apart after the collision, and no serious damage was done. The Olga steamed ahead towards the mud-flats in the eastern part of the bay, and was soon hard and fast on the bottom. Not a life was lost, and several weeks later the ship was hauled off and saved.

"The Trenton was not able to get out into the bay again after her collision with the Olga. She was now about two hundred feet from the sunken Vandalia, and was slowly drifting towards the shore. A new danger seemed to arise. The Trenton was sure to strike the Vandalia, and to those on shore it seemed that the huge hull of the flag-ship would crush the Vandalia to pieces and throw into the water the men still clinging to the rigging. It was now after five o'clock, and the daylight was beginning to fade away. In a half-hour more the Trenton had drifted to within a few yards of the Vandalia's bow, and feelings hard to describe came to the hundreds who watched the vessels from the shore.

"The memory of the closing incidents of that day will cling to me through life, for they were a spectacle such as few have ever seen. No American can recall those patriotic features without feeling a glowing pride in the naval heroes of his country. I was standing with others as far down on the beach as it was safe to be, watching the ships through the gathering darkness, and every incident that occurred came under my personal notice.

"Presently the last faint rays of daylight faded away, and night came down upon the awful scene. The storm was still raging with as much fury as at any time during the day. The poor creatures who had been clinging for hours to the rigging of the Vandalia were bruised and bleeding; but they held on with the desperation of men who were hanging between life and death. The ropes had cut the flesh on their arms and legs, and their eyes were blinded by the salt spray which swept over them. Weak and exhausted as they were, they would be unable to stand the terrible strain much longer. They looked down at the angry waters below them, and knew that they had no strength left to battle with the waves. The final hour seemed to be upon them. The great black hull of the Trenton could be seen through the darkness, almost ready to crash into the stranded Vandalia and grind her to atoms. Suddenly a shout was borne across the waters. The Trenton was cheering the Vandalia. The sound of four hundred and fifty voices broke upon the air, and was heard above the roar of the tempest. 'Three cheers for the Vandalia!' was the cry that warmed the hearts of the dying men in the rigging.

"The shout died away upon the storm, and there arose from the quivering masts of the sunken ship a response so feeble it was scarcely heard upon the shore. Men who felt that they were looking death in the face aroused themselves to the effort, and united in a faint cheer for the flag-ship. Those who were standing on the beach listened in silence, for that feeble cry was the saddest they had ever heard. Every heart was melted to pity. 'God help them!' was passed from one man to another. The cheer had hardly ceased when the sound of music came across the water. The Trenton's band was playing 'The Star-Spangled Banner.' The thousand men on sea and shore had never before heard strains of music at such a time as that. An indescribable feeling came over the Americans on the beach who listened to the notes of the national song mingled with the howling of the storm.

"Men who had exhausted every means, during the whole of that awful day, of rendering some assistance to their comrades, now seemed inspired to greater effort. They ran about the beach eager to afford help, even at the risk of life itself. They looked despairingly at the roaring torrent of water that broke upon the shore, and knew that no boat could live in such a sea. Bravely as the Samoans had acted, there was not one of them who would again venture into the surf, where certain death would befall them.

"Persons on shore were simply powerless, and there was nothing to do but remain on the beach ready to lend assistance in any possible way which might present itself.

"But the collision of the Trenton and Vandalia, instead of crushing the latter vessel to pieces, proved to be the salvation of the men in the rigging. When the Trenton's stern finally struck the side of the Vandalia, there was no shock, and she swung around broadside to the sunken ship. This enabled the men on the Vandalia to escape to the deck of the Trenton, and in a short time they were all taken off.

"By ten o'clock at night the natives and nearly all the white persons who had watched the storm seemed to be satisfied that no further harm could come to the two ships; and the shore, which had been thronged with people all day, was soon deserted. The three Nipsic officers and myself patrolled the beach all night in the hope of rescuing some one who might not have escaped to the Trenton. We found but one man, Ensign Ripley, who had jumped from the Vandalia before the Trenton touched her, and had reached the shore. He was lying on the beach exhausted and about to be washed out by the undertow when we came upon him and carried him to the consulate. The storm had abated at midnight, and when day dawned there was no further cause for alarm. The men were removed from the Trenton and provided with quarters on shore.

"During the next few days the evidences of the great disaster could be seen on every side. In the harbor were the wrecks of four men-of-war,—the Trenton, Vandalia, Adler, and Eber; and two others, the Nipsic and Olga, were hard and fast on the beach and were hauled off with great difficulty. The wrecks of ten sailing vessels also lay upon the reefs. On shore, houses and trees were blown down, and the beach was strewn with wreckage from one end of the town to the other.

"Above the whole scene of destruction the stars and stripes and the flag of Rear-Admiral Kimberly floated from the shattered masts of the Trenton, as if to indicate that America was triumphant even above the storm. The American naval forces took entire control of the town, and a guard of Marines, under Captain R. W. Huntington, of the Trenton, and Lieutenant Fillette, of the Nipsic, was stationed in every locality to prevent any trouble which might arise on account of the great confusion which prevailed on shore.

"A muster showed that one hundred and forty-four lives had been lost in the storm. Of these, ninety-one were from the German ships, Eber and Adler. The Vandalia had lost four officers and thirty-nine men, and the Nipsic had lost seven men. One man was killed on the Trenton by a piece of flying timber, and two victims from the schooner Lily were added to the list.

"Not more than one-third of the bodies were recovered. The others were either swept under the coral reefs in the harbor, or washed far out to sea."

---

## CHAPTER XXXVII.

### THE PARIS EXPOSITION.

**1889.**—On the 13th of April a detachment of Marines, under the command of Captain H. C. Cochrane, United States Marine Corps, assisted by First Lieutenant P. St. C. Murphy, United States Marine Corps, sailed on the steamer La Gascoigne for duty at the Paris Exposition. The detachment reached Havre on the 21st and Paris on the 22d.

The exposition opened on the 6th of May with imposing ceremonies, and the splendid detachment of American Marines received not only the applause of their countrymen, but the admiration of the foreigners present.

The duty assigned the Marines was performed with faithfulness, soldierly exactitude and firmness, and one of the interesting features of the Exposition was the American quarter, where could be seen daily these soldiers from the New World.

On the Fourth of July the detachment marched from the *Ecole Militaire,* the barracks where they were quartered, to the grave of Lafayette. There, under the direction of Captain Cochrane and Lieutenant Murphy, and in the presence of a large multitude, the American Marines decorated with flowers the modest tomb of *America's hero.*

The following description of the ceremony is taken from Galignani's *Messenger,* of the 5th of July of that year:

"This ceremony took place on Thursday morning, July 4, at ten o'clock, at the small cemetery, attached to the convent of the Sœurs Blanches, Rue Picpus, where General Lafayette is buried. Dr. Thomas W. Evans, acting as chairman of the organizing committee, received at the gate of the convent the persons who came to take part in the ceremony. Among those present were the American Minister, Mr. Whitelaw Reid, General W. B. Franklin, Consul-General Rathbone, Dr. J. A. Tonner, Hon. Andrew D. White, Colonel W. Cody, Mr. N. Salisbury, Rev. Rowland B. Howard, Rev. Dr. Thurber, Mr. Ward De Lancey, General Meredith Read, Lieutenant H. D. Bomp, United States army, military attaché to the American legation, etc. A little after ten the United States Marines from the exhibition grounds, under the command of Captain Cochrane, made their appearance, whereupon the procession was immediately organized. The Marines, with the stars and stripes floating, opened up the march, followed closely by the representatives of the Lafayette family, Minister Whitelaw Reid, General Franklin, Dr. Thomas W. Evans, Consul-General Rathbone, and all the members of the organizing committee, who were present, carrying flowers and beautiful wreaths, who were followed by the whole assembly. The procession slowly marched to the tomb of General Lafayette, which is situated at some distance from the entrance of the convent,—the funeral march being played by the buglers of the Marines. The Marines surrounded the modest tombstone, and reversed their rifles. The wreath-bearers then advanced, one by one, to deposit their floral offerings on the tomb. When the decoration was completed, M. Edmond de Lafayette, the grandson of the general, stepped forward, and in a few but touching words expressed, in the name of the Lafayette family and in his own, to all present, the sincerest and warmest thanks for this beautiful demonstration of the gratitude of the American people to their well-beloved relative. Captain Cochrane then drew up his command in line, and the United States Marines fired three volleys of musketry over the grave of the Franco-American hero. The whole ceremony in its simplicity was of a very imposing character, and will be remembered by all present as one of the most impressive demonstrations which has ever been made in France by foreigners in memory of a deceased French citizen. It will be an interesting fact to the many Americans who attended the ceremony, to know that they were very near—some even upon—American soil; for at the time of the death of General Lafayette, the United States government, as a testimony of gratitude to the deceased, sent to France some American earth, which, it was desired, should be mingled with the French soil, on the place where the body of the French general of the United States army was interred."

On the 19th of November the detachment was honored by the presentation of medals by the American exhibitors; the officers' medals being of gold and those of the men of bronze. Captain Cochrane, the commander, also had conferred upon him the decoration of Chevalier of the Legion of Honor, and complimented in general order upon return.

(From *Army and Navy Register,* December 14, 1889.)

" The following letter has been addressed by Secretary Tracy to the committee of American exhibitors at the Paris Exposition:

"NAVY DEPARTMENT, WASHINGTON, D. C., December 6, 1889.

"SIR,—Colonel C. G. McCawley, commandant, United States Marine Corps, has sent to the Department a copy of a communication addressed to him under date of the 26th ultimo, by Captain H. C. Cochrane, United States Marine Corps, who commanded the detachment of Marines on guard duty in connection with the American exhibit at the Universal Exposition of 1889, at Paris, in which he gives an account of the presentation of commemoration medals, one of gold to each of the officers and one of bronze to each of the men of the detachment, by the American exhibitors.

"Captain Cochrane refers to the presentation as an incident of unusual interest and as an event of the most gratifying character. That the members of the Marine detachment should have been so highly esteemed by all with whom they came in contact, during their term of duty at the Exposition, is not only very gratifying to each and every one of them, but also to the Department and the Corps they represent.

"They will bring with them the emblem of that esteem in the beautiful and appropriate medals bestowed on them by their highly-intelligent and appreciative countrymen, and will long cherish the recollection of the kind words for and generous courtesies to them at all times, and especially on the brilliant occasion when they were recipients of such marked honors. The Department avails itself of this means of making known to the committee its appreciative sense of the distinction conferred on the detachment by the American exhibitors.

"Very respectfully, "B. F. TRACY,
"*Secretary of the Navy.*

"JAMES ALLEN, ESQ.,
"*Representative Pennsylvania Railroad Company, Secretary of Committee American Exhibition.*"

(From *Galignani's Messenger,* November 20, 1889.)

"As a token of regard for the detachment of United States Marines, who have been on duty at the Paris Exposition a *fête* was given in their honor last night by the American Exhibitors' Committee, who had struck commemorative medals for distribution among the officers and men as a *souvenir* of their stay in this city. The ceremony was presided over by Mr. Somerville Tuck, United States assistant commissioner-general, in the spacious rooms of the Salons Delmonico, which were suitably ornamented with the stars and stripes and the French tricolor. It was hoped that the United States Minister, Mr. Whitelaw Reid, would be present, but at the last moment he wrote to Captain Cochrane as follows:

"DEAR CAPTAIN,—The physicians still think it best to keep me shut up, and as I could not go out to-day it would seem to be defying them to go out to-night. I am sorry not to be able to witness the deserved decoration of your splendid little body of men, of whom we have all been so proud during the last six months.

"The Marines, who, as usual, presented a splendid appearance, were under the orders of Captain Cochrane and Lieutenant Murphy. There were present, besides the chairman, United States Consul-General Rathbone, Captain Maigret, representing the President of the republic; Lieutenant-Colonel Delorme, on behalf of the Minister of Marine; M. Crosier, for the Minister of Foreign Affairs, and some officers representing the Twenty-eighth Regiment of French Infanty, which served in American during the Revolution; Senator Lafayette, General Meredith Read, Messrs. Hammer, Miles, W. Stiassny, and James Allan, members of the committee; General Young, Mr. Amory Austin, and Mr. Washington I. Adams, of the United States Commission; Mr. J. B. Jackson, late of the United States Navy; Dr. Tonner; Mr. Williams, United States Consul at Rouen; Captain Zalinski of the United States Fifth Artillery; Mr. Ellis, private secretary to the United States Minister in Paris; Messrs. Jeffrey and Chanute, the deputation sent from Chicago to make a special report on the exhibition; Messrs. Gunnell, Bickford, Collins, Butterworth and Philip Franklin, nephew of General Franklin, United States commissioner-general, who has returned to America; M. Armand Dumaresq, the artist; Mr. J. H. Hobson, Mr. Henry Haynie, Mr. Hatch, Mr. George Strong, and Mr. Small, of Baltimore, and many others.

"Mr. Somerville Tuck, in opening the proceedings, addressed the officers and men as follows:

"'Among the many pleasant experiences which have fallen to my lot this summer, none equals this, when I have the honor to make a presentation which is the object of our being here to-night. A number of the exhibitors in the United States section have had the happy thought of uniting themselves for the purpose of showing their approval of the services done by you, and determined that it should take the form of a diploma and commemorative medal. You represent a Corps, the most ancient in the military and naval organization of our country, and now well into the second century of its existence. Beginning with a battalion it now equals a brigade, and bears the same proportion to our fortunately small warlike organization as the similar bodies of the Old World. Military in form, you serve almost entirely with the Navy, of which you form part, and to which you impart a military character. Yours is a corps which has also served with the army in the field, in our campaigns with foreign foes, against the Indians, in political and labor riots, at large conflagrations, and scarcely any event in our history can be recalled where the Marines have not been actors, and always to their renown and credit. I can recall the Tripolitan and Corean wars, the only occasions the United States troops captured a fort in the Old World, China, the Japan expedition, your landing at Alexandria, the Panama expedition, when within three hours every depot had its draft *en route* for New York to sail within twenty-four hours, and lastly Samoa, in this present year, when Marines volunteered for the fire-room, shovelled coal into the furnace of a sinking ship, the Vandalia, and those who reached the land instantaneously took up their duty.

" 'Peace has its victories as well as war. Your Corps has always been selected upon the occasion of civic processions and solemnities. In 1878 you served here in a similar capacity. Let me quote the words of General Bullock, of Massachusetts, who was present on the occasion of the closing celebration in that year: "At the recent distribution of prizes, when the flag of our distant country came into the building under the guardian custody of those manly representatives of American discipline, you saw and heard and felt the plaudits which ran over the vast building with a freshness and heartiness which it seemed to me was accorded in no equal degree to any other nationality. It was the tribute of Frenchmen and Frenchwomen. I saw thousands of the women of France around me who rivalled the enthusiasm of the colder and sterner sex. It was the spontaneous tribute of Frenchmen and French women to the fond recollections of the past. The memories of D'Estaing and Franklin, of Lafayette and Washington, floated in the very air."

" 'History repeats itself, and no words of mine could more fully describe your reception and appearance and the applause given you on the 20th of September last.

"'With such a record as yours, it was natural that the commission, of which I was a member, should have applied for your detail at this exhibition. Your commandant responded with the alacrity characteristic of your Corps. His choice of your officers was particularly happy. Your captain I have had the privilege of knowing for many years, a modest gentleman and gallant soldier, who will bravely bear always the cross which came to him from the executive of this our sister great republic, and which has the great and exceptional merit of having been conferred and not sought. Your lieutenant I met for the first time when I had the pleasure of receiving you at Havre; but he was not unknown to me, and the coincidence was a most singular one. At the time the Navy Department requested your detail, I met a lieutenant of the Navy, an old friend, and, saying to him I had outgrown or lost sight of my old officer friends in your Corps, and had no idea of the selection that might be made, he replied that there was a gentleman and officer who had been so courteous and so considerate to him on his last court-martial (my friend has a record of about two in every five years) that he would like very much to see him get the billet. The name was mentioned; I could do nothing, knowing the etiquette which forbids the applying for any particular officer for details of this kind; but I was most happy to know that my friend's desire had been fulfilled when I was presented to your lieutenant upon the steamer in the roads of Havre.

"'The gentlemen who have made this testimonial preserve their anonymity to such a degree that I know only the secretary of the manifestation, and must to him, therefore, and for you, express my official and personal gratitude for this graceful recognition of services well and faithfully done. No medal was ever more worthily bestowed, and in years to come it will be to you all a beautiful *souvenir* of the important part borne by you in this great triumph of labor and industry, which has allayed party strife at home, promoted international conciliation, and by assembling the nations of the earth in friendly intercourse brought so much nearer the time when there will be peace on earth and good will to men.

"'In conclusion, permit me to express my pleasure at welcoming here to-night the officers of the Twenty-eighth of the Line, a regiment which stood shoulder to shoulder with our line in the days that tried men's souls, which helped us to become the old republic of the New World, and brought back from America principles which have expanded until France is to-day the new but eternal republic of the Old World.'

"The medals and diplomas having been distributed, the kindness which prompted the testimonials was acknowledged in a spirited and humorous speech by Captain Cochrane, who also paid a tribute to the magnificent hospitality dispensed by the Paris municipality.

"Senator Lafayette, who spoke in English, was enthusiastically cheered. He thanked the committee for inviting him to be present on such an interesting occasion, and congratulated the officers and men upon the rewards which they had so thoroughly deserved. He hoped they would keep in their hearts the best remembrance of France and the French; and he wished them a happy and safe return to their country,—that land of liberty, prosperity, and progress,—the land of Washington."

The following is the general order of the commandant of the Corps, publishing the letters of General W. B. Franklin, commissioner-general of the United States, and of the Hon. B. F Tracy, Secretary of the Navy, in relation to the detachment:

GENERAL ORDER No. 1.

HEAD-QUARTERS UNITED STATES MARINE CORPS,
WASHINGTON, D. C., January 10, 1890.

It is with great pleasure that the following letters of General W. B. Franklin, commissioner-general of the United States to the Universal Exposition of 1889, at Paris, and the Hon. Secretary of the Navy, referring to the soldierly bearing and excellent conduct of the detachment of Marines lately on duty at that Exposition, are published to the officers and enlisted men of the United States Marine Corps. C. G. MCCAWLEY, *Colonel-Commandant.*

NAVY DEPARTMENT, WASHINGTON, January 9, 1890.

SIR,—The department takes pleasure in enclosing a copy of a communication, addressed to it under date of the 3d instant, by the Secretary of State, accompanied by a copy of a letter from General W. B. Franklin, commissioner-general to the Paris Exposition, commendatory of the detachment of Marines at the Paris Exposition. Very respectfully, B. F. TRACY,
*Secretary of the Navy.*

COLONEL C. G. MCCAWLEY,
*Commandant Marine Corps Head-quarters.*

OFFICE OF THE UNITED STATES COMMISSIONER-GENERAL,
PARIS EXPOSITION OF 1889,
HARTFORD, CONNECTICUT, December 28, 1889.

SIR,—I have been officially informed that the detachment of United States Marines, under the command of Captain Henry Clay Cochrane, United States Marine Corps, assisted by Lieutenant P. S. C. Murphy, of the same Corps, which you were so kind as to detail for service at the Paris Universal Exposition of 1889, has returned to the United States, arriving at New York on the 24th instant. I make the following statement concerning them.

Their behavior during the Exposition and after its close was in all respects what it ought to have been; they were admirable in discipline; their conduct was irreproachable; their performance of duty was excellent.

Their soldierly bearing and military appearance were far ahead of these qualities of any troops which I saw at the Exposition or in Paris. When they appeared as a body they were enthusiastically received, and it was the unanimous opinion, not only of those connected with the United States Commission, but of the citizens of the United States who visited the Exposition, that their presence added dignity to the United States exhibit, and that they did honor to their country and their flag.

Captain Cochrane and Lieutenant Murphy not only performed their duties admirably as the officers of the detachment, but were of great assistance to me in various other duties that they were kind enough to undertake.

The French authorities of the War Department were very kind in their reception of them, and in transporting them and in furnishing their quarters in the barracks of the Ecole Militaire, with French troops. In fact, all French troops treated them as comrades, and the relations between these soldiers and them were of the pleasantest kind.

Dr. J. A. Tonner, a citizen of the United States, resident in Paris, and medical attaché of the United States Commission, attended the detachment professionally without compensation. It gives me pleasure to acknowledge his efficient and patriotic service.

I thank you for having given me the opportunity of close acquaintance with such excellent troops.

Respectfully yours,
(Signed) W. B. Franklin,
*United States Commissioner-General.*

Hon. B. F. Tracy,
*Secretary of the Navy, Washington, D. C.*
Through Department of State.

---

# CHAPTER XXXVIII.

1891.—The troubles at Nevassa Island—The Marine Corps Detachment at Valparaiso, Chile, During the Revolution—The Cruise of Captain Cochrane, U. S. M. C., and Detachment in Bering Sea—The Foundation of the School of Application for Officers and Men of the Marine Corps.

1891.—During the period of trouble with the negro laborers on Nevassa Island a detachment of Marines from the U. S. S. Kearsarge, under command of Lieutenant George T. Bates, was landed to protect American lives and property; remaining about three weeks on shore and rendering such soldierly and efficient service as to elicit a letter of high commendation from Commander Horace Elmer, U. S. N., commanding the Kearsarge.

When, in the month of August, the city of Valparaiso, Chili, was captured by the forces of the Congressional party, the Marines from the warships on the South Pacific Station, under command of Captain W. S. Muse, were landed to aid in protecting the American consulate. They remained on shore three days (August 28, 29, 30), and by their presence and soldierly bearing did efficient service in preventing any forms of violence, as was fully acknowledged by the commander-in-chief of the station.

During the months of June and July, and a part of August, a detachment of forty enlisted men of the Marine Corps, under the command of Captain H. C. Cochrane and Lieutenants J. A. Turner and J. H. Pendleton, cruised in Bering Sea, on the steamer Al-Ki, specially chartered by the Navy Department in the work of suppressing seal poaching. The success of the expedition, and the important part played by the American force are set forth in Captain Cochrane's letter of report to Colonel Charles Heywood, commandant of the Marine Corps. This letter, which abounds in interesting incidents, is as follows:

U. S. Marine Barracks,
Mare Island, Cal., October 16, 1891.

Sir: I have the honor to report that the Marine detachment, consisting of three officers (First Lieuts. Turner and Pendleton and myself) and forty enlisted men, embarked by your telegraphic order of June 19th, last, for special duty in Bering Sea, in connection with the agreement between Great Britain and the United States for the arrest of seal poaching, returned on the 14th instant, having been absent nearly four months.

We went on board the steamer Al-Ki, a chartered vessel belonging to the Pacific Coast Steamship Company, at Mare Island, on the 21st, and sailed from San Francisco on the 22d of June. Arrived in Bering Sea July 2d, and at Iliuliuk, Ounalaska, Aleutian Islands, the same day, in advance of the other vessels ordered.

The U. S. S. Thetis arrived on the 3d, the Mohican on the 8th, and the Alert on the 9th of the same month, and the Marion on the 14th of August, making the combined force of marines present 5 officers and 113 men. H. M. S. Nymphe, from Victoria, British Columbia, joined the fleet on the 7th of July, the Pheasant on the 12th, and the Porpoise, from the China station, on the 27th, each provided with a guard of marines.

While the men-of-war, together with the United States revenue cutters Rush and Corwin, were engaged in cruising and furnishing all sealing, whaling, and fishing vessels with notice of the President's proclamation and the orders of the British Government relative to the fur seal fisheries, the Al-Ki acted as harbor and prison ship at Ounalaska. The crews of vessels seized were promptly transferred to our custody upon being brought into port and were uniformly well treated.

Owing to the determined attitude of the combined governments, and the custom of giving each vessel found a preliminary warning, but four seizures were necessary. These were the schooners E. B. Marvin, British, July 6th, the La Ninfa, American, July 14th, the Ethel, American, July 30th, and the

Of these vessels, the first and the last were sent to Victoria for adjudication, and the others were towed to Sitka, 1,200 miles, by the Al-Ki, and turned over to the United States marshal for Alaska. A prize crew of 1 corporal and 3 privates were put aboard of the La Ninfa, and 2 privates on the Ethel, each crew containing an expert signalman, by whom communication was maintained with the towing steamer. This unusual duty was exceedingly well performed.

While at Sitka a public drill was given on shore, in the Plaza, August 14, in the presence of the governor of Alaska, the judge of the United States Court, the collector of customs, and other distinguished citizens, which elicited many compliments and gratified the inhabitants, Russian, American and Indian, of that quiet town.

August 15th the Al-Ki left for Ounalaska, arriving on the 20th. After September 1st, no sealing vessels were found in Bering Sea, and after the 15th the weather became generally stormy, and continued so until our departure, October 5th, when the mountains were all snow-clad.

The entire summer was cool and moist, with much fog, and either overcoats or rubber clothing were in use at some time each day.

The men were as comfortably quartered as circumstances would permit, and had liberal rations, save as to fresh meat, a privation common to all of the ships in the expedition. With a hauling seine we caught great quantities of fish, salmon, haddock, cod, trout, herring and flounders, which partly supplied the absence of meat.

Very good clams were also obtained near the anchorage. This region furnishes neither vegetables nor fruits, except a few berries late in the season. These, salmon-berries, and blueberries, grow wild and are very abundant and good. Daily liberty was given the men, who preserved excellent relations with the natives (Aleuts) as well as with the white citizens.

Throughout our absence drills and instructions were unremitting, and every officer and man is trained in boating, in the Army signal code, in skirmishing, the bayonet exercise, and target practice, in addition to his routine duties. The men were organized into boats' crews from the first day, and did all of the boarding and pulling required. An excellent rifle range up to 300 yards and four signal stations were established on Amnak Island, and great interest was taken in the competitions.

The exhibition drill given in the village of Iliuliuk was highly praised by the United States officials and others present.

Much trouble was occasioned during the cruise by mutinous manifestations on the part of the sailors of the ship, who were foreigners, and "union men," and by some of the engineers' department.

The aid of the civil law was invoked twice, and on the third occasion the master of the vessel came to my office (September 9), and said: "Captain Cochrane, if the worst comes to the worst, can the Marines help me out" (referring to the attitude of his crew)? I replied: "Yes; you can set every one of them on the beach, firemen and all, and we'll take the ship to San Francisco." He has reported to the steamship company that this declaration ended the troubles. In this connection I inclose extracts from the San Francisco press.

Otto, British, August 31st. The total number of prisoners received was 48, including a dozen nationalities, and 7 Nootka Sound Indians, hunters. The crew of the Otto was not transferred to the Al-Ki.

I am thankful to report the detachment of full strength and all well, and to be able to say that its conduct has been most creditable.

The total distance traveled is about 6,600 miles.

I am, sir, very respectfully yours, HENRY CLAY COCHRANE,
*Captain U. S. Marine Corps, late Commanding Detachment.*

The COLONEL COMMANDANT, U. S. MARINE CORPS,
*Headquarters Washington, D. C.*

The year 1891 was further made memorable by the foundation of the School of Application of the Marine Corps, at the Headquarters, Washington, D. C. The object sought was to supply a long-felt want for a school of practice in which officers and enlisted men of the Corps might receive such systematic instruction as is needed "to keep pace with the progress made in the methods of warfare, and thereby insure for the Navy greater utility and continued efficiency on the part of its military arm." The direction and supervision of the work of this institution was vested in the commandant of the Corps, with such assistance from the officers of his command as might be designated by him. The general objects and scope contemplated in the school are well set forth in Colonel Heywood's General Order No. 1, dated from headquarters, May 1, 1891, as follows:

HEADQUARTERS U. S. MARINE CORPS,
General Order No. 1. WASHINGTON, D. C., May 1, 1891.

1. The colonel commandant takes pleasure in formally announcing to the Marine Corps the establishment of a School of Application at the headquarters of the Corps, with the approval of the Secretary of the Navy, as contained in the following communication:

NAVY DEPARTMENT, WASHINGTON, April 18, 1891.

SIR: The department is in receipt of your communication of the 13th instant, submitting for its consideration an outline of the course of instruction for a school of application, at the headquarters of the Marine Corps, for officers, non-commisioned officers, and privates of the corps; and the course of instruction for the school, as proposed, is, in accordance with your recommendation, approved. Very respectfully, B. F. TRACY,
*Secretary of the Navy.*

The COMMANDANT U. S. MARINE CORPS,
*Headquarters, Washington, D. C.*

2. Outline of course of instruction:

First. *Infantry tactics and small-arm instruction.*—To include the school of the soldier; school of the company; school of the battalion, and evolutions of the brigade; bayonet exercise; instructions for skirmishers; trumpet signals; ceremonies, and guard duty; position and aiming drills; target practice; estimating distances, and pistol practice.

Second. *Gunnery instruction.*—To include machine and rapid-fire gun drills; naval great-gun exercise; nomenclature and description of guns, carriages, mounts and gun implements; description and mode of manufacture of gun powder, projectiles, cartridges, fuses, and primers; pointing, sighting, sights, and range finding.

Third. *Torpedoes.*—The various kinds and their uses; how exploded; how used offensively and defensively.

Fourth. *High explosives.*—Gun cotton, nitroglycerine, dynamite, fulminates—how prepared, preserved, used; care necessary in handling.

Fifth. *Electricity.*—How produced; conductors and insulators; heating effects of currents, how applied to fire guns and torpedoes; simple electrical tests and measurements; telegraphy.

Sixth. *Field service and modern tactics.*—Exercises in application in summer camps; marches; duty of advanced guards; camps and bivouacs; outpost duty; patrols; signalling; street, bush, and open country fighting; formations for attack and defense; importance of cover; care of wounded; application of the tourniquet.

Seventh. *Field intrenchments.*—To embrace the making on the ground of rifle pits, shelter trenches, gun pits, obstacles, and improvised field defenses.

In addition to the foregoing, non-commissioned officers and other enlisted men of sufficient intelligence will receive careful instruction in the use of the logarithms, solution of plane triangles, and the practical use in the field of angle-measuring instruments in making military reconnoisances and hasty surveys.

Non-commissioned officers will be instructed also in the duties of guards when embarked, landing and campaigning with the Naval brigade, and the best formation for fighting against superior numbers armed with inferior weapons.

Additional instruction for commissioned officers is left to the discretion of the colonel commandant.

DESIGNATION.

3. The school is officially designated the "School of Application of the United States Marine Corps," and it will be under the direct care and supervision of the colonel commandant.

ORGANIZATION.

4. The School of Application shall consist of a director of instruction, who shall have the immediate command of the school and post; instructors and assistant instructors, and such officers and enlisted men as may be assigned to it for duty or instruction.

5. The school shall have two divisions, one for commissioned officers and one for enlisted men.

ADMINISTRATION.

6. Instructors and assistant instructors shall, as far as practicable in the judgment of the commanding officer of the school, be exempt from all duties that will in any way interfere with their preparation for and proper performance of duty as instructors and assistant instructors.

7. Instruction will be given according to a regular programme, and in such manner and by such methods as the colonel commandant may hereafter prescribe, and which will be duly announced in general orders.

8. The instruction as prescribed shall be obligatory for all commissioned officers and non-commissioned officers, and for such other enlisted men as may be assigned to the school for duty or instruction.

9. The course of instruction in military hygiene will be under the direction of the medical officer of the post.

10. Instructors and assistant instructors shall keep marks of all recitations and exercises, and shall submit the same at the end of each week to the director of instruction, who will forward them to the colonel commandant as a part of the weekly report of progress.

These marks will be used as aids in determining questions of merit and proficiency.

11. Maximum marks given for perfect recitations and exercises shall be 3; for a total failure 0; and between these limits according to quality, fractions expressed in tenths.

12. A board of three officers to be designated by the colonel commandant shall be appointed to attend and witness the examinations at the close of each course of instruction at the school.

It shall be the duty of the board to examine into the actual condition of the school respecting its discipline, the methods of instruction employed, etc. The board shall report its observations and make any recommendations it may deem advisable to the colonel commandant.

13. The commanding officer of the school will make to the colonel commandant an annual report of the progress and wants of the school, and he will recommend such alterations in and additions to the programme of instruction and code of regulations as he may from time to time consider necessary or advisable.

14. With the course of instruction as outlined for the School of Application, and a modified course for branch schools at each post, together with the instruction, officers are able to impart to their guards on board ship, the Marine Corps will be enabled to keep pace with recent progress in the profession of arms, and to this end the colonel commandant confidently relies upon the co-operation of all the officers of the corps.

CHARLES HEYWOOD, *Colonel Commandant.*

The School of Application was under command of Captain D. Pratt Mannix, U. S. M. C., until his death, February 6, 1894. He was succeeded by Captain Paul St. C. Murphy, U. S. M. C., who was commandant for two years (1894-1896), and finally, by Captain F. H. Harrington, U. S. M. C., who served until the opening of the Spanish-American War. After the war, the practical work of instruction was conducted in the various branches at different posts of the Marine Corps and the Navy, and separate reports were made by the several officers in charge.

The scope of the work originally contemplated and very thoroughly carried out in the school is well stated in Captain Mannix's report on the first year's work (1891-1892):

## REPORT OF CAPTAIN D. P. MANNIX, COMMANDING SCHOOL OF APPLICATION.

HEADQUARTERS SCHOOL OF APPLICATION, U. S. MARINE CORPS,
MARINE BARRACKS, WASHINGTON, D. C., October 1, 1892.

SIR: In compliance with paragraph 13, General Order No. 1, Headquarters U. S. Marine Corps, series of 1891, I have the honor of making the following report of the operations of the School of Application of the United States Marine Corps for the past year, the first of the existence of the school:

The School of Application of the U. S. Marine Corps was established at the headquarters of the corps by General Order No. 1, Headquarters U. S. Marine Corps, May 1, 1891, with the view of supplying a want long felt by the corps for a school of practice, by which it might be enabled to keep pace with the progress made in the methods of warfare, and thereby insure for the Navy greater utility and continued efficiency on the part of its military arm for all duties it might be called upon to perform, ashore or afloat, in peace or war, requiring the disciplined soldier, the skilled marksman, and the expert artillerist.

It was the intention also that this school should supplement the course of the Naval Academy by instruction of the academic graduates assigned to the Marine Corps in the administrative and military duties of the corps. They are, therefore, sent here when assigned to the corps, for instruction before joining other stations.

The headquarters of the corps has special advantages not possessed by any other place for a school of practice. Here student officers and men are comparatively free from the care of guard duty, except so much as may be necessary for instruction, and in consequence display more interest in their drills and studies and have more time to devote to them. The naval experimental ground for ordnance at Indian Head is of easy access, as is the naval magazine with its excellent range for target practice, and a few miles back of the Potomac's eastern branch the country is well adapted for field training. The navy-yard and gun foundry are conveniently at hand, combining a water front admirably suited for instruction in boat pulling and sailing, with the advantages afforded by the gun shops of acquiring a practical knowledge of the manufacture of guns, carriages, etc., not to be obtained elsewhere.

The school has also the services of the corps band, which it could enjoy at no other station, and it has moreover at headquarters the personal care and supervision of the commandant of the corps.

#### PROGRAMME OF INSTRUCTION.

The school has two divisions, one for commissioned officers and one for enlisted men.

### DIVISION FOR COMMISSIONED OFFICERS.

For the purpose of administration and convenience in instruction the course is arranged into departments as follows:

I. Department of Infantry.
II. Department of Artillery.
III. Department of Administration and Sea Service.
IV. Department of Law.
V. Department of Torpedoes.
VI. Department of Engineering.
VII. Department of Military Art.

The course in each department is both practical and theoretical, preference being given to the former wherever it is possible to do so.

The class of student officers report for instruction and duty September 1 of each year. Five days are allowed to settle and arrange their affairs. The instruction then commences in the department of infantry.

## Department of Infantry.

The course of infantry is divided into four parts, embracing the drill instructions, guard duty, small-arms firing regulations, and infantry fire discipline, and consists of recitations, drills, problems, and field exercises.

### DETAILED PROGRAMME OF STUDIES.

#### Part I.—*Infantry drill instructions.*

Definitions; general regulations; school of the squad; school of the company; school of the battalion; evolutions of the brigade; extended order drills; formation for street riots; ceremonies; signals.

#### Part II.—*Guard duty.*

Rosters; the officer of the day; the officer of the guard; the sergeant of the guard; the corporal of the guard; the soldier of the guard; orders for sentinels on post.

#### Part III.—*Small-arms firing regulations.*

Care of rifle; nomenclature; dismounting and assembling; sighting drills; position and aiming drills; gallery practice; general regulations for range practice; details for individual practice; suggestions for riflemen; skirmish firing; volley and file firing; targets; ranges; marking and scoring; classification; motion of bullets; variations in the trajectory; the trajectory as affected by atmospheric conditions; the effects of fire; measuring distances by pacing; estimating distances by sight; estimating distances by sound; pistol practice.

#### Part IV.—*Infantry fire discipline.*

Long-range *versus* short-range fire; supply of ammunition on the battlefield; observations on supplying ammunition; uncontrolled and controlled fire; advantages and disadvantages of controlled and uncontrolled fire; fire discipline and the control and direction of fire.

1. The distance at which fire should be opened and the number of cartridges to be used to attain a definite object;
2. Determination of the force required;
3. On choice of ground;
4. The selection of the objects;
5. Determination of range, sights;
6. The kind of fire to be used; advantages and disadvantages of mass and volley firing;
7. On the intensity;
8. On the attitude of the men;
9. The observation of results;
10. When men in movement may fire;
11. Number of rounds to be fired at each halt;
12. When the bayonet should be fixed; fire units. Résumé.
13. Tactical deductions.

## Department of Artillery.

The course of artillery is divided into two parts, embracing artillery drill instruction and naval gunnery, and consists of recitations, drills, and practical exercises.

### DETAILED PROGRAMME OF STUDIES.

Part I.—*Artillery drill instructions.*

General rules; school of the section, horsed, and with the drag; school of the battery.

Part II.—*Naval gunnery.*

Care, nomenclature, and description of guns, carriages, mounts, and gun implements, drills of main and secondary batteries; dismounting and assembling the mechanism of machine and rapid-fire guns; description and mode of manufacture of projectiles; charges, fuses, and primers; the theory and practice of pointing; explanation of sights, concentration of fire, and when to use the various kinds of projectiles; employed against torpedo boats and in field service; mounting and dismounting; subcaliber practice; exterior ballistics; the practical use of instruments in determining pressure, velocity, time of flight, and range.

## Department of Administration and Sea Service.

The course in administration and sea service is divided into two parts, military administration, and sea service, and consists of lessons and exercises in applicaton.

### DETAILED PROGRAMME OF STUDIES.

Part I.—*Military administration.*

General principles of administration in the U. S. Marine Corps; the system of recruitment; the system of purchases; care of property and supplies; system of accountability for money and property; transportation of men, munitions, and supplies by rail, steamer, or wagon; military and naval regulations; orders; reports; correspondence in general, and returns.

Part II.—*Sea service.*

Embarking and disembarking; interior discipline; care of small-arms and accouterments; routine; mess arrangements; official intercourse; military etiquette; duties of non-commissioned officers; special duties of first sergeant, of non-commissioned officer on guard, of orderlies, of sentinels; nomenclature of ships; parts and fittings of boats; boat-pulling and sailing; swimming; slinging and lashing hammocks; making knots, splices, and hitches, and the application of pulleys, blocks, and tackles.

## Department of Law.

For the present the course of law is limited to one part, military law and courts-martial, and consists of lessons supplemented by lectures.

### DETAILED PROGRAMME OF STUDIES.

Military law proper; the subject defined and divided; constitutional provisions; the written military law; the unwritten military law; the court-martial; the constitution and composition of general courts-martial; the jurisdiction of general courts-martial; the procedure of general courts-martial; arrest; the charge; assembling and opening of the court; the president and members; the judge-advocate; challenges; organization; arraignment; pleas and motions; the trial; evidence; the finding; sentence and punishment; action on the proceedings; the reviewing authorities; summary courts-martial; courts of inquiry.

### Department of Torpedoes.

The course of torpedoes is divided into four parts, torpedoes offensive and defensive, torpedo fuses, electricity, and explosives, and consists of lessons and exercises in application.

#### DETAILED PROGRAMME OF STUDIES.

##### Part I.—*Torpedoes offensive and defensive.*

Description of the various kinds of torpedoes, mechanical and electrical, contact and ground, with their advantages and disadvantages, and how used offensively and defensively; name and use of the instruments and gear embraced in a ship's torpedo outfit; preparation of the service spar torpedo and contact gun-cotton torpedo for use; method of defending a harbor by torpedoes, and where the different kinds should be placed in the channel; how to find the size of charge, the depth at which effective, and the space between torpedoes; the strength of torpedo case, how tested and made waterproof; filling and planting torpedoes, and how the cables, buoys, etc., should be arranged during the operation; junction boxes and their use; how the cables should be introduced into the firing station; firing the torpedo, and how to determine the position of a vessel with reference to any ground torpedo; shutter apparatus; firing and observing arcs and position finders; circuit closers and their use; nature of enemy's attack on torpedoes and how to defend them; improvised electrical and mechanical torpedoes; improvised circuit closers; description of various kinds of land torpedoes; the use and application of hand torpedoes; torpedo boats; description of the Whitehead and Howell torpedoes; manner of discharging the Whitehead and Howell torpedoes.

##### Part II.—*Torpedo fuses.*

Mechanical and platinum wire fuses and detonating fuses for gun cotton; how made, tested, and used; the number of fuses necessary and their position in a charge of gunpowder or gun cotton.

##### Part III.—*Electricity.*

How electricity is produced; conductors and insulators; heating effect of currents; its general application to torpedo work; how applied to fire guns and for lighting purposes; how used in telegraphy and for ringing bells; the Leclanche and other batteries, how made and how taken care of; testing room instruments; the Wheatstone bridge and its practical use in measuring electrical resistance; electrical tests of cables for insulation, conductivity, and resistance.

##### Part IV.—*Explosives.*

Gunpowder: composition, kinds, manufacture, care; principles of explosion; size and form of grains.

High explosives: Gun cotton, nitro glycerine, dynamite, fulminates; how prepared, preserved, and used; peculiarities of each; effects of explosion; care necessary in handling.

### Department of Engineering.

The course of engineering is divided into four parts: Topography, field engineering, signaling and telegraphy, military gygiene, and consists of lessons, lectures, and exercises in application.

#### DETAILED PROGRAMME OF STUDIES.

Part I.—*Topography.*

Construction of scales; copying, reducing, and enlarging plans; the conventional signs.

Terms used in describing hills; the representation of hills; the scale of horizontal equivalents.

Sections.

Making use of plans on the ground; measurement of distances; the plane table illustrated in intersection and resection.

The prismatic compass and the protractor; their uses illustrated in intersection and resection; the magnetic variation.

Trasversing with compass and field book; plotting; trasversing and plotting at sight.

Trasversing with the plane table.

Hill and mountain sketching.

Sketching without instruments; general remarks on sketching.

Reading contoured plans.

Reconnaissance of a defensive position; sketch and report.

Sketching on horseback. The Vernier; the pocket sextant, adjustment and use; the theodalite, its construction and adjustment; levelling; trignometrical surveying.

Part II.—*Field engineering.*

General principles and definitions; clearing the ground; hasty intrenchments, obstacles; defence of posts; field level and field geometry; deliberate intrenchments; field casemates; brushwood; revetments; working parties; siege trenches; roads; cordage and use of spars; bridges; floating bridges; camping arrangements; hasty demolition.

Part III.—*Signaling and telegraphy.*

Instrument in signaling and telegraphy as prescribed by General Order No. 380, Navy Department, January 2, 1890; use of flag and torch; use of signal disks in transmitting messages.

Part IV.—*Military hygiene.*

Definition; clothing; lodgment of troops; camps; bivouacs; exercise; cleanliness; water tests for the common impurities and methods for purification; early aid to the injured; the best and most expeditious manner of temporarily treating gunshot wounds, poisoned wounds, frost bite, bruises, dislocations, hemorrhage and fracture of bones; application of the tourniquet, and the most approved methods and mode of application in the restoration of the apparently drowned.

### Department of Military Art.

The course of military art is divided into three parts: Minor tactics, grand tactics, and strategy, and consists of lessons and exercises in application.

DETAILED PROGRAMME OF STUDIES.

Part I.—*Minor tactics.*

General principles and definitions; security and information; reconnoitering duties; time and space; advanced guards and outposts; principles of attack and defence; employment of infantry; employment of cavalry; rear guards and marches; rivers and defiles; villages and woods; convoys; night attacks.

Part II.—*Grand tactics.*

General considerations upon tactics; its progress and epochs; division of armies into arms; infantry, its rôle, its armament, its tactical units, its formations, and maneuvers; cavalry, its rôle, its armament, its tactical units, formations, and maneuvers; artillery, its rôle, armament, tactical units, formations, and maneuvers; combined action of the three arms in battle.

Influence of the ground upon tactics; positions, their flanks, their front, their approaches, their interior, their rear, detached advanced posts, occupation of ground and proportion of troops to space, tactical marches and the element of time in military movements, orders of battle and engagements offensive and defensive, principal tactical combinations, different phases of battle, surprises, attack of strong places, and *coups de main.*

Part III.—*Strategy.*

Definition and object of strategy and general principles of warfare, system of operations, the theatre of operations, bases of operations, strategic lines and points, fronts of operations, strategic fronts, lines of defense and strategic positions, lines of operations, strategic lines of maneuver, depot of supplies and their relation to marches, rôle and strategic importance of railways and telegraphs, marches of concentration, strategic front marches, forward and retreat, strategic flank marches, principle strategic combinations, indispensable conditions to be fulfilled by all strategic combinations, defense of frontiers and sea coasts by forts, etc.

## DIVISION FOR ENLISTED MEN.

*(Non-commissioned officers' school.)*

Non-commissioned officers and such privates as show sufficient aptitude are taught the theory and principles of the subjects outlined in the programme for their division, but the main object is to give them a thorough practical knowledge of these subjects.

All enlisted men are required to attend the course of practical exercises at the school. Attendance upon the course of studies in the division for enlisted men is compulsory only on the part of all non-commissioned officers.

The course of studies and of practical exercises for enlisted men proceeds with that of the commissioned officers.

DETAILED PROGRAMME OF STUDIES.

The course for non-commissioned officers in infantry, artillery, administration and sea service and torpedoes is essentially the same as for the commissioned officers' division, and consists of recitations, lectures, drills, and field exercises.

FIELD ENGINEERING.

The course of engineering is divided into five parts: Military sketching, elementary field fortifications, signaling, military hygiene, preliminary course in mathematics, and consists of recitations, lectures, and exercises in application.

Part I.—*Military sketching.*

General principles; definitions; scales; comparative scales; diagonal scales; prismatic compass; plane table; field book; protractor; north point; finding true north; variation of compass; dip; plotting; eye sketching; contours; copying and enlarging; how to proceed to make a sketch with field book and prismatic compass.

Part II.—*Elementary field fortifications.*

General observations; nomenclature; normals for battle trenches, to fire standing, kneeling, and lying down; skirmishers' rifle pits; covering trenches; revetments; obstructions; wire obstructions; abatis; removal of obstructions; hasty demolitions; utilization of natural cover for defense; holes and sunken roads; embankments; hedges; plank and board fences; walls; making loop-holes; defense of woods, buildings, farm yards, and villages.

Camping: To unpack, pitch, strike, and pack tents; telling off and exercise of cooking, latrine, water, ration, and wood parties. Construction of field kitchens and lantrines, trenching of camp, and water supply.

Part III., *Signaling,* and Part IV., *Military hygiene.*

Same as for commissioned officers' division.

Part V.—*Preliminary course in mathematics.*

[Arithmetic, algebra, geometry, and trigonometry.]

The study of arithmetic, elementary algebra, geometry, and plain trigonometry for those not sufficiently proficient in these subjects.

Field Service.

The course of field service consists of lessons, drills, and field exercise, as follows:

General observations on field tactics; marches; reconnoitering; advanced guards; rear guards; outposts; attack and defense of positions.

*Text-books used at the School of Appliance, U. S. Marine Corps.*

DIVISION FOR COMMISSIONED OFFICERS.

| Departments. | Text Books. |
|---|---|
| Infantry | Instructions for Infantry, U. S. Navy; Manual of Guard Duty; Blunt's Firing Regulations for Small Arms; Mayne's Infantry Fire Tactics. |
| Artillery | Instructions for Artillery, U. S. Navy; School of the Section, Light Artillery; Gunnery Drill-Book, U. S. Navy. |
| Administration and sea service | Accountability Instructions; Duties of Marines Embarked. |
| Law | Winthrop's Military Law and Courts-Martial; Instructions for the Administration of Law and Justice, U. S. Navy. |
| Torpedoes | Sleeman's Torpedoes; Spar Torpedo Instructions, U. S. Navy. |
| Engineering | Manual of Field Engineering; Richard's Military Topography; Notes on Signaling. |
| Military art | Shaw's Modern Tactics; Hamley's Operations of War. |

DIVISION FOR ENLISTED MEN.

| Departments. | Text Books. |
|---|---|
| Infantry | Instructions for Infantry, U. S. Navy; Manual of Guard Duty; Blunt's Firing Regulations for Small Arms; Batchelor's Infantry Fire, its use in battle. |
| Artillery | Instructions for Artillery, U. S. Navy; School of the Section, Light Artillery; Radford's Hand-Book of Naval Gunnery. |
| Administration and sea service | Accountability Instructions; Duties of Marines Embarked. |
| Torpedoes | Catechism of the Spar Torpedo, U. S. Navy; Catechism of Explosives, U. S. Navy; Notes on Submarine Mines. |
| Field engineering | Notes on Elementary Field Fortifications and Signaling; Davies' Practical Arithmetic, Elementary Algebra, Elementary Geometry. |
| Field service | Notes on Minor Tactics. |

It has been found necessary to prepare notes and circulars at the school for use as text-books in some parts of the course of instruction. These notes and circulars are intended to meet such changes in the subject taught as will keep pace with the progress of the times and to embody only what is really needed. If the school was in possession of a medium-sized hand printing press and a small lithographic press, with means of doing photo-lithographic work, these professional notes and circulars could be disseminated throughout the Corps.

The first class to take the course reported on September 1, 1891, and was assigned to duty in April last. Seven months was too short a period in which to accomplish all that it was desired to teach, but the lack of sufficient quarters at the post for officers and the appropriation for hire of quarters becoming exhausted, together with the demand from other stations for the services of officers, necessitated the limited time.

In addition to the drawback arising from want of time, the need of proper facilities for practical instruction was greatly felt. The class, however, was very attentive and zealous and showed great interest in the course. With the permission of Commodore Folger, Chief of the Bureau of Ordnance, and the assistance of Commander O'Neil, Inspector of Ordinance at the Navy-yard, and his assistant inspectors, the class availed itself of the advantages afforded by the gun shops in obtaining information regarding the manufacture of ordnance, etc., including the heating and assembling of gun hoops and jackets; boring guns and jackets; means of measuring expansions and interior and exterior diameters; rifling and sighting guns; different types of breech mechanism; gravity return carrages; methods of controlling recoil, of training and elevating, rapid-fire guns and carriages; banding projectiles; method of manufacture and inspection of primer tubes; the preparation of fuses and primers for issue; manufacture of De Bange gas check pads, and of modern guns and carriages from the rough forgings as received from the manufacturers to the finished gun and gun-carriage.

The thoroughness of the instruction given the class was indicated in the parade, written, and oral examinations held before the board of examiners, composed of the colonel commandant and Major H. B. Lowry, Quartermaster of the Corps. Written questions had been prepared covering the course, and when answered by the class were submitted to the board, who, as a further test of efficiency, assigned each officer selected questions from the list to answer orally and demonstrate at the blackboard.

The board was most favorably impressed with the proficiency displayed by the officers under instruction, as well in the parade and practical exercises as in the theoretical course. I append a copy of the questions for the written part of the final examination, officers' division. The division for enlisted men has only been organized during the present year. The facilities for practical instruction are much better for the present class than they were for the last. During the year the Bureau of Ordnance and other bureaus of the Department gave strong proof of their interest in the advancement of the school.

Through the kindness of Commodore Folger and his great interest in all that tends toward the improvement of the service the school has been supplied with an excellent torpedo outfit from the torpedo station, including a liberal supply of the hand-books for seaman gunners on the spar torpedo and explosives, for the use of the non-commissioned officers' school, and with a quantity of drill cartridges from the Washington Navy-yard for using calibre .45 ammunition with the 37-millimeter Hotchkiss gun. By direction of the Department the school has also been furnished with a condemned launch, No. 19, which has been refitted and answers very well for practical instruction in the spar torpedo and spar torpedo boat fittings. A small plat of ground at the post has also been utilized during the summer for an engineering camp, wherein have been constructed some model battle-trenches, rifle and gun pits, fascines, gabions, etc., with a short piece of wall, showing method of defense by loop-holing, notching, etc. This camp, although small, will aid materially the course of practical engineering in the construction of intrenchments, skirmishers' pits, gun pits, and in the principles of the defense of banks, hedges, ditches, and walls, also in the practical details of camping.

The school is still very much in need of a small supply of electrical, surveying, and drawing instruments, materials, etc. We have no fund upon which to draw for their purchase, and the school is dependent wholly upon such small sums as the quartermaster willingly lets us have when he can spare them from his contingent fund.

The personnel of the school consists of the commanding officer, who also commands the post, and three first lieutenants of the corps, who act as instructors. These officers continue to add to their knowledge, and consequently their effectiveness as instructors, the duties of which are laborious and important. During the first school term only one instructor, Lieutenant Prince, was available. His services were most efficient and valuable. Dr. Dickson, U. S. N., of the post, also rendered valuable service to both divisions of the school by his interesting lectures on military hygiene and early aid to the injured.

The officers for instruction are those appointed from the naval cadets, graduates of the year, at the conclusion of their six years' course, to fill such vacancies as occurred during the year in the Marine Corps.

The garrison consists of the force of enlisted men ordinarily maintained at this post. The increased demand for Marines on board ship and for duty at the other stations has rendered it impracticable to order a sufficient number to the school for instruction or to grant the requests of non-commissioned officers and privates who have applied for the privilege of going through the course. With even the ordinary small force the course is frequently and seriously interrupted during the year by the necessity of detaching for service elsewhere non-commissioned officers and men when only fairly started and interested in their studies.

At the present time all the available school force is rendering service at

Camp Low, Sandy Hook, where it was dispatched on the 13th ultimo at forty minutes' notice.

The importance of such a school as this to the corps and the service and the advantage of an uninterrupted course are apparent to all.

But if the corps would reap the benefits of the school to the extent originally contemplated its enlisted force should be sufficiently increased to permit of at least one battalion of four instruction companies, thirty-two non-commissioned officers, and one hundred and twelve privates being assigned to the school, to be relieved only upon completion of the course and then replaced by an equal number.

It should be remembered that while undergoing training such a battalion would be ready at the hand of the Department for any temporary emergency. That such emergencies arise we know, and we know also that when they do the Marines are always called upon to take a leading part.

It is time it was clearly perceived that the corps should be prepared to meet these sudden demands without unduly weakening the force engaged in guarding and protecting the public property at the Navy-yards and stations throughout the country. Very respectfully, D. P. MANNIX,

*Captain U. S. Marine Corps, Commanding.*

THE COLONEL COMMANDANT U. S. MARINE CORPS.

Approved October 15, 1892. CHARLES HEYWOOD,

*Colonel Commandant U. S. Marine Corps.*

The officers studying in the school enjoyed the advantage of opportunities to visit the ordnance shops at the Washington Navy-yard, where they witnessed all the operations connected with the manufacture and assembling of modern heavy naval guns, gaining thereby a thorough understanding of their construction and operation They were also allowed to visit the naval equipment grounds at Indian Head, through the courtesy of the chief of the Bureau of Ordnance, witnessing the trials of guns and enjoying opportunities for full explanations from the officer in charge.

At the beginning of the second school year, on September 5, 1902, the organization of the Enlisted Men's Department, for the practical training of non-commissioned officers and intelligent privates, was completed and a class was formed in the subjects outlined for presentation in their division. An apprentice school was also founded to supplement the regular educational work, and in this division regular courses were given to prepare boys, between the ages of 14 and 16 years, for the positions of drummers and buglers in the Corps. In addition to musical instruction, these apprentices were also given the rudiments of a good English education.

The work of the school for the first year is outlined by Colonel Heywood, in his report for the year 1892, as follows:

"The School of Application, established at the Marine Barracks, Washington, D. C., under orders of May 1, 1891, has completed the first year of its existence, and the most excellent results have been obtained in the course of instruction to both officers and enlisted men, and show in a very gratifying manner the advisability of its establishment, and the benefit that will result to the corps. The first class of officers consisting of the seven young graduates of the Naval Academy who were appointed as second lieutenants in the U. S. Marine Corps July 1, 1891, completed their course April 1, 1892, and were assigned to duty at the different shore stations. The next class of five officers, appointments of July 1, 1892, is now under instruction here. Owing to the limited number of officers and men, and the duty that is required of them, it has been impossible to order as many here for instruction as I would like. Many officers and enlisted men have applied to be detailed, and I hope, if the corps is increased, to see large classes here in future, and gradually to have every one in the corps instructed at this school."

---

## CHAPTER XXXIX.

**1892.**—The Cholera Scare and the Camp of Marines at Sandy Hook—Fire in the New York Navy-yard—Marine Detachment at Opening of the Columbian Exposition, Chicago—Changes in the Discipline and Equipment of the Corps.

1892.—In the months of September and October, 1892, the arrival in the harbor of New York of several ships carrying cholera patients created intense excitement throughout the United States, at the same time awakening the Federal authorities to the necessity of adopting decisive measures to prevent the spread of the contagion A quarantine camp was, accordingly, established at Sandy Hook, whither large numbers of immigrants were sequestered, under guard of a detachment of U. S. Marines. In speaking of this incident in his annual report, Colonel Heywood, commandant of the Marine Corps, writes as follows:

During the cholera excitement in New York, and upon the establishment of a quarantine camp at Sandy Hook for the detention of emigrants, by order of the Department, a battalion of Marines, under command of Major R. W. Huntington, consisting of ten commissioned officers and 201 enlisted men, was ordered there for patrol duty to prevent the escape of any of those detained. These men were housed under canvas and established a regular camp on this sand spit. They rendered the most efficient service to the country in helping to prevent the introduction of cholera. The battalion was formed from detachments drawn from the small force at the Navy-yards at Boston, Brooklyn, League Island, Annapolis, Norfolk, Washington, D. C., and the Marine Barracks, Washington, D. C., the United States receiving ship Vermont, and the U. S. S. Atlanta and Minnesota. In some cases, notably that of the Marine Barracks, Washington, the men were prepared to start for Brooklyn, in heavy marching order, in forty minutes after the order to proceed was received. In Boston they were en route in less than one hour after the receipt of the order, and all reported in New York not later than the morning after the order was issued, and some the same day. This shows in a marked degree the discipline and efficiency of both officers and men, and the cheerfulness and ability of Marines to perform, on the shortest notice, all and any duty that may be required of them, particularly in this case, which was attended with some danger and great exposure and discomfort. These men were isolated for over three weeks, having no communication with the outside world at all during that time, and they deserve the highest credit for the work they performed."

The following extracts from the report of Captain R. W. Huntington, U. S. M. C., on the behaviour of the Marines at Camp Low, Sandy Hook, during this memorable period are worthy of preservation:

REPORT OF MAJOR R. W. HUNTINGTON, COMMANDING MARINE DETACHMENT, SANDY HOOK.

MARINE BARRACKS, BROOKLYN, N. Y., October 11, 1892.

SIR: I have the honor to report that on the 14th ultimo, in obedience to orders received from the commandant of this Navy-yard, dated September 13, I proceeded by tug Traffic, in command of a detachment consisting of Captain F. H. Harrington, First Lieutenant C. G. Long, and 56 enlisted men, to Camp Low, Sandy Hook, and there relieved Captain S. Mercer from command. He, with First Lieutenants C. A. Doyen and B. S. Neumann, Assistant Surgeon Henry La Motte, U. S. Navy, and 61 enlisted men, remained there under my command.

Captain Mercer had encamped his men on the line of the railroad, about 500 yards west from the end of the wharf upon which the buildings of Camp Low are partly placed. I approved his selection of site for camp and put my men under canvas there. The 15th and 16th September were fully occupied in arranging the camp. On the 17th a detachment, consisting of Captain M. C. Goodrell, Second Lieutenants J. Prochazka and R. McM. Dutton, with 83 enlisted men, reported for duty, and were tented alongside the railroad. The command was then organized as follows:

Major R. W. Huntington, commanding; First Lieutenant B. S. Neumann, adjutant and quartermaster; Assistant Surgeon Henry La Motte, U. S. Navy, surgeon; Sergeant A. Bernstein, S. M. and Q. M. sergeant; Company A, Captain F. H. Harrington, First Lieutenant C. G. Long; Company B, Captain M. C. Goodrell, Second Lieutenant R. McM. Dutton, Second Lieutenant J. Prochazka; Company C, Captain S. Mercer, First Lieutenant C. A. Doyen, and to each company about 67 enlisted men.

Our Camp was named Camp Henry Erben, in honor of the commandant of this Navy-yard.

On the 18th emigrant passengers to the number of about 900 were landed at Camp Low, and we communicated with them by telephone, considering them as in quarantine. From that day to the 28th of September there continued to be a number of passengers held at Camp Low, and our sentinels prevented egress from that camp. The daily guard consisted of one company. The main guard was at our camp, and an outpost, consisting of the officer of the guard, two sergeants, three corporals, and eighteen privates, was tented north of Camp Low. This guard was connected with our camp by a plank walk across a water course through the swamp. The walk and bridge were built, in the first instance, by Lieutenant Neumann. The tide rising higher than we had expected, the whole walk was made into a footbridge by Captain F. H. Harrington. This bridge very materially shortened our line of sentinels. The line was, as thus established, about 900 yards long.

The passengers landed were emigrants, and no trouble wes experienced in keeping them from wandering. It was sufficient in every instance for the sentinels to indicate that they had orders to fire upon persons persisting in advancing toward our line, to make them fall back. Eleven posts over Camp Low were kept. This guard would in any case be found enough to restrain steerage passengers. Probably cabin passengers would be harder to handle.

During the stay of the emigrants at camp there were a few cases of cholera, of which only one was fatal.

The battalion was frequently drilled, although the ground is poorly adapted for drill, and the woods are full of poison oak. Twenty rounds of ball cartridge per man were expended in target practice, with fair results. No extended range was practicable on account of the lay of the land and the ordnance proving ground. The men were comfortably tented. In addition to the tents we took with us, others were borrowed from Camp Low, so the men had more than regulation space. The tents were well floored. Floors for our tents were made at the Navy-yard, and for those tents we borrowed lumber was also given from Camp Low. Our thanks are due to the officials there for many acts of kindness. Kitchens were built by the companies. The third company used, as mess room and kitchen, a house belonging to the railroad company.

The supply of water was from pipe wells pumped by steam into the tank formerly used by the locomotive engines. We furnished Camp Low with water. Private James McGantlin, of the guard of the U. S. S. Atlanta, who, a short time ago, for some weeks was waiting discharge (bad conduct) in these barracks, under sentence of summary court martial, ran the pump and engine and did honest and valuable service.

On the 18th Captain F. L. Denny, assistant quartermaster, reported for duty, and was ordered to attend to the purchase and shipment of necessary stores in New York and Brooklyn. This duty he did well. The rations of the men were uniformly good. Occasional issues of fresh beef and bread were made as opportunity allowed.

Quarantine against the whole Sandy Hook reservation was declared by the authorities of New Jersey. Dr. Woodward, of the State Board of Health, who was in charge of this quarantine, proved himself a kind-hearted and reasonable man, and placed no unnecessary restrictions upon our intercourse.

Through his kindness we received occasional supplies and mail by railroad train.

The sick list was generally very small. There was a slight epidemic of diarrhœa when we were first in camp, due probably to the change in the manner of living from the barracks to camp. Several men were disabled for a time by oak poisoning.

On the 21st Private Joseph McMahon, of the Vermont's guard, died from purpura hæmorrhagica. He was buried on the right side of the railroad, going toward the main line, on the 22d. On the 28th Second Lieutenant Julius Prochazka was transferred to the Naval hospital, Brooklyn, with typhoid fever.

The battalion was withdrawn on the 5th of October, with the exception of First Lieutenant C. G. Long and eleven enlisted men. These were withdrawn on the 8th.

The conduct of the enlisted men was very good generally. The guard duty was done carefully and absolutely well.

The officers were attentive and zealous in performance of duty. I consider that Captain F. H. Harrington and First Lieutenant B. S. Neumann deserve special mention for the energy and intelligence they showed.

I am, very respectfully, your obedient servant,

R. W. HUNTINGTON,
*Major U. S. Marine Corps, Commanding Marines.*

THE COLONEL COMMANDANT, U. S. MARINE CORPS,
*Washington, D. C.*

The necessity of withdrawing so large a number of men from the forces on shore duty, at the various navy yards and stations, of course, involved a corresponding reduction of the effective guards for United States ships and stores. The knowledge that the forces were thus diminished encouraged certain miscreants to kindle fires in the Brooklyn Navy-yard, with the result that several buildings were partly or wholly destroyed, and several ships, notably the U. S. S. Cincinnati, greatly injured. Such an incident, which was quite unprecedented in the history of the Station, gives an example of the high estimate of the efficiency and usefulness of the Marine Corps existing in the public mind.

A battalion organized in Brooklyn, with the band of the U. S. Marine Corps, participated in the Columbian parade, in New York city, on October 12th, marching with detachments from the Army and Navy forces, and from the militia of the various States. Another battalion, consisting of fifteen commissioned officers and four companies of thirty-six men each, under command of Major R. W. Huntington, with the band and field musicians, an additional fifty men, specially detailed by order of the Navy Department, participated in the opening exercises of the Columbian Exposition, Chicago, Ill., on October 19th, 20th, 21st, and 22nd, being quartered temporarily in one of the completed buildings in Jackson Park.

During the year 1892 several improvements were made in the discipline and uniform of the Marine Corps. Among these may be mentioned the adoption of the "Instructions for Infantry and Artillery, U. S. Navy," and the change in the uniform of the Corps, in accordance with the recommendations of the Board of Inspectors of the Navy Department. The recommendations, as carried out at this time, included canvas fatigue suits, and the substitution of black helmets for the shakos always previously worn by the Marines.

## CHAPTER XL.

1893.—Marines in the Hawaiian Islands—Disaster at Port Royal, S. C.—Marine Detachment at the Columbian Exposition.

1893.—Early in this year the ancient native kingdom in the Hawaiian Islands was overthrown; the queen, Liliuokalani, deposed, and a provisional government formed, preparatory to the establishment of the republic. During the period of threatened disorder succeeding this event a detachment of U. S. Marines, under command of First Lieutenant Herbert L. Draper, was landed from the U. S. S. Boston, in order to afford protection to the lives and property of American residents at Honolulu. The eminent part played by the Marines in preserving order at this time, when violence was constantly fomenting, was recognized in the following handsome resolution, adopted by the Provisional Government and presented to Lieutenant Draper:

*Resolved,* That the sincere thanks of the executive and advisory councils of the Provisional Government of the Hawaiian Islands be, and they are hereby, tendered to Lieutenant Herbert L. Draper, U. S. Marine Corps, and to the men composing the Marine guard of the U. S. S. Boston, under his command, for efficient, faithful, and valuable services rendered to the Provisional Government while in charge of the Government buildings and grounds at Honolulu from February 1 to April 1, 1893.

During this period the watchful care over valuable Government property, the orderly conduct and discipline maintained by Lieutenant Draper and his men are deserving of the highest praise and commendation.

*To Lieutenant Herbert L. Draper and the officers and men of the U. S. Marine Corps, U. S. S. Boston:*

GENTLEMEN: I have the honor hereby to inform you that the above resolution was unanimously passed by the executive and advisory councils of the Provisional Government of the Hawaiian Islands in session this day.

JAMES B. CASTLE,
*Secretary of the Executive and Advisory Councils.*

HONOLULU, HAWAIIAN ISLANDS, April 3, 1893.

Another eminent instance of the readiness of Marine Corps men to render assistance on any occasion when their services may be required, occurred at Port Royal, S. C., when, during the night of August 27th-28th, a cyclonic tidal wave inundated the town and Naval Station, bringing death and destruction in its wake. This service is recorded by Captain L. A. Beardslee, U. S. M. C., in the following letter addressed to the Colonel Commandant:

COMMANDANT'S OFFICE, U. S. NAVAL STATION,
PORT ROYAL, S. C., September 6, 1893.

SIR: I take pleasure in informing you that during the night of the 27-28 of August, while a cyclone was sweeping over this station and a tidal wave inundated it, the conduct of every man in the Marine guard was most praiseworthy. Almost without exception they waded to and fro through the flood of water which was driven by the wind into the seas so dangerous that many people were overthrown and drowned by them, doing their utmost to preserve life and property. The influence of First Sergeant Gallagher, and of the non-commissioned officers, who are all excellent men, has brought the tone of the guard to a most creditable condition. Respectfully,

L. A. BEARDSLEE,
*Captain Commandant.*

THE COLONEL COMMANDANT U. S. MARINE CORPS,
*Washington, D. C.*

During the period of the Columbian Exposition at Chicago, Ill., a detachment of three officer and sixty enlisted men, commanded (at different times) by Captains B. R. Russell and Paul St. C. Murphy, were in constant attendance. This detachment, consisting of picked men drawn from the various stations of the Marine Corps, served as a guard for the valuable collections, loaned by the Department of State and several foreign governments, and for the Naval exhibits; a specially chosen party also acted as an escort in conveying these treasures to Washington after the close of the exposition. However, the most conspicuous service rendered by men of the detachment occurred on July 10th, when an obstinate and fatal fire destroyed the Cold Storage Building on the fair grounds. This service was acknowledged by George R. Davis, Director-General of the Exposition, in a letter, dated July 12th, 1893, and addressed to Captain Russell. The same official again referred feelingly to the incident in a letter to Captain Murphy, under date of December 18th.

A detailed account of the service of the Marine detachment at the Exposition is given in Captain Murphy's letter of report to the Colonel-Commandant herewith annexed:

WASHINGTON, D. C., February 7, 1894.

SIR: I have the honor to enclose herewith a copy of a report submitted by me to Lieutenant Commander E. D. Taussig, U. S. Navy, representative of the Navy Department, World's Columbian Exposition, in relation to the services of the Marines in connection with the Naval exhibit, World's Columbian Exposition.

Very respectfully, your obedient servant,

PAUL ST. C. MURPHY, *Captain, U. S. Marine Corps.*

THE COLONEL COMMANDANT, U. S. MARINE CORPS,
*Headquarters, Washington, D. C.*

Colonel GREEN CLAY GOODLOE, Paymaster

U. S. M. C. **Plate 6**

WASHINGTON D. C., February 5, 1894.

SIR: I have the honor to submit, agreeably to your instructions, the following report in relation to the services of the Marines in connection with the Naval exhibit, World's Columbian Exposition.

A detachment consisting of three commissioned officers, Captain B. R. Russell, First Lieutenant George Barnett, and Second Lieutenant C. S. Radford, and 50 enlisted men were assigned to that duty, and arrived on the 6th of May. Later ten additional men joined the detachment, being sent at the request of the Department of State, for duty as a special guard over the exhibits of that Department.

The detachment moved at once into the encampment which had been prepared for it by the quartermaster of the Marine Corps.

The camp, named in honor of the Secretary of the Navy, "Camp Herbert," was located on the Government plaza adjoining the naval observatory, and in close proximity to the model battleship Illinois. It was laid out regularly in streets, with boarded walks, and the grounds bordering upon it and the intervening spaces between the tents were sodded. The grass thrived and added greatly to the attractive appearance of the camp. A tent was assigned to every two men, and was furnished with bunks, bedding, washstands, lockers, and other articles necessary to their comfort. Later, when the weather became cold, an oil stove was supplied to each tent. By an arrangement made by the quartermaster of the Marine Corps, the men were subsisted at a restaurant outside the Exposition grounds, at a convenient distance from the camp, and marched to their meals.

Passed Assistant Surgeon A. C. H. Russell, U. S. Navy, attached to the model battleship Illinois, gave the necessary medical attention to the men, and displayed at all times an interest in their health and in the sanitary condition of the camp.

The duty performed by the detachment was mainly that of guarding the Naval exhibit on board the Illinois, the encampment and the Naval observatory, together with certain exhibits belonging to the Department of State, located in the Government building. During the hours when visitors were admitted thirteen posts were maintained on board the battleship, two in the Department of State, and one over the camp and observatory. At night, after the exhibits were closed to the public, these posts were reduced to the number necessary to guard against fire, usually four. The duty performed by these sentinels was in many ways more trying than ordinary guard duty, and was both difficult and delicate. The handling of great crowds, made up of people of all nations and of every age and condition of life, required in an extraordinary degree the exercise of patience, courtesy, and firmness. The average daily number of visitors to the battleship was about 20,000, a number greatly increased on State and other special days, as, for instance, on Chicago day, when it was estimated that 70,000 persons visited the ship. The sentinels and non-commissioned officers of the guard managed the crowds on these occasions with so much judgment and tact that no injury to persons or property worthy of mention occurred at any time.

Besides doing the regular guard duty, the detachment took part in parades and ceremonies of special importance, such as the parade to celebrate the arrivals of the Caravels, the reception and parade in honor of his highness the

Maharajah of Kapurtalah, and the parade on Grand Army Day, when it acted as escort to the Battalion of Naval Veterans.

On the occasion of the cold-storage fire, which resulted in so deplorable a loss of life, the men of the detachment were present and the service rendered by them was of such a character as to merit recognition by the director general of the Exposition in a letter commending their efforts and thanking them.

At the close of the Exposition the detachment was charged with the additional and responsible duty of guarding certain relics of Columbus that had been on exhibition in the Monastery of La Rabida. Among these were the papers belonging to the Government of Spain, to the Duchess of Alva, and the Duke of Veragua, together with certain other exhibits, the property of the Vatican and the French Government.

The most valuable of these exhibits were brought by the detachment to Washington and delivered under guard to the Department of State. The remainder (the Vatican exhibit) was, by the direction of the Navy and State Departments, left in Chicago in the custody of his grace Archbishop Ireland.

Capt. B. R. Russell was relieved from command of the detachment at the end of July, 1893, and Lieut. Barnett was left in charge until August 4, at which date I relieved him and assumed command.

The detachment remained intact until the close of the fair, when its strength was gradually reduced. This reduction took place as follows:

(1) November 2, 20 men were sent to the navy yard, New York, for duty on board the U. S. S. New York.

(2) November 30, myself and Lieut. Radford and 22 men were detached. Lieut. Radford and 12 men were sent to the marine barracks, navy yard, New York, and I with 10 men to guard the Columbus relics, with the custody of which I was charged, proceeded to Washington.

(3) Lieut. Barnett with the few men remaining continued on duty in connection with the naval exhibit until December 18, 1893, when all were withdrawn.

The camp equipage in use by the detachment, except the bedding and four tents, had been purchased out of the funds assigned to the naval exhibit, and when the men were finally withdrawn this property was turned over by your authority to the quartermaster of the Marine Corps.

The good behavior and soldierly bearing of the men were the subject of frequent and favorable comment, some of which expressed in official letters I have the honor to lay before you. These letters are the following:

World's Columbian Commission,
Office of the Director-General of the Exposition,
Administration Building, Jackson Park,

Chicago, Ill., July 12, 1893.

Dear Sir: You will permit me to express my profound gratitude for the prompt and heroic service which the marines under your command rendered during the awful catastrophe which visited these grounds on Monday, and I would be greatly obliged if you would have this letter read in the hearing of your men.

I have the honor to be, sir, your obedient servant,
George R. Davis,
*Director-General.*

Capt. B. R. Russell, U. S. Marine Corps,
*Camp Herbert, Jackson Park, Chicago, Ill.*

DEPARTMENT OF STATE,
WASHINGTON, October 23, 1893.

SIR: The representatives of this department on the government board of managers, World's Columbian Exposition, having called my attention to the behavior of the marines under the command of Capt. Paul St. C. Murphy, U. S. Marine Corps, detailed to guard the Department of State exhibit, I desire to express my appreciation to you, and through your department to thank the men, whose names are Privates Patrick Kane, Allen J. McDonald, Egbert W. Thomas, John MacDonald, Charles Masche, and John A. Dixon, for their exemplary conduct. They have been faithful, discreet, and courteous, and their fine personal appearance has gratified all visitors to the department's section.

I have the honor to be, sir, your obedient servant,

The SECRETARY OF THE NAVY. W. Q. GRESHAM.

DEPARTMENT OF STATE,
WASHINGTON, December 8, 1893.

SIR: I have the pleasure in informing you that the U. S. marines, under the commmand of Capt. Paul St. C. Murphy, which you kindly put at the disposal of this department to guard and bring to Washington certain valuable documents lent by foreign governments to the United States to be exhibited at the World's Columbian Exposition, have performed the duty assigned them to the complete satisfaction of this department, and that the objects under their care have been safely delivered here.

I desire to express to you my high appreciation of the service which you have hereby rendered this department, and request that you will convey to Capt. Murphy my thanks for the efficient way in which he performed this duty.

I have the honor to be, sir, your obedient servant, EDWIN F. UHL,

The SECRETARY OF THE NAVY. *Acting Secretary.*

WORLD'S COLUMBIAN COMMISSION,
OFFICE OF THE DIRECTOR-GENERAL OF THE EXPOSITION,
CHICAGO, Ill., December 15, 1893.

SIR: Shortly after the cold-storage fire I took occasion to thank the U. S. marines for their praiseworthy conduct at that unfortunate catastrophe, and I now desire to attest high appreciation of their general service during the Exposition. The detachment, numbering, as I understand, 3 commissioned officers and 60 men of the U. S. Marine Corps, was here during the entire period of the Exposition, and not only most efficaciously discharged its duty of guarding the Navy and State Department exhibits, but was of material assistance on many occasions of ceremony, acting as escort, and taking a prominent part in the marine functions which were given from time to time. The behavior of the marines throughout was excellent, and their officers zealous in furthering the common interests of the government and the Exposition, as well as ever ready to extend proper courtesies. I have the honor to request that you convey the thanks of the Exposition to the members of the detachment.

Yours very truly, GEORGE R. DAVIS,
*Director-General.*

Capt. PAUL ST. C. MURPHY,
*Commanding U. S. Marines, Jackson Park.*

I would add that the zeal and ability shown by Lieuts. Barnett and Radford in connection with their duties contributed in no small degree to the efficiency of the detachment.

In closing this report I would take the opportunity to express my appreciation of the kind interest shown by you in all matters relating to the officers and men of the detachment, and in their name and my own to thank you.

I have the honor to be, sir, very respectfully,

PAUL ST. C. MURPHY,
*Captain, U. S. Marine Corps, late Commanding Detachment,*
*U. S. Marines, Camp Herbert, Naval Exhibit.*

Lieut. Commander E. D. TAUSSIG, U. S. Navy,
*Representative of the Navy Department, World's Columbian Exposition.*

---

## CHAPTER XLI.

1894.—The Railroad Strike Troubles in California.

1894.—The summer of 1894 saw sundry uprisings and threats of violence, as a consequence of the prolonged labor strike in California. To aid in the preservation of peace, as well as to protect trains and railroad property, with a view to keeping open the transmission of the United States mails, a detachment of Marines, from the Mare Island Navy-yard and the receiving ship Independence, was detailed to coöperate with the army troops. The Marines, under the command of Major Percival C. Pope, U. S. M. C., and with Captain O. C. Berryman and First Lieutenants C. M. Perkins, R. C. Draper and R. McM. Dutton, were detailed for duty, with the troops of the Department of California, under the immediate command of Colonel William M. Graham, Fifth Artillery, U. S. A. In addition to the Marines, Major Pope's battalion consisted, at various times, of Company E of the First Infantry and Batteries A, K and L of the Fifth Artillery, being stationed during most of the troublous period in and in the immediate vicinity of Sacramento. In spite of the intense heat and the numerous obstacles encountered, the Marines performed their duties in the most creditable manner, as was testified by the Department Commander, General T. H. Ruger, U. S. A., and Colonel Graham. In his report on the episode, Major Pope describes the men of the Marine detachment as "ready, cheerful, full of zeal, respectful in language and deportment, so much so as to call forth praise and admiration from all officers with whom they came in contact."

The following are reports on the action of the Marines from the officers of the coöperating military forces:

[Orders, No. 20.]

OFFICE OF COMMANDING OFFICER, U. S. TROOPS,
SACRAMENTO, Cal., August 8, 1894.

[Extract.]

III. In accordance with telegraphic instructions of the 7th instant from the department commander, Major P. C. Pope, Marine Corps, is relieved from further duty with this command.

In relieving Major Pope the commanding officer desires to express his high appreciation of the zeal and judgment shown by him in the exercise of his many difficult duties, under trying circumstances, and to thank him for his uniform courtesy and promptness in the execution of orders. He also desires to compliment Major Pope on the efficiency shown by the marine battalion while on duty here. The cheerfulness of officers and men, their soldierly and exemplary conduct, and their disregard of personal comfort in the performance of duty, made them, as a command, deserving the highest praise, and of which he may feel justly proud.

Major Pope will proceed to the Mare Island navy yard and report to the commandant thereof.

The travel directed is necessary for the public service.

By order of Col. Graham: W. W. GALBRAITH,
*First Lieutenant and Adjutant, Fifth Artillery, Adjutant.*

HEADQUARTERS DEPARTMENT OF CALIFORNIA.
SAN FRANCISCO, Cal., August 20, 1894.

SIR: In accordance with the purpose stated in my letter of July 27 last, I desire to testify to the very efficient and important services rendered by the battalion of marines under the command of Major Percival C. Pope, Marine Corps, as a part of the force under command of Col. William M. Graham, at Sacramento, Cal., and on the line of the Central Pacific Railroad, during the month of July and the present month.

Very respectfully, your obedient servant, THOS. H. RUGER,
*Brigadier General, U. S. Army, Commanding Department of California.*

The COMMANDANT MARE ISLAND NAVY YARD, CALIFORNIA.

[General Orders, No. 18.]

HEADQUARTERS DEPARTMENT OF CALIFORNIA
SAN FRANCISCO, Cal., September 4, 1894.

* * * * * * * *

It is with satisfaction that the department commander alludes to the services rendered by the naval forces in conjunction with the troops. The battalion of U. S. marines from the Mare Island navy yard, under immediate command of Major Pope, rendered essential service as part of the command of Col. Graham, at Sacramento and on the line of the Central Pacific Railroad. Its performance of duty was very efficient and its conduct excellent.

Important service was also rendered by the force of sailors and marines under command of Lieut. Commander W. H. Reeder, detached for service with the Army from the U. S. steamers Charleston, Monterey, and Thetis, and the ship Independence. This force was organized as a naval "brigade" for duty ashore, and performed valuable services at Oakland; it also constituted a reserve relied upon for effective service elsewhere in any emergency that might arise.

* * * * * * * *

By command of Brig. Gen. Ruger: O. D. GREENE,
*Assistant Adjutant General.*

[Indorsement on Major Pope's report to commandant navy yard, Mare Island, Cal.]

U. S. NAVY YARD,
MARE ISLAND, Cal., August 23, 1894.

Approved and respectfully forwarded to the honorable Secretary of the Navy. The officers and enlisted men of marine battalion have been prompt, energetic, and faithful during the whole of the time from the receipt of my order to "stand by for a call" until their return here with the report of "well done" from Col. Graham.

H. L. HOWISON,
*Captain, U. S. Navy, Commandant Navy Yard and Station.*

Another detachment of Marines from the U. S. S. Charleston was under the command of Lieutenant-Commander W. H. Reeder, U. S. N., in connection with the naval brigade at Oakland.

---

## CHAPTER XLII.

**1894-1897.**—Detachments of the Marine Corps in Korea and China.

**1894-97.**—Although, on several occasions in the years immediately preceding, detachments of the U. S. Marine Corps had seen heavy and arduous service, the events in Korea and China in the year 1894 gave eminent opportunity to know the American sailor and Marine under test conditions. The progress of the war between China and Japan determined the U. S. Government to dispatch the Cruiser Cincinnati, later also the Monocacy and Yorktown, of the Asiatic Station, to protect American interests in the seat of hostilities.

Among the most important tasks facing Admiral McNair, Commander-in-Chief of the U. S. Naval force, was the protection of the American consulate at Seoul, the capital city of Korea, as well as the numerous missionaries and foreign residents collected there. The force, consisting of twenty-one Marines and twenty-nine sailors, under command of Captain G. F. Elliott, U. S. M. C., made a night march from the post at Chemulpo to the city of Seoul, a total distance of thirty-one miles, part of the way through submerged rice fields and in the midst of intense heat. Arrived at the Capital with very few mishaps the detachment performed important services in guarding the American and several foreign consulates and in helping to preserve order. Later Captain Elliott was ordered, with his force and the Marine detachment of the Monocacy, to embark at Tientsin for service, in case of need, in protecting American interests in the City of Peking against riots occasioned by the approach of the Japanese army. This expedition was later found to be unnecessary.

Captain Elliott's report on the services of his men, from the time of their detachment for duty at Seoul, is full of most interesting material, and is consequently quoted *in extenso.*

## REPORT OF CAPT. G. F. ELLIOTT.

### MARINES IN SEOUL, KOREA.

The treaty between China and Japan, known as "the treaty of Tientsin," contained an agreement that neither country should land troops in Korea without due notification each to the other.

In the month of June, 1894, the U. S. flagship Baltimore was ordered to Korea to watch proceedings and give a feeling of security to the numerous missionaries living in the capital, Seoul, some 26 miles from its seaport, Chemulpo.

Chinese and Japanese fleets were constantly steaming into the harbor, anchoring, and moving out, evidently watching each other's movements.

Early in July the Chinese landed 4,000 soldiers 16 miles south of Chemulpo, at a small village called Ahsan, and went into intrenched camp, which was inland sufficiently to be out of range of navy guns.

As soon as this move on the part of the Chinese was positively known, the Japanese landed in light order 300 sailors from the fleet, hired Korean coolies to carry supplies and baggage, and attemptel a night march to Seoul. The roads being bad, they did not reach there until evening of the next day. They remained in charge of their legation in Seoul for about ten days, when they were relieved by the Japanese troops. Why they did not go by the river in their boats was never satisfactorily answered.

Within a week after the sailors had occupied Seoul, transports, conveyed by Japanese men-of-war, disembarked 6,000 infantry, a single horse battery, rapid fire guns, and 240 cavalry. All were landed from sampans on the single mole rapidly, without noise, confusion, or loss of any kind. They were beautifully equipped in all the necessities for field service. The cavalry were a picked body of men; their mounts, stallions, about 15 hands, vicious and not up to hard work. All had a peculiar flatness of the flanks that I have never noticed in horses of any other country.

On July 23 telegrams urgent in tone were received on board from our minister in Seoul asking for a guard to protect the legation, also stating that the Japanese had seized the palace and king the night before after a small skirmish, in which 40 Koreans and 3 Japanese were killed or wounded.

A detail of 2 ensigns, a cadet midshipman, and assistant surgeon, the paymaster's clerk, 21 marines, and 29 sailors was made to proceed to Seoul by boat at 4 p. m. the next day. At 7 p. m. the only pilot procurable had not returned from up the river, and for this reason we did not expect to be able to start the next morning.

The weather was extremely hot and the atmosphere filled with humidity. It was the opinion of officers on board that the march to Seoul could not be made in a night and that a march by day would endanger the lives of men.

Believing at the time that a guard was urgently needed, I stated my views and said that I felt certain that the marines could make a night march of it with 60 rounds of ammunition and three days' rations.

At 7 p. m. received an order from the commanding officer to proceed to Seoul. At 8 p. m. left the ship; landed at 8.45 p. m. As the Japanese occupied and patrolled the roads, their consul was asked to furnish a guide, which he willingly did, and we received only courtesy from their many small posts on the way.

The road was little more than a footpath for pack animals and varied in every degree of wretchedness, over clay hills, a mountain pass, three miles of deep sand, with two good-sized streams to ford. The night was extremely dark, hot, and sultry. The distance was added to in order to avoid a Japanese division on the move for Ahsan, where the first land fight of the war took place three days later. This divergence took us through three miles of rice fields without a road, and the entire distance of thirty-one miles was made in eleven hours. As the men had not drilled on shore for a year, they were severely taxed, but there was no straggling.

After crossing the river three miles from Seoul we met many fugitive Koreans fleeing from the city with their women and children. Over 100,000, it is said, deserted the city for the country. On seeing us they always hastily deserted the road. On entering the south gate not an inhabitant was to be seen. The houses were all closed, and the market place, usually thronged at this hour, was entirely deserted. The gates of the city were heavily guarded by Japanese infantry. The gates of the legation were also closed, and there were some Korean refugees of rank housed in the compound.

At 2.30 a. m. Ensign Heyward reported for duty, having with him 2 subalterns, an assistant surgeon, an acting paymaster, and 29 sailors. They had marched some three miles from the river. Two of the men were in the hands of the surgeon from sun collapse. One of these returned to the ship two weeks later still on sick report.

Two old quarter-deck awnings, two boat stoves, and the paymaster's stores having arrived, the men were soon sheltered in two tents. A cook house was erected and sinks dug. Later board floors, mosquito nets (purchased by the men), a bath house, and other luxuries made the camp as comfortable as the extreme heat would allow.

The legation compound is very small—about 140 yards long by 100 yards wide, is surrounded by a wall varying in height from 6 to 10 feet, contains two good wells, and is situated between the English and Russian legations, the three occupying a ridge which slopes gently to the south and abruptly to the north. The legations face south and are entirely surrounded by the one-story Korean hovels. Buildings in the American compound, viz: Minister's house, brick, of one story; secretary's office, same, but smaller; a small Korean cottage occupied by the officers; legation office, old Korean cottage, large enough for 20 men in summer; a new building put up at the minister's expense, 45 feet by 16 feet, called marine barracks, and a few sheds for the many coolies employed as servants.

In spite of the hot weather and contracted space the sinks were kept free from all disagreeable odors by the free use of quicklime and very little earth thrown in twice a day.

Sentinels were established at both gates and at advantageous points to protect the legation, also in the lane connecting the English and Russian legations with the American. At the earnest request of missionaries a patrol at night made rounds through their compounds, which were situated within hail of the guard in the legation.

Through courtesy, needed supplies were sent up from the Baltimore in a small steamer purchased from a Chinese merchant by the British chargé d'affaires. He requested through our minister that I furnish a quota of sentinels to guard a storehouse where the boat landed. This I was obliged to refuse, and thereafter our own steam cutter made necessary trips.

Up to the 26th of August no man left the compound. All thoroughly understood the situation, that the Japanese garrison had been depleted to 300 men in order to send troops to Pingyang, and that in case they were defeated their legation would probably be attacked in Seoul. An event of this kind would endanger the lives of all foreigners, for there is a strong party of Koreans whose cry is Korea for Koreans; also a religious sect of fanatics calling themselves Tonghaks, organized to do away with Christianity; but the term is now generally applied to rebellious Koreans. This belief proved to be true, for a letter written by the king's father, the old regent, to the commanding Chinese general in Pingyang, stated: "You gain a victory and I will stir the foreigners up here in rear." The regent was confronted by this letter in the hands of the Japanese minister in Seoul later, and admitted he wrote it, and for this lost most of the power to which they had restored him.

The conduct of the men up to August 26 was exemplary in the extreme. Money was served out after this, and they were allowed liberty until 6 p. m. (when they could be spared). Temptations to smuggle liquor were offered by Koreans constantly. Still there were but few cases of drunkenness or overstaying liberty. Their general conduct was good, and, although they would have been glad to get back to the ship, I never heard a complaint.

I was relieved on the 26th of September by Lieutenant Gill, U. S. N., from U. S. S. Concord, in command of a detachment of marines. Ensign Heywood and Assistant Surgeon Morris were detached from my command and reported to Lieutenant Gill for duty. We joined the ship on the evening of the 27th of September, making the trip down the river in ships' boats towed by steam launches in eight hours.

### PORT ARTHUR.

Pingyang having been taken, the Chinese garrison fled to the northern portion of the peninsula of Korea. The roads leading north through the country were so contracted and difficult that it was almost impossible for the commissariat to supply the Japanese army, its means of transportation being an organized corps of coolies.

As the Japanese paid cash—600 to the silver yen—for everything used or taken in the country, even to firewood, the transportation of money became a very serious difficulty. To lessen this in a degree they forced the Koreans to make use of their idle mint and coin silver and copper.

The Chinese army had halted in its flight from Pingyang on the Yalu River, and it was evidently the intention to collect a larger land force here than they had yet put into the field to recover Pingyang and eventually to capture Seoul. Stores and transports were to be conveyed by their fleet to the mouth of the Yalu.

A counter stroke by the Japanese on Port Arthur would keep this army from moving south, or at least from receiving succor by water. This could only be done by landing a fresh corps in transports within easy striking distance of Port Arthur and attacking on its land face.

The naval engagement known as the battle of Yalu was risked and fought to accomplish this. Although not actually decisive, the Chinese fleet was so crippled that a corps was embarked in transports, landed, and took the fortifications protecting Taliwan Bay, 25 miles from Port Arthur. This place was given up almost without defense, and its splendid armament of modern Krupp guns left intact.

General O Yama, having established a base at Taliwan Bay, moved on Port Arthur, 25 miles distant. A large Chinese column attempted a sortie on the 18th or 19th of November, but was turned back by the fire of a single battery. It is probable that an assault by the Japanese would have been entirely successful at this time, but their attack was not hurried.

The Japanese army, numbering 20,000 men (about 15,000 actually participating in the fight), was divided into two columns. The right, the main assaulting division, under the direct command of General O Yama, and a mixed division of 3,000 infantry and a battery, under command of a lieutenant colonel, the left. The two divisions were not actually in touch, but were separated by more than a mile.

The outer line of defense, about four miles in length, stretched across this narrow peninsula, and was about the same distance from Port Arthur, the inner line about two and a half miles distant.

The contour between these lines was a mass of steep, rocky hills, varying from 200 to 600 feet in height, separated by narrow, tortuous valleys, deeply seamed with dry gullies, the prevailing color presented to the eye being a grayish brown. Stunted bushes of oak and small pines sparsely dotted the hillsides.

A chain of ten forts on the crests of the hills had mounted modern ordnance. The forts supported each other in flank; had lines of rifle pits along the ridges connecting them.

A magazine in one of the principal forts was blown up by shell fired from a field battery, and the infantry were said to have assaulted in fine style. The second Nangoya regiment broke through on the Japanese right and poured over the hills into the town. They were closely followed by the head coolies of the commissariat department, who were armed with Japanese swords. Some hundreds of Chinese soldiers were within the village; they had shed their uniforms, and were dressed in the blue blouse of the coolies. The second regiment had seen a day or two before the heads of a lieutenant and two soldiers (belonging to them) stuck on poles by the wayside, and their bodies horribly mutilated. I was told by a Japanese captain that the Chinese had fired on them when they entered the village. However this may be, the Japanese soldiers shot and the coolies slashed with the old sword of Nippon about 240 in the streets and houses of the village.

The mixed division, forming the left, did most of its fighting, carrying the inner line of rifle pits on their left, broke through, but were halted before reaching Port Arthur.

The Chinese army, some 13,000, were allowed to retreat around the Japanese left, crossing their line of communication with Taliwan Bay, and escaped up the Laotong peninsula.

The allowing Chinese armies to escape was a policy pursued throughout the war; to parole them would be a farce, feeding and holding them as prisoners an expense. As Chinese soldiers were little more than coolies, the number captured could not affect numerically the fighting force of three hundred millions of people.

News reached the U. S. S. Baltimore at Chefoo on the 23d of November that Port Arthur had fallen on the 22d of November at 2 p. m., and that night we sailed, anchoring 3 miles from Port Arthur at daylight on the 24th.

At 7.30 a. m. a whaleboat containing three naval officers and myself left the ship with directions to ask General O Yama's permission to visit points of interest within the captured port.

Japanese steam launches were engaged in removing torpedoes, one or two of which exploded while we were en route, and though not in close proximity, they gave a nasty shock to the boat. The wind having died out one of the dragging launches towed us into port, avoiding buoys, either placed on discovered mines, or dummies placed by the Chinese.

The entrance to the harbor is very narrow. On the starboard hand a hill 656 feet high, rising almost like a bluff from the water, was crowned with a heavily armed citadel, and on a lower level several batteries in barbette; the guns were modern Krupp. This hill breaks off in almost a perpendicular cliff, forming the starboard shore of the narrow entrance. Its reverse side from the sea is so steep that a fine military road, cut and walled with faced stone, V's to the citadel. It is a natural traverse for the dockyard and the little village situated directly behind it.

The entrance is little over 100 yards wide; the port shore slopes more gradually from the water, but its angle not far from 45 degrees. Its summit, some 500 feet high, was fortified, having mounted in barbette two 24-centimeter Krupp guns. Numerous batteries dotted the hills on the right and left.

The channel bears to starboard just after entering, so that the dockyard is masked from gun fire at its entrance. The harbor is about half a hill long; at its head, cut at right angles to it on the starboard shore, is a basin faced with cut granite. This basin is about 600 yards long by a little more than 100 wide. Facing its head on the right is the dockyard proper, having machine and work shops of varous kinds, a fine pair of iron shears, connected with them a hoisting engine; also numerous storehouses. Opening into this basin near its head is a fine stone dry dock for large vessels; also a small one for torpedo boats. On the left of the basin was the coal pile, about 20,000 tons.

Landing in the dockyard, it was noticeable that nothing had been destroyed by the retreating Chinese. A torch properly applied would have left nothing except the dry dock and coal pile.

General O Yama courteously received us, and ordered his chief of staff to give us such aid as we needed in visiting points of interest. Wishing to see the land defenses rather than the heavy sea batteries which had not been engaged, a captain of artillery was sent with Lieutenant Halsey and myself to visit the battlefield. We moved out to the northeast, over a dusty, barren plain, much cut up with dry watercourses. On the road we passed a few Chinese soldiers who had been slashed with the sword; the neck was usually severed by a single blow. In some instances half-starved hogs and miserable dogs were feeding on these dead.

About two and one-half miles out the plain ended. A rather steep range of hills formed its boundary. Following, and just below the crest of these, we could see a long line of the red and white uniforms of the Chinese infantry. For a time we believed the soldiers of the Celestial Empire had marked a single modern battlefield with glory, but the uniforms had all been shed before the fight began.

On the crest of the hill was a line of rifle pits, peculiar in construction. It consisted of a series of holes about two feet deep by two in circumference, about a foot apart, the earth being thrown to the front. Planted in this was a flat slab of stone, when it could be procured, or else a number of smaller pieces were set together. The earth covering it prevented the splashing of lead when a ball struck. The new earth was hidden as well as possible by twigs cut when the leaves were on.

A few men had fallen in their pits, but the majority that stood had evidently huddled through fear or that desire that men have for company in time of danger. It was not from discipline or drill, and little groups had fallen together as the Japanese broke through.

Five kinds of ammunition of as many calibres were found on the line. Cartridges were laid loosely by the pits, and were scattered all down the hill, thrown away in flight. Empty shells were few in proportion.

The Japanese had assaulted up hills so steep that it was difficult even to walk up unencumbered, and they must have been exposed to the fire of 37-millimeter Hotchkiss and rapid-fire guns, well placed in earthworks having commanding positions.

The Japanese killed had been buried and their wounded removed to hospitals. Captured prisoners under sergeants' guards were digging shallow graves in the hard soil and forcing the stiffened bodies of their dead comrades into them by the free use of spade and poles

Whether the Chinese wounded were removed to hospital or not I do not know. There were none but dead on the field, but they had all fallen by the bullet or shell.

The entire line was not assaulted, but the point described answers for a description of all the points taken. I do not think there could have been more than 400 dead; probably 1,200 killed and wounded is a fair estimate.

Crossing the main road leading to Taliwan Bay we met a body of Chinese coolies willingly carrying the loads of Japanese coolies for a very small stipend; the fight was not theirs. On our return we visited the village and inspected the dockyard, leaving for the ship about half an hour after sundown.

### GUARD DETAILED FOR PEKING.

The Baltimore arrived in Nagasaki on November 29; there found a telegram ordering the Marines to Tientsin, to be held at the call of the United States Minister in Peking as a guard for the legation there if needed.

The Baltimore was coaled, and we sailed for Chefoo on December 2, arriving at 7 A. M. on the 4th. Entering the harbor just before us was the British steamer bound for Tientsin. The paymaster boarded her and secured passage for three officers and forty-two men. The accommodations for the latter were poor in the extreme. We left the Baltimore at 8 P. M. and sailed at 10 P. M., arriving off Taku Bar the evening of the 5th.

There were several steamers before us unloading in lighters, and our steamer was obliged to wait her turn. We had left Chefoo hurriedly, so that there was a lack of rations for the men if detained. Everything had been eaten up on board at breakfast the morning of the 6th except flour in barrels, and there were no means of cooking this for so large a number.

There was every indication of a blow coming up: in that case it would be three or four days before we could communicate with the shore. Added to this was the danger of the river freezing up, which usually occurs by the 10th of December. Taking this into consideration, I determined to take the men to Tientsin by rail, thirty-two miles.

I was well aware I had no right to land an armed force without permission in a foreign country, but it was impossible for me to get this permit within necessary time, so I embarked the men and baggage on a tug at 1 P. M. After crossing the bar and steaming about ten miles, we entered the Kiho River, the mouth of which is fortified by large concrete and mud forts on both sides.

As soon as we were well in sight of the forts numerous trumpets sounded "to arms," and the Chinese garrison manned the guns. They went through the form of loading them—may have actually done so. However, we were allowed to pass unquestioned, and landed six miles above the forts at a coal dock, 500 yards from the railroad station. There I left Ensign Belknap with ten men to guard the baggage, and marched the remainder to the depot. Curiosity had collected a motley crowd of dirty natives, coolies, and soldiers about us, numbering not far from 500. They pressed in so closely that it was necessary to place the men back to a wall and put in front numerous sentinels to keep off the press. The sentinels, losing patience, would now and then give vicious blows with the butts of their pieces, which soon had a good effect.

The men stood fully equipped from 3.30 P. M. until 6 P. M., when the train pulled up, having next to the engine two coal floats for our transportation. Ensign Belknap had his detail and baggage on board. The thermometer was much below freezing and the wind coming over the ice fields biting.

We were obliged to use the open floats, as Chinese soldiers traveling in American style of passenger cars had made their floors a foul sink. The men danced and thrashed themselves to keep from freezing, and after an hour and a half we arrived at Tientsin, crossed the river, and reported on board the U. S. S. Monocacy.

Moored alongside of the bund was a gunboat of each of the following nations, viz., American, German, English, Russian, and French; also guards of Marines or sailors held for duty in Peking if wanted, viz., fifty-one American Marines, fifty English Marines, fifty German soldiers, fifty Russian sailors, fifty Italian sailors, thirty-two Spanish Marines.

Preparations were being made by the commanding officers of these guards to get their men properly equipped for the extremely cold weather of winter. Long sheepskin coats, fur caps, and gloves were purchased at Government expense by the various nationalities for their men. In case of the American Marines they had to pay for their coats themselves $3.50 in gold, and these were absolutely worthless to them in the spring. The strong smell of wool prevented their being stored on board ship.

All guards were expected to be ready to start for Peking, eighty-six miles distant, by December 15, but an edict forbidding them to enter the city was issued by the Emperor.

Every one believed that this edict would be rescinded, and in order that the men might be in proper physical condition to make this march in four days or less, the American Marines were marched daily without regard to weather (there being no rain) about seven miles, drilling in the afternoon when the weather was good. The German sailors were the only body of men who marched regularly, and were trained systematically.

When the river froze and steamer communication was cut off, the streets of Tientsin were barricaded, and the commanding officers of vessels had at least an understanding what was to be done by each command in case of riot, fire, or disturbances. Rumors varying in the sensational was the daily order of things.

Knowing if the Japanese saw fit to make an attack on the Taku forts, thirty-two miles down the river, or at Shanhaiquan, 170 miles to the west, connected by rail with Tientsin, that guards should be called to Peking, and that it would be impossible to procure carts for the transportation of stores, three months' rations for fifty men were sent and stored in the legation.

In February I received an order from the commanding officer of the ship, by direction of the Admiral, to proceed to Peking to confer with our minister, Colonel Denby, in regard to making arrangements for housing, feeding, etc., the men, and to see what the missionaries intended to do, provided the Japanese landed in the Pichili district and moved on Peking.

After these orders were issued the commanding officer held them for a week, as foreigners had been molested en route, and Chinese soldiers had robbed the banks (money lenders) in the town through which I had to pass. Twenty-five thousand Chinese soldiers were encamped along the route, beginning twenty miles from Tientsin and ending sixteen miles from Peking.

It took two days to make this disagreeable trip on horseback. Chinese camps lined the road; they were usually surrounded by a wall made of mud mixed with straw, about sixteen feet high, castellated for ornament, having a single arched entrance. Their use was to break the cold winds and keep men from wandering out of camp—did sentinel duty. Bodies of troops met on the move were armed with all manner of weapons. Tower muskets, gingals, Remington rifles .50 and .45 calibre, German carbines, Grass magazine rifle calibre 30.03, bamboo poles with a small flag and spike, tridents, swords with handles four feet long, etc. And I saw 1,000 muster to pay honor to Li Hung Chang, who was supposed to be on his way to Peking, who were unarmed. I do not believe that these troops so near Peking had a single cartridge, but of this I am not certain. Cartridges would be served out when they started for the enemy, so straggling privates of different regiments stated.

I called on our minister, Colonel Denby, and showed him my instructions. He was not only personally courteous, but interested himself in aiding me to carry out my orders.

The American legation compound is less than an acre in area and is badly cut up with buildings, none of which was fit to house fifty men. Besides, the great wall of the city, sixty feet high and parapeted, was within 100 yards of the compound. This allowed it to be enfiladed by rifle fire, which made it necessary to look for other quarters.

The Methodist mission was visited, a mile distant, a compound of nine acres, but this was situated under the wall and removed from support of the other legations, which were grouped. I found two strong houses on Legation Street well suited. They were in charge of Sir Robert Hart. Called on him with our minister in regard to them; received a note the next day regretting that it would be impossible to let us have them, as they were Chinese property. Finally secured the Dutch legation, which was vacant at the time, a compound suitable in every way. Paymaster Cowie made all arrangements necessary in case of occupation, even for having a distiller put up, as all water in the city is strongly alkali, cistern water being very limited.

At Colonel Denby's request I visited the six American missions, which necessitated a cold ride of sixteen miles within the walls of the city. A report has been made to Admiral Carpenter in regard to the situation in Peking at this time and the probable results of a move on Peking by the Japanese.

The Emperor of China for the second time gave an audience to the foreign ministers, the occasion being New Year's. Paymaster Cowie and myself acted as military attachés to the American legation.

The day before I returned to Tientsin Prince Lung, the Emperor's uncle, called on Colonel Denby, begging him to act in the name of the Emperor to the extent of drawing up a request to the Japanese authorities asking them to receive a peace commissioner. Pending these negotiations our minister did not wish a guard in Peking, and I left with the understanding that the guard would not leave Tientsin unles the Japanese transports appeared off the coast or the commission failed.

Had the Japanese landed in the district of Pichili, 150,000 Chinese soldiers would have been defeated, and an immense rabble of armed men swarming through the country would have made the best of their way home, and there was not a haversack in the Chinese army.

Peace having been declared on May 10, we were ordered back to the Baltimore May 16. I left five men behind to fill vacancies on the U. S. S. Monocacy. Proceeded by merchant steamed to Chefoo, and there embarked the Marines on the U. S. S. Yorktown and U. S. S. Concord, and arrived in Nagasaki May 20 and joined the U. S. S. Baltimore on that date.

Very respectfully, C. F. ELLIOTT,
*Captain, United States Marine Corps.*

ADDENDA TO CAPT. G. F. ELLIOTT'S REPORT.

The quiet, soldierly discipline of the Japanese troops astonished all foreign officers who had the opportunity of observing them while disembarking, on the march, occupying camps near, or while billeted in cities not their own.

In 1870, in obedience to an edict of the Mikado, the Diamios mustered their Samuri, and each and all of the 2,000,000 hereditary soldiers of Japan laid by their swords in a day, and the privileged class became of the people. The feudal system which had defended the country, closed it to foreigners, made invasions and filled Nippon for hundreds of years with chivalry and oppression, ended.

The country was defenseless, having neither army nor navy, but the progressive party immediately took steps to create both, modeled after those existing among enlightened nations.

Army officers of good standing from America, France, and Germany were employed to instruct, organize, and drill the new defense, but no system was adopted until from observation the German was preferred, and the Japanese army is now modeled with slight and suitable changes after that country.

Casual observers, though military men, are apt to credit the iron discipline adopted from the Germans with creating in less than twenty years an army thoroughly equipped, proficient in drill, and possessing in the highest degree that discipline which gives unquestioning obedience and patient endurance; but these latter qualities already existed in the sons of the Samuri, and soldiers not belonging to this class were leavened by them; besides, for the first time in the history of their country, they were allowed the privilege of fighting for her. These two conditions had more to do with their high standard as soldiers, in my opinion, than the system adopted.

Japanese infantry regiments are homogeneous physically; the men do not vary more than 2 inches in height, seven years in age, or more than 20 pounds in weight. In the field they are not burdened with over-fleshy men, neither with the "faithful old soldier," whose pride has outlived his strength. Forced marches did not leave a fourth of the regiment straggling in the road, and the endurance of the men could be counted on nearly as a unit.

Undoubtedly they are brave, but have not been tried in large bodies up to a demoralizing loss, although small parties on one or two occasions were cut off and fell together like true soldiers. Whether they will take a panic under heavy reverses is not known.

They are fairly well set up, but do not show it as much as the regular foreign soldier, good weight carriers, but I believe slow marchers, although for short distances they get over ground very rapidly on a trot, having what is known as "good wind." Drill has in a great degree eliminated the native peculiarity of being pigeon-toed. In marching they step too much from the knees and do not move out from the hips; this I think is due from the use of the Kimona in early life.

Uniform and equipment.—The cap is of dark-blue cloth, German in pattern, having a small drooping visor, a 2-inch band of yellow braid and the imperial chrysanthemum crest for ornament; it is heavy and ugly, but is the universal headgear. The coat is of dark-blue cloth, of roundabout pattern, single-breasted, having one row of good-sized buttons, a standing collar faced in colors to denote the regiment; a shoulder strap of the same color has worked upon it in silver bullion the regimental number. For winter dark-blue cloth trousers, and for summer white cotton duck. Overcoat of German pattern, dark-blue cloth.

All rifles are made at the arsenal in Tokyo, and are called by the inventor's name Murata. They are of two patterns. The one used by the great majority of troops was a single-loading bolt gun of .44 caliber, sighted to 1,400 meters, without wind gauge; the other a magazine gun of .315 caliber. The stocks are very short and the cheek wide. The cartridge box is of leather, holding about 45 rounds, about 4 inches deep, so as to allow the hand to enter it readily. A small bottle with screw stopper is made into it at one end to contain oil. The belt, of heavy black leather with a tongue buckle. The knapsack is of russet leather; a pocket at one end holds the blade of the Linnamen intrenching spade, the handle of which projects above the soldier's left shoulder when the knapsack is slung. The blanket was of sky blue and when rolled was strapped on top and down each side of the knapsack. An extra pair of shoes was also strapped on each side, soles out. Just below the flap on the back was strapped a tin box shaped like the case of a pair of binoculars; this was the haversack, and was used for eating from and carrying boiled rice. There was a tray for tea, etc. The water bottle was of glass covered with leather.

Cooking in camp was done for the company at kitchens, large iron pots being the only utensils. Boiled rice took the place of bread, tea, beef, when it could be purchased, and dried fish brought from Japan formed the ration. I think the cost of a single ration per day was 3 cents, gold.

While in Tientsin, China, I visited several times the hospital under charge of Dr. Irwin, surgeon to Viceroy Li Hung Chang. Wounded Chinese soldiers were treated here during the winter. Surgeon Major James, of the British army, had charge of some of the wards, and was much interested in wounds made by the small-caliber enveloped bullet. He kindly gave me every opportunity to see, and much information in regard to these wounds. I saw about thirty of the wounded; in no case were bones shattered to any extent, even at the exit of the ball; the puncture was clean; no bullet was found in the patient, so the casing was unknown, whether of steel or a softer metal. They did not seem to deflect on striking, but cut their way through in the line of flight. Thirty patients testified they had received their wounds at distances varying from 150 to 400 yards. They healed readily, often by first intention. None of the wounds were in the leg, for the men had made their own way from the battlefield on foot, and in carts, to the railroad, more than 200 miles, and were several weeks on the road. All suffered more from frost-bitten feet than the bullets. Out of the thirty cases I saw, I think probably the old .45-caliber lead ball would have left on the field nine of them dead.

The services of the Marines from the U. S. S. Yorktown are also recognized in the following documents:

U. S. S. YORKTOWN,
CHEMULPO, KOREA, July 7, 1896.

SIR: I have the honor to bring to your notice the excellent behavior of the Marine guard of this vessel during the eight weeks they were under my command at the United States legation in Seoul.

The men early evinced the intention to conduct themselves with propriety. This led me to allow them much latitude and they proved themselves worthy of the confidence so placed in them.

All work was well and cheerfully performed; no man had to be excused from his tour of duty, and not one punishment was necessary. Many complimentary things were said to me of the men, and I am sure they left Seoul with the respect and good will of all its residents. It is difficult to conceive of better behavior.

I am, sir, very respectfully, your obedient servant,

R. R. BELKNAP, *Ensign, United States Navy.*

THE COMMANDING OFFICER U. S. S. YORKTOWN,
*Chemulpo, Korea.*

**(First Indorsement.)**

U. S. S. YORKTOWN (THIRD RATE),
CHEMULPO, KOREA, July 8, 1896.

Respectfully forwarded for the information of the Colonel Commandant, United States Marine Corps.

C. H. STOCKTON,
*Commander, United States Navy, Commanding.*

U. S. S. YORKTOWN (THIRD RATE),
CHEMULPO, ÆOREA, July 8, 1896.

SIR: Herewith I enclose a report of Ensign R. R. Belknap, concerning the Marine guard of this ship while it was stationed at the United States legation at Seoul, Korea, which is considered a matter worthy of your notice.

From personal knowledge I can second the report of Ensign Belknap, and can further say that the United States minister to Korea, Mr. Sill, speaks of the guard as a model one, and expressed in an official letter to me his very high appreciation of its good conduct and faithfulness.

I am, sir, very respectfully, C. H. STOCKTON,
*Commander, United States Navy, Commanding.*

THE COLONEL COMMANDANT,
*Headquarters, United States Marine Corps, Washington, D. C.*

(First Indorsement.)

U. S. FLAGSHIP OLYMPIA,
YOKOHAMA, JAPAN, July 20, 1896.

Approved and forwarded. F. V. MCNAIR,
*Rear Admiral United States Navy, Commanding Asiatic Station.*

UNITED STATES NAVAL FORCE ASIATIC STATION,
UNITED STATES FLAG-SHIP OLYMPIA,
YOKOHAMA JAPAN, July 7, 1896.

SIR: Please communicate to Ensign R. R. Belknap, United States Navy, the commanding officer of the guard from the vessel under your command recently maintained at the United States legation at Seoul, Korea, and also to the men who composed the said guard, that the commander in chief has been highly gratified by the expressions of appreciation of the creditable manner in which their duty was performed which have come to him from the minister resident of the United States at Seoul and from the commanding officer of the Yorktown.

Very respectfully, F. V. MCNAIR,
*Rear Admiral, United States Navy,*
*Commanding United States Naval Force, Asiatic Station.*

THE COMMANDING OFFICER U. S. S. YORKTOWN,
*Chemulpo, Korea.*

---

## CHAPTER XLIII.

**1898.**—The Destruction of the U. S. S. Maine—Heroism of Private Anthony.

**1898.**—The story of the Spanish-American War would be most incomplete and fragmentary with no mention of the distinguished services rendered by men of the United States Marine Corps. The many opportunities accorded to this branch of the service to display the qualities of bravery and of efficiency in the performance of duty were nobly improved, winning high praise and commendation, and adding to the glories of American arms. The first sad episode in the chain of events leading to the memorable conflict—the destruction of the Battleship Maine by torpedo explosion in Havana harbor—gave occasion for a display of courage and coolness in a private of the Corps that would be worthy the emulation of the bravest and greatest.

Realizing at his post of duty that the ship had been destroyed and was rapidly sinking, this brave man, William Anthony by name, thought first of his commander's safety, and risked his own life, amid the smoke and darkness, to inform him of the occurrence. The story was gratefully told by Captain Sigsbee himself in the following letter to the Secretary of the Navy:

NAVY DEPARTMENT, WASHINGTON, D. C., April 8. 1898.

SIR: I desire to call the attention of the Department to the soldierly conduct of Private William Anthony on the occasion of the explosion of the Maine. At the time of the explosion I was in the captain's cabin of the Maine. The lights of the vessel were instantly obscured and the apartments were filled with smoke; there was immediate and intense darkness. On leaving my cabin through the usual passage forward, feeling my way along, I was met near the outer door of the superstructure by Private Anthony, who was coming into the cabin to fulfill, on that dangerous occasion, the precise duties of his position by notifying me of the explosion. He ran against me in the darkness, apologized hastily, and reported to me that the ship had been blown up and was sinking. We then proceeded together to the quarter deck.

The special feature in the case of this service performed by Private Anthony is that, on an occasion when a man's instinct would lead him to seek safety outside the ship, he started into the superstructure and toward the cabin, irrespective of the danger. The action was a noble one, and I feel it an honor to call his conduct to the attention of the Navy Department with the recommendation that he be made a sergeant.

Very respectfully, C. D. SIGSBEE.

*Captain. United States Navy, lately Commanding the U. S. S. Maine.*

THE SECRETARY OF THE NAVY, *Washington, D. C.*

Captain Sigsbee's letter called forth the following commendation from Secretary Long, addressed to Private Anthony:

NAVY DEPARTMENT,
WASHINGTON, D. C., April 9, 1898.

SIR: The Department is in receipt of a letter from Captain C. D. Sigsbee, U. S. N., of date April 8, 1898, calling attention to your conduct on the occasion of the disaster to the Maine.

In transmitting to you a copy of this letter the Department is glad to add its expression of commendation. Your courage and fidelity to duty at the time of the explosion, as detailed by Captain Sigsbee, was most commendable. Your conduct was a credit to the Naval service, and entitles you to the hearty thanks and approval of the Department.

Very respectfully, JOHN D. LONG, *Secretary.*

PRIVATE WILLIAM ANTHONY,
*United States Marine Corps.*

Colonel Charles Heywood, Commandant of the Corps, also refers to the incident in the following terms:

"The discipline and instruction of the Marine Corps have been maintained at a high standard, and to this is attributed in a large measure the efficiency of the services rendered by the Marines in the war between the United States and Spain. One of the instances of discipline connected with the war, which attracted public attention, was the conduct of Private William Anthony in performing the very letter of his duty as orderly on the occasion of the destruction of the battleship Maine in Havana Harbor by going below to the captain's cabin, irrespective of danger, and informing him that the ship had been blown up and was sinking. For his action on this occasion Private Anthony received commendatory letters from Captain C. D. Sigsbee, of the Maine, and the Secretary of the Navy, and was promoted to the rank of sergeant by myself. The letters mentioned are appended, and I request that they be printed with this report."

---

## CHAPTER XLIV.

**1898.**—The Work of the Marine Corps in the Battle of July 3, 1898.

**1898.**—In the ever-memorable naval battle off Santiago, on July 3, 1898, the men of the Marine Corps attached to the various ships of the fleet distinguished themselves by their skillful and effective handling of the secondary batteries and rapid-fire guns, to which the Navy Department had determined they should be assigned. In view of the fact that the wrecks of several of the Spanish ships showed more serious damage from these batteries than from even the heavy guns, no further comment on the skill and endurance of the Marines is required. The details of this and other services of the men of the Corps are given in the reports of the various commanding officers.

The report of the services of the detachment on the battleship Oregon, during her historic run from San Francisco to Jupiter Inlet, Fla., around Cape Horn, and afterward through the weary weeks outside the Harbor of Santiago, Cuba, to the battle of July 3d, are given in the following letters of Captain Dickins, U. S. M. C., and of Captain C. E. Clark, U. S. N., Commander of the Oregon:

"U. S. S. Oregon,
"New York Navy-yard, August 29, 1898.

"Sir:—I have the honor to acknowledge the receipt of your letter of August 9, and in reply would respectfully state that this ship took a most prominent part in the bombardments of Santiago on June 6, 16, and July 1 and 2, and in the battle of the 3d of July, and was also under fire during the sinking of the Reina Mercedes.

"The Marines were stationed at the secondary battery and had the two 1-pounders in the fighting top, the four 6-pounders on the bridge, and four 6-pounders on the superstructure, which were manned by six non-commissioned officers and thirty-eight privates. The remainder of the guard were stationed as sharpshooters.

"From the time the Oregon left Callao, Peru, until after the destruction of the Spanish fleet off Santiago de Cuba the Marines kept a constant watch, as sharpshooters, forward and aft, and a look-out for every pair of 6 and 1-pounders which they manned, the remainder of the guns' crews sleeping at their guns with ammunition at hand.

"We went to general quarters frequently, at times twice in one night. The men were thoroughly drilled at their guns, and within thirty seconds after general quarters sounded every gun manned by them was ready to open fire.

"This was the result of constant and careful drill and instruction. When the Marines went into action they went in coolly, every man doing his duty, even though they occupied the most exposed position on the ship, with no armor of any kind to protect them.

"For the way the Marines manned and fought their guns, and for their good behavior under fire in every battle that their ship took part in, I would respectfully refer you to Captain C. E. Clark, the commanding officer, and Lieutenant Commander J. K. Cogswell, the executive officer.

"The destructive effect of the fire of the secondary battery of the fleet is shown by the report of the Board ordered to report on the effect of the gun fire on the Spanish fleet.

"The Marine guard of this ship also formed three-fourths of the force which made the first landing of armed troops on the Island of Cuba at Guantanamo Bay on June 10, 1898.

"I respectfully submit the station bill of the Marine guard at general quarters, which was faithfully carried out. "Very respectfully, "R. DICKINS,

*"Captain, United States Marine Corps, Commanding Guard."*

THE COLONEL COMMANDANT UNITED STATES MARINE CORPS,

*Washington, D. C.*

In indorsing this report, Captain Clark says:

"The within report is a modest statement, by a brave and most deserving officer, concerning the highly meritorious part taken by himself and Lieutenant A. R. Davis and the non-commissioned officers and privates of the Marine guard of the Oregon during the days and nights of anxious waiting for the Spanish fleet on the Atlantic, and with searchlights at the entrance of Santiago harbor, during three engagements with the enemy's batteries, and in the battle of the 3d of July. Every man in the guard had an exposed station, and the only reluctance ever shown by any of them promptly to obey was when ordered to take shelter behind the turrets, while the alacrity with which they ever sprang to their posts showed that they were all animated by the spirit that has given the Marine Corps its reputation for bravery and faithfulness during a full century. "C. E. CLARK, *"Captain, U. S. N."*

A similar story of duties creditably performed is told by Lieut. Radford, U. S. M. C., Commander of the Marines on the Battleship Texas:

U. S. S. TEXAS (FIRST RATE),

NAVY-YARD, NEW YORK, August 11, 1898.

SIR: I have the honor to submit the following report of the stations and services of the Marine guard of this vessel on July 3, and other engagements participated in by the Texas:

Stations.—In the fore-top, six men, two 1-pounders; on forward superstructure, six men at two Hotchkiss revolving cannon; on forward superstructure, five men at 3-millimeter rapid-fire gun; in port and starboard waists, six men at two 1-pounders; on after superstructure, six men at two Hotchkiss revolving cannon; on after superstructure, six men at two Colt automatic guns; in maintop, six men at two 1-pounders; one man at central station; two orderlies for commanding officer; two sentries in engine room.

In all the bombardments the men went to and remained at their stations. The Texas was in the bombardments of Santiago of June 6, 16, and July 2. On June 15 forced the mined entrance to Guantanamo (14-100 pound guncotton mine afterwards recovered) and reduced the fortifications. Without assistance silenced the Socapa battery on June 22, which had successfully withstood the combined fire of the western squadron on the 6th and 16th. Played a conspicuous part in the destruction of the Spanish fleet on July 3, engaging the Infanta Maria Teresa, the first to leave the harbor, and was present at the successive surrender of the remaining vessels, including the Cristobal Colon, fifty miles to the westward of Santiago. Total secondary battery fire, 730 shots, the Marines firing 330.

As all secondary battery guns are manned every night with two men of each crew, one man of each gun always on lookout, the service was hard but cheerfully performed. The regular post duty was in no way neglected. The guard of this vessel, by direction of Captain J. W. Philip, U. S. N., was landed at Guantanamo on June 12 (taking ashore two Colt automatic guns), and assisted in the defense of Camp McCalla on June 12 and 13, the men behaving well under fire. The funeral escort, for the burial of Dr. Gibbs and two privates, remained at parade rest and perfectly cool under the stray firing of the Spanish sharpshooters.

The guard has done all that was required, and in a cheerful and satisfactory manner. Very respectfully, CYRUS S. RADFORD,

*First Lieutenant, United States Marine Corps.*

In indorsing Lieut. Radford's report, Capt. Philip of the Texas writes:

The performance of all duty of the Marine guard under command of Lieutenant Radford met with my approval and commendation.

Besides their work at the secondary battery in all engagements, I desire to call attention to special instances:

During the chase on July 3 it was reported to me that the firemen and coal heavers were giving out, and the engineers desired more men from the deck. The main battery having already been drawn upon for this extra work, I directed Lieutenant Radford to detail fifteen or twenty men to go in the fire room to shovel coal. Immediately, and with a rush to be first, all the Marines started for the fire room to aid the Texas to maintain her speed in the chase.

On arrival at Guantanamo Bay, June 12, Colonel Huntington asked that the guard of the Texas be sent ashore to reinforce and assist his command. It was landed at once, and on arrival on the hill I noticed it was stationed on picket duty immediately and under fire at once.

The valuable service rendered by Lieutenant Radford on the 12th and 13th was later especially commended to me by both Commander McCalla and Colonel Huntington; and in this connection I desire to call attention of the colonel commandant not only to the gallant conduct of Lieutenant Radford, but to the fact that he has the distinction of being the only officer in the Marine Corps who has done service both ashore and afloat during this war, a fact that should be brought to the attention of the Navy Department for its consideration.

J. W. PHILIP,
*Captain, United States Navy, Commanding.*

On the Cruiser Brooklyn the Marines handled their batteries with equal skill and effect, as is shown in the following interesting account of the action by Captain Paul St. C. Murphy, U. S. M. C., commander of the detachment:

U. S. FLAGSHIP BROOKLYN,
GUANTANAMO BAY, CUBA, July 6, 1898.

SIR: I have the honor to submit the following report of the part taken by the Marine Guard of this vessel, on the 3d instant, in the action which resulted in the destruction of the Spanish squadron.

At the moment the alarm was given that the enemy's ships were coming out of the harbor, the guard was at quarters, ready for inspection. It was immediately dismissed and the men sent to their stations for battle.

The distribution was as follows:

| | |
|---|---|
| Six-pounders: 2 sergeants, 1 corporal, 13 privates | 16 |
| One-pounders: 1 sergeant, 1 corporal, 10 privates | 12 |
| Colt automatic guns: 1 sergeant, 1 corporal, 10 privates | 12 |
| Signals: 2 music, 4 privates | 6 |
| Battle orderlies, 8 privates | 8 |
| Flag orderly, 1 private | 1 |
| Commanding officer's orderly, 1 private | 1 |
| Ammunition and fire party, 10 privates | 10 |
| Commanding and supervisory: 1 captain, 1 lieutenant, 1 first sergeant | 3 |
| Total (full strength of guard) | 69 |

The men were full of enthusiasm, but there was no excitement or disorder, and apparently no concern for personal safety.

The battery was handled with admirable coolness and deliberation. Greater care could not be taken in setting sights and aiming if the men had been at target practice and each striving to make a record score.

Considering the fact that the enemy was within effective range during the greater part of the action the fire of the secondary battery must have been most destructive to his men and material, and contributed its full share to bringing the battle to an end so speedily and with so little loss to ourselves.

It is reported that Spanish officers have stated that so deadly was the effect of our secondary battery fire it was impossible to keep their men at the guns.

Where all did their duty manfully it is a difficult matter to select individuals for special mention. There are some, however, who deserve to be brought to your notice by name for conduct that displayed in a conspicuous manner courage, intelligence, and devotion to duty.

During the early part of the action a cartridge jammed in the bore of the starboard forward 6-pounder, and in the effort to withdraw it the case became detached from the projectile, leaving the latter fast in the bore and impossible to extract from the rear. Corporal Robert Gray, of the port gun, asked and received permission to attempt to drive the shell out by means of the rammer. To do this it was necessary to go out on the gun, and the undertaking was full of difficulties and danger, the latter due in a great measure to the blast of the turret guns firing overhead. The gun was hot, and it was necessary to cling to the jacob ladder with one hand while endeavoring with the other to manipulate the long rammer. After a brave effort he was forced to give up, and was ordered in. Quarter Gunner W. H. Smith then came, sent by the executive officer, and promptly placed himself in the dangerous position outside the gun port, where he worked and failed as the corporal had done. Neither had been able to get the rammer into the bore, and there seemed nothing left to do but dismount the gun. At this juncure Private Macneal, one of the crew, volunteered to go out and make a final effort. The gun was so important, the starboard battery being engaged, that as a forlorn hope he was permitted to make the attempt. He pushed out boldly and set to work. The guns of the forward turret were firing, the blast nearly knocking him overboard, and the enemy's shot were coming with frequency into his immediate neighborhood. It was at this time that Chief Yeoman Ellis was killed on the other side of the deck. Macneal never paused in his work. The rammer was finally placed in the bore and the shell ejected. The gun was immediately put in action and Macneal resumed his duties as coolly as if what he had done were a matter of everyday routine. The battle orderlies well merit a place among those whose conduct is worthy of special mention. They were on the move constantly bearing battle orders to all parts of the ship, and in no instance did they fail in the prompt and intelligent performance of their responsible duty. The signalmen occupied very exposed positions during the action and rendered excellent service. Signal halyards and numbers, battle flags and speed cones, were riddled by small projectiles and fragments of bursting shell, casualties that show in what a zone of danger the signalmen performed their duties. Signalmen Coombs and McIntyre and Battle Orderlies Rall and Davis were so near Yeoman Ellis when he was killed that they were bespattered with blood.

The following are the names of the battle orderlies: To the flag officer, Privates Rall and Davis; to the commanding officer, Privates Kelly, Smith, and Sanjule; to the executive officer, Privates Brennen, G. Wismer, and P. O'Donnell.

The flag orderly, Private Richmond, and commanding officer's orderly, Private Woodsun, were on deck and rendered good service. The following are the names of the signalmen: Forward, Privates Coombs and McIntyre; aft, Privates Shaw and A'Hearne.

The music boys, Drummer Weisenberg and Fifer Stewart, were stationed on the main and gun decks, respectively, to sound trumpet calls, and behaved manfully. None showed more unflinching courage than the men in the military tops, who stood by their guns delivering their fire with unerring precision, undismayed by the projectiles flying about them and striking in their immediate vicinity. Private Stockbridge, the only man on the sick list, climbed into the main top at the signal for battle, where he remained to the end of the action, doing good work at his gun.

The non-commissioned officers, First Sergeant Manning, Sergeants Bristow, Montair, Ingalls, and McDevitt, Corporals Dittmeier, Doyle, and Gray, showed excellent soldierly qualities in the management of the men and battery.

First Sergeant Manning rendered valuable aid in supervising the widely separated detachments of the guard.

I cannot speak too highly of the conduct and bearing of Lieutenant Borden. His courage and excellent services proved him a valuable officer.

There were no casualties of a serious nature. Two men were slightly injured, one, Private Flynn, in the back by a splinter; the other, Private Barfield, in the leg, cause unknown. In neither case was it necessary for the men to leave the station.

It is a matter of greatest pride to the guard, officers and men alike, to be able to claim a share in the splendid work done by the Brooklyn on the 3d instant.

A report nearly identical with this has been made to the commanding officer.

Very respectfully, PAUL ST. C. MURPHY,
*Captain, United States Marine Corps, Commanding Guard.*

In indorsing this report, Admiral Schley writes:

"U. S. FLAGSHIP BROOKLYN,
"OFF SANTIAGO DE CUBA, July 15, 1898.

"The conduct of the Marine guard under Captain Murphy's command on the occasion of the destruction of the Spanish squadron on July 3, 1898, was in every way worthy of the accompanying report. W. S. SCHLEY,
*Commodore U. S. N., Commander in Chief Flying Squadron."*

Although the Cruiser New York was not present with the other ships of the fleet during the engagement with the Spanish, on July 3d, the services of the Marines in other actions, and during the extended investiture of the port of Santiago, are made the subject of the following report:

U. S. FLAGSHIP NEW YORK,
NAVY YARD, NEW YORK, August 27, 1898.

SIR: In obedience to your directions, I have to report on the stations and services of the Marine guard of this vessel on the occasion of the destruction of the Spanish fleet, July 3, 1898, as follows:

On July 3, 1898, the men went to their stations with much enthusiasm and were delighted at the prospect of coming to close quarters with the enemy. Privates William Rapp, William O'Neil, and Charles W. Berthold were on the sick list, but went to their stations for general quarters and performed their duties there.

The men at all times showed a commendable coolness while under fire.

Following is a list of stations of officers and men July 3, 1898, during the action with the enemy:

| Name. | Rank. | Station. |
|---|---|---|
| Meade, R. L. | Major, | Commanding marine guard. |
| Lane, R. H. | First lieutenant | In charge two 6 pounders No. 14. |
| Kinne | First sergeant | Rifleman, upper foretop. |
| Olinger Charles | Sergeant | In charge marine ammunition (passers first division). |
| Buerger | ....do | In charge marine ammunition (passers second division). |
| Shire, James H | ....do | In charge marine ammunition (passers third division). |
| Erbs, A. | Corporal | No. 1, starboard 6 pounder No. 14. |
| Ludlow, Thomas | ....do | Rifleman, upper foretop. |
| Givens, William | ....do | Rifleman, maintop. |
| Walters, G. H. | ....do | No. 1, port 6 pounder No. 14. |

| Name. | Rank. | Station. |
|---|---|---|
| Shea, John J. | Fifer | With executive officer. |
| Cassidy, J. P. | Drummer | Forward spar deck. |
| Anders, C. C. | Private | Rifleman, lower foretop. |
| Barksdale, M. S. | ...do | No 3, starboard 6 pounder No. 14. |
| Bartlett, C. H. | ...do | Rifleman, upper foretop. |
| Benson, John B. | ...do | Ammunition passer, first division. |
| Bernard, A. H. | ...do | Orderly to commander in chief. |
| Berthold, C. W. | ...do | No. 4, port 6 pounder No. 14. |
| Bolster, M. | ...do | Ammunition passer, fourth division. |
| Bonner, P. A. | ...do | Ammunition passer, third division. |
| Bracken, E. | ...do | Rifleman, maintop. |
| Brownlee, W. | ...do | Aid to wounded, berth deck. |
| Demerais, C. | ...do | Ammunition passer, third division. |
| Donohue, E. | ...do | Ammunition passer, second division. |
| Donohue, T. J. | ...do | No. 4, starboard 6 pounder No. 14. |
| Donovan, J. L. | ...do | Ammunition passer, fourth division. |
| Doyle, Alfred | ...do | Rifleman, lower foretop. |
| Doyle, John | ...do | Ammunition passer, fourth division. |
| Foley, James J. | ...do | Ammunition passer, third division. |
| Gill, Patrick | ...do | Rifleman, upper foretop. |
| Haisman, F. W. | ...do | Ammunition passer, fourth division. |
| Haistings, J. C. | ...do | Do. |
| Heath, F. W. | ...do | Rifleman, upper foretop. |
| Hendershot, N. V. | ...do | Ammunition passer, first division. |
| Hesse, John | ...do | Do. |
| Horton, J. T. | ...do | Ammunition passer, second division. |
| Hutchinson, J. C. | ...do | No. 3, port 6 pounder No. 14. |
| Johnson, Benjamin | ...do | No. 2, starboard 6 pounder No. 14. |
| Kane, John | ...do | Rifleman, maintop. |
| Kearins, M. | ...do | Ammunition passer, fourth division. |
| Kelly, John | ...do | Ammunition passer, third division. |
| Kelly, Thomas | ...do | Aid to wounded, berth deck. |
| Kenny, William | ...do | Ammunition passer, third division. |
| Knip, Adolph | ...do | No. 2, port 6 pounder No. 14. |
| McGovern, James | ...do | Ammunition passer, fourth division. |
| Morgan, Daniel | ...do | Rifleman, lower foretop. |
| Newton, C. E. | ...do | Ammunition passer, third division. |
| O'Hara, R. | ...do | Do. |
| O'Neil, William | ...do | Ammunition passer, first division. |
| Penny, Edw. | ...do | Ammunition passer, second division. |
| Petitclerc, Edw. | ...do | Do. |
| Power, M. | ...do | Orderly to commanding officer. |
| Prescott, H. G. | ...do | Rifleman, maintop. |
| Putney, B. C. | ...do | Ammunition passer, third division. |
| Rapp, William | ...do | Ammunition passer, first division. |
| Ross, Albert | ...do | Ammunition passer, fourth division. |
| Sauer, William | ...do | Ammunition passer, second division. |
| Shea, John | ...do | Ammunition passer, first division. |
| Smith, F. M. | ...do | Ammunition passer, third division. |
| Smith, James | ...do | Rifleman, lower foretop. |
| Smolt, M. | ...do | Ammunition passer, first division. |
| Sullivan, James | ...do | Do. |
| Sprowls, M. | ...do | Do. |
| Vaughn, R. H. | ...do | Ammunition passer, third division. |
| Walter, R. F. | ...do | Rifleman, upper foretop. |

The ammunition passers were subject to call as riflemen.

The stations during the engagement at San Juan with the shore batteries, May 12, 1898, were the same as given above, with the exception of the following men, who had the stations given opposite their names:

| Name. | Rank. | Station. |
|---|---|---|
| Meade, R. L. | Major | Marine officer of the fleet. |
| Goodrell, M. C. | Captain | Commanding marine guard. |
| Kinne, F. A. | First sergeant | In charge of top ammunition. |
| Bernard, A. H. | Private | Ammunition passer, first division. |
| Power, M. | ...do | Ammunition passer, second division. |

Private William Rapp was slightly wounded by a fragment of a shell at San Juan.

The stations during the different engagements with shore batteries were practically the same as at San Juan.

The services of the Marines of this vessel were on each occasion efficient.

In addition to their regular duties the Marines furnished parts of prize crews, patrol launch crews, stood watch at the guns, and were posted as armed lookouts for torpedo boats.

Very respectfully, RUFUS H. LANE,
*First Lieutenant, United States Marine Corps, Commanding Marine Guard.*

Capt. Chadwick adds to this report the following high commendation:

The officers and men of the Marine guard on board this ship at all times in action did their duty most creditably. I have nothing but praise for their conduct and bearing at any time when engaged or when there was a prospect of engagement. Very respectfully, F. E. CHADWICK,
*Captain, U. S. N., Commanding.*

Capt. Littleton W. T. Waller gives this highly spirited account of the behavior of the Marines under his command on board the Battleship Indiana:

U. S. S. INDIANA (first rate),
OFF TOMPKINSVILLE, N. Y., September 1, 1898.

SIR: In accordance with the request contained in the letter of the colonel commandant, United States Marine Corps, under date of August 9, I have the honor to submit the following report:

The Marines of this ship are stationed for battle as follows: Twenty-four manning port 6-pounder rapid-fire guns on superstructure deck; 20 in powder division passing rapid-fire ammunition; 2 orderlies to commanding officer; 1 signal man; 2 orderlies passing ranges on orlop deck; 1 central telephone station; 2 in top, assisting in range-finding and indicating; 8 on search lights (at night); 15 in reserve as riflemen and as supports and reserves for rapid-fire guns.

First Lieutenant W. C. Dawson is stationed on the bridge with the commanding officer, acting as signal officer.

By direction of the commanding officer I have charge of all the rapid-fire guns on the superstructure and bridge decks with the exception of 2, making, in all, 16 guns.

The Marines took part in all the engagements in which the ship participated: San Juan de Puerto Rico, May 12; Santiago de Cuba, June 22, July 2, July 4, and the destruction of the Spanish fleet on July 3. The bombardments of the city of Santiago de Cuba were carried on by the great guns, and the Marines took no part in the firing. In all picket boats on duty at the mouth of Santiago Harbor there were detachments of Marines.

Practically, since a day or two after the destruction of the Maine, this ship has been on a war basis, and all precautions observed in time of war have been carried out. In all of the wearing and wearying watch duty, all war work, both in preparation for and during actual warfare, the Marines willingly and cheerfully took part, performing the many duties required of them so well that, during the time of preparation and war, there were only two cases of dereliction of duty; and at all times meeting with the approval and gratification of the commanding officer, who has stated to me that he regarded our men as the best in the squadron.

I cannot mention specific cases of duty performed in a markedly superior manner when all did so well.

On July 2, during the attack on the forts at Santiago, the Marines did all the firing at the eastern battery and Morro. This firing was so accurate and controlled as to draw strong expressions of admiration and approval from the commanding officer and the officers of other ships. For three days prior to this fight all the Marines except the orderlies and non-commissioned officers were, with the crew, coaling ship, only completing the work at midnight on the 1st of July, in time to get underway, leave Guantanamo Bay and reach Santiago in the early morning, to go into action without time to wash the coal dust off their persons. They went into the fight with such spirit and showed such qualities of discipline and precision as to draw forth a special order from the commanding officer commending seamen and Marines for their fine work.

On this occasion the Marines fired 570 shots with splendid accuracy.

On the occasion of the destruction of Cervera's fleet, the rapid-fire battery of this ship fired 1,744 shots in about sixty-five minutes; of this number, 1,534 were fired by the guns under my charge, a little more than a third of this number being fired by the Marines.

While the Marines were stationed at the port battery, and the starboard battery was engaged on that occasion, the fire was so rapid and sustained, the shock of explosion and the blinding smoke from the 8 and 6-inch guns were so great, that it was necessary frequently to relieve the crews of the starboard with the Marines from the port. The only trouble experienced at the time was the difficulty in keeping the men actually engaged under cover; they would creep up to the guns, waiting for the chance to take part in the action.

The condition of the Oquendo and Teresa after the action, together with the rapid destruction of the torpedo-boat destroyers, attest the accuracy of the fire of the rapid-fire batteries. The statement made at Norfolk by Captain Conchas, of the Teresa, to the effect that he could not keep his men at the guns or send messages and orders on account of the terrific fire from the rapid-fire guns of the three eastern ships of the squadron is a further proof of the accuracy and rapidity of fire of these guns.

Captain Conchas stated in conversation with civilians that his ship was destroyed by the fire of the three first ships a few moments after leaving the channel and turning to the westward.

As I have previously stated, I cannot mention any special instances of extraordinary conduct on the part of the men during the several engagements in which they participated; they have at all times and under all circumstances performed their duties faithfully and well, meeting with the approval of all.

In the case of Lieutenant Dawson, I cannot speak of his conduct during any of the action from personal observation, as he was on the bridge with the commanding officer.

In all the preliminaries and preparations for war Lieutenant Dawson has been responsive and indefatigable in the discharge of his duties. I cannot speak with too much praise of his capacity and interest.

I must leave to my seniors to say whether or not I performed my duties satisfactorily and well. Very respectfully,

LITTLETON W. T. WALLER,

*Captain United States Marine Corps, Commanding Marines.*

LITTLETON W. T. WALLER, LIEUTENANT-COLONEL

U. S. M. C. Plate 7

Captain Waller's report is supplemented by Captain Taylor's high praise of the conduct of the men of the Marine Corps on the Indiana, giving them full credit for the part played by their batteries in the great fight:

> The conduct of the Marines equaled in excellence that of the other divisions of the ship. All divisions of the ship, including the Marines, behaved admirably in all the various engagements of the war.
>
> In the destruction of Cervera's fleet the Marines fired about 500 shots from the secondary battery of the ship, about 1,200 being fired by the seaman division. The smaller number fired by the Marines was caused by their having the port battery of 6 pounders, while the starboard battery was the one engaged. The accuracy of the 6-pounders fired both from the seamen and the Marines was exceptionally good.
>
> With reference to the Marine officers of the Indiana, their conduct was equally deserving high praise and commendation with the officers of the other divisions of the ship.
>
> H. C. TAYLOR,
> *Captain, U. S. N., Commanding.*

Admiral Cervera, Commander of the Spanish squadron destroyed on July 3, 1898, was confined at the U. S. Naval Academy, Annapolis, Md., with the captains of the several Spanish vessels and other officers of high rank. Here every reasonable liberty was allowed to the distinguished prisoners, although the regulations of war were strictly maintained by a guard of the Marine Corps, under command of Maj. W. S. Muse, U. S. M. C. This guard also served to protect the grounds and buildings of the Naval Academy until the departure of the prisoners on September 8.

---

## CHAPTER XLV.

1898.—Departure of the First Marine Battalion from Cuba to its Camp at Portsmouth, N. H.—The Prison Camp at Portsmouth.

1898.—On its departure from Camp McCalla, on Guantanamo Bay, the First Marine Battalion, under command of Lieut.-Col. Robert W. Huntington, embarked on the U. S. S. Resolute, which had served as a transport during the war. The Resolute sailed on August 9, bound for the Isle of Pines, but the destination having been altered by commands subsequently received, the course was changed to Manzanillo, where they arrived on the 12th. Here, as was expected, an engagement between the U. S. vessels and the shore batteries was imminent, but the bombardment was suddenly interrupted on the 13th by information that the Peace Protocol had been signed.

Consequently, on the following day, the Resolute sailed for Playa del Este, where a detachment of U. S. Artillery was taken on board. On the 18th, she sailed for Montauk Point, N. Y., going thence to Portsmouth, N. H., where the battalion was disembarked and encamped, on August 26.

The last engagement of the war is vividly described in the following report of Capt. Cooper F. Goodrich, U. S. N., Commander of the U. S. S. Newark, which accompanied the Resolute on the voyage to Manzanillo:

U. S. S. NEWARK,
Off MANZANILLO, CUBA, August 13, 1898.

SIR: I have the honor to submit the following report of the movements and operations of this vessel and her consorts up to 8 o'clock this morning.

On the afternoon of the 9th the Newark left Guantanamo and was joined shortly after off the entrance to that port by the U. S. S. Resolute, carrying the battalion of marines under Colonel Huntington. We proceeded to Santiago de Cuba, where we communicated with the St. Louis and then continued to the westward. On Wednesday afternoon, the 10th instant, we fell in with the Hist and Suwanee off Cape Cruz. Lieutenant-Commander Delahanty, of the Suwanee, having proceeded us to this point, communicated with the Hist and learned from her commanding officer, Lieutenant Young, that the condition of affairs of Manzanillo was such as to warrant the belief that an attack by the force under my command would result in a speedy capitulation of the garrison and city. This he reported to me as being in entire accordance with a letter addressed to you by Commander Todd of the Wilmington, which he had been permitted to read on board your flagship. Lieutenant Young, who had on board a competent pilot, assured me that it was entirely practicable to approach to within two miles of Manzanillo in a ship drawing as much water as does the Newark.

Inasmuch as the force detailed by you for the contemplated operations at the Isle of Pines was not all on hand, and as the Wompatuck could, in all probability, not leave Guantanamo until the 12th, or possibly the 13th, it appeared to me well to occupy this time of waiting in an attempt at securing Manzanillo and its garrison.

We waited off Cape Cruze that night in order to be joined by the Alvarado and Osceola, and then on the morning of the 11th started for Cuatro Reales Channel, the following ships accompanying the Newark: Resolute, Suwanee, Hist, Osceola, and Alvarado. In order to minimize the chances of accident that would be incurred in navigating waters only imperfectly charted, I sent the Hist with her pilot ahead. On her starboard beam was the Suwanee. In rear of these came the Osceola. In rear of the Osceola came the Resolute, and lastly the Newark, with the Alvarado close aboard, all keeping the lead going constantly. By a preconcerted system of signals the presence of shoal water or other danger could be instantly communicated from the leading ships in ample time to stop the progress of the Resolute and Newark, heavy-draft vessels.

We experienced no difficulty whatever in getting through Cuatro Reales, the least water found by the Newark being 5½ fathoms. At dark that day we anchored inside of the Great Barrier Reef, in 10 fathoms of water, about 40 miles distant from Manzanillo.

Yesterday morning, the 12th instant, my little flotilla got under way at half past 4 and proceeded to the vicinity of Manzanillo. The Resolute, Suwanee, Hist and Osceola anchored well inside of the northern entrance. I hoisted a flag of truce on the Newark and proceeded to an anchorage about 2 miles distant from the town, whence I sent the Alvarado, also bearing a flag of truce, to present to the military commandant a demand for surrender, a copy of which I have the honor to inclose. This demand was placed in his hands by Lieutenant Blue at thirty-five minutes past noon. The reply was to the effect that the Spanish military code forbade a surrender except as the consequence of a siege or other military operation.

The town, being fortified, is exempt from the privileges and immunities attached to defenceless places. Nevertheless, as you will perceive from my demand, sufficient time was given to permit noncombatants to leave the city. At 3 o'clock I signaled to the outlying vessels to take the stations off the town which had been assigned, and at 3.35 hauled down the flag of truce on the Newark and proceeded toward Manzanillo until the shoalness of the water forbade her further approach. At 3.40 fire was opened from this ship on the batteries, and was maintained with tolerable steadiness until 4.15 o'clock, with an accuracy surprising in view of the short time during which she had been commissioned, the other vessels following shortly after.

At 4.15 P. M., having seen supposed white flags hoisted on the Spanish gunboat Cuba Espanola and the commandant's quarters, I made signal "Cease firing," and sent the Alvarado in under flag of truce. At the same time the Suwanee, Hist, and Osceola, all under the immediate orders of Lieutenant-Commander Delahanty, were approaching the town from the southward through the middle channel. When these vessels were within 1,000 to 1,500 yards of the batteries, the Spanish authorities opened fire on them at 4.35, paying no attention to the flag of truce on the Alvarado, which (as I have since been informed) they failed to perceive. The Alvarado hauled down her flag of truce and joined the other gunboats in returning the fire. At 4.50 opened fire again from the Newark. The Cuban forces at this time appeared to the northward of the town and began discharging volleys, which were apparently returned by Spanish artillery. The Newark threw a number of 6-inch shells in this direction, in order to assist the Cubans. The Suwanee, Osceola, Hist, and Alvarado soon returned to the neighborhood of the flagship, and we all anchored at about 5.30 P. M. for the night. From that time until daylight this morning 6-inch shells were fired from the Newark at the batteries at irregular intervals, one shot being fired during each half hour. Daylight revealed a large number of white flags flying over the block-houses and batteries of Manzanillo and the approach of a boat from the shore bearing a flag of truce. The captain of the port came off and delivered to me a dispatch from the Secretary of the Navy, reading as follows: "Protocol of peace signed by the President; armistice proclaimed." My disappointment was, as may be imagined, very great, for I had every reason to believe that the garrison was entirely ready to surrender. I had hoped that the fleet under your command might have won one more laurel and gained one more important victory before the conclusion of peace.

A few projectiles fell close to this ship, but the enemy's attention was naturally directed chiefly against the gunboats. I am happy to report no casualties or injuries beyond three shots from Mauser rifles through the Suwanee's ensign. What was possibly the last shot of the war was a 6-inch projectile fired from the Newark at 5.20 A. M. to-day. It gives me great pleasure to speak in the highest terms of the officers of this ship and of the remarkable gun practice she displayed.

I inclose reports of the commanding officers. It is impossible for me to exaggerate their loyalty, zeal, and bravery, which have been too often proved during the war to render eulogy on my part necessary. Subjected as they were to close fire from guns of various calibers, from old-fashioned smoothbores to Krupp 14-pounders, and volleys of Mauser bullets, they stuck to their post and upheld the honor of the flag. I earnestly commend to your favorable consideration Lieutenant-Commander Delahanty, Lieutenants Young, Purcell, and Blue, commanding the Suwanee, Hist, Osceola, and Alvarado, respectively.

A part of the contemplated plan of operations was the landing of some or all of the Marines of Colonel Huntington's command. This officer's regret at the loss of an opportunity to win additional distinction for his corps and himself was only equaled by his careful study of the necessities of the case and his zealous entrance into the spirit of the enterprise.

Commander Eaton was, as is his wont, most ready and efficient, and with his ship was extremely helpful toward others. It was only the nature of his ship and her personnel which, under my positive orders, kept him from a more prominent place in the action.

Very respectfully, your obedient servant, CASPER F. GOODRICH,
*Captain, U. S. N., Commanding.*

The COMMANDER IN CHIEF,
*Flagship New York, Guantanamo, Cuba.*

At Portsmouth, N. H., the First Marine Battalion went into camp, remaining for several weeks, in order to rest and recuperate after the arduous services of the Cuban campaign. This camp—called Camp Heywood, in honor of the Colonel Commandant of the Corps—was located on Seavey's Island. On the upper part of the same island, something over a month before the arrival of the First Battalion, a camp had been prepared for the Spanish prisoners taken in the battle of July 3. The latter encampment was named Camp Long, in honor of the Secretary of the Navy.

With the humaneness characteristic of American methods in warfare, Camp Long was carefully provided with every convenience for the comfort and well-being of the prisoners, while its location, on the banks of the river, overlooking several pleasant summer resorts, was one of the most picturesque on the coast. The camp, however, was enclosed by a high board fence, outside of which, at a distance of twenty feet, was another of barbed wire. In order to provide against possible uprisings, four Gatling guns were planted on an overlooking eminence, and two others, on the bridges leading to the Navy-Yard. Galleries and sentry boxes were also built on the outside of the fence, giving the sentinels full view of the interior of the stockade.

The Marine guard at Camp Long, established on July 7, consisted of 114 men and 4 officers, under the command of Col. James Forney, U. S. M. C. Later, by detachments of Marines from several ships of the fleet investing Santiago harbor, who were sent north with the prisoners, it was increased to 220 men and 8 officers, and continued at this strength, until the close of the war. Late in August, Col. Forney was superseded in command by Lieut.-Col. Meade.

"On July 11, 10 officers and 692 men, Spanish prisoners, were received from the St. Louis, and on July 16, 2 officers and 961 men were received from the Harvard; also, on August 24, 8 officers and 8 enlisted men, captured on the Argonaut at the beginning of the war, arrived from Fort McPherson, Ga., where they had been confined, making a total of 20 officers and 1,661 men. Capt. B. R. Russell, from the Marine battalion, and Second Lieut. T. S. Borden and 28 Marines from the Brooklyn, and the guard of the Marblehead, came north with the first detachment of prisoners on the St. Louis, in addition to the regular guard of the ship, arriving at Portsmouth July 10; and Capt. A. C. Kelton and First Lieut. F. J. Moses and 60 men from the First Marine Battalion came north as a guard with the second detachment of prisoners on the Harvard, in addition to the guard of this vessel, arriving at Portsmouth July 15. Captain Russell, Second Lieutenant Borden, 28 men of the Brooklyn's guard, and Captain A. C. Kelton, First Lieutenant F. J. Moses, and 60 men from the battalion were transferred to the Marine barracks, Portsmouth, N. H., after the arrival of the vessels, the Marblehead's guard returning on the St. Louis."

Immediately upon their arrival, the captured Spaniards were given clean, warm clothing, comfortable bedding, and excellent food. Such of them as had money were supplied with tobacco and other luxuries at nominal prices; no efforts being spared to mitigate, as far as possible, their condition of captivity. The sick and disabled among them, also, were cared for at the Naval hospital.

The general conditions of life at Camp Long, and the accommodations provided for the prisoners, are interestingly set forth in the following extracts from the report of Col. James Forney, Commander of the camp:

"I assumed command of the Marine barracks and the Spanish prisoners at this station on the 8th of July, 1898.

* * * * * * * * * *

"Many of the prisoners, when they landed, were in a sick and enfeebled condition, a number of them falling down and being unable to move until picked up and placed in an ambulance, in which they were carried either to the camp or to the Naval hospital.

"I found upon my arrival here that the buildings being erected for the prisoners were entirely inadequate for the purpose and, upon my representation to Admiral Carpenter, the commandant of the Navy-yard, he authorized other buildings to be built from time to time, and finally accommodated all the prisoners comfortably.

* * * * * * * * * * *

"The Marine camp was located outside the stockade, and the Marines to the number of 130 were quartered there. They had a large building to sleep in and another to mess in. The Marine officers had a building of their own, situated on the highest point of Seavey's Island, where I also had my quarters. A flagstaff was erected in front of these quarters and a flag hoisted at the regular times, a sunset gun also being placed there and being fired at sunrise and sunset, which regulated the camp into a regular military encampment.

"At the northeastern part of the stockade I had a guardhouse built for 50 Marines, and 14 cells built for refractory Spanish prisoners. On this part of the island I discovered a well that had been covered up with rock and dirt, which, upon cleaning out, I found to contain most excellent water.

"On the northeast part of the island I laid aside a plot of ground for a graveyard for the prisoners who have died, and have their graves marked so that they can be identified at any time. I gave all the dead military funerals, as I considered that, being prisoners of war, they were entitled to it. I have had three volleys fired over their graves, the Spanish flag wrapped around their coffins, and a Catholic priest to read the services.

* * * * * * * * * * *

"The discipline of the camp has been most excellent in every respect. I was compelled to confine some of the prisoners from time to time for breaking the rules of the camp, but the offenses were at no time of a very serious nature.

"Captains Allen C. Kelton and Benjamin R. Russell rendered me great assistance and were always vigilant and attentive to their various duties. First Lieutenant Theodore P. Kane, especially, was of great benefit to me, helping me with the correspondence and getting the camp into shape. Second Lieutenants Thomas S. Borden, David D. Porter, Charles S. Hatch, and Don A. Baxter were all careful in regard to their duties. The three latter came in for the war and I trust that they may be retained.

"Medical Inspector Remus C. Persons was always on hand to look out for the sick. He distributed many articles of clothing to the prisoners with his own hands. Passed Assistant Surgeon Lewis Morris, who was detailed to accompany me every morning on my tour of inspection, was always on hand and careful to detect any trouble in regard to the sanitary condition of the camp. Pay Inspector Joel P. Loomis's reputation as a good caterer followed him here, and in clothing and feeding the prisoners and attending to his duties, in connection with his clerk, Mr. John A. Kelly, worked most excellently. Carpenter Joseph B. Fletcher did excellent work in superintending the construction of the buildings and was of great assistance to me.

"The system adopted of marching the prisoners down in three columns of twos to take their meals from the three tables near the pavilion worked well, the men coming up in single file and taking their food in a bowl and plate and then going inside the pavilion where I corralled them with sentinels.

* * * * * * * * * * *

"Seavey's Island, upon which the camp is situated, is rocky, hilly, and rough, with bold ledges of rock running up to a considerable elevation, displaying the finest views of the river and harbor. The officers' building is situated over the most dangerous point, called Pull-and-be-damned-Point. The camp is at one end of the island, surrounded at one side by a stockade of boards 10 feet high, outside of which, at a distance of 20 feet, is a high barbed wire fence. Sentry boxes are at each angle, and there are twelve sentinels inside the camp, patrolling along the dead line, the prisoners' quarters, sinks, and the pavilion.

"Every facility is given the prisoners to wash their clothes, and on the river side down the steep slope to the swiftly running tides they were paroled from 8 A. M. until sunset each day. They never broke their parole, always coming back on time, and were of great service to me in managing the prisoners.

"The Spanish commissioned officers have a separate house of their own and have servants from the prisoners to wait on them. They have much better fare than the rest of the prisoners, and claret wine is served to them. I have given them all the respect and assistance due their rank, and I feel that they have appreciated it. I utilized them by putting them on duty as officer of the day, assisting in the policing, and looking out for the buildings; and they have fully co-operated and assisted in carrying out the routine of the camp with my own officers.

"I had 36 warrant officers, consisting of engineers, sergeants of marines, etc., put in a separate building, with a mess table of their own and bunks to sleep on instead of hammocks. The 125 petty officers from the different ships I had kept together in a building separate from the rest. All the prisoners I had divided up into ships' companies according to their respective ships, and had the ten buildings lettered A, B, C, etc.

"The sick prisoners at the Naval hospital were from the Maria Teresa, Vizcaya, and Oquendo. They suffered from insufficient nutriment and a pernicious malarial fever. Some of the crew of the Cristobal Colon were convicts from the Canary Islands, but gave me no trouble. The prisoners all had fresh beef, coffee, fish, butter, hash, etc.

* * * * * * * * * * *

"There was a system of signals arranged between the camp and the Navy-yard barracks by rockets, and there was telephone connection between the two places, making security doubly sure in case of an outbreak among the prisoners.

"Second Lieutenants Frank A. Kinne and Robert E. Devlin were on duty here a short time and rendered efficient service.

* * * * * * * * * * *

"I had the camp for the battalion of Marines under the command of Colonel Robert W. Huntington laid out on the northern part of the island, near Camp Long, water pipes put in from the reservoir, sinks built, floors laid for all the tents, etc.

"Camp Long assumed historical importance in the Spanish War and was visited by thousands of people, who were always received courteously and given every attention. The orders had to be very strict in regard to actually going inside the camp, and the exceptions were very rare in this respect.

"Admiral Cervera visited the camp on the 15th of August and received a cordial reception from his men and from the people of the city of Portsmouth, as well as those in the adjacent country, who flocked to see him during his visit. The paymaster who accompanied him paid out about $35,000 to the Spanish prisoners.

"I established a store inside the camp that was ably managed by Mrs. Ida N. Gulick, the post trader, where the prisoners could obtain small stores at a fixed price, which added greatly to their comfort and benefit.

"On August 24 more prisoners arrived—8 officers and 8 enlisted men. They were captured on the Argonaut at the beginning of the war, off the coast of Cuba, and have since been confined in Fort McPherson, Ga. They were brought here under the charge of First Lieutenant A. P. Buffington, Thirteenth Infantry, United States Army."

## SERVICES OF THE FIRST MARINE BATTALION IN CUBA.

The preparation of the Marine Corps for service in Cuba is described graphically by Colonel Heywood in his annual report for 1898, as follows:

"In accordance with the verbal instruction of the Department of April 16, 1898, to organize a battalion at New York for service in Cuba, I issued orders on the 17th and 18th of April for the immediate assembling at New York of detachments of men from all the Eastern posts of the Corps and receiving ships. On the night of April 18, by direction of the Secretary, I proceeded to New York for the purpose of organizing the Marine battalion service. The battalion, as organized, consisted of 23 commissioned officers of the Marine Corps, 1 surgeon of the Navy, and 623 enlisted men, all under command of Lieutenant-Colonel R. W. Huntington, U. S. M. C. The battalion was divided into six companies, one of which was an artillery company, having four 3-inch rapid-fire guns, received from the ordnance department, Navy-yard, New York, and was composed of young, strong, and healthy men. The following is the organization of the battalion:

Lieutenant Colonel R. W. Huntington, commanding.
Major P. C. Pope.
Major H. C. Cochrane.
First Lieutenant H. L. Draper, adjutant.
Captain C. L. McCawley, A. Q. M., quartermaster.
Surgeon John M. Edgar, United States Navy, surgeon.
First Sergeant Henry Good, sergeant-major.
First Sergeant W. J. Limerick, quartermaster-sergeant.

Company A: Captain Allan C. Kelton, First Lieutenant F. J. Moses, Second Lieutenant L. J. Magill.

Company B: Captain B. R. Russell, First Lieutenant C. L. A. Ingate, Second Lieutenant M. J. Shaw.

Company C: Captain G. F. Elliott, First Lieutenant L. C. Lucas, Second Lieutenant P. M. Bannon.

Company D: Captain W. F. Spicer, First Lieutenant W. C. Neville, Second Lieutenant Newt. H. Hall.

Company E: Captain H. K. White, First Lieutenant J. E. Mahoney, First Lieutenant A. S. McLemore.

Company F (artillery): Captain F. H. Harrington, First Lieutenant C. G. Long, First Lieutenant W. N. McKelvy.
Color guard: One Sergeant, two corporals.

"Each company consisted of 1 first sergeant, 4 sergeants, 4 corporals, 1 drummer, 1 fifer, and 92 privates; total 103.

"Total in battalion, 23 commissioned officers, 623 enlisted men.

"Before leaving Washington for New York, I was informed by the Department that the commandant, Navy-yard, New York, had been directed to fit out the Panther, formerly the Venezuela, for the transportation of a battalion of 400 men, the number decided upon by the Department, and that he had been instructed to render me all possible assistance in fitting out the ship as a transport, having regard for the health and comfort of the men. Upon my arrival at the Navy-yard, New York, I reported to the commandant of the station, Rear Admiral F. M. Bunce, United States Navy, who afforded me every facility, and accepted every suggestion looking to the proper fitting out of the ship. Everything was done to make the men as comfortable as possible in the way of providing bunks and other conveniences, although the ship was not well adapted for use as a transport, as there were no air ports between decks, and the only ventilation was from the hatches used for loading freight and two small ventilators in the after part of the ship.

"The vessel was ready in two days for the battalion of 400 men, which could have sailed then. When the battalion was ready to sail, two days after the arrival of the men at New York, orders were received from the Department directing that two companies be added to the battalion, and accommodations for these additional men had to be immediately provided. Work was proceeded with night and day to make the necessary provision for the increased number of men, and two days later, on April 22, the Panther sailed, with the battalion of 24 commissioned officers and 623 men, for Cuba.

"As the men marched from the barracks to the ship they were greeted with great enthusiasm by the officers, sailors, and others on the vessels at the Navy-yard, as well as those on shore. The band of the yard was loaned by the commandant to escort the battalion to the landing. As the Panther left the Navy-yard and proceeded down the river she was repeatedly greeted with cheers and whistles from the vessels passed.

"The greatest care was exercised in fitting out the battalion by the quartermaster of the Corps, Major F. L. Denny, U. S. M. C., the quartermaster of the battalion, Captain C. L. McCawley, U. S. M. C., and myself, and when the Panther sailed the battalion was thoroughly fitted out with all the equipments and necessities for field service under the conditions prevailing in Cuba which experience and careful consideration could suggest, including mosquito netting, woolen and linen clothing, heavy and light weight underwear, three months' supply of provisions, wheelbarrows, push carts, pickaxes, shovels, barbed-wire cutters, wall and shelter tent, and a full supply of medical stores. Campaign suits of brown linen and campaign hats were ordered, but owing to the great demand for these articles at the time by the army it was impossible to send them with the battalion. They were shipped later, however, and proved a great comfort to the men. Tent floors were purchased at Key West.

"After orders were received to increase the strength of the battalion by two companies, making in all 623 men, it was found that the Panther would be very much crowded with this number on board. I reported the fact to the commandant of the station, and was informed by him that he had received orders to fit out the Resolute, formerly the Yorktown, as a permanent transport for the use of the battalion. This vessel was particularly well adapted for use as a transport, as she had a large number of staterooms for the accommodation of officers, and more than enough accommodations for a thousand men, and her upper decks were open fore and aft, making a clear, unobstructed deck the length of the ship, fitted with air ports throughout and a system of artificial ventilation by steam blowers. This deck is large enough to permit the erection of standing bunks sufficient to accommodate with ease about 850 men, and if all the space in the ship were utilized, bunks for 1,000 men could be erected. After my return to Washington, I suggested to the Department that distillers and a machine for manufacturing ice be installed in the vessel, which was done. I also recommended to the Department that, as there is frequently occasion to transport Marines and the crews of vessels, the Resolute be retained in the service as a permanent transport, and I respectfully renew this recommendation. After the Resolute was fitted out and ready to sail, and provisions placed on board for the battalion, the exigencies of the service required that she be taken for other purposes, and she was not available for the use of the battalion until it embarked at Guantanamo for the Isle of Pines.

"After leaving New York the Panther proceeded to Hampton Roads for the purpose of awaiting a convoy to Cuba, arriving on April 23, 1898. Major P. C. Pope and First Lieutenant J. E. Mahoney, who had been ordered to the battalion, joined it at Hampton Roads. The Panther left Hampton Roads April 26, under convoy of the U. S. S. Montgomery, arriving at Key West April 29. During the time the Panther remained at Key West, from the date last mentioned to June 7, the men were landed and went into camp there. The battalion received orders at 5.30 in the afternoon of May 24 to land,, with all stores, by 3 o'clock the following morning, which was accomplished. Just before the Panther sailed from Key West, Major P. C. Pope was detached from the battalion. The Panther sailed from Key West for Cuba on June 7, 1898, and arrived at Santiago de Cuba on the morning of the 10th. On the same day, at 1 P.M., the ship arrived at Guantanamo Bay, Cuba, and at 2 P. M. of that day the battalion landed, with stores, and prepared to go into camp. On the 11th the camp was attacked by a much superior force of Spaniards, and from that time until the 14th the battalion was constantly under fire, and repulsed the enemy on every attack. The holding of the position at Guantanamo Bay was of the utmost importance to the Navy, as it was the only harbor where the vessels could seek shelter during the hurricane season. Owing to the dense undergrowth, affording safe shelter to the Spanish sharpshooters, it would have been impossible for the vessels, by shelling the shore, to keep the enemy from harassing those on board the ships with their Mauser rifles to such an extent as to make it dangerous for them to remain there.

"Captain Geo. F. Elliott, of the battalion, was sent on June 14 with a detachment of two companies of the battalion and 50 Cubans for the purpose of destroying a well at Cuzco, about 6 miles from the camp, which was the only water supply of the enemy within 12 miles. This small force attacked and defeated a body of about 500 Spaniards and accomplished the destruction of the well.

"About 1 A. M. on the morning of the 12th of June, during a very severe attack on the camp, Assistant Surgeon John Blair Gibbs, U. S. N., was killed by a Mauser bullet, reported by Surgeon John M. Edgar, of the battalion, to have been fired at a range of from 600 to 800 yards. The death of Assistant Surgeon Gibbs cast a gloom over the whole command, as he was a most popular officer, liked by all, and his services were very much missed and the battalion could ill afford to lose them.

"I regret to have to report the following list of enlisted men of the Corps who lost their lives in the brave defense of the flag at Guantanamo Bay:

"Killed: Sergeant Major Henry Good, Sergeant Charles W. Smith, Private Goode Taurman, Private William Dumphy, and Private James McColgan.

"The following men of the battalion were severely wounded: Corporal William B. Glass, Private Bartholomew McGowan, Private

James D. Bourke, Private Robert J. Fleming, Private Albert E. Halvosa, Private Patrick Long, Private Charles C. Marley, Private Lewis L. Noonan, Private James Roxbury, Private Thomas Wallace, and Private Arthur Walker."

The following detailed account of the services on shore of the detachment of the Marine Battalion, under command of Lieutenant-Colonel Huntington, is embodied in the report of Captain Charles L. McCawley, Quartermaster of the First Marine Battalion. Taking his story from the date of departure from New York City, he says:

"At first it was the intention of the Navy Department to have the battalion sail on April 20, but that plan was found impossible of execution owing to the fact that the Transport Panther, which had been prepared for a battalion of 450 men, could not be made ready for an increase of over 200 men to that strength, which was ordered added to the command. The battalion actually sailed on the 22nd of April. The delay in sailing was fortunate as it enabled me to purchase and receive some necessary articles which in the great haste of preparation had not been obtained. The work of loading the supplies on the Panther began on the morning of the 22nd, the work continuing throughout the day, and the battalion embarked at 5 P. M., amid a scene of marked enthusiasm on the part of a great crowd of people assembled at the Brooklyn Navy-yard. The Panther sailed at 8 P. M., arriving at Fort Monroe the following evening, where she waited until the morning of the 26th for the Montgomery, which vessel conveyed her to Key West. This port was reached on the morning of the 30th. There the Panther anchored to await instructions from the commander in chief of the North Atlantic Station.

"The battalion remained at Key West over a month. While there I purchased a number of articles, including tent floors, deemed necessary by the commanding officer of the battalion. Excepting the tent floors the purchases were of trivial articles, as the battalion had been so completely and efficiently fitted out under your direction before leaving Brooklyn. At 5 P. M., May 23, orders were received for the battalion to disembark and go into camp, it having been determined that the Panther should tow a monitor to Havana. The commandant of the Naval station stated that he wished to put the Marines 'on their mettle,' and that the command and all of its stores must be out of the ship by 4 A. M. of the following morning. These orders necessitated the engaging of a lighter, teams, and wharfage, also a storehouse in which to place the stores. These were promptly secured and the work of unloading the ship commenced. By dint of the hardest work on the part of both officers and men the stores were finally landed and the battalion disembarked at 4 A. M. After this night's work my labor began afresh, for the tents, tent floors, cooking outfits, and all required equipage had to be transported by teams to our camp, which was distant about two miles from the docks upon which the stores were landed. I completed sending these stores to camp and placing in the storehouse those not required there at 3 P. M. Upon my arrival at camp I found all the tents pitched, floors laid, latrines dug, supper prepared and, in fact, the whole camp in running order.

"The battalion remained in camp for two weeks, and the experience gained thereby by the officers and men was most valuable. The location of the camp was not all that could be desired, but was the best obtainable. Cistern water had to be hauled to it in casks from the town, but, this water not being good, its use in its natural state was discontinued, and thereafter it was only used, both for drinking and cooking, after being boiled.

"It was while at this camp that there arrived campaign suits of brown linen which you had procured. Their issue was hailed with the greatest satisfaction by the officers and men, who had been sweltering for weeks in blue uniforms under a tropical sun. The entire battalion was fitted out, including the officers, who were authorized to wear these suits, and the appearance of the men in this comfortable, businesslike uniform excited favorable comments from army and Navy officers who came in contact with the battalion. The main purpose of these suits being to afford comfort to the men serving in a tropical clime, it is considered that they fully filled the purpose, and while frequent service in Cuba showed that the color of the material is not the best for campaigning, they were of decidedly practical benefit to the command.

* * * * * * * * * * *

"Conforming to orders, the battalion again embarked on the Panther on June 6 and sailed the following day to join the commander in chief of the North Atlantic Station at Santiago, Cuba. This port was reached without special incident on the morning of the 10th of June. Upon reporting to the commander in chief the battalion was immediately ordered to proceed to Guantanamo, about forty miles to the eastward, there to be landed under order of Commander B. H. McCalla, U. S. N., commanding the U. S. S. Marblehead. With the aid of all the boats and steam cutters of the vessels in the harbor, four companies and most of the equipage were landed that evening. Camp was promptly pitched in a place previously selected on the top of a hill where stood the remains of a Spanish blockhouse, destroyed by the fire of the ships before the battalion landed, and outposts were established. It was a very laborious task carrying the tents, tentpoles, cooking outfits, ammunition, intrenchment tools, etc., up this hill, which was quite steep, and when night fell the men were completely tired out. They had little or no rest that night, as an attack by the Spanish was expected, they having been heard by the outposts stealthily moving through the paths in front of the camp. The next afternoon the enemy did make an attack and continued it during the entire night, and in the morning it became necessary to strike the tents and get them out of the way in order that the command might intrench itself. This was completed under fire in an orderly manner, and the whole camp outfit was carried over the hill facing the harbor. The trenches were rapidly dug, the men working superbly. Some of the tents had to be used as breastworks in places, and their appearance shows the effects of the Spanish fire, they in some cases being riddled with bullets.

"For several days and nights the command was exposed to almost incessant firing, and it was not until after the battle of Cuzco, when Companies C and D. under Captain Elliott's command, drove the Spanish away, that we had any rest at all, and even then it was several days before we felt assured that there was to be freedom from further annoyance, it being known that there were 7,000 Spanish troops at Guantanamo, only fifteen miles away. For about ten days the men spent all their time in the trenches, which from day to day were perfected, going down the hill for meals by detachments. When more confident of not being further attacked, I gradually brought over from the Panther clothing sufficient to meet all demands, and from time to time made requisitions on you and the assistant quartermaster at Philadelphia for such articles as were needed. It is with much pleasure that I record the fact that these requisitions were always filled in the promptest and otherwise satisfactory manner. A quantity of supplies was received by me on the transport Resolute, which reached Guantanamo June 21. In these supplies were gray felt campaign hats, which were required and which proved quite comfortable, undress caps being illy adapted to service in a tropical climate. A few days after her arrival the Resolute was ordered away, and it became necessary to send ashore all stores belonging to the battalion and to place in tents all those which required protection.

"Rations for the enlisted men were procured from the Panther and Resolute present, and at other times from the supply ships Celtic and Supply. Usually I received a ten days' issue. The rations were kept in a rudely constructed storehouse, which served to protect the perishable provisions from the weather. Fresh beef and vegetables were also obtained from these ships to serve out to the command according to the Navy allowance. I had an ice house made by digging a hole in the sand and boarding it on the sides, in which the meat was kept, and this enabled me to obtain a two or three days' issue at a time. We received ice with each issue of fresh beef. The meat was of a very superior quality and kept well. First Lieutenant James E. Mahoney, United States Marine Corps, had charge of the messing of the men and was particularly zealous and efficient in the discharge of his duty. I turned over to him the daily rations for the men and he made the issues to the companies. The command drank distilled water, which was also used for cooking purposes. This was obtained daily from the Panther and Resolute most of the time and for quite a period from the Vulcan. I had anticipated that water might not be obtained on shore and had purchased empty wine casks in Key West. The casks were conveyed to the vessels and returned to the dock in a large sailing launch. The water was distributed from the boat to the various companies of the battalion, whose cooks came to the landing to receive it. The men also filled their canteens at the same place. Under my direction, Sergeant Richard Silvey had charge of the distribution of the water and he performed his duty most satisfactorily. The Cuban officers and soldiers also received their food and water supply from us.

"It is my opinion that much of the excellent health of the battalion while in Cuba was due to the fact that distilled water only was used for drinking and cooking. There were other elements that entered into the good health of both officers and men, chief of which were the excellent sanitary arrangements, the use by officers and men of the light-weight woolen underwear, and the absence of tropical fruits from the vicinity of our camp and the fact that the men were at all times supplied with proper and sufficient food and clothing, and it was these that enabled us to bring home 98 per cent. of the battalion, fit for duty. Not a single man of the command died from disease."

Under date, June 17, 1898, Lieutenant-Colonel Huntington reports as follows on the operations of the Marines under his command at Camp McCalla, Guantanamo Bay, Cuba:

"The stores of this battalion were sent to the dock at Key West from Camp Sampson, on Sunday, June 5. We broke camp at 2 A. M. on June 6, and went on board the Panther, Major Pope going to Key West hospital.

"On June 7, at 7.10 P. M., we sailed from Key West and arrived off Santiago de Cuba on the morning of the 10th; on the same day, at 1 P. M., we arrived in Guantanamo Bay; at 2 P. M., the battalion landed with stores. Company C was landed and deployed on the hill near the beach on the right of the entrance to the harbor. This hill is about 150 feet high and on top was formerly occupied by the Spanish troops, but when the position was vacated the day before our landing, the blockhouse on the top of the hill was burned.

"On landing all houses and huts lately occupied by the Spanish forces were burned.

"The hill occupied by us is a faulty position, but the best to be had at this point. The ridge slopes downward and to the rear from the bay; the space at the top is very small, and all the surrounding country is covered with thick and almost impenetrable brush. The position is commanded by a mountain, the ridge of which is about 1,200 yards to the rear.

"On the afternoon of landing, tents were pitched and outposts established.

"On the 11th, about 5 P. M., an attack was made upon one of the outposts and two privates, McColgan and Dumphy, of Company D, were killed, each receiving more than eight wounds, each of which would have caused death. These two men were patrols. A detachment was sent out from camp to support the outpost, and we found only faint traces of the enemy. After nightfall fire was opened upon our camp by small parties from different directions on five different occasions. The men turned out each time under arms with promptitude and courage. About 1 A. M. a more combined attack was made, and noisy fire from south, southeast, and southwest, was opened. During this attack Acting Assistant Surgeon John Blair Gibbs, United States Navy, was killed. From the best information attainable about 160 men were engaged in this attack.

"On the morning of the 12th Sergeant C. H. Smith was killed and Corporal Glass, Privates McGowan and Dalton, all of company D, were wounded—not dangerously.

"On the morning of the 12th all tents and material were removed from the position and taken on the bay side of the hill, and a trench was dug on the south front, about forty yards across, and a barricade made around the position, which would enable us to hold it, as I was informed that more troops were being assembled by the enemy in this immediate vicinity.

"On the night of the 12th many persistent and trifling attacks were made, in reply to which we used a good deal of ammunition. About 2 A. M. Sergeant Major Henry Good was killed. On the 12th we were joined by sixty insurgent troops, and they, being acquainted with the country, and excellent woodsmen and fearless, were of the greatest assistance.

"On the 13th, about 8 P. M., fire was opened upon the camp and subdued without loss or difficulty. About 8 A. M. on the 14th a rather smart fire was opened for a few moments on the camp and easily repelled. About twenty Cubans came from below the hill at this alarm, but their help was not needed. They opened fire.

"At 9 A. M., 14th, a force consisting of companies C and D, the native troops above mentioned, with about twenty-five more from Guantanamo, all under the direction of Colonel Tomas, Cuban army, proceeded through the hills about six miles and destroyed a well, said to be the only available water supply within nine miles.

"From the best information I can gather, this force was opposed by four regular companies of Spanish infantry and two companies of guerrillas, making a total of a little short of 500 men.

"The engagement between these forces lasted from about 11 A. M. until 3.30 P. M. Our troops drove the enemy at every point, being obliged to make the first advance under fire, which, owing to the lay of the country, they could not return.

"Captain Elliott reports that the men in many cases coolly estimated distances, borrowed his field glass to pick up parties of the enemy, and at a distance of 1,000 yards often inflicted damage and caused withdrawal.

"Second Lieutenant Magill, with 50 men and 10 Cubans, joined Captain Elliott, climbing the mountain through cactus and brush; this advance was intended to cut off the retreat of the Spaniards, which unfortunately failed of its principal object, owing to the fact that his advance was stopped by the fire of the U. S. S. Dolphin.

"Being apprehensive for the success of the movement, I ordered First Lieutenant Mahoney to be joined by First Lieutenant Ingate—these officers each having 50 men with them on picket—this combined force to proceed to Captain Elliott's assistance. Lieutenant Ingate failed to find his way to Lieutenant Mahoney, and Lieutenant Mahoney advanced alone, arriving too late to take an active part in the affair.

"Our losses were 2 Cubans killed, 2 wounded, and 3 privates wounded, not dangerously; after the affair, while descending the mountain, Lieutenant Neville wrenched his hip and will probably be unfit for service for a month; about 10 or 12 of our men and 2 Cubans were overcome by the heat.

"From information received from prisoners, which I believe to be reliable, about 60 of the Spanish force were killed and something more than 150 wounded, and 1 lieutenant and 17 privates were captured. The force returned to camp at 8 P. M., exhausted by the long, hard march through this mountaineous and tropical country.

"This affair was planned by the Cubans, but too much praise cannot be awarded to the coolness, skill and bravery of our officers and men, by which alone its success was achieved.

"Captain Elliott's cool advance up a rocky, steep mountain path, under fire for twenty minutes without being able to return it, and the gallantry and skill displayed by him throughout this affair were essential to the great success attained by the expedition, and are worthy of and I earnestly recommend that he be advanced in rank one grade. Captain Elliott mentions in terms of high praise, the conduct of First Lieutenants Lucas and Neville and Second Lieutenants Magill and Bannon."

In referring Huntington's report to Rear Admiral Sampson Commander Bowen H. McCalla, of the U. S. S. Marblehead, under date July 19, 1898, gives a revised version of several incidents. He says:

"This report requires several corrections.

"The blockhouse referred to on page 2 was burned by the gun fire from the Yankee on the 7th instant.

"The position referred to on the same page was not occupied again after a small Spanish force had been driven away, when the Marblehead took permanent possession of the bay on the 8th instant.

"Early on the morning of the 10th instant Captain Goodrell, with 40 Marines from the Oregon and 20 Marines from the Marblehead, examined the locality occupied by the Marines, who arrived shortly after he had completed this duty. On the arrival of the Panther, Captain Goodrell was sent on board to give Colonel Huntington the benefit of his observations.

"Referring to paragraph 4, page 2, the position occupied by the Marines has been pronounced by Major-General Perez, of the Cuban army, on the 17th instant, to be the only tenable position on the bay which could be successfully held by a small force. He also stated that 5,000 Spaniards could not take it.

"If the Marine position is commanded by a mountain ridge, that mountain ridge is commanded in turn by the ten 5-inch rapid-fire guns of the Marblehead, and of such other ships as may be here.

"The mistake of locating the camp between the main position and the outpost was corrected on the 11th instant, at my suggestion.

"The expedition was suggested by Colonel La Borde, and the Dolphin was sent to cover the sea front of our force.

"Twenty-three Marines overcome by the heat were brought back by the Dolphin.

"This exhaustion was due, I believe, mainly to the fact that the campaign hats of the Marines were on the Resolute, and not in the Marine camp.

"The behavior of the officers and men of the Marine battalion generally has been most gallant, and is in general worthy of all praise."

HEADQUARTERS FIRST MARINE BATTALION.
PLAYA DEL ESTE, CUBA, July 31, 1898.

SIR: I have the honor to make the following report: After the action of June 14 the enemy retreated farther up country and has never since annoyed us.

On June 25, at 3 a. m., Companies C and E and about forty Cubans, under my command, crossed to the west side of Guantanamo Bay in small boats for the purpose of cutting off a body of the enemy who had been annoying small boats from the Marblehead in their search for mines. A landing was made, and the troops disposed to cut off any retreat of the enemy on the point, while the Marblehead watched the isthmus leading from the mainland to our position.

A heavy patrol was then sent to search the point, but none of the enemy were found, although unmistakable signs showed that a force of 100 or 150 had occupied this point a day or two before.

This force re-embarked at 7.30 A. M. and returned to the camp.

The regular pickets have been maintained—fifteen men by day and a full company with all its officers by night. This line of observation is about 800 yards to a mile from our position. One-half of this line—the left—is the same as that established on the 10th day of June, when we first landed. The right half of this line has been drawn back to easier supporting distance.

Sentries on each face of the fortified position occupied by us are maintained, but I have reduced these materially from the number which were kept on duty from the 10th to the 30th of June, inclusive.

Strong scouting parties, in addition to those sent out by the Cubans, have been sent out frequently to examine the surrounding country for the enemy.

During the past few days water has been reported in the well at Cuzco, which was filled up by our force after the affair on the 14th ultimo, as reported to you in my communication of June 17, but inasmuch as rations have been sent from here to the Spaniards in Caimanara it does not seem necessary to fill up the well, but it is being closely observed by scouting parties from this camp.

The graves of our dead have been appropriately marked with headstones and a record placed in a bottle beneath the headstone in each case.

The strength of the battalion at this date is 515. Of this number 23 are commissioned and 482 enlisted; deducting 21 sick, leaves 484 available.

Your attention is invited to the reduction in the strength of the battalion as shown by the muster rolls forwarded herewith.

R. W. HUNTINGTON,
*Lieutenant-Colonel, U. S. Marine Corps, Commanding Battalion.*

THE COLONEL COMMANDANT UNITED STATES MARINE CORPS,
*Headquarters, Washington, D. C*

Captain G. F. Elliott, whose conduct is so highly praised in Lieutenant-Colonel Huntington's report, renders the following account of the fight at Cuzco:

CAMP MCCALLA,
GUANTANAMO BAY, June 15, 1898.

SIR: I have the honor to submit the following report:

In accordance with your verbal directions, I left camp at 9 A. M. yesterday with two companies of the battalion, C and D, commanded respectively by First Lieutenant L. C. Lucas and Captain William F. Spicer, with an aggregate of 160 men and fifty Cubans, under command of Lieutenant-Colonel E. Eugene Tomas. Colonel Laborde, Cuban Army, was also present, but without command.

My orders were to destroy the well at Cuzco, about six miles from this camp, which was the only water supply of the enemy within twelve miles of this place, and the existence of which made possible the continuance of the annoying attacks upon our force in camp here.

Two miles and a half from Cuzco half the Cubans and the first platoon of C Company, under Lieutenant Lucas's command, passed over a mountain on our left, hoping to cut off the enemy's pickets. In this we failed, and our force was discovered by the Spanish outpost, which retreated immediately and gave the alarm to the main body, whose headquarters were in a house at Cuzco.

A high mountain separated the two forces at this point, and each attempted to gain its crest as a point of advantage. In this we were successful, but were fired on heavily by the enemy from the valley, at a distance of 800 yards. This fire was replied to by the Cubans of the main body. Lieutenant Lucas, with 32 men of his platoon and the remaining Cubans, came into the fight at 11.15. The other nine men of his platoon becoming exhausted were obliged to return to Camp McCalla. Lieutenant Bannon conducted the second platoon of C company just below the crest of the hill, out of fire from the enemy, leaving the narrow path, which was the only road, and making their way through the cacti. Just in rear of this platoon and following in single file was D company. The crest of the hill was in the shape of a horseshoe, two-thirds encircling Cuzco Valley and the well. The Cubans, C and D companies occupied one-half of this horseshoe ridge, while Second Lieutenant L. J. Magill, with one platoon (50 men) of A company, came up from the valley on the opposite side, where he had been stationed as an outpost from Camp McCalla, having been attracted by the heavy fire, and believing his force necessary to our assistance, and occupied the left center of this horseshoe ridge. As soon as he saw our position he sent one of his men around the ridge to report to me. For fifteen minutes we were marching under a heavy fire, to which no reply was made, to gain this position. By the use of glasses and careful search by the men, individuals were discovered here and there, and fire being opened upon them, they would break from cover to cover, and we were thus enabled to gain targets at which to fire, which had been heretofore impossible, owing to the dense chapparal in which the enemy sought successful cover.

Many of the men fired as coolly as at target practice, consulting with each other and their officers as to the range. Among these were Privates Carter, Faulkner, and Boniface, all of whom did noticeable execution. This movement of the enemy gave Lieutenant Magill an opportunity to get in a cross fire, which was well taken advantage of.

Having reduced the enemy's fire to straggling shots, the U. S. S. Dolphin, Commander H. W. Lyons, U. S. N., which had been sent along the coast to co-operate with us if possible, was signalled to shell the house used as the enemy's headquarters and also the valley, but she was so far to the front, having mistaken the valley intended, that her fire was in Lieutenant Magill's direction, driving him to the reverse side of the ridge.

However, this shell fire started the enemy from his hiding place, which gave the other companies the opportunity to fire on them on the move.

Signal was made to the Dolphin to cease firing, and Lieutenant Magill was directed to form skirmish line and move down the valley in front of him toward the sea. This was defeated by renewed shell fire from the Dolphin.

The fight, which began at 11 A. M., was now drawing to a close, being over at 3 P. M. The enemy began a straggling retreat at 2 P. M., getting out of the valley as best they could.

The fire of the force under my command was at all times deliberate and aimed, sights being adjusted, and volleys were fired when sufficiently large bodies of the enemy could be seen to justify it. The two platoons of Company C. under First Lieutenant Lucas and Second Lieutenant P. M. Bannon, were handled with the best of judgment. D company overcrowded on the firing line and men needlessly exposed themselves by standing in groups. First Lieutenant W. C. Neville, commanding the first platoon, did his best with the men in front of him. Captain Spicer, commanding D company, was overcome by the sun on the top of the hill and had to be sent on board the Dolphin. Lieutenant Neville injured his hip and ankle in catching his foot and falling down the mountain side after the fight was over. These accidents left Second Lieutenant M. J. Shaw in command of D company, which he handled with entire satisfaction. Forty men left the crest of the hill at 3.15 P. M., under Lieutenant Lucas, and destroyed the well and burned the house lately occupied by the enemy. Canteens were taken from the men still holding the crest and filled with water required by signal from the Dolphin.

The Marines fired on an average about sixty shots each, the Cubans' belts being filled during the action from the belts of the Marines, each having to furnish six clips or thirty cartridges.

The loss to our force was one private of D company wounded slightly and ten or twelve overcome by heat. These latter were kindly taken on board the Dolphin and cared for. The ship rendered every possible assistance to the expedition. Two Cubans were wounded during the fight on the hill, one being accidentally shot by Colonel Laborde by a pistol.

While destroying the well the Cubans were placed up the valley from which the enemy retreated and began a noisy and hot fight with guerrillas who had not been dislodged. In this fight the Cubans lost two killed and two wounded, but killed five of the enemy.

The march home began at 5.30 P. M., camp being reached at 8 P. M.

From the best of information since obtained, which is believed to be reliable, sixty of the enemy, among whom were two officers, were killed. The wounded were numerous, but the wounds were probably light, owing to the range of 600 or 1,000 yards, at which distance all the explosive effects of the bullets are lost. Eighteen prisoners, including one lieutenant, were captured; about thirty Mauser rifles and a quantity of ammunition.

Lieutenant Magill also captured a complete heliograph outfit and destroyed the signal station. This had been used since our arrival here and could be seen at all times. Before closing I desire to commend Lieutenant Magill's good judgment in coming up and the excellent manner in which he handled his men.

Sergeant John H. Quick was obliged to stand on the open ridge under fire to signal the Dolphin, which he did with the utmost coolness, using his rifle with equal judgment while not thus engaged. My only regret is that E company, under the command of First Lieutenant James E. Mahoney, which had been sent to us from an outpost near Camp McCalla, when the heavy firing was heard there, was unable to report to me until 4 P. M. Had he been an hour and a half sooner, I am satisfied that the entire force of the enemy, which was about 500 men, would have been captured. This delay was not due to any lack of zeal on his part.

I have the honor to be, sir, very respectfully, your obedient servant,

G. F. ELLIOTT,
*Captain, United States Marine Corps,*
*Commanding C Company.*

Lieut. Col. R. W. HUNTINGTON,
*Commanding First Battalion of Marines,*
*Camp McCalla, Guantanamo Bay, Cuba.*

HEADQUARTERS FIRST MARINE BATTALION,
GUANTANAMO, CUBA, June 18, 1898.

SIR: I desire to make the following supplementary report: Upon leaving camp you asked me if I wanted an adjutant. I declined to take one, the command being short of officers for duty; but having been notified that a Mr. Stephen Crane would be allowed to accompany the expedition, I requested him to act as an aid if one should be needed. He accepted the duty and was of material aid during the action, carrying the messages to fire volleys, etc., to the different company commanders. Very respectfully, G. F. ELLIOTT,
*Captain, United States Marine Corps, Commanding Company C.*

Lieut. Col. R. W. HUNTINGTON, U. S. M. C., *Commanding Battalion.*

U. S. S. MARBLEHEAD, THIRD RATE,
GUANTANAMO, CUBA, June 16, 1898.

SIR: I have the honor to inform you that on the 14th instant, at the suggestion of Colonel Laborde, the Cubans under the command of himself and Lieutenant-Colonel Thomas, supported by two companies of Marines, under the command of Captain Spicer and Lieutenant Elliott, routed the force of about 300 Spaniards stationed in the pass between the Marine camp and the south coast.

One portion of the command advanced by the cliffs so far as the well and blockhouse, which I referred to in my No. 88, supported by the Dolphin.

The other portion diverged from the coast line and advanced up the valley to the southeast, the two forces eventually uniting on the sides of the mountain in the vicinity of the blockhouse and well.

In this vicinity the Spaniards, numbering about 300, were encountered and driven from their position, sustaining a loss of between forty and sixty killed and one officer and seventeen soldiers captured.

As the day was well advanced, it was not possible for our force to make a search for the Spanish wounded, and I fear that many were left on the field uncared for.

We suffered a loss of two Cuban soldiers killed; six wounded, four of whom were Cubans. In addition, twenty-three Marines were prostrated by the heat and, with the wounded, were transferred to the Dolphin, from which ship the force was also supplied with ammunition during the engagement.

The well and blockhouse referred to, on the south coast, were destroyed and a set of heliograph instruments taken.

The object of the movement was for the purpose of relieving the pressure on the Marine camp by an offensive movemnt, and it was, I believe, entirely successful.

I need hardly call attention to the fact that the Marines would have suffered much less had their campaign hats not been on the Resolute.

I desire to call particular attention to the devotion of the Cubans to the cause of freeing their island, shown in so many ways, by stating that the last words of the Cuban who was shot through the heart and buried on the field were, "Viva Cuba Libré."

Inclosed, marked "A," is a list of the Spanish soldiers captured.

The second lieutenant, also captured, is Francisco Batista, of Guantanamo City.

The Marines who were prostrated by the heat were nearly all able to return to their camp early in the evening.

Very respectfully, B. H. McCalla,
*Commander, United States Navy, Commanding.*

The Commander in Chief, *North Atlantic Station.*

The story of the Marine Battalion is thus continued in the Annual Report of the Colonel Commandant for 1898:

"On the 5th of August the battalion embarked on the Resolute, which had been previously carefully fitted out as a transport as stated above, and on the 9th of the same month sailed for the Isle of Pines. After sailing, the destination of the vessel was changed to Manzanillo, where the ship arrived on August 12.

"On August 13, news having been received of the signing of the peace protocol, the town surrendered, and on the 14th the Resolute, with the battalion on board, sailed for Playa del Este. On the 18th of the same month the Resolute, having taken on board certain officers and men of the artillery of the army, sailed for Montauk Point, at which place she arrived on the 23d. Having landed the army detachment, and getting a clean bill of health, she proceeded to Portsmouth, N. H., where the battalion disembarked on August 26."

---

## CHAPTER XLVI.

1898.—Disbandment of the First Marine Battalion—Post-Bellum History of the Corps—The Centenary of Its Organization.

**1898.**—The men of the First Marine Battalion, after return from Cuba, as already stated, went into camp at Portsmouth, N. H., instead of being returned to the various naval stations. Here they remained from August 26 until September 16, when orders to disband the battalion were received. The detachments from the barracks at New York, Philadelphia, Norfolk, Washington, and Annapolis passed through Boston in a body, going thence to their several stations. The Washington detachment, on its arrival in that city, on September 22, was reviewed by President McKinley, who took occasion to mention their excellent condition and military bearing in the highest terms.

Among the honors awarded to the officers and men of the Battalion, the following conspicuous list is given:

"A resolution giving the thanks of Congress to the officers and enlisted men of the First Marine Battalion was introduced on the last day but one before adjournment, but failed to pass through lack of time.

"The President recognized the services of the First Marine Battalion by advancing or brevetting a number of its officers, as follows:

"Lieut. Col. R. W. Huntington, advanced one number and appointed colonel, for eminent and conspicuous conduct in battle.

"Capt. George F. Elliott, advanced three numbers on the list of captains, for eminent and conspicuous conduct in battle.

"Capt. Paul St. Clair Murphy, appointed major by brevet, for gallant service in the naval battle of July 3, off Santiago.

"First Lieut. W. C. Neville, appointed captain by brevet, for conspicuous conduct in battle at Guantanamo, Cuba.

"Second Lieut. Thomas S. Borden, appointed first lieutenant by brevet, for distinguished service in the naval battle at Santiago, Cuba.

"Second Lieut. Louis J. Magill, appointed first lieutenant by brevet, for good judgment and gallantry in battle at Guantanamo. Appointed captain by brevet for good judgment and gallantry in battle at Guantanamo, Cuba.

"Second Lieut. Philip M. Bannon, appointed first lieutenant by brevet, for distinguished conduct in the battle at Guantanamo, Cuba."

Recommendations in the cases of a number of other officers of the corps were submitted to a specially appointed board of the Navy Department, which awarded brevets and promotions in several cases, according to services.

Just as the first incident in the train of events leading to the war served to demonstrate the high standard of discipline of the Corps, in the person of Private William Anthony, so the whole record of its service only served to further enforce this fact. In view of the exceptionally severe conditions experienced at Camp McCalla, where the men were under constant fire, often for several days together, exposed to all kinds of weather and performing exhausting services, all the while undergoing the hardships of a tropical climate, the fact that only 2 per cent. were stricken with disease, while none died from such cause, is remarkable evidence of the complete and careful preparation for service, and the rigid discipline, as well as of the vigilance of the quartermaster and medical officer. After going into camp at Portsmouth, N. H., the average for indisposition from disease was only nine-tenths of 1 per cent. Neither of these records was equaled in the army or navy forces serving at the same period.

Continuing the post-bellum history of the Corps during 1898 from the Commandant's report for the following year, we read:

"By order of the Secretary of the Navy the Marine Band and two battalions, consisting of four companies each, aggregating 26 officers and 451 enlisted men, under command of Col. Robert W. Huntington, U. S. M. C., were ordered to Philadelphia, Pa., to participate in the Peace Jubilee parade in that city on October 27, 1898. The men for the battalions were ordered from the Marine barracks, Navy-yards, League Island, Pa., New York, N. Y., Washington, D. C. (navy-yard and headquarters), and Annapolis, and the Marine guards of the following ships: U. S. F. S. New York, U. S. S. Columbia, Dixie, Prairie, Yankee, Texas, Richmond, New Orleans, and Michigan. The battalions were composed almost entirely of men who had seen active service with the First Marine Battalion in Cuba. The men who took part in the parade wore undress uniform, campaign hats and leggins, and were equipped with knapsacks, haversacks, canteens, web belts, and Lee rifles. They presented a fine appearance and elicited much applause along the line of the parade.

"During the time it was necessary for the battalions to remain in Philadelphia the men were quartered on the U. S. S. Yosemite and the officers on that ship and the U. S. S. Minneapolis. Subsistence was provided and mattresses loaned by the Navy, and the men were extremely comfortable. Commodore Silas Casey, U. S. N., commandant navy-yard, League Island; Capt. Theodore F. Jewell, U. S. N., commanding the U. S. S. Minneapolis, and Lieut. Frank E. Sawyer, U. S. N., commanding the U. S. S. Yosemite, did all in their power to contribute to the comfort of the marine officers and men, and their efforts in this direction were very much appreciated by the officers and men of the battalions and the commandant of the corps."

The year 1898 is also notable as being the centenary of the Marine Corps as an organization; the act of Congress establishing it as a permanent branch of the national service having been passed July 11, 1898. The anniversary was appropriately noticed by the Secretary of the Navy in the following order:

"NAVY DEPARTMENT,
"General Order No. 494. "WASHINGTON, D. C., July 30, 1898.

"On the 11th day of the present month the United States Marine Corps, which has been a part of the Naval establishment of the Government for one hundred and twenty-three years, completed the one hundredth year of its existence as a corps. During this period the many occasions on which it has received the thanks of Congress for distinguished conduct in conflict with the enemy, and the numerous other instances in which its duties have been performed in such a manner as to bring it prominently into notice, have served to impress upon the public, and especially upon the Naval service, the great worth of the Corps. This order is issued in recognition of the centennial anniversary of this important and useful arm of the Government, which, occurring as it does in the midst of the war with Spain, and at a time when the Marine Corps has added new glory to its already gallant record, affords the Department an added pleasure in announcing this anniversary to the the service.

"JOHN D. LONG, *Secretary.*"

---

## CHAPTER XLVII.

**1898-1899.**—Changes in the Organization and Internal Affairs of the U. S Marine Corps by the Congressional Acts of May 4, 1898, and March 3, 1899.

**1898-1899.**—The Spanish-American War, in addition to furnishing conspicuous and nearly unprecedented opportunities to test the Marine Corps to the fullest extent, also marked a period in the history of its internal organization. The opening of the war found the Corps splendid in discipline and equipment, but sadly deficient in numbers. So greatly was this the case that, as the Colonel Commandant himself confesses in his annual report for 1898, referring to the organization of the First Marine Battalion:

"In order to organize this battalion, to furnish guards for the auxiliary ships, and to make the increases in the guards of the regular vessels requested by their commanding officers, it was necessary to very greatly deplete the strength of the shore stations of the Corps, leaving most of them in charge of non-commissioned officers, and in some instances with a strength of only six or seven men."

Accordingly, Congressional action was necessary to provide, as far as possible, to supply the deficiency, within the limited period at disposal. By the Naval Appropriation Act, passed May 4, 1898, provision was made for the enlistment of 473 men, for permanent service, thus bringing the Corps to its full authorized strength, 3,073 men, and, for 80 corporals and 1,500 men, for service during the war. As provided further in the same Act, 60 gunnery sergeants were to be appointed for the war, but since, by some strange oversight, no appropriation was made for their pay, none were enlisted. Speaking of these provisions, Col. Heywood remarks:

"If the additional 473 men had not been appropriated for, the Corps would have been unable to meet the demands for men required as guards on board ship and men for the battalion and at Key West, and even after these 473 men were added to the Corps, there were but 71 men of the permanent establishment available for duty at the different posts. In addition to the men required at the navy-yards, guards war had not been provided, the Corps would have been unable to furnish adequate guards for the various navy-yards and stations, where millions of dollars worth of public property is stored. As the men enlisted for the war became sufficiently drilled, some of them were distributed among the various Marine guards on board ship, relieving older men for positions as non-commissioned officers at the different posts. In addition to the men required at the navy-yards, guards composed of selected men were ordered established at the magazines at Norfolk and Philadelphia, as attempts had been made by Spanish spies to blow them up."

With the provision temporarily increasing the strength of the Corps, the Commandant's rank was raised from colonel to brigadier-general. Previous to this enactment the Marine Corps had enjoyed the rather anomalous distinction of being the only force in the world, of 3,000 men or over, commanded by a colonel. Consequently, in raising the rank of the commandant, Congress did no more than confer on him a dignity commensurate with the duties and responsibilities of his position.

By the further provisions of the same Act, the appointment of forty-three additional second lieutenants was authorized, forty of whom were to be drawn from civil life and three from the capable non-commissioned officers of the Corps These newly-appointed officers were hurriedly drilled and prepared for duty, and assigned to posts on several regular and auxiliary war-ships, to the various shore stations and to the First Marine Battalion, organized for service in Cuba. During the war several of them attained distinction by performance of conspicuous services, although, by the terms of enlistment, all were discharged shortly after the cessation of hostilities—the last one being mustered out on March 16, 1889.

The end of the war, however, found the Corps sadly in need of additional officers for permanent service, as well as of a larger number of enlisted men, to meet the constantly increasing demands for naval guards, both afloat and ashore Consequently, several very radical changes were authorized by a law passed on March 3, 1899, and known as the Naval Personnel Act. This act provided for making appointments to vacancies in the authorized number of first and second lieutenants, from four classes: First, from graduates of the U. S. Naval Academy; second, from those who had enlisted as second lieutenants in the Corps, during the Spanish-American War, under the Act of May 4, 1898; third, from meritorious and competent non-commissioned officers of the Corps; and, fourth, from civil life.

Among the appointees under these provisions were thirty of the forty-three second lieutenants enlisted for the war, all of whom were immediately commissioned as first lieutenants. One of these, Lieutenant Charles Go. Andresen, was formerly a non-commissioned officer of the Corps. Fifteen others were appointed from civil life, after passing satisfactory examinations, thus completing the total number of first lieutenants to be appointed before January 1, 1900.

The reason for thus adding to the number of commissioned officers of the Corps was that this same Act had, very wisely, provided to increase the permanent force to 211 officers and 6,000 enlisted men, exclusive of the Marine Band. Thus the Corps was brought to a strength consistent with all needs.

The Commandant having been elevated to the rank of brigadier-general by the Act of May 4, 1898, the specified officers of the Corps, according to the Naval Appropriation Act, were: One brigadier-general commandant; one adjutant and inspector; one quartermaster; one paymaster, each with the rank of colonel; one assistant adjutant; one assistant paymaster; two assistant quartermasters, with the rank of major; three assistant quartermasters, with the rank of captain;

five colonels; five lieutenant-colonels; ten majors; sixty captains; sixty first-lieutenants, and sixty second-lieutenants. Provisions were also made for commissioning competent non-commissioned officers, after satisfactory examination, thus greatly stimulating ambition and inducing a desirable class of men to enlist. Further opportunities for advancement were afforded in the authorization of four sergeant-majors, nineteen quartermaster-sergeants, seventy-two gunnery-sergeants, and ten additional first sergeants, all of whom were to be competent to assume the responsibilities of command on occasion.

Reporting upon the new conditions of the Corps required by the provisions of the "Naval Personnel Act," Colonel George C. Reid, Adjutant and Inspector of the Corps, writes as follows under date September 20, 1899:

"Since my annual report of September 20, 1898, the status of the Marine Corps, with regard to the number of its officers and enlisted men, has been increased by the act of Congress approved March 3, 1899, to a total of 211 officers and 6,000 enlisted men, exclusive of the band. This puts the Corps upon a footing, when its full strength has been attained, which will enable it more fully and efficiently to comply with the requirements of the Naval service, and afford a relief from the extreme ardurousness of the duties as they have existed heretofore, and will give opportunity for the more thorough and really necessary detailed instruction of both officers and enlisted men. It will enable each post to have a sufficient number of officers and enlister men to perform all the duties required of them, and yet give them time for study and for that theoretical and practical instruction required of all officers in the interest of thorough efficiency, and at the same time more nearly assimilate the periods of duty of both officers and men to that which has been for some time in vogue by Army Regulations, and which in cases of emergency is limited to one day in five.

"The appointment of an assistant adjutant and inspector, provided for by the recent legislation of Congress, has furnished a much-needed aid to the adjutant and inspector in the jerformance of the constantly increasing duties of his office, and particularly important are the duties of the superintendent of small-arms firing, which have been confided to his charge. To this officer also properly belong the duties of the Marine Corps member of the Naval Board of Inspection and Survey. Though the adjutant and inspector himself has for the past five years been, by order, a member of this Board, and until the past year has performed his duties as such, he no longer is notified when the inspections of ships are to be made, and the inspections of the Marine guards of ships have of late been made by a Naval officer of the Board. I am of the opinion that better results would obtain if the military inspection of these guards were confided to a Marine officer, and that officer one of the adjutant and inspector's department who is especially qualified for such duty. The good effects from a rigid technical inspection must necessarily result to any guard, but more especially necessary are such inspections in the cases of the guards of small ships which are under the charge of non-commissioned officers, who have not the authority and influence of commissioned officers in charge of men, and who have to do with their comfort and discipline as well as their proper instruction and training for their various duties.

* * * * * * * * * * *

"The addition of 4 sergeant-majors, 72 gunnery sergeants, 10 first sergeants, and 20 quartermaster-sergeants has done much to encourage and stimulate the ambition of the enlisted men. The grade of gunnery sergeant is one of the most important, and will, without doubt, produce a class of non-commissioned officers most valuable to the service in the future. The increase in the number of quartermaster-sergeants will supply a demand that has long been urgent, and will afford a very necessary assistance to the commanding officers of the different posts in the matter of the care of and responsibility for public property and in the rendering of accounts. A thorough course of both theoretical and practical instruction for gunnery sergeants in ordnance and gunnery is most important, and its early establishment is intended by the Brigadier-General Commandant.

"Since my last report there have been established in the new possessions acquired by the United States the following stations: Cavite, Philippine Islands; Guam, Ladrone Islands; Havana, Cuba, and San Juan, Porto Rico; and during the past year there have been transferred to these stations one battalion consisting of 14 officers and 260 enlisted men, under command of Colonel P. C. Pope, to Cavite, Philippine Islands, and another one, later, consisting of 14 officers and 362 enlisted men, under command of Lieutenant Colonel G. F. Elliott; a battalion to Guam, Ladrone Islands, consisting of 5 officers and 121 enlisted men, under command of Major A. C. Kelton; a detachment of 5 officers and 75 enlisted men to Havana, Cuba, under command of Major (now Lieutenant-Colonel) W. S. Muse, and another consisting of 2 officers and 40 enlisted men to San Juan, Porto Rico, under command of First Lieutenant (now Captain) L. C. Lucas, making in all 40 officers and 858 enlisted men. The detachment at Havana has been temporarily withdrawn on account of the breaking out of yellow fever at that station, resulting in the death of two privates. The post at Pensacola, Fla., was re-established December 12, 1898, when 1 officer and 46 enlisted men were sent to that station, under command of First Lieutenant (now Captain) B. S. Neumann.

"Immediately after the publication of the proclamation of peace between Spain and the United States, as issued by the President, 41 officers who had been appointed for service in the Marine Corps during the war were mustered out, and the discharge of men enlisted for the war has progressed as rapidly as the exigencies of the service will permit, until the present time, when there are 18 still in the service.

* * * * * * * * * * *

"The demands upon the Corps for foreign service have necessitated a reduction in the strength of the different posts to the extent that requires the performance of guard duty day on and day off in connection with the various other duties of an enlisted man at all the different stations. While the strength of the Corps in the past has compelled this arduous duty, it is earnestly hoped that the near future will see the Corps recruited to its full strength allowed by law; that this old and wearisome routine of day on and day off will become a thing of the past; that the enlisted man will have at least sufficient time between his tours of guard duty to enable him to give the necessary attention to the care of his arms, accouterments, etc., required to keep them in the best condition, and that he will be able to have more than one night's uninterrupted rest. I am fully of the opinion that this excessive guard duty, coupled with the numerous other duties of a soldier in garrison, is one of the principal causes of that discontent and dissatisfaction, especially on the part of the recruit, which conduce to desertion."

---

## CHAPTER XLVIII.

1899.—The Disorder in Samoa—The Skirmish at Tagalii Between the British-American Force and the Hostile Natives.

1899.—During March-April, 1899, there was considerable disorder on the island of Samoa, owing to a dispute over the right of succession between the two native chieftains, Mataafa and Malietoa. Since the island was at that time recognized as a quasi-independency, under the combined suzerainty of the United States and Great Britain, the naval forces of both nations were actively drawn into the dispute, interfering to establish the claims of Malietoa, and to put down the uprising against their property authority. Although the affair was of short duration, it gave opportunity for some sharp fighting; also calling forth several acts of conspicuous bravery in American Marines, sailors and naval officers.

The most important operation during this brief period of trouble was the disastrous expedition on shore, composed of sailors and Marines from the American and British warships, under the command of Captain L. C. Stuart, R. N., of H. M. S. Tauranga. Its object was the breaking up of a camp of Mataafa natives in the vicinity of Vailele. The American force, consisting of sixty officers and men, including twenty Marines, was under the command of Lieutenant F. V Lansdale, U. S. N., of the U. S. S. Philadelphia, who was assisted by Ensign Monaghan, U. S. N., Lieutenant C. M. Perkins, U. S. M. C., and Passed Assistant Surgeon G. A. Lung. The British force, consisting of sixty-two officers, seamen, and Marines, was commanded by Lieutenant Cave, R. N., of H. M. S. Porpoise; the combined forces being under the direction of Lieutenant Freeman, R. N., of H. M. S. Tauranga, specially detailed for the occasion by Captain Stuart.

In this encounter occurred a sad but glorious event, the deaths of Lieutenant Lansdale and Ensign Monaghan, with several of the men comprising the crew of a Colt's automatic gun. Captain White, of the Philadelphia, reports on the incident as follows:

"Several incursions have been made into the bush back of Apia during the last ten days of March. Very few natives had been observed. When seen, however, they were always fired upon by the Colt automatic gun, and they fled in terror. Lieutenant Lansdale set great store by the gun and frequently operated it himself. He appears to have depended greatly upon the gun when the party was in ambush on April 1. Twice it did not function and time was lost in overhauling it, and great delay was consumed in trying to get it through the wire fence. Lieutenant Lansdale was loath to abandon it, but the fire was so galling that before he was wounded he was compelled to scatter the important parts and leave it behind. Prudence led him to deploy his men in open order. The thicket was so dense that after the order for retreat was sounded it was not possible for the groups to render each other mutual support.

"Lieutenant Lansdale was wounded below the knee soon after the Colt was abandoned and rendered incapable of marching. He was assisted by his men, one of whom, N. E. Edsall (ordinary seaman), was mortally wounded while doing so. It is not clear when Mr. Lansdale received the wound in his chest. It is evidence most clear that when Ensign Monaghan discovered that the lieutenant was wounded he used his best endeavors to convey him to the rear, and, seizing a rifle from a disabled man, made a brave defence, but undoubtedly he fell very shortly after joining him, and the hostile natives, flushed with success, bore down on our men in his vicinity. The men were not in sufficient numbers to hold out any longer, and they were forced along by a fire which it was impossible to withstand. Ensign Monaghan *did* stand. He stood steadfast by his wounded superior and friend—one rifle against many, one brave man against a score of savages. He knew he was doomed. He could not yield. He died in heroic performance of duty.

"Private Henry L. Hulbert, U. S. M. C., informs me that he saw both Lieutenant Lansdale and Ensign Monaghan after they were killed; that he had been near them a few moments before, and that he had assisted in covering their retreat. He is mentioned by Lieutenant Perkins for bravery and good conduct.

"James Butler (coxswain) was killed instantly while standing and firing against the enemy. The surgeon's report covers the cases of the wounded men.

"I have to commend the steadiness of F. D. Fisher (gunner's mate, first class), in charge of the Colt automatic gun, for coolness and steadiness in twice dismounting and overhauling the gun under fire. J. S. Ranlett (ordinary seaman) assisted Ensign Monaghan in binding up Lieutenant Lansdale's leg with a neckerchief belonging to B. Callaghan (blacksmith) and assisted in carrying him when first wounded.

"The number of cartridges fired by each of the bluejackets averaged 50. From what appears to be reliable authority, 40 of the assailants were killed and 50 wounded. This information was obtained several days after the engagement. The last to arrive were H. M. Wagner (coxswain), E. D. Myrick (ordinary seaman), F. Walker (ordinary seaman), P. Kensie (oiler), A. E. Gustafsen (seaman), Henry L. Hulbert (private, U. S. M. C.), and two British sailors, whose names I have not learned."

The following report of Lieutenant Perkins to Rear-Admiral Kautz gives an interesting detailed description of this unfortunate affair, which seems to have given occasion for several acts of heroism:

CAMP HOLLOWAY, APIA, SAMOA, April 2, 1899.

SIR: In obedience to instructions I make the following report of the engagement of yesterday:

Pursuant to verbal orders of Lieutenant P. V. Lansdale, U. S. N., the senior Naval officer on shore, I joined at 1 P. M. a column of combined British and American troops.

The British detachment included stretchermen, signalmen, etc. With the column went interpreters Mackie, Scanlon, George Reid, Macdonald, Missionaries Wright and Henrickson, of the Mormon mission, and Mr. Stanley Osborne, secretary to the consul-general, who acted as volunteer aid, and a native contingent numbering, I should judge, about 100. Lieutenant Freeman, R. N., was the senior officer of the combined column present, and I was directed by Lieutenant Lansdale, U. S. N., to obey generally his instructions.

Upon taking up the march to the eastward along the beach road we formed in the following order: British marines and signalmen, Colt gun detachment, Marine guard United States Marine Corps, seamen infantry, natives, and civilians. Lieutenant Freeman and Mr. Lung marched near the head of the column; Lieutenant Lansdale with the gun detachment; Ensign Monaghan with his company; I kept, generally, with my own detachment, occasionally going to the head to speak to Freeman and Lansdale as to the route, having gone over a part of it the day before as far as Faglii.

We first halted on the west bank of the river Vaivase, to the westward of Faglii village, when signal was made to H. M. S. Royalist to cease firing. This was at 2.15 P. M. The column then took skirmishing order and advanced cautiously, after burning the enemy's huts near Mormon camp (village of Faglii). Proceeded along road skirting German plantation beyond Vailele to village of Tonga, which natives fired. Here we made a halt. A consultation was held between Lieutenant Freeman and Lieutenant Lansdale at which I was not present, but each afterwards informed me that we would retire inland to Apia by main road skirting Copre plantations. This is known as the road from Vailele, half a mile from the seaboard, and for a long distance it is straight, bordered by groves of cocoanuts and inclosed by barbed-wire fence. Upon taking up the return march I was directed by Lieutenant Lansdale to preserve our then formation—inverse order of march—the marines becoming the rear guard or left of line. A short distance along the road the enemy was first seen crossing a transverse road about 200 yards to left. Here we halted, formed line, and fired for five or six minutes. The Colt gun, which was hurried to the head of column (right), failed to work. I was informed by Lieutenant Lansdale that it had been wrongly assembled by the armorer after being cleaned before starting. It was overhauled and a round fired to test it. We then formed in double column of files on each side of the road—two lines of flankers and the friendly natives sent into the bush ahead and on left flank.

After marching about a mile further we descended a defile and forded a river, the road zigzagging. On opposite side climbed to higher plateau and struck level road again. Proceeded along this a short distance, when we were suddenly attacked in force from the left, the hostiles being entirely concealed and firing on the column from the grass, ground slightly rising toward them. Returned fire without apparent effect; friendly natives firing excitedly and recklessly, endangering our own men. Enemy's fire slackened a few moments and then began more spiritedly from all along the left and from the left flank and rear. Being the only officer on the left of the line I urged (through interpreters Mackie, Wright, Hendrickson, and Scanlon) the friendlies to prevent our being flanked and to protect our rear, but to no avail. The fire growing hot all around our left and rear and two men being wounded, I ordered the left flank to fall back to the right toward a wire fence, about 300 yards distant. While falling back to the fence I met Lieutenant Cave, R. N., and Dr. Lung, who were accompanying the wounded to the rear. Both agreed that a retreat to the shore line was the only course to pursure to prevent being cut off entirely. The right was then engaged firing to the front (left). I could not tell if the Colt gun was working. Having reached the wire fence with several of my men, the doctor with wounded, together with Lieutenant Cave and British marines and signal party having already crossed, I directed the trumpeter, Fifer Tietze, to sound the signal for the guard to form and retire. The enemy had then approached within 50 yards to left and rear, firing boldly. I made a stand at the fence, directing Sergeants McNally and Forsterer to rally the men and protect retreat of others as they fell back. This was the most dangerous spot of the field, the bluff descending abruptly beyond into a deep ravine covered with dense shrubbery, banana plants, etc., and as our men withdrew and entered this hollow we were fired upon continuously from the bluff to the left and by sharpshooters perched in the cocoanut trees directly overhead. I saw one of them fall from a tree, shot.

The greater part of my men had then crossed, being preceded by the British under Lieutenant Cave, Dr. Lung, and the wounded. Sergeant Bruno A. Forsterer and three men remained at the fence till firing ceased. Had we been a few minutes later in crossing this ravine it is my opinion we would have been cut off and killed to a man. I did not see the Colt gun in action, nor Lieutenants Lansdale and Freeman, or Ensign Monaghan after the battle began. Sergeant Forsterer and Private Hulbert saw both officers shot. Facing to the rear and firing at the sharpshooters as we went we finally reached the shore, where we had originally halted, at the village of Faglii. I requested Lieutenant Cave to signal to the Royalist, which was lying offshore directly opposite, to send boats with reinforcements, which was done, and used my utmost endeavors to throw up intrenchments. Here the marine guard and native contingent halted while the main body of British sailors moved to the westward in the direction of the consulate, halted about 300 yards distant. I directed Mr. Stanley Osborne to return to the consulate with the wounded and a message asking for reinforcements.

At Dr. Lung's suggestion I assigned him to command of the company of bluejacket infantry. It was then that I learned for the first time that Lieutenant Landsdale and Ensign Monaghan were not with us, though I was informed by Dr. Lung that he believed they were both safe. I continued to cover the rear with the marines, British and American, and tried to induce the natives to go into the bush on the left to protect our left flank, which was exposed. Happily, however, we were not fired upon after reaching the beach.

When I directed the left to fall back we were almost surrounded, and had I delayed a moment later I believe the column would have been massacred.

I can not close this report without commending in the highest terms the behavior of the marine guard under my command, who, to a man, acted with the utmost coolness and intrepidity, particularly Sergeants Michael J. McNally and Bruno A. Forsterer, the latter holding the stand at the fence until surrounded, and afterwards, when we reached the sea, volunteering to lead the friendly natives into the bush to protect our flank.

The gallantry of Private Henry L. Hulbert, who remained behind at the fence till the last, and who was with Lansdale and Monaghan when they were both killed, I desire especially to mention. His behavior throughout was worthy of all praise and honor.

Dr. Lung acted with the greatest coolness and bravery and rendered valuable services outside his profession.

Private McCarthy, U. S. M. C., was the only one of my command wounded—shot through the wrist. Very respectfully, C. M. PERKINS,

*First Lieutenant, United States Marine Corps, Commanding Guard.*

The COMMANDING OFFICER, *U. S. Flagship Philadelphia.*

Lieutenant Perkins' recommendations of Sergeants Forsterer and McNally and Private Hulbert called forth the following letters from the Secretary of the Navy:

NAVY DEPARTMENT,
WASHINGTON, D. C., May 23, 1899.

SIR: The Department is in receipt of the report of Rear-Admiral Kautz and the reports forwarded through him concerning the reconnoissance made by the combined American and British forces at Apia, Samoa, April 1, 1899.

The Department notes with pleasure the high terms in which Lieut. C. M. Perkins, U. S. M. C., speaks of your conduct during the trying events of that day. That part of his report concerning you is as follows:

"I cannot close this report without commending in the highest terms the behavior of the Marine Guard under my command, who, to a man, acted with the utmost coolness and intrepidity, particularly Sergeants Michael J. McNally and Bruno A. Forsterer, the latter holding the stand at the fence till surrounded, and afterwards, when we reached the sea, volunteering to lead the friendly natives into the bush to protect our flank."

Your conduct as outlined above reflects the greatest credit upon you and the Marine Corps. JOHN D. LONG, *Secretary.*

Sergt. BRUNO A. FORSTERER, U. S. M. C.

NAVY DEPARTMENT,
WASHINGTON, D. C., May 23, 1899.

SIR: The Department is in receipt of the report of Rear-Admiral Kautz and the reports forwarded through him concerning the reconnoissance made by the combined American and British forces at Apia, Samoa, April 1, 1899.

The Department notes with pleasure the high terms in which Lieut. C. M. Perkins, U. S. M. C., speaks of your conduct during the trying events of that day. That part of his report concerning you is as follows:

"I cannot close this report without commending in the highest terms the behavior of the Marine Guard under my command, who, to a man, acted with the utmost coolness and intrepidity, particularly Sergeants Michael J. McNally and Bruno A. Forsterer, the latter holding the stand at the fence till surrounded, and afterwards, when we reached the sea, volunteering to lead the friendly natives into the bush to protect our flank."

Your conduct as outlined above reflects the greatest credit upon you and the Marine Corps. JOHN D. LONG, *Secretary.*

Sergt. MICHAEL J. MCNALLY, U. S. M. C.

NAVY DEPARTMENT, WASHINGTON, D. C., May 22, 1899.

SIR: The Department is in receipt of the report of Rear-Admiral Kantz and the reports forwarded through him concerning the reconnoissance made by the combined American and British forces at Apia, Samoa, April 1, 1899.

The Department notes with pleasure the high terms in which Lieut. C. M. Perkins, U. S. M. C., speaks of your conduct during the trying events of that day. That part of his report concerning you is as follows:

"The gallantry of Private Henry L. Hulbert, who remained behind at the fence till the last, and who was with Lansdale and Monaghan when they were both killed, I desire especially to mention. His behavior throughout was worthy of all praise and honor."

Your conduct as outlined above reflects the greatest credit upon you and upon the Marine Corps. Very respectfully, JOHN D. LONG, *Secretary*.

Private HENRY L. HULBERT, U. S. M. C.

---

## CHAPTER XLIX.

**1899-1901.**—The Operations of Battalions of the U. S. Marine Corps in the Philippine Islands.

**1899-1901.**—The acquisition of the new possessions—Guam, the Ladrones and the Philippines, not to mention Porto Rico and, temporarily, Cuba—of course created unusual demands for both army troops and Marines to perform guard and garrison duties. As a consequence, the Congressional Act of March 3, 1899, increased the effective force of the Marine Corps none too soon.

In the Philippine Islands detachments of the Corps saw considerable active service, which constantly necessitated draughts upon the available force at home, and furnished the subject for considerable correspondence, much of which is given in the succeeding pages.

The first move toward establishing a post of the Marine Corps in the Philippines was made in obedience to a cablegram from Admiral Dewey, dated March 9, 1899, and asking for a battalion of 250 Marines to garrison the naval station at Cavite. Immediately upon receipt of this request a force of fifteen officers and 260 enlisted men was assembled at the New York Navy-yard, under command of Colonel P. C. Pope, U. S. M. C. Among the other officers were Major W. F. Spicer, Major C. L. McCawley, the latter being assistant adjutant, and Assistant Surgeon J. R. Waggener, U. S. N.

It was the original intention of the Department to transport this battalion in the U. S. S. Glacier, but the accommodations of that vessel proving inadequate to requirements, the men were sent overland to San Francisco, Cal., by special train, leaving Jersey City April 13,

1899. The train arrived at San Francisco on the 19th, and on the following day the battalion was embarked on the army transport Newport, which immediately put to sea.

As described in Colonel Pope's reports, the voyage was uneventful and free from trouble, except for a slight outbreak of measles among the men. On arrival at Cavite, the battalion was quartered in the navy-yard adjoining Fort San Filipe, where the best available living accommodations and sanitary arrangements were prepared. The rigors of the tropical climate and the constant danger of fevers necessitated repairs and improvements from time to time, and, finally, by a special appropriation from the reserve funds of the Navy Department, the fort—which had been turned over to the Marine Corps—and its neighboring buildings were remodeled and renovated for the safe accommodation of a force adequate to all probable demands.

The conditions found by the battalion on its arrival at Cavite are set forth in the following extract from a report of Colonel Pope, dated May 27, 1899:

"The battalion strength at present of 282 men and 15 officers is not sufficient to be prepared for every call that might be made upon it here. At the same time, according to present arrangements, there is no available space for any more troops in the yard and no place within the necessary distance where the men can go into camp. Joining the yard is old Fort Filipe, which is in a very dilapidated and unsanitary condition, and is uninhabitable. There is an old, small one-story wooden building in the fort inclosure, now occupied by a company of volunteers. There are no cooking, messing, or washing facilities. It is surrounded by high fort walls, is unhealthy, and the heat most trying. When the army troops are withdrawn this could be made a comfortable barracks by taking down the old building and erecting a light frame two-story building, with mess hall, kitchen, etc., below the quarters, and this would furnish quarters for one more battalion. There are also two sets of officers' quarters in bad condition, which could be pulled down or repaired and enlarged for officers. It would take about $5,000 for the necessary repairs. The remainder of the volunteer regiment stationed here is quartered in various places through the town of Cavite wherever they find a vacant place in the bungalows. These arrangements for the volunteer troops are bad in every way. They are continually changed around, so that the same troops are not kept here for any length of time. No one in this climate sleeps on the ground floor, even the poor native bungalows being raised well above the ground. When people sleep near the ground they are continually ill with fever and soon break down, and the space thus cleared should be utilized as a site for building a small barracks. Otherwise, at the present time, every available space and living room are occupied, though possibly a limited number of bungalows could be rented outside. No more troops could be taken care of until some additional arrangements are made for their being quartered.

"If more Marines are to be sent here, and if it is to be arranged to occupy Fort Filipe after the withdrawal of the volunteers from it and its renovation, I think the most economical plan would be to increase the battalion to six companies of 100 men each. With such a force we could be prepared to respond to any call with an organization that could be easily handled, quartered and subsisted.

"I think that after the officers and men are well settled they will be very comfortable here."

When the battalion first went into quarters at Cavite, the sick list was very large, on account of the prevalence of climatic fever among the officers and men; but this condition of affairs proved of short duration, the fever soon wasting itself and leaving those affected by it in excellent condition After this brief epidemic, the sick-list was reduced to less than one-half of one per cent. of the total command.

Previous to making the repairs necessary to secure thoroughly sanitary quarters for the men as many of them as possible were lodged in the building formerly occupied by the Spanish Inspector General. This place, rather humorously called "the palace," was described as "filthy in the extreme." The process of repair, which involved the virtual rebuilding of this structure, consisted in laying cement floors on arches to raise the first story above the ground, and in arranging the drainage in an approved fashion. The deplorable ignorance of the Spanish on all matters connected with sanitation could be no better demonstrated than by the deplorable condition of this building intended for their officers. The arrival of the American troops at once marked a perceptible change in these matters.

As Colonel Pope had predicted soon after landing, the needs of the Philippine War quickly rendered a larger body of Marines in the islands a practical necessity, and these conditions, coupled with the demand for an available force in China, led to the dispatch of five more battalions during the next year and a half. In compliance with a cabled request from Rear-Admiral Watson, dated July 26, 1899, a second battalion of Marines, consisting in all of 367 enlisted men, under command of Major (later Lieutenant-Colonel) George F. Elliott, U. S. M. C., was organized at the Navy-yard, New York, dispatched by special train to San Francisco, and there embarked on the army transport City of Sydney, on August 18. The arrival of this additional force enabled the relief of the army guards, leaving the station at Cavite in possession of the Marines. The Third Battalion, consisting of 325 men and fifteen officers, under command of Major Littleton W. T. Waller, sailed from San Francisco, Cal., November 11, 1899, and arrived at Cavite December 15. The Fourth Battalion, although organized for service in the Philippines, in compliance with the cabled request from Rear-Admiral Remey, was finally sent to China, as was also the Fifth Battalion. Another Marine detachment, known as the Sixth Battalion, was organized in compliance with the cabled request of Rear-Admiral Louis Kempff, dated September 4, 1900, for service in the Philippines. It consisted of two companies, each composed of three officers, one gunnery sergeant, eight sergeants, six corporals, and ninety privates. Like the previous battalions, it was taken overland to San Francisco, where Company B was permanently stationed; Company A, under command of First Lieutenant F. M. Eslick and Second Lieutenants P. E. Chamberlain and W. H. Pritchard, being embarked on October 1 and arriving at Manila October 28. These battalions, together with the men on detachments of the various ships, gave a total of fifty-eight officers and 1,547 enlisted men of the Marine Corps serving in the Philippines at the end of 1901.

The record of service of the Marines in the Philippine campaign is characterized by the same qualities of fine discipline and personal courage that have ever been conspicuous in the history of the Corps. Never before had the men of this branch of the service a better opportunity to display real soldierly qualities, and never was an opportunity more nobly improved.

Conspicuous among the early services ashore may be mentioned the very material assistance rendered by a squad of these Marines—Corporal Thomas F. Prendergast and Privates Howard M. Buckley and Joseph Melvin—the crew of a Colt automatic rapid-fire gun, operating under command of Ensign Cleland Davis, U. S. N., in connection with the soldiers of the Eighth Army Corps. Their efforts served not only to fully test the capabilities of this weapon, but also to render valuable service in several successive engagements. In transmitting Ensign Davis' report to the Navy Department, Admiral Dewey uses the following laudatory expressions:

"While this crew was not composed of volunteers, none being asked for, the men performed their duty under most trying conditions of war in the most exemplary manner, and deserve high praise. I hope the Department will reward their services in a suitable manner."

The story of this interesting incident is well narrated in Ensign Davis' own words:

"With my detachment I was assigned to the divisional artillery under Major R. W. Young, U. S. V., chief of artillery, and was stationed on the firing line at Caloocan until March 23, when, in company with the artillery, I proceeded to La Loma Church.

"On March 25 operations commenced. The general plan of advance was as follows: General McArthur was in command. His division consisted of the first brigade, Brigadier General H. G. Otis, on the left, composed of the Third Artillery, Kansas and Montana regiments; and the second brigade, Brigadier-General Hale, on the right, composed of the Pennsylvania, South Dakota, and Nebraska regiments. The divisional artillery was in the center, which rested at La Loma Church. This general formation was maintained until Malinta was reached, the center advancing along the Caloocan-Noveleaches road to Cabalahan, thence along the Malinta-Noveleaches road to Malinta, the right wing swinging so as to preserve the front. At Malinta the division was joined by Brigadier-General Wheaton's independent brigade, consisting of the Third and Twenty-second Infantry, the Oregon and part of the Minnesota regiments, which had advenced along the railroad from Caloocan. From there on this brigade was in reserve, guarding the railroad communications. The front was now contracted and the advance continued with the center along the railroad track.

"The character of the country was extremely favorable for defensive warfare. The fields were rice land, covered with numerous copses of dense bamboo thicket. There was a network of tide-water rivers, mostly unfordable. In addition, the enemy had built strong intrenchements from ten to twenty-five feet thick, at short intervals along the roads, on the river banks, and especially along the railroad. These trenches were of the most modern type. The advance of the army was so rapid that the enemy had no time to destroy the iron railroad bridges, and the unfordable streams were crossed on these with little delay, the mules and horses swimming.

"The detachment under my command went into action in the following engagements:

"Near Cabalahan on March 25, covering with the artillery the advance of the Montana and Pennsylvania regiments against strong intrenchments on the Malinta-Noveleaches road. In the afternoon of the same day a platoon of thirty men from the Fourth Cavalry found the enemy strongly fortified on the opposite bank of the Tullahan River and engaged, with heavy loss to themselves. The Colt gun, with one piece of artillery, went into action under heavy fire on the left of the road, and the enemy shortly fled from their intrenchments. At the Marialo River, March 28, the detachment advanced under cover to within seventy-five yards of the enemy's trenches, strongly thrown up on the bank across the river, and by a sweeping fire covering the trench, which was about 150 feet long, silenced the enemy's fire, and enabled the artillery to come up on the open road to within seventy yards of them. Twenty-three of them surrendered in this trench, though a deep river was between. Of some twenty odd who attempted to escape nearly all were shot down. As an instance of the accuracy of the Colt gun, Colonel Funston, of the Twentieth Kansas, and Assistant Surgeon Smith, attached to the artillery, reported that one man was found dead with five holes in his body in a space that could be covered by the hand, all made by six-millimeter bullets from the Colt gun as he attempted to escape. At Guiguinto, on March 29, the enemy were encountered in force on the opposite bank of the river, retreating before the advance of our troops to a fringe of woods about 1,500 yards distant, from which they poured in a heavy and destructive fire as we crossed the river on the railroad bridge. Our troops were here under a great disadvantage, their Springfield rifles not being effective at this range. My detachment crossed the bridge under this fire and opened up at a range of from 1,600 to 1,900 yards, with, it is believed, good effect. Near Malolos, on March 31, the artillery and the Colt gun commenced the action, and in a few minutes the enemy retreated from behind strong intrenchments. After the artillery had driven them from their works the Colt gun kept up a fire on the retreating enemy up to a range of 2,000 yards. Malolos was then occupied with little resistance. On April 4 I took part with my detachment in a reconnoissance northward as far as the Quinqua River, where the enemy were encountered in some force, fortified on the opposite bank.

"I returned to the ship on April 5, in obedience to your orders of the 3d instant.

"In my opinion the efficiency of the automatic gun in operations on shore was amply demonstrated in this campaign. The light weight of the gun and ammunition and its simplicity of handling makes it available for various uses. As an adjunct to artillery, especially as the modern tendency seems to be toward closer ranges, it would seem to be invaluable. A gun, tripod, and 2,500 rounds of ammunition, the whole weighing less than 250 pounds, could be readily carried on the limber of each piece. But two men would be required to set it up and operate it, and it would be equivalent to the support of a company of infantry, with the additional advantage of being able to fire over the heads of advancing troops with perfect safety, as was done at Cuiguinto. Its portability is such that it could form part of the equipment of each infantry company or cavalry troop, and it is so small and compact that it can be taken with its tripod almost anywhere a man can go. Another point is its value for high angle fire. The value of a battery of such guns to a regiment is obvious.

"During the campaign about 4,500 rounds were fired from the gun. An examination of the barrel and mechanism shows the whole to be in excellent condition after a total of over 7,000 rounds had been fired from it. The Winchester ammunition furnished proved to be defective and not fit to be used in the gun. The U. M. C. ammunition was satisfactory in every respect. The last 2,500 rounds were fired without a single jam.

"The conduct of the detachment is deserving of commendation."

In transmitting Ensign Davis' report Major Richard W. Young, Chief of Artillery, Second Division, Eighth Army Corps, writes:

"Ensign Davis usually fired the gun himself, and always with accurate aim and unflinching courage. Were he an army officer I would unhesitatingly recommend him for a brevet for gallant and meritorious services."

General MacArthur writes: "I have personal knowledge of the matters referred to by Major Young, and take great pleasure in expressing strong concurrence in all he says of the professional skill and personal gallantry of Ensign Davis, which came repeatedly under my personal observation."

The accounts received of the various actions in which Marines played leading parts, if not operating entirely without assistance from the army troops, show that some of the most important works in the lines of subduing and Americanizing the islands must be credited to them.

On October 3, 1899, a detachment of eighty-four men and five officers, under command of Captain H. C. Haines, U. S. M. C., with a Colt's gun squad of four men and a sergeant, were ordered to report to Brigadier General Grant at Bacoor. At this place they were joined by another detachment of twenty Marines from the U S. S. Baltimore, under command of Captain J. T. Myers, U. S. M. C., and twenty-four sailors, under command of Ensign A. W. Marshall, U. S. N. This combined force accompanied the army troops in the march toward Imus, through an exceedingly uneven country, the greater part of the way being through thickets, swamps and small streams. After about a mile and a half out of the total four miles had been traversed, straggling parties of the enemy were encountered, firing from behind bamboo clumps. The American troops advanced upon their positions, without firing a shot, and succeeded in capturing eight prisoners. Both the army officers and their own commander, Captain Haines, speak in high praise of the bravery and implicit obedience to orders displayed by the Marines in this skirmish, the first occasion in which a regular detachment of the Corps co-operated with the army in the Philippines.

Five days later, on October 8, a force of 356 Marines, under command of Lieutenant Colonel George F. Elliott, attacked the insurgents entrenched in and around the town of Novaleta, dislodging them and capturing the town. The result of this expedition was to effect a junction with the army troops, under General Theo. Schwan, which had been the object in view. The real significance of this victory may be understood, in view of the fact that the Spanish had made numerous unsuccessful attempts to take the place, on one occasion, according to an apparently well-founded report, having once lost an entire regiment there. In the words of Colonel Elliott, "the fight was a dogged, steady rush through mud and water, with heavy firing from the left flank and lighter firing from the right flank," in which only ten men were seriously wounded, two of them fatally.

The details of the action are given in the following spirited report of Colonel Elliott to Rear Admiral Watson:

HEADQUARTERS MARINE BRIGADE, CAVITE, P. I., October 12, 1899.

SIR: I have the honor to report as follows in regard to the fight of Sunday, October 8, at Novaleta, in order to make a diversion for the army in its attack on Cavite Viejo:

In obedience to your verbal order I reported to Brigadier-General Grant at Bacoor in order to form some preconcerted plan by which a battalion of 400 Marines and the U. S. S. Petrel could co-operate in a simultaneous attack on Cavite Viejo by the army, and Novaleta by the Marines, whose right flank should be protected by shell fire. I also conferred with Captain Cornwell of the Petrel, and it was agreed as follows: A steam cutter placed opposite the mouth of the Imus should give notice when the army had crossed that river; the Marines to leave the navy-yard by 9 A.M., carrying haversacks, canteens, ponchos, and 200 rounds of ammunition and proceed to the outpost at Caridad, some 3 miles distant; the Petrel to lie off Caridad until signaled to begin firing. While marching to the outpost it was learned from the flagship that 4 cascoes, supposed to be loaded with armed men, had landed in San Roque in rear of the outpost. On receipt of this information, I detached Captain Bisset and 50 men in order to make the outpost absolutely secure; this reduced the two battalions to 356 men and 20 officers.

The country being so rough, I deemed it best to divide the command into two small battalions, which were commanded by Captain Haines on the left and Captain Fuller on the right, with two officers to the company. I had as a staff, volunteer aids, Lieut. Walter S. Crosley, U. S. N., and Second Lieutenant Rifelberick of the Sixteenth Infantry, U. S. A.; Asst. Surg. C. D. Langhorne, U. S. N., being surgeon in charge. We arrived at the outpost at 9:45 A.M. sharp; firing being heard from the army advance at about 10:15. Having received the message that the army had crossed the Imus, the Petrel was signaled 'commence firing,' and she shelled our advance across the causeway. Lieut. A. E. Harding moved out first with 16 scouts, followed by the advance guard under Captain Davis 400 yards in the rear; following this company came Captain Fuller's battalion of two companies, commanded, respectively, by First Lieut. Henry Leonard and First Lieut. G. C. Thorp, at 200 yards' distance. The rear battalion, commanded by Captain Haines, was separated from the first by an interval of 24 paces, in the following order by companies: First Lieutenant Butler, First Lieutenant Hill, and Captain Borden.

After passing over the causeway a dense thicket of thorn bushes was encountered, absolutely impassable except by a single narrow road, the ground being low and marshy and intersected by tide-water runs. When the scouts had advanced about a mile, sharp fire was opened on them—Captain Davis's company had pushed into the stream in order to work to the front. I ordered First Lieutenant Butler's company to push up the road, following the advance of the scouts, as heavy fire had been opened upon them. Hardly had this road been well entered before a heavy fire came from the chaparral at a range of about 400 yards, on the left flank, and two men were wounded here. The men were opened out and hurried along on the run, in single file, in order that we might reach the open field some three-quarters of a mile ahead. Having arrived at the rice field, I ordered First Lieutenant Leonard to debouch to the right and move forward, as a very heavy fire was coming from the front from a small fort with flanking intrenchments. This was done in fine style, but the companies following received a severe fire from the left flank, when the men dropped without orders, faced to the left, and opened a rattling fusillade, soon quelling this fire, and although they were formed flank to the principal fort, it was with difficulty that they were made to cease firing and move out of the road into the rice fields, deploying toward the works.

Lieut. Leof M. Harding was conspicuous in his endeavor to compel deployment. Captain Haines coming up with his two companies stopped all annoyance from the left flank, then deployed his companies on that side of the road. Captain Fuller managed to get his line formed to the right of the road. Lieutenant Gilson's section of Captain Davis's company having broken through the chaparral, came up about this time. The ships could no longer fire in our front and a forward movement was made by rushes through rice fields from knee-deep to armpit deep in mud and water. When we arrived within 250 yards of the intrenchments, a slight dike was found on the left and a line of old rifle pits on the right. Here the men were gotten fairly well straightened out, but were unable to charge immediately, as they were absolutely 'blown' from the fast pace and heavy ground.

A tremendous fire was kept up from both sides. From a blockhouse on the left excellent shooting was done. This was made untenable by Captain Haines's orders to Borden's company, while the companies of Lieutenants Leonard and Butler and Lieutenant Gilson's section fired some excellent volleys on the right. During the heat of this fire Private Bartholomew O'Shea, who has since died of his wounds, reported to me in the road that Dr. Langhorne, Lieutenant Hill, and two men were wounded, some half mile to the rear; that they were being fired at at short range by a small party of the enemy, and asked that 'I pull them out.' Lieut. A. E. Harding was close beside me, and I ordered him to take 20 men and proceed down the road to the rear to their rescue. The men were so busy firing that it was with difficulty he gathered together 14 men and carried out this duty, driving away some eight or nine Filipinos who would have eventually killed this small wounded party. Lieutenant Harding's actions during the entire day were those of an excellent soldier.

Four buglers sounded the charge repeatedly, the officers without exception calling on their men to respond, and they finally broke forward in a dogged advance, without cheering, as they were breathless, and the enemy abandoned the entire length of the trenches, but kept up a short fire from the nipa huts from the farther side of the narrow, unfordable river which was directly in front of their works. The men of the right wing were the first in the works, as those on the left were blocked by lagoons and thorn bushes. The blockhouse was burned, as were all nipa huts from which firing was seen, and which were used by the garrison as barracks or shelter. A great deal of personal bravery among officers and men was shown, even up to reckless bravado, of which I highly disapprove, and I believe they will fight as well but with better judgment in the future. Dr. Langhorne, while attending Lieutenant Hill, was wounded through the arm, but continued looking out for the injured as well as for nearly fifty men prostrated by heat and over-exertion. The two aids, Lieutenant Crosley, who was hit by a spent ball, and Lieutenant Rifelberick, did excellent work in helping men, many of whom were in their first action, to the proper performance of their duties.

Connection was made in Novaleta with advance guard of General Schwan, which was the object of the expedition. The fort, having a clear field of fire for 800 yards down the road and over the fields, was pierced for rifle fire, and the enemy was but little exposed behind the parapet. It was necessary to carry this place by direct assault; a flank movement could not have been made toward the left on account of dense thicket and the fact that it would eventually form my line with the rear across the road leading to old Cavite, down which the enemy were retreating before General Schwan. Had the attack been made on the right flank it would have blanketed the fire of the Petrel, which would have been my base of safety in case assault failed.

Every endeavor was made to provide for wounded men, stretchers and "Chinoes" for body bearers, but it was impossible for me to get Chinese to perform this labor; three stretchers were provided without bearers, and we are indebted to the captains of the Wheeling and Petrel for sending aid to us by boat. This provision was known to be so deficient that Assistant Paymaster Sanford, U. S. N., and Pay Clerk Hunt, U. S. N., followed to the outpost with vehicles and proceeded to the field during the fight. Had this not been done I should have been obliged to hold the fort during the night until communication could have been made with Cavite, so that I could remove my wounded and exhausted men.

I had a private in my ranks who was born in Bacoor, near by, and who served in the Spanish army. He alleges that on several occasions the Spaniards endeavored to take this place, but were on every occasion repulsed.

I returned to the garrison at 6 P.M., after having rested for a short time at the outpost. Very respectfully, G. F. ELLIOTT,

*Lieutenant-Colonel, Commanding Marines.*

THE COMMANDER IN CHIEF, UNITED STATES NAVAL FORCE,
*Asiatic Station.*

In forwarding the above-quoted report, Rear-Admiral J. C. Watson, U. S. N., commander in chief United States Naval force on Asiatic Station, stated:

"I take pleasure in commending Lieutenant-Colonel Elliott, with his aids, and the whole command, for the very creditable performance of the duty assigned. The statement made, relative to the unsuccessful attempts of the Spaniards to take Novaleta on several occasions, is confirmed by information in the possession of the military governor."

Capt. B. H. Fuller, who commanded the other battalion in the engagement, makes the following report relative to First Lieuts. George C. Thorpe and David D. Porter:

"Lieutenant Porter, followed by Lieutenant Thorpe, were the first to enter the enemy's position, followed by two or three men, and I wish to commend them to you as especially deserving of recognition for bravery."

This report is indorsed by Lieutenant-Colonel Elliott, as follows:

"Respectfully forwarded. First Lieutenants George C. Thorpe, R. M. Gilson, and David D. Porter, U. S. M. C., when within 150 yards of the insurgents' fort, stood upon the parapet of the first entrenchments taken while their men were firing, and I respectfully request that the Admiral will admonish these young officers for bravado which might have caused a failure in carrying the fort provided these officers had been killed or wounded before the charge had been sounded. I approve of the report of Captain B. H. Fuller, and his recommendation for personal bravery shown after the charge was sounded."

In indorsing this report, Rear-Admiral Watson stated:

"The commander in chief is glad to commend Lieutenants Porter and Thorpe for bravery, but has admonished them and Lieutenant Gilson for bravado, and for being out of their proper position while the men were on the firing line."

In recognition of the services of the Marines in this battle Major Gen. E. S. Otis, commanding Department of the Pacific and Eighth Army Corps, addressed a letter to Admiral Watson, Commander-in-Chief of the Asiatic Station, as follows:

Headquarters Department of the Pacific and Eighth Army Corps,
Manila, P. I., October 25, 1899.

Admiral: For the prompt assistance rendered by the body of marines under the command of Lieutenant Colonel G. F. Elliott, which in accordance with my request you caused to give assistance at Novaleta and vicinity to our column of troops commanded by General Schwan and marching westward from Bacoor, permit me to thank you and, through you, Colonel Elliott and his body of marines for the able and efficient service and aid rendered us.

I have the honor to inclose an appreciative communication submitted by General Schwan and which he desires to be forwarded.

I am, Admiral, with great respect, your most obedient servant,

E. S. Otis, *Major-General, U. S. Volunteers, Commanding.*

Brigadier-General Schwan's letter is as follows:

Headquarters Department of the Pacific and Eighth Army Corps,
Manila, P. I., October 23, 1899.

Sir: I beg to bring to your attention the action of a body of Marines under the command of Lieutenant-Colonel G. F. Elliott, who in accordance with a preconcerted plan moved on the 8th instant over the causeway at Caridad on Novaleta, about the same hour that an expedition under my command advanced in the direction of that town from old Cavite, which latter the insurgent troops evacuated early in the morning, upon the approach of a battalion that had been sent forward by me from Binacayan.

The demonstration of Lieutenant-Colonel Elliott's battalion, supported by the fire of the gunboat Petrel, which shelled the beach, doubtless drew off a considerable number from or prevented the material re-enforcements of the enemy's troops occupying the strong position at Putol, which was captured by my troops about noon.

I respectfully request that you convey to the commander in chief of the United States Naval forces on Asiatic Station the thanks of my command for the very timely and most efficient aid rendered to it by Lieutenant-Colonel Elliott's troops and the Petrel.

Very respectfully, THEO. SCHWAN,
*Brigadier-General of Volunteers.*

MAJOR-GENERAL COMMANDING DEPARTMENT OF THE PACIFIC AND EIGHTH ARMY CORPS, *Manila, P. I.*

While the Bay of Manila, where the greater portion of the U. S. fleet continued, was the base of operations for a large number of the Naval and Marine Corps expeditions, considerable attention was given to the vicinity of Subig Bay. Here several ships were stationed, and systematic operations begun.

In September, 1899, a large rifled gun was discovered in a position near the mouth of the Kalakan River, at Port Olongapo. This the Naval authorities determined to destroy, in order to avoid the annoyance it might cause to ships and landing parties. Consequently, on the 23d of the month, a detachment of 34 Marines from the Baltimore, 18 from the Concord and 18 from the Charleston, together with a number of sailors, the whole force under the command of Lieut. J. D. McDonald, was landed to effect this purpose. The landing was supported by the guns of the ships, and the gun was speedily destroyed, in spite of a heavy fire from parties of insurgents armed with Mauser rifles. Squads, under command of Sergts. George Herbert and Louis Sample, were dispatched to points nearer the enemy, and by their effective marksmanship soon silenced them. In addition to the gun, a number of boxes of brown prismatic and black powder were found and destroyed; also a small muzzle-loading cannon of antique pattern was taken aboard the Baltimore.

Systematic operations around Olongapo were begun in December, 1899, when a force of 117 men, with First Lieuts. G. C. Thorpe, Logan Feland, and J. W. Lynch, under command of Capt. H. L. Draper, was sent from Cavite to Subig Bay. From Olongapo expeditions were sent through the neighboring districts, for the purpose of clearing the country of insurgents and robbers. Very few encounters of importance took place, the enemy seeming to prefer attacking small squads, and generally keeping hidden when larger detachments appeared.

On Feb. 16, 1900, a small squad, under commandof Corporal Wallace A. Sullivan, had a skirmish with a party of the enemy at Banictican, whither they had been sent for water. Corporal Sullivan and Private C. M. Welsh were killed, and it would undoubtedly have gone ill with the rest of the party, had not Sergt. H. Harvey and seven men been dispatched to their relief. Sergeant Harvey's gallantry in this affair was highly praised by Captain Draper.

The story of the pacification and Americanization of his district is interestingly told by Captain Draper in the following selected reports on the operations of his command. In a report to Col. Elliott, dated at Olongapo, Feb. 10, 1900, he writes:

On January 5 I thoroughly scouted the valley passing Santa Rita.

On January 6 Lieutenant Feland, with scouting party, burned the insurgent signal station to the northwest of the town of Olongapo; on the following day another party was sent to complete the destruction of this place.

On January 13 I sent Lieutenant Thorpe up the river leading to Boton with scouting party. This town had never been visited previously by American troops. He surrounded the eastern half of the town and captured 2 insurgents armed with war bolos. I have been informed that these two Filipinos, one of whom was an officer, were recruiting for the rebel Major Alba.

On the 15th ultimo Lieutenant Thorpe scouted the country to the northward, but none of the enemy were encountered.

Having received reliable information that at least 50 insurgents were in Boton, at 3.30 A. M., January 25, I took 40 men in four boats and proceeded to that place, two boats going up the Boton River with me and two up the Opac River with Lieutenant Thorpe. By this plan it was hoped that the town might be surrounded and the enemy captured. Owing to the length of time occupied in getting up the muddy rivers and the great difficulty encountered, all the men being obliged to jump out of the boats and push them over the mud banks, at least every 50 yards, the enemy escaped from the town. On our return, one detachment, under Corporal W. A. Sullivan, was fired upon from the thick chaparral lining the banks of the Boton River. On hearing the firing I hastened with our boat's crew to assist the detachment fired upon through almost impenetrable swamps, abandoning the boat on a sand bar 200 yards from the beach. On my arrival the enemy had retreated, and it being found impossible to pursue, we returned to the arsenal. There were no casualties.

On the 25th ultimo I was informed that this same party of insurgents would leave in the early morning to join the forces of the enemy concentrated near Castulejos. Accordingly, at 2:50 A. M. on the 26th, I left this post with Lieutenants Thorpe, Feland, and Lynch and Acting Assistant Surgeon Haas, to ambush all the trails leading from Boton to Castulejos. No guides could be obtained. Lieutenant Lynch was posted at Santa Rita, in ambush with 25 men; Lieutenant Thorpe and 15 men at the junction of the Benictican-Castulejos trail, and with Lieutenant Feland, Dr. Haas, and 30 men I ambushed the Bacbac trail, this being the one over which the insurgents' main body were to pass. These ambuscades were held until about 10 A. M., when, the enemy not having been encountered, the expedition returned to the arsenal. The enemy, in some manner, received notice of my movement and retreated precipitately to Moron.

Many small scouting parties have been sent out from time to time to see that this district is clear of the enemy.

A patrol system has been inaugurated in and about Olongapo. This insures peace and tranquillity to the town. No more robberies by ladrones have occurred since this was established, and the population of the town is increasing daily by the ingress of men from the insurgents and families from the mountains.

To promote the general welfare and secure the regular routine of peaceful life for the Filipinos in Olongapo, I held an election for municipal officers on the 28th ultimo. This election was held with the usual Filipino ceremonies, secret ballot, and resulted in the election of men in whom I have some confidence for president, vice-president, and secretary of Olongapo, for alcalde of Benictican, and alcalde of Santa Rita. The officers were installed by me in their offices with due and appropriate ceremonies.

After announcing the result of the election I made a speech to the newly elected officials and electors to the effect that my government guaranteed to every man the fruit of his own toil, the rights of life and liberty and the pursuit of happiness, and the right to worship God as he saw fit, and only demanded in return obedience to the laws.

I find that the new government works excellently. I have issued through them 204 cedulas. The municipal government having recommended, I have appointed 5 native policemen, uniformed in machetas and old full-dress helmets, so that they may be distinguished by the patrols at night.

Since my arrival here forty families have moved into Olongapo, where no person lived before; government has been organized, the peaceful people are protected; an English school has been started, with Lieutenant Thorpe as teacher; rations have been issued to save from starvation some of the natives; medical attendance and medicines have been supplied when needed, and a constant scouting of the surrounding country maintained.

I respectfully request that I may be appointed governor of the district, which includes Olongapo, Santa Rita, and Benictican, together with their small barrios; also that blank cedulas and blank passes may be furnished me.

Very respectfully, H. L. Draper, *Captain, Commanding.*

The Commanding Officer, *First Regiment of Marines, Cavite, P. I.*

In a later report he continues his narrative, as follows:

United States Arsenal, Olongapo, P. I., March 14, 1900.

Sir: I have the honor to submit the following report:

It was my intention to submit in this report a full statement regarding the skirmish at Benictican in which Corporal W. A. Sullivan and Private C. M. Welch were killed on the 16th ultimo. This is now unnecessary, as a full account may be gotten from the records of the court of inquiry. I most earnestly recommend that Sergeant H. Harvey may be promoted to the rank of gunnery sergeant for great gallantry in the presence of the enemy. Sergeant Harvey, with only 7 men, went at once to the relief of his comrades in obedience to orders previously issued by me. He carried out these orders zealously, efficiently, and gallantly.

Having received reliable information that most of the party who made the attack of the 16th had retired on Morong, at 1.30 on the morning of the 17th I took 107 men to that town by water and surprised it. About 400 rounds of Remington ammunition was destroyed, and the houses containing it were burned. The gunboat Manilano was broken down at the time. It being absolutely essential to strike Morong while the enemy were yet congratulating themselves on the killing of my men, the master of the native steamer Balayan was requested to tow the Manilano, which had my attacking force on board. This request the said master complied with. The Manilano rendered very effective assistance on this occasion. No formidable resistance was made by the enemy. Morong was a large town and an insurgent headquarters. It contained a large blockhouse, which was burned, and a strong system of intrenchments.

The command returned at 1 P. M., when the construction of coffins for the dead was continued, the same having been necessarily stopped in order that the men could be available for duty with the expedition. The dead were buried at 5 P. M.

On the afternoon of the 17th all inhabitants of Benictican and Baton were ordered into the town of Olongapo, and given three days to get in on pain of being declared outlaws. This order was complied with by all the inhabitants with the exception of six insurgent families, who removed to another town.

On the 23d of February I took 100 men and officers to Benictican. The Nashville having bombarded the town, I entered it, found nobody, and destroyed the town entirely, giving the good people who owned houses in Benictican houses in Olongapo. The surrounding mountains, valleys, and swamps were thoroughly searched, and the enemy could not be found.

On the 5th instant I embarked 45 men on the U. S. S. Nashville to co-operate with the army commanded by Colonel Wilder, of the Maccabebe scouts, in an attack on Bagac. The plan agreed upon between Colonel Wilder and myself was that I should keep the enemy occupied on the water side of the town while he with his force of 300 Maccabebe scouts should march to the rear of the town and gather in the enemy. Colonel Wilder was not able to find the trail leading to the rear of the town, but advanced by the beach. I covered his advance into the town with a line of skirmishers. One native, who was endeavoring to run away, and acting suspiciously, and who refused to halt after being repeatedly told to do so, was shot by my advance guard. A small quantity of arms and ammunition was captured by members of my command on this occasion. The detachment returned to this post at 7.30 A. M. the next day. I respectfully ask that Colonel Wilder may be thanked for his courtesy to myself and men.

Having received reliable information that several insurgent officers and men were living in or near the town of Calapacuan, which is within three-fourths of a mile of the nearest detachment of the army at Subig, that this town was being used as a storehouse by insurgents, and owing to the fact that an insurgent officer had written letters from this town to the civil officials of Olongapo levying contributions for the insurgent cause, I asked the permission of the commanding officer at Subig to raid the town, which permission was given. At 5 A. M. of the 8th instant First Lieutenant Logan Feland, with 80 men, surprised the town and, in obedience to my orders, brought the male able-bodied population to this post. Several war bolos were captured, and many tons of rice were found. After thoroughly investigating the cases of the prisoners, all were released except insurgent Captain Calixto Mindogorin; the local presidente, Augustin Mindogorin, and insurgent Sergeant Galope. The said captain had only left his command in the adjacent mountains two days before, and is considered the most dangerous and cruel leader in the immediate vicinity, and all the well-inclined natives here expressed gratification and relief at his capture.

The said presidente acknowledged to me that despite the fact that he had a commission as a United States official he received and stored rice and money contributions to the insurgent cause. He stated he did this through fear of the insurgent leaders, and at the same time acknowledged that he had never stated these facts to the commanding officer at Subig, who was only three-fourths of a mile from the town. I know this man to be very dangerous and the greatest civil prop to the insurrection in this part of the country. As further evidence of Augustin Mindogorin's support of the insurgent cause, I would state that he received and guarded as a prisoner an insurgent first lieutenant turned over to him by the rebel Colonel Arce. This first lieutenant, Marcelino Reyes, is detained here by me until I can send him to Cavite, a free man, he being a native of that town and entirely converted to the American cause.

Insurgent Sergeant Galope, aside from being in active service, has repeatedly denounced the good men of Olongapo to the insurgent leaders, and has acted in this capacity much as a spy. The two Mindogorins and Galope will be held by me as prisoners until further orders. I earnestly request that severe measures be taken in the cases of these men, both on account of their aggravated cases of treason and as an example to others.

In this connection I desire again to commend to your notice the efficient, zealous, and courageous qualities displayed by First Lieutenant Logan Feland.

The commanding officer at Subig generously acknowledged the services rendered on the above occasion by the marines.

Owing to the fact that the government established by me in Olongapo has been working smoothly and efficiently since the 29th of January, and that the town is rapidly growing, solely because perfect security and justice reign in the town, I respectfully request that the port may be opened. If it is thought fit to order me in charge of same, I respectfully request that I may be furnished with the necessary blanks and the orders issued by the military governor bearing upon the same.

I beg to renew my request for blank cedulas. Over 400 cedulas have been issued, all written. My object in issuing these written cedulas was to help bring order out of chaos. This has been accomplished. No charges have been made for these cedulas, so that when I am furnished with the regular blank cedulas the government collections on the same can be made.

Small taxes have been levied for the support of the police, the cleaning of the streets, and sanitation of the town.

On auditing the town books on the 6th instant, I found the total amount on hand 52 pesos. Very respectfully, H. L. Draper, *Captain, Commanding.*

The Commanding Officer, *First Regiment Marines, Cavite, P. I.*

Among the other notable naval operations, in which Marines had a prominent part, may be mentioned the expedition from the U. S.

Flagship Brooklyn, sent to rescue a number of prisoners, Spaniards and Americans, who were detained by the insurgents at a town on the Gulf of Ragoy. The force detailed for this work, including a detachment of 20 Marines, under the command of First Lieut. Philip S. Brown, U. S. M. C., sailed from Manila Bay to San Narcisco, and thence to the vicinity of Piris. There they found and rescued 522 persons, including 10 Americans, 460 Spanish officers and soldiers, and a number of priests. Although the Marines encountered no opposition in landing and in rescuing these captives, the task undertaken was difficult and arduous. In the work of bringing their charges in boats from the shore, Lieut. Brown and his men were on duty for sixteen consecutive hours. In his report, Lieut. J. H. Gibbons, U. S. N., Commander of the General Alava, says:

> Lieutenant Brown and the 20 marines returned on board in the last boat at 5 A. M., February 28, 1900, having been on duty continuously since 1 P. M. the previous day, twelve hours of which time was spent knee-deep, and sometimes waist-deep, in the water. About 400 prisoners were brought in during the night. Rations, including hot coffee, were served to them as they came aboard. I cannot commend too highly Lieutenant Brown's conduct, while in charge of the embarkation on shore.

In recognition of Lieut. Brown's exceptional zeal in the performance of this duty, Rear-Admiral Watson wrote to the Navy Department:

> I would suggest that the work done by First Lieutenant P. S. Brown, U. S. M. C., and the men under his command be called to the attention of the brigadier general commandant of the Marine Corps.

The Admiral's recommendation called forth a special letter from the Secretary of the Navy to Lieutenant Brown, ending with these words:

> The Department is pleased to note the commendation with which the commander-in-chief speaks of you and the men of the Marine Corps under your command. Very respectfully, JOHN D. LONG, *Secretary.*

During 1899-90 detachments of Marines from the headquarters at Cavite were detailed as guards and garrisons at the following points in the Philippines: Olongapo, Port Isabela, Balabac Light-House, Malabrigo Light-House, Cabras Light-House, Capones Light-House, Cape Melville Light-House, Caridad, Dalahican, Sangley Point, San Roque, Canacoa, and Santiago Light-House. However, on account of the large draughts of men necessary for service in China, the last six posts were temporarily abandoned. During 1900-01, new posts were established at Novaleta, Las Pinas, Rosario, Cavite Viejo, Bacoor, and Paranaque.

From the date of the arrival of the First Marine Battalion at Cavite, until Oct. 15, 1899, when he was retired for ill health, Col. P. C. Pope was Commandant of the Marines in the Philippines. He was succeeded by Lieut.-Col. George F. Elliott, who held the command until September 11, 1900, when he was succeeded by Col. H. C. Cochrane. Colonel Cochrane continued at Cavite until detached and ordered home on May 8, 1901, being then succeeded by Lieut.-Col. M. C. Goodrell.

Lieut. Col. Goodrell is notable as being the officer who received the surrender of the insurgent colonel, Alba, with 13 officers and 83 men of his command.

## CHAPTER L.

### 1900.—The Siege and Capture of Tientsin.

1900.—In the spring of 1900 the "Boxer" uprising in China created a demand for United States troops and Marines, to protect the lives and property of American citizens at various points in the Empire. This need was so immediate that all available forces, possibly to be spared, were shipped from the Philippines. Among these were several companies of the 9th U. S. Infantry and a detachment from the 1st Regiment of Marines, quartered at Cavite. The Marine detachment, consisting of 6 officers and 101 enlisted men, was embarked on the U. S. S. Newark, for transportation to Taku, China, on June 14, 1900. Arriving at its destination four days later, its strength was augmented by the addition of another 30 men and 2 officers, who had been brought from Cavite in the U. S. S. Nashville. The officers of these detachments were: Major L. W. T. Waller, commanding; First Lieuts. S. D. Butler, Henry Leonard, George C. Reid, R. F. Wynne, W. G. Powell, A. E. Harding, and Second Lieut. Wade L. Jolly. The combined force, under the command of Major Littleton W. T. Waller, landed on the 19th, and immediately began the forward movement by marching to Tong-Ku, a distance of about 12 miles from Taku.

At this point was the terminus of a railway to the interior, but the track was found torn up in various places and the rolling stock entirely deserted. It was, consequently, necessary to make numerous repairs to both, in order to prepare a train for transporting the Marines. This work was speedily accomplished under the direction of Capt. F. M. Wise, U. S. N., Commander of the U. S. S. Monocacy, under whose supervision the train was also run to a point some 18 miles from Tong-Ku and 12 miles from Tientsin. After the dispatch of the Marine detachment on the 20th, the railroad was continuously operated by Captain Wise for forwarding troops of several different nationalities. In his report of the 22d he describes this work as follows:

"On Wednesday at 1 P. M. I dispatched by a train toward Tientsin 440 Russians, 130 United States Marines, the latter having one Colt and one 3-pounder. The next day (Thursday) I dispatched 500 Welsh Fusileers and 240 English bluejackets. I sent with the latter the Monocacy's 3-pounder, with directions to turn it over to Major Waller when encountered. Commander Craddock, R. N., was in command of the train. That same day (yesterday) about 6 P. M., I dispatched a second train with 900 Russian troops and four pieces of artillery with their four guns and four light Maxims, with about 100 horses. This morning at 8 A. M. I dispatched a train with about 200 Russian soldiers, 240 German, and two heavy English guns on fixed mounts, the latter being secured to a box car, constituting movable field artillery on the railway line. There were with the Russians this morning about 100 horses and provision wagons. I have also sent out two platform cars containing large fresh-water tanks, each with a capacity of about 5,000 gallons, and I have been especially careful that every train and even car carried drinking water in small utensils, breakers, buckets, cracker tins, etc., as there is no potable water between here and Tientsin. I am now preparing another train to take out the remaining Russian troops (Cossacks) who arrived this morning; and have been informed that bluejackets from the Terrible have arrived, and they will also be dispatched. There will thus have been dispatched about 3,000 men. Of the trains sent out, that of Wednesday, with our Marines, returned the same night. They had to repair track and bridges as they went along, but they had no mishap. Train No. 2, under Captain Craddock, had two cars derailed by an open switch about four miles out. They got past the obstruction and went on.

"At 11 P. M. last night an engine with two cars came back from the force farthest advanced containing wounded men, Russians and Americans. I inclose copy of Major Waller's report. He had retired to Chung Liang Cheng. I am about at the end of my resources with cars until some empty ones return from the front. Mr. Tuckey, Mr. Sherriff, and Mr. Clark, of the late imperial service, have given me valuable assistance in running this makeshift train service, but the employés are all my own, with some English and German firemen and stokers."

About 8 miles from Tong-Ku, the Marine detachment, under Major Waller, was met by a battalion of Russian infantry, about 440 strong, which joined forces with them, being conveyed over the remainder of

COLONEL FRANK L. DENNY QUARTERMASTER

U. S. M C. Plate 8

the route in the same train. They arrived at a point about 12 miles from Tientsin at 11 P. M., going into bivouac immediately, with the intention of awaiting the arrival of reinforcements, before attempting to resume the forward movement. The Russian commander was so anxious, however, to march on the beleagured city at once, that Major Waller yielded to the force of superior argument, and joined the advance at 2 P. M.

The combined column of nearly 600 men, accordingly, marched toward Tientsin; the crew of the American Colt gun, commanded by First Lieut. W. G. Powell, U. S. M. C., going ahead, the Russians following and the Marines bringing up the rear. At about 6:30 A. M., on June 21, they reached the Chinese part of the city. At about 7 A. M., when nearly opposite the imperial arsenal, the enemy began firing from the wall, lightly at first, but presently, at the front and flank, heavily and with considerable accuracy, at the ranges of 300 and 900 yards. The Colt gun being brought into action, kept the firing at the front well under control, until it finally jammed and had to be disabled. The enemy's force numbering between 1,500 and 2,000 men, the Americans and Russians were finally obliged to withdraw.

In his report of June 22, Major Waller continues the story from this point, as follows:

"I threw my men back and with some of the Russians (enlisted men) formed a line to the right and opened on the enemy on our right. These men on our front and right were imperial troops. A few boxers came on our left and annoyed us exceedingly with their fire, but were driven off by our men. The fire from the front was from hidden trenches. The enemy began to push troops to our right rear, and my line then became long and thin. The Russians fell back from the front and formed on my right flank, moving by that flank under the protection of the railroad embankment. The Colt, at this time, was in the original front with the Russian support of about 17 men. This dwindled to 2. The gun's crew lost 1 man killed and 2 wounded, when it jammed. The enemy's fire being unbearable, the lieutenant disabled the Colt gun and retreated. At about 8:15 the enemy again attacked my left flank, and as the Russians were some distance on my right flank I began my retreat by the same flank. During this retreat my force formed the rear guard. We were followed for four hours, but kept down the enemy's fire and brought off our wounded. We moved to this point, arriving a little before 2 o'clock in the afternoon. Our losses were 1 corporal and 2 privates killed, 1 corporal, 1 sergeant and 5 privates wounded. The killed were Corporal Lannigan, Privates J. K. Miller and W. H. Morris. The seriously wounded were Privates

Carter and Francis, and Corporal Kates. The other wounds were slight. The entire march was 30 miles. The fight occupied altogether about five hours.

"English and Russians arrived last night and bivouacked here. This morning more Russians with artillery arrived, and also a small party of Germans. I have sent a body of 16 men forward, to scout in front of the English force, and shall join them later with my whole body, co-operating with them.

"I can not say too much for the officers and men under my command; but I must speak especially of the conduct of Lieutenants Butler, A. E. Harding, Leonard, and Wynne. At a later time I shall take occasion to mention certain non-commissioned officers and men.

"The first relieving column reached Tientsin, but failed to relieve the place. I have done the best I could, sir, under the conditions surrounding me. Our men were the front to begin with and formed the rear in retreat. I have lost two guns, one from military necessity, the other by capture. We are footsore and weary, but will go forward now. The condition at Tientsin is almost hopeless. If we can not attack to-morrow, I fear the worst. An American escaped from Tientsin informs me that there were 2 killed and 4 wounded among the Americans before he left—six days ago. He didn't know the men."

The retreat was made to Chin Liang Cheng, the point from which the combined forces had started on the march to Tientsin, the entire distance marched within 12 hours being 30 miles, with 5 hours consumed in continual fighting. Upon the arrival of the British force, under Commander Craddock, R. N., Major Waller determined to co-operate with them, and continued in their company.

In his report to the Brigadier-General Commandant of the Marine Corps, under date, Tientsin, June 28, he thus continues the narrative:

"On the following day we moved as far as railhead, our bivouac of the day before, where we camped for the night. The force amounted at this time to about 2,000 men, 1,000 being Russian and the rest English, German, American, Italian and Japanese, in strength in order mentioned, the British being about 600 strong. It was agreed that we should advance in two columns on the next day at 4 A. M., my force occupying the advance of the British column and the right of the firing line. We struck the enemy at about 7 A. M. and drove them steadily until about 12:30 P. M., when we entered Tientsin, relieving the besieged Europeans, our losses being for the day 1 killed and 3 wounded. The Russian column deflected from the scene of our defeat two days previous and followed us into Tientsin. We rested for the remainder of the day, and at 12:30 A. M. the next morning, June 25, moved to the relief of Vice-Admiral Seymour, Captain McCalla, and the men of the powers who had been forced back from their march on Pekin to the relief of the ministers and intrenched at a point about 8 miles from Tientsin. We met very little opposition, and succeeded in relieving at 12 A. M., our casualties being 2 wounded from shell fire, 1 bullet wound. The force relieved had, by a brilliant charge and without knowing what they had opposed to them, succeeded in capturing one of the most important arsenals in China, the place being filled with a plentiful supply of all sorts of munitions of war. Capt. B. H. McCalla was in command of our forces, and was still on his feet, although suffering from three wounds. We moved the sick and wounded from the arsenal on the evening of the 25th, and encamped on the opposite side of the river for the night. At 4 A. M. the following day, June 26, we moved back to Tientsin with the sick and wounded of the besieged, the march being very slow on account of the large number of disabled men. Captain McCalla left me in charge of the combined force of American seamen and Marines.

"At noon on the 27th, the Russians having attacked the arsenal, the scene of my repulse on the 21st, and which had not been captured, asked for re-enforcements. I sent out Second Lieutenant Jolly with 40 men, Mr. Harding, my adjutant, going as a volunteer, and placed the whole under the command of Commander Craddock, R. N. This force was about 1,800 strong, and succeeded in driving the enemy from the parapets out of their fortifications and in full flight. It was developed that the enemy had about 7,000 men at this point. Our men charged over the parapet with a British company, being the first in this part of the fight. Our loss here was one wounded and Lieutenant Jolly overcome with heat, but not until after he had brought his men back to their quarters. Lieutenant Harding acted as a volunteer and captured an imperial flag, which he has presented to me. To-day we are resting, sending away the wounded, and getting ready for a march on Pekin. My effective strength is now about 89. Having given you the bare facts, I now wish to invite attention to the incidents of the busy week.

"Our men have marched 97 miles in five days, fighting all the way. They have lived on about one meal a day for about six days, but have been cheerful and willing always. They have gained the highest praise from all present, and have earned my love and confidence. They are like Falstaff's army in appearance, but with brave hearts and bright weapons.

* * * * * * * * *

"Captains Myers and Hall and the Marines under them are besieged at Pekin. As soon as sufficient forces have arrived with train the forces will move on Pekin. Our force is disgracefully small, considering our interests.

"I have to earnestly recommend to your notice for such reward as you may deem proper the following officers: Lieutenant Smedley D. Butler, for the admirable control of his men in all the fights of the week, for saving a wounded man at the risk of his own life, and under a very sever fire; Lieutenant A. E. Harding, for conspicuous gallantry in action, for saving wounded at the risk of his own life under a heavy fire; Second Lieutenant W. L. Jolly, for the same risk and for leading a fine charge over two parapets in the face of a heavy fire; Lieutenant Leonard, for saving life under fire and for admirable control and direction of the fire; Lieutenant Powell, for working and managing the Colt gun under a fierce fire and without support, after the crew had been shot down; Lieutenant Wynne, for his steadfast courage and encouragement of his men. As for the men, I feel that I cannot do them justice. They have made history, marked with blood, if you please, still glorious and brilliant. They were the first in the field, and, please God, they will remain until the last man, woman, and child is relieved from the toils of these barbarians. I shall send you the names of special instances in these cases, hoping that a suitable reward may be given them as far as the law allows. For myself, sir, I have only to say that I did my best. I have carried the colors you surrendered to me through each fight. * * * I tried to get into Tientsin to help the besieged. I failed. I lost a gun. If there is any fault it is mine. I only remark that it took 2,000 men with 6 guns to do what I failed to do with 530 without guns. I am awaiting re-enforcements asked for, and shall move forward as soon as the combined forces are ready.

"I append a list of the casualties to date, not including sick.

"I have also to ask that you urge the Department to thank the British surgeons for their care on the field and in hospital of our wounded; especially do I wish to recommend to the Department's notices the services of Robley H. J. Brown, R. N., H. M. S. Alacrity. So sure was his service and search of the field that we were enabled to get all rifles on the firing line with the sure knowledge that the dead and wounded would be attended to. We had no surgeon or medical supplies. The operations under Commander Craddock, R. N., were admirably planned and executed.

"List of casualties to date (inclosed with letter): Killed—Privates Lannigan, J. K. Miller, Morris, Provensal, and Hunter. Wounded—Corporals Francis, Kates and Hetrick, Sergeant Sullivan (slight),

Privates C. S. Smith, Cork, Sullivan, J. J. Bailey, Pennington, Carter; the most serious wound, the thigh being badly fractured.

"We need several carts for transportation. The report received from Pekin to-day is very bad, the runner having left there five days before. All the Europeans were in the English legation, ammunition very short, only three of the legations left standing Artillery fire would probably be directed against the British legation, in which case the suffering would be terrible.

"There seems small chance of any movement toward Pekin for three weeks."

In endorsing this report, Rear-Admiral Kempff writes, under date, U. S. S. Newark, July 4, 1900, as follows:

"Respectfully forwarded approved, with the request that the valuable and able service of Major L. W. T. Waller, U. S. M. C., commanding detachment, receive due attention and proper recognition at the hands of the Department.

"I would suggest a suitable medal for Major Waller and 5 per cent. additional pay for life in various grades he may reach. The other officers and men should receive medals and such other recognition as may be considered suitable by the Department, except to a few men not deserving, and who will be specially reported. Major Waller has been requested to submit a list of the names of these undeserving men.

"I was delighted when the Marines arrived in the Solace to find that Major Waller was in command, feeling certain that the men would be well cared for and render creditable service. It is with our Marines under Major Waller as with the force under Captain McCalla—foreign officers have only the highest praise for their splendid fighting qualities."

In a report to Rear-Admiral Kempff, dated at Tientsin, July 2, Major Waller announces the approach of a Chinese army of 10,000 men and four batteries, under command of General Ma San Yuen, which had started from Tsai Tsun, twenty-eight miles distant, on the previous afternoon. The allied forces, with 9,500 men, felt little fear from this Chinese horde, their only movement being a reconnoissance by the Russian cavalry. Regarding their plans Major Waller says:

"It was decided in council to-day for the Japanese, American, English and French to attack the west arsenal to-morrow morning and capture Tientsin City. If the reconnoissance develops the fact that the Chinese column advancing is not within striking distance of the town, the Russian and German forces will advance by the left bank of the river and capture the Chinese camp and fort on that side.

"Tientsin, Chinese, held by us with a garrison of 2,000 men. We apprehend no trouble from Chineses army."

In another report bearing the same date, Major Waller thus describes an heroic action of a party of Marines in the course of a reconnoissance of a Chinese fort:

"We made a reconnoissance in the direction of the fort near Tientsin City (Chinese). Forty of our men took part, supporting the English left. The firing was very heavy and sustained, but the enemy were driven out of the villages and the houses burned. Some of the British troops being in a hot corner and unable to retire, Lieutenant Butler volunteered to protect them, and moved to their position under a hot fire, permitting the British to retire. He then fell back, forming the rear guard of the column and protecting them thoroughly.

"Our men were thanked by Colonel Barr of the British service for their great assistance to-day. The villages destroyed have given us much trouble, forming cover for 'snipers.'"

In spite of the fact that, as he reports, the firing from the Chinese city continued "very bad," Major Waller was obliged to curb his impatience. While strongly advising an immediate attack on the city, before any further Chinese troops could arrive, he was outnumbered in the council of the allied commanders, who decided not to begin offensive operations until July 4. The principal opposition seems to have come from the Russian General, Stessel, who expressed his unwillingness to co-operate at present. At this council, he says, he presented the following information, "received from private and reliable sources": "There are four gates to the city of Tientsin, the roads therefrom intersecting the city and dividing it into four equal parts. The city has no troops in it at present. The suburbs on the west side are thick and occupied by 'Boxers,' the headman living on the north side just north of the viceroy's yamen. With him is General Nieh, whose troops we have so often defeated. This house is on the opposite side of the grand canal from the walled city. The guards in the city are composed of men employed by the Chinese merchants. They are in number about 5,000, but are not in favor of the 'Boxers.' The Chinese merchants are very anxious to save the town from destruction. I have proposed to them that when we attack they direct their guards not to fire on us, also to open the south gate, so we may gain entrance to the town and walls. In return we will promise them immunity from trouble. The Chinese forces there are about 3,500, located near the viceroy's yamen. There are several thousand 'Boxers,' all armed with rifles. The plan for attack is for the Anglo-Americans to form in two columns, consisting of troops and sailors. The

Americans with the troops pass on the south side of the mud wall and attack the arsenal; the sailors follow on the north side with field guns; these in turn followed by the French with six guns, small, and the Japanese forming three echelons on the north side of the wall.

"Reaching and carrying the arsenal by storm, the echelons make a partial change of front to the right and move on the city. The march will probably begin on July 4 at 2.30 A. M., so as to be in position by daybreak."

To this he adds:

"I believe the attack should be made at once, but the Russians will not move to-day.

"The Russians, with the Germans, are to advance by the left bank of the river, engaging the forts and capturing the yamen and barracks. As the Russians have all, or nearly all, the artillery, it is risky to move without them; still, if the Chinese get 10,000 more men in the town, it means so much more difficulty for us.

"There is hardly a chance for a move on Pekin for the next two or three weeks."

The delay in attacking the Chinese city was evidently due to the real lack of harmony between the various commanders, the unwillingness to yield to arguments looking toward immediate active co-operation being particularly noticeable in the Russion general, whose troops still formed the larger part of the allied forces, being also the most completely equipped for an effective campaign.

Owing to this delay in bringing about a concerted action, the Chinese force, under General Ma San Yuen, arrived at Tientsin on the evening of July 2, and entered the city without opposition. Early on the following morning they began shelling the allies, keeping up an incessant fire during the greater part of the day and ceasing only with the nightfall. Major Waller reports that their aim was accurate, several houses adjoining his quarters being repeatedly struck. As a result of this annoying bombardment, the Russian and Japanese artillery went into action, and, although their fire made a perceptible effect on the Chinese batteries, each lost four men killed, besides numerous lesser casualties. One of the Russian field guns was dismounted by a well-aimed shell.

Although, on this day, July 3, Major Waller reports that the assault on the city had been indefinitely postponed, the situation was considered sufficiently grave to warrant the order that all women and children be removed from the town occupied by the allies, and sent to the coast as soon as possible, along with his wounded. He also adds that the "river communication may be cut at any moment. All the other wounded capable of being moved will be sent down as soon as possible. The shell and bullet fire is almost incessant." To add to the complications of the situation, the relations between the powers, although outwardly friendly, were "not as cordial" as they had been; the report of disturbance in the German concession, Kiau Chou, threatened the withdrawal of the German forces, while the Russians still persisted in their refusal to join a concerted assault. Major Waller reports, however, that he was "in close touch" with the English and Japanese, and "very friendly with the Russians."

A probable explanation of the extraordinary attitude of the Russian general is that he considered himself entitled to recognition as the commander-in-chief of the allied forces, and, because this post was not formally accorded him, he was unwilling to co-operate with the other powers. His eulogistic order to the detachment that captured the East Arsenal on June 27, bearing his signature as "Chief of the Detachments," shows his readiness to attribute the highest glory to the Russian arms. A translation of this order is as follows:

ORDER TO THE DETACHMENT OF THE EXPEDITION.

CAMP NEAR TIENTSIN, June 15-18, 1900.

By the aid of God, and the bravery of my troops and those of other nations, viz., German, English, American, and Japanese, we yesterday succeeded in taking by storm a stronghold of Tientsin known by the name of the east arsenal.

Neither the open plain, covered by a frightful, hostile fire, nor the ditches filled with water and soft mud, nor the steep walls, were able to stop the advance of the brave storming party which only would be ordered to cease the pursuit of their task when the enemy fled in all directions. Hurrah!

To you, brave comrades, there is nothing impossible. The Lord, our protector, will show us the way to other victories and glory. On my part, as chief of the expedition, I wish to express to you my heartfelt thanks and my congratulations on the wreath of laurel with which you have decked your glorious colors anew.

CHIEF OF THE DETACHMENTS,
STESSEL, *Major-General.*

Between July 4 and 12 little was attempted in the line of engaging the enemy beyond posting pickets and batteries to neutralize hostile movements. Several skirmishes were fought in which the Marines figured conspicuously.

In his report of July 7 Major Waller writes that he had just dispatched a party of sharpshooters, composed of Marines and men of the Royal Welsh Fusileers, who should occupy the tower of the English College and pick off "snipers," or Chinese sharpshooters, and also to "watch for and locate any signals or flag flying from the French concession." This party was under the command of Lieutenant R. F. Wynne, U. S. M. C. After a day in the station their report was that the "snipers" had been driven off and that the Chinese had extended their lines well to the westward and were busy digging trenches in that direction. The evident plan in this movement was to surround the position of the allies and cut off river communication.

Meantime the forces of the powers were being steadily augmented; 300 Japanese and 200 British-Chinese troops arriving on the 7th, as well as several British 4.7, 9 and 12 pounders. Major Waller relates, however, that the 4.7 pound guns were not immediately mounted, owing to objections from the Russians that the chasm emplacements were too near their camp. At noon on the same day shelling of the Chinese city, the West Arsenal and the viceroy's yamen was begun, but, although the enemy's fire had been steadily maintained until that hour, it ceased almost immediately, to be resumed only on the cessation of the British fire in the evening. Then numerous casualties resulted among the British and other allied troops.

In his report of July 7 Major Waller also relates the following conference regarding the proposed advance on Pekin. He says:

"The aid to Admiral Seymour has come to me from the admiral, asking me to state what force of men it would take to make the march on Pekin, holding the line of communication. I replied to this, Not less than 40,000 seasoned troops, with not less than 25,000 in the attacking column. The Japanese general has stated that it would take 55,000 troops.

"I was then asked if I would be willing to serve under a commanding general, if one could be selected. I replied that for the purposes of the expedition I would gladly do so, as I recognized the absolute necessity for a common head for the military operations. I was then asked if I would serve under a prince of the blood if selected. To this I replied in the same manner, that as far as the military operations for the purposes of this expedition I would do so. I was asked if I would serve under a Japanese field marshal, and replied in the same manner. I was asked if I thought the other nations would agree to this, and replied that I did not think the Russians or French would so agree."

On July 9 (Monday) an assault was made on the Chinese right flank with a force of 2,000 men—American, English and Japanese. The plan was to advance beyond the Chinese forces entrenched to the westward; to curve in upon and drive them into the West Arsenal, where a general attack could be made. Major Waller's duty was to cut off retreat from the arsenal toward the city. Regarding his allotted part in the battle, he says, "My force and its engagement is really a more or less independent action on our side, and only joining in the general line when it has caught up and advanced beyond our left."

The result of this fight is well set forth in his own words, in his report of June 9:

"We drove the enemy steadily to the westward, and finally the flanking party, our Marines and some sailors (Japanese) entered the arsenal together. We were subjected to a heavy shell fire, but had no casualties, although the shells exploded overhead and among men for an hour.

"Chinese losses about 500—I think more. Japanese cavalry and artillery did excellent work. Our losses were about 45, of which 7 were killed.

"The arsenal was so badly wrecked that no garrison was left in charge.

"Three miles of the line to the westward entirely clear of Chinese. My prisoners say that Boxers lost so heavily last rain that they fell back to the westward that day. Last rain was on Friday.

"Report as to the capture of Nieh's family by Boxers is confirmed by prisoners. At the same time, they say the troops we fought to-day were his (Nieh's).

"Ma's troops on the other side of the river confronting Russians.

"I think it more than probable that we will make a general attack on the city to-morrow.

"The firing and fire discipline of our men was expert to-day. Fifty of them controlled and overpowered the fire of the enemy while our artillery was passing over a bridge I was thanked by three different nations for the work of our men. I am glad to say that they are well and fit as fiddles for any work.

"We still need artillery and cavalry very much.

"Some political questions may arise in a day or two and I shall report them immediately. At the same time I am being governed entirely by the Secretary's instructions.

"Chinese shelling from the east did great damage. I have secured quarters for the Ninth Infantry entirely out of line of fire. Everything else under fire. My barracks struck again to-day."

In his report of June 10 Major Waller speaks of the excellent behavior of the Marines under his command, in the following terms:

"I take the liberty of sending you the copies of two letters received by me concerning the part taken by my men in yesterday's fight. I hope, sir, that these men may be mentioned to our Government. They have done magnificent work since the 20th of June. To-day there is not a man on the sick list except the wounded. Yesterday their fine discipline was the admiration of all nations, and their admirable fire directed against the enemy kept down the Chinese riflemen so that all the trains, artillery and baggage were permitted to pass over an exposed bridge without one casualty.

"If the shelling from the Chinese guns continues as bad as yesterday I shall be obliged to move my men. My barracks have been hit three times, and the compounds on either side are plowed up with shells. The quarters I have reserved for the Ninth Infantry are out of the line of fire.

"We expect to have the 4-inch guns from the Terrible in position to-day. These guns use the lyddite shells."

As a result of the successful movement of June 9, another assault, on the east and north of the city, was planned for the 11th, in which Major Waller was to have commanded the Marines and Royal Welsh Fusileers, occupying the head of the column and the right of the firing line. He adds also:

"I have arranged to relieve the Centurion [British] men and Marines in charge of the railroad station to-morrow (10th). This place is a very hot corner, and I shall use only the men I now have, as they are thoroughly accustomed to the shelling and 'sniping.'"

The contemplated attack of the 11th did not take place on account of the destruction of the bridges, and, by the arrival of Col. Meade on the 12th, Major Waller was relieved of command.

Col. Meade had been detached from the Marine station at Cavite, P. I., on June 30, and, with a force of 300 men and 18 officers, embarked for Taku on the U. S. S. Brooklys, Rear-Admiral Remey sailing at the same time, to assume command of the naval forces and operations in China.

The officers of this detachment were: Col. R. L. Meade, Major George Richards, Capts. M. J. Shaw, W. B. Lemly, A. R. Davis, C. G. Long, B. H. Fuller, P. M. Bannon, First Lieuts. J. H. A. Day, C. G. Andresen, R. H. Dunlap, A. J. Matthews, D. D. Porter, W. H. Clifford, J. F. McGill, William Hopkins, Second Lieuts. F. M. Wise, Jr., W. McCreary, L. M. Little, United States Marine Corps.

Arriving at Tientsin about midnight on July 11, Col. Meade was just in time to lead the Marine battalion in the long-delayed assault on the walled city of Tientsin, which at a conference of the Powers, held on the 12th, was determined for the following day.

While it is evident that none of the allied commanders expected the battle to be as long and sharply contested as it actually proved, the result was a complete victory, the city being occupied and for some time administered by the Powers. The men of the Marine regiment bore the full brunt of the fight on the firing line, doing their full share in routing the enemy before their position, and being the first to enter the city by the south gate. The Americans, English and Japanese co-operated, under the English commander, Brig.-Gen. A. R. F. Dorward, but the Russians and French acted more or less independently, each national force being controlled only by the general lines of agreement decided on at the conference of the previous day.

The details of this historic battle are well set forth in Colonel Meade's report, as follows:

HEADQUARTERS UNITED STATES FORCES,
TIENTSIN, China, July 16, 1900.

SIR: I left Tongku on July 11 at about 8.15 A. M. and arrived at Tientsin after midnight. I found Major Waller and his force quartered in the European concession in houses which were nightly under shell fire of the enemy. Small bodies of the enemy also controlled the streets with rifle fire at night, this fire generally beginning about 10 o'clock and lasting until about daylight.

Only the foreign concessions were held by the allied forces when I arrived, the French concession being nearest to the walled city, the English next, and the German lowest down the river. The walled city, strongly fortified, and all other portions of Tientsin, also strongly fortified, were held by the Chinese imperial troops and the "Boxers."

Our force took turns with the other troops of the alliance in guarding the railway station, which was an exposed place, almost continually under shell fire and a very dangerous duty.

On the 12th instant, at a conference held at the English general's headquarters (Brig. Gen. A. R. F. Dorward), it was decided to attack the city at about daybreak the next day (13th), and I was called upon to furnish a quota of 1,000 men, our marines (22 officers and 326 men) and a force (673 men) from the Ninth United States Infantry, who had preceded us to Tientsin by a few hours. One battalion of the Ninth Infantry was still at Taku or en route.

At 3 A. M. I marched out of barracks with a force of 22 officers and 326 men, in four companies, Companies A, D, C, and F, commanded respectively by Lieut. S. D. Butler, Capt. C. G. Long, Capt. A. R. Davis, and Capt. B. H. Fuller. Company F was an artillery company of three 3-inch rapid-fire guns and three Colt's automatic guns, and this company was supported by Company D (Captain Long, who was also the commanding officer of the Second Battalion of the temporary organization I have with me in China).

We marched through the Taku gate of the walled city in two columns, the Japanese forces being to the right and the English and American forces on the left. The column in which the Americans were was distributed as follows: Two companies of the Royal Welsh Fusileers leading, followed by the marines (infantry and artillery), the English naval brigade, and finally the Ninth United States Infantry (673 men). The road was very heavy for artillery such as we had, and I do not advise the naval gun to be used as a field piece until some device is gotten up as a limber, because the trail wheel plows into the ground, and the dikes and ditches, which were frequent, necessitated all the united force of the two companies to get the guns across, costing much strength which should have been reserved for the fatigue of the battlefield.

Our verbal orders (we had no written ones) were to march on a line parallel to the city wall, about 1,000 yards in rear and to the southward of the bridge at the south gate, and there the commanding officers were to receive their final instructions. No such meeting, however, was held, and my orders for the marines were to advance along the mud wall in a northerly direction with two infantry companies, leaving the artillery company and its infantry support to act in connection with the British field artillery and to open fire at a point where the Chinese had some 4.7-inch (or 6-inch) guns mounted, which had been particularly obnoxious. We arrived at the south gate at 5 o'clock A. M.

The naval battery of the Terrible, under command of Captain Bayly, of the royal navy, had opened fire on the forts and guns of the enemy just before our arrival, and they were responded to by the enemy vigorously. This battery was so accurate in its fire that every shell landed in the place intended for it, and at about 5.45 o'clock A. M. the Chinese magazine was exploded with a shock which was almost like an earthquake shock, and was distinctly felt by all of us, who were standing fully one mile and a half from the point of explosion.

At about 6.30 A. M. I received orders from the British general to support the Royal Welsh Fusileers in an attack on the extreme left, and we crossed the wall in skirmish line, having an extensive swamp to cross. The country was a flat, level one, with grave mounds and dikes and ditches in great numbers; and these already dug trenches were a very considerable help to us, as in such an open, fire-swept plain we would have had difficulty in advancing, and would have been compelled, with only the bayonet, to throw up hasty intrenchments. The fire of the Chinese, both in artillery and infantry, was fearfully accurate, as the casualty list will evidence; and I thank God for the mounds and dikes.

We advanced by rushes to a line of trenches about 800 yards from the enemy. We found that in our front there were very bad swamps and a stream of water, which would render it impossible for us to have reached the city at that point; but I believe it was not intended that we should advance farther, as the Royal Welsh Fusileers were then in the same skirmish line with us. We reached the advanced position about 8 a. m. I took 180 rounds per man with me—100 rounds in the belts and 80 in the haversacks. This is not sufficient for an all-day fight, and as it grew toward night I began to be apprehensive of being left in an advanced position in a fight where no prisoners were taken on either side with only the bayonet to fight with.

On the firing line the action was especially hot and the enemy's fire especially rapid and accurate, and about 8.30 a. m. the enemy appeared in large numbers upon our left and among the grave mounds of the field in which we were, with the evident intention of flanking us. I made a turning movement to the left and rear, and we drove them away. Later in the day, about 2 p. m., they again made a flanking effort, but at this time the infantry support of the artillery company was on the mud wall of the city and aided us by a cross fire. The company was commanded by Capt. C. G. Long. The effort of the enemy proved a failure, and we drove them in.

We remained in the trenches until about 8 p. m., when we received an order from the brigadier general commanding to withdraw, which was probably the most difficult action of the day, since the enemy had so well covered our position that their shots struck the crests of the trenches and threw dirt in our faces, many being hit. I ordered the withdrawal in small parties of 8 or 10 men, to rush from mound to mound or trench to trench. I had previously sent the wounded to the rear under particularly unfortunate circumstances. I had also to send one dead officer to the rear.

The withdrawal was successful, only one man being hit, and we were in safety under the mud wall near the south gate.

General Dorward ordered that the troops should sleep upon their arms that night and on the following morning to enter the city, the south gate to be blown in by gun cotton.

The troops had had nothing whatever to eat on the 13th save the small luncheon (if it may be so called) which each man carried in his haversack. It was not expected when we started that the action would prove so long, but General Dorward, knowing the situation, kindly sent to the reservation for food and other necessaries, and the bivouac proved a success, and the men, although very fatigued, were ready for duty.

On the 14th instant, the south gate having been blown in, we moved into the walled city at about 6 o'clock A. M.

We found the city filled with dead Chinamen and animals. No resistance was made to our occupation in the walled city ifself, but an infantry fire was kept up by the Japanese infantry upon the enemy, who responded from the suburbs. Since then we have had undisturbed possession of all Tientsin.

During the day of the 13th instant my force of marines stationed at the railway station were vigorously attacked and suffered heavily. I respectfully append the report of the commanding officer of the detachment.

The conduct of my officers and men I cannot praise too highly. I had them for the most part under my personal eye. I desire especially to call your attention to the conduct of First Lieut. Charles G. Andresen, whose fearless conduct excited the admiration of all; First Lieut. S. D. Butler, who, at the risk of his life, went out of the trench to bring in a wounded man and was shot while doing so; First Lieut. Henry Leonard, my adjutant, who brought First Lieut. Butler in in safety and was dangerously wounded. All conducted themselves well, and I cannot commend them too highly. I append a list of the names of the officers who were engaged in the battle.

Since the armed force of the Cinese have been driven away one conference of the representatives of the eight powers has been held with reference to city government, and this conference is to meet again to-morrow to elect a president.

* * * * * * * *

I had almost forgotten to state that the Russians, in force, attacked the north side of the city while we engaged on the south. One part of my force was at the railway station during the battle and were driven back by the shell fire. They did not retire far, however, and they guarded what they were sent to guard.

I regrt to report the death of Capt. A. R. Davis, who was killed at my side in the advanced trench. He was killed almost instantly. I had his body brought in with the wounded, and he is buried here in Tientsin, his grave being marked. This was all I could do. Col. E. H. Liscum, commanding the Ninth United States Infantry, was killed in the action, being twice wounded, once through the lungs and again through the foot. He is buried here in Tientsin.

It being impossible to bring in all the dead, they were buried in the trenches where they fell.

All the forces engaged spent the night of the 13th on the ground near the south gate, provisions and water having been sent for by the British brigadier general commanding.

It has been utterly impossible for me to make a report before this. Every moment since my return from the front has been fully occupied by important matters. I desire to call attention to the work of Capt. M. J. Shaw, acting commissary and quartermaster, after Captain Lemly was wounded. His untiring activity in keeping the command supplied with ammuniton, food, water, and all other necessaries merits commendation from me.

Major George Richards, assistant paymaster, and Capt. W. B. Lemly, assistant quartermaster, on the regimental staff, volunteered to act as my aides, and they accompanied me during the day of the battle (13th). Captain Lemly was wounded in the leg very early in the action and before we had reached the arsenal.

The forces engaged are estimated to be about 5,650 of the allied forces. The Chinese had about 60 guns, and their forces are variously estimated—nothing being correct—but there was a large army of imperial troops and Boxers.

Very respectfully, ROBT. L. MEADE,
*Colonel, U. S. M. C., Commanding United States Forces in Tientsin, China.*

The BRIGADIER GENERAL, COMMANDANT,
*United States Marine Corps, Washington, D. C.*

The following supplementary extracts from reports of the various company commanders furnish additional details regarding the services and behavior of the Marines.

Capt. P. M. Bannon, U. S. M. C., commanding Company N, writes:

"At 7 P. M. on July 12 I relieved First Lieutenant Butler at the railroad outpost. The detachment consisted of First Lieutenant McGill and 50 enlisted men. We remained on this duty until 12:30 A. M., July 14, being relieved by the Ninth Infantry, U. S. A.

"The American position was on the right of the French, and extending to the engine house. On the right of the engine house were the English troops.

"The engine house was not occupied, though it was properly prepared for defense against infantry attacks. Traverses made with cross-ties were also built on left flank.

"The lines were subjected all night to a very heavy infantry fire from front and left. The enemy opened fire with their artillery about 4:30 A. M., July 13. At 5 A. M. the artillery fire became so severe that all the men were ordered out of the trenches, except an observation point. Good shelter was obtained by covering the ash pits of the engine house with cross-ties. The enemy having advanced and occupied a burnt village about 600 yards in our front, it became necessary to again occupy the trenches. This was done promptly under a severe fire. After driving back the enemy, the men were again ordered under shelter. At this point 50 men of the Ninth United States Infantry, under command of Lieutenant Brown, arrived, having been ordered out as re-enforcement. The fire continued severe until about 11 A. M., when it became less intense. At 5 P. M. the artillery fire ceased, and we had only the infantry fire to contend with.

"This was the state of affairs when I was relieved, at 12:30 A. M., July 14."

To this report Capt. Bannon adds, under date June 24:

"I respectfully invite the attention of the commanding officer to the action of that part of the company (N) compelled to remain behind as a barrack guard. While the fighting was going on, July 13, these men made trip after trip to our trenches with water and ammunition. Considering the distance and the severe fire to which they were subjected, I consider them to be deserving of the highest praise."

Reporting on the conduct of the men of the Second Battalion of Marines, under his command, Capt. Charles G. Long writes:

"The Second Battalion, under my command, consisting of one battery of artillery, three rapid-fire guns, and three Colt's automatic; and one company, Company D; the former commanded by Capt. B. H. Fuller and the latter by First Lieut. R. H. Dunlap. The other officers of the battalion were First Lieut. A. J. Matthews, adjutant; Lieutenants Clifford, Jolly, Little, and McCreary, U. S. M. C., and Assistant Surgeon Thompson, U. S. N. We started from our barracks at about 3 A. M. with the First Battalion, all under the command of Col. R. L. Meade, U. S. M. C.

"When about one and one-half miles from the western arsenal, I was ordered to go into action on the right of the Japanese artillery, who had one battery of light field guns. The First Battalion moved to the front and the Second Battalion inclined to the right and moved across the marsh to a position just to the right of the Japanese and outside the outer wall. This was about 2,200 yards from the south gate of the Chinese city, the point to be forced.

"The infantry helped to place the guns in position, and about 75 or 100 shells were fired over the outer wall and into the city; Lieutenant Matthews indicating from the top of the wall the fall of the projectiles. Seeing that a better view of the south gate could be obtained from inside the mud wall, the battalion was moved by the left flank through the arsenal gate and took a position inside and to the left. This was at about 6:30 A. M. From this position the fire was continued, and some very good shots were placed in the Chinese city. One of the Chinese batteries on our left located us and tried to drive us from our position. One 3-inch gun was turned on this battery, also the Colt's guns, but the range was too great for the latter to have any effect.

"Our ammunition supply was small, as we had to move it over ditches by hand and keep pace with infantry. After exhausting our ammunition the guns were moved outside the wall and left in charge of Captain Fuller and about 60 men. Lieutenant Porter and the Colt detachment joined Company D, commanded by Lieutenant Dunlap, the Colt guns being left with Captain Fuller.

"We then moved with this force of about 100 men to the extreme left flank of the allied forces, which was noticed to be open and liable to a flank attack by the Chinese, which would enfilade on men already on the line about 600 yards from the inner wall. Arriving there we opened fire on the enemy who appeared to harass the flank. Shortly afterwards, having received a signal to move forward, we went over the wall and advanced in extended order a distance of about half a mile across the marshes, coming up on the left of the First Battalion.

"During the advance the fire against us was continuous and quite heavy, coming from the front and left flank, but we advanced without halting, the last 200 yards in double time. One man, Private Kellerher, was hit in the shoulder just as we arrived on the line, and was sent to the rear later. Just before moving to reinforce the line, Lieutenant Wynne and about 40 men joined my command. This was about 8:45 A. M. About 9:15 A. M. the enemy tried to flank us, so I moved my command about 50 yards to the rear and 100 yards to the left, thus protecting the flank, and also sent Lieutenants Wynne and Jolly with 40 or 50 men well to the left.

"The fire was received and returned all day, and several attempts at a flank attack frustrated. My command remained in position until about 7 P. M., when, after the other companies had moved to the rear for a night position, we followed them. Lieutenant Wynne's detachment from the outer wall covered the movement to the rear.

"The officers and men during the engagement, which lasted thirteen or fourteen hours, displayed coolness and in every way the qualities of good soldiers and Marines. The artillery, under Captain Fuller and his officers, was handled well, and the fire was effective in spite of the poor ammunition. I inclose a report from Captain Fuller as to the further movements of part of his company. Lieutenant Dunlap, in command of Company D, kept his company under excellent control, and by well-directed volleys and individual fire well protected the extreme left flank of the allied forces. Sergeant Kolock was shot and instanty killed shortly after arriving on the line. He was buried on the field.

"After moving to the rear we took up a night position, and early next morning, the south gate having been forced, the companies moved into the Chinese city of Tientsin, returning to the barracks in town about 1:30 P. M."

The gallant behavior of the Marines, acting with the men of the 9th U. S. Infantry, called forth the following generous and highly commendatory letter from Gen. Dorward, Commander of the British forces:

*From the general commanding British forces, North China, to the officer commanding United States forces.*

TIENTSIN, CHINA, July 15, 1900.

SIR: I desire to express the high appreciation of the British troops of the honor done them in serving alongside their comrades of the American Army during the long and hard fighting of the 13th instant, and the subsequent capture of Tientsin City, and of my own appreciation of the high honor accorded to me by having them under my command.

The American troops formed part of the front line of the British attack and so had more than their share of the fighting that took place. The ready and willing spirit of the officers and men will always make their command easy and pleasant, and when one adds to that the steady gallantry and power of holding on to exposed positions, which they displayed on the 13th instant, the result is soldiers of the highest class.

We all deeply sympathize with you in the heavy losses you have suffered, especially with the Ninth Regiment in the loss of their gallant colonel, E. H. Liscum, while at the head of his men, and with the First Regiment of Marines in the death of Captain Davis, who met a soldier's death in the very front of the fight.

I blame myself for the mistake made in the taking up of their position by the Ninth Regiment, not remembering that troops wholly fresh to the scene of action and hurried forward in the excitement of attack were likely to lose their way. Still the position they took up and gallantly stuck to all day undoubtedly prevented a large body of the enemy from turning the right of the attacking line and inflicting serious loss on the French and Japanese.

Among many instances of personal bravery in action I propose especially to bring to notice in dispatches the conduct of First. Lieut. Smedley D. Butler, United States Marine Corps, in bringing in a wounded man from the front under heavy and accurate fire. Leutenant Butler was wounded while so doing, but I am glad to learn not seriously. The regimental adjutant, First Lieut. Henry Leonard, as Lieutenant Butler was suffering severely, volunteered to carry him out of the firing line. This gallant feat he successfully accomplished, but I regret to say was very dangerously wounded in so doing.

The Ninth Regiment were firing somewhat outside my sphere of action, so I am to bring forward only one instance of personal gallantry in that regiment, although circumstanced as they were, fighting for about twelve hours almost alone and unsupported, and never giving back a foot of ground until directed to retire under cover of night and fire of the naval guns, such instances must have been very numerous. The one I would refer to was the bringing back to me by the acting regimental adjutant, Captain Lawton, of the account of the position of the regiment across a wide and fire-swept space, and returning with reenforcements to guide them to his regiment, when he was severly wounded.

The withdrawal of the regiment was a delicate military operation finely carried out, on which I congratulate Lieutenant-Colonel Coolidge and the officers and men under his command.

I have the honor to be, sir, your obedient servant,

A. R. F. DORWARD, *Brigadier-General.*

The list of officers of the Navy and Marine Corps engaged in the battle of Tientsin, July 13, 1900, is as follows: Col. R. L. Meade; Majors George Richards, L. W. T. Waller; Capts. W. B. Lemly, A. R. Davis, C. G. Long, B. H. Fuller, P. M. Bannon; First Lieuts. S. D. Butler, H. Leonard, G. C. Reid, W. G. Powell, J. H. A. Day, R. F. Wynne, R. H. Dunlap, D. D. Porter, J. F. McGill, C. G. Andresen, A. J. Matthews, W. H. Clifford, A. E. Harding; Second Lieuts. F. M. Wise, W. McCreary, L. McC. Little, W. L. Jolly; Surg. C. D. Norton, U. S. N.; Asst. Surg. Edgar Thompson, U. S. N.

In addition to the high commendation expressed for the services of a number of his officers and men, Colonel Meade writes, as follows, of the conduct of Major Waller, whom he had superseded in command, just previous to the memorable conflict of July 13:

HEADQUARTERS UNITED STATES FORCES,
TIENTSIN, CHINA, July 18, 1900.

SIR: 1. I desire to make a separate report of the services of Maj. L. W. T. Waller, U. S. M. C., of my command, his conduct being of such meritorious character as demanding it from me.

2. Major Waller left Cavite with 6 officers and only 100 men. He reached Taku at about the same time First Lieut. Henry Leonard did, who left Cavite before Major Waller, and the 2 officers and 32 men of First Lieutenant Leonard were added to Major Waller's command.

3. A full report of his operations while on shore in China before my arrival has already been sent to you by him, and I know of this work only from the high praise everywhere bestowed upon him by our own people and the British forces, by whom, and especially by Vice-Admiral Seymour and Gen. A. R. F. Dorward, he is held in great esteem.

4. Since my arrival here Major Waller has shown untiring zeal. His conduct in the battle of Tientsin on the 13th instant and in the occupation of the city on the 14th was that of a fine soldier, and since then he joined in an expedition outside of the city in which 16 guns and an immense amount of rifles and ammunition were seized. These guns did not have to be fought for, but the possession of them aids our cause materially.

5. I respectfully recommend Maj. L. W. T. Waller for promotion to the next higher grade. Very respectfully,

ROBT. L. MEADE, *Colonel, U. S. M. C., Commanding U. S. Forces.*

The BRIGADIER-GENERAL, COMMANDANT,

*United States Marine Corps, Washington, D. C.*

In endorsing this well-deserved letter of praise, Rear-Admiral Remey writes:

UNITED STATES NAVAL FORCE ON ASIATIC STATION,
FLAGSHIP BROOKLYN, OFF TAKU, CHINA, July 20, 1900.

Approved and respectfully forwarded. Nothing but praise is heard of the conduct of Major Waller during the recent disturbances and engagements about Taku and Tientsin. All reports indicate that he is an officer of high merit, reflecting much credit on the Marine Corps and the whole United States service, and I would be glad to see him rewarded.

GEO. C. REMEY, *Rear-Admiral, U. S. N., Commander in Chief.*

As given by Colonel Meade, the foreign forces before Tientsin numbered, all told, about 5,650 men, of which the Russians had 2,300; the Japanese, 1,600; the Americans, 900; the French, 600, and the Germans, 250, at the time of the battle. The losses among these forces on July 13 were American, 24 killed, 98 wounded, 1 missing; English, 17 killed, 87 wounded; Japanese, 320 killed and wounded; French, 13 killed, 50 wounded; Russians and Germans, 140 killed and wounded, making a grand total of 750.

The following report of Major Waller is particularly interesting, as giving a number of details and letters relating to the Marine regiment, not otherwise included:

HEADQUARTERS FIRST REGIMENT, U. S. MARINES,
TIENTSIN, CHINA, July 30, 1900.

SIR: 1. It has been suggested by Colonel Meade that I make a report to you covering the operations of our men since my last report, and up to the capture of Tientsin city, including an expedition under my command on July 16. On July 3 a small party of our men, subsequently increased to 80 in all, went under the immediate command of Lieutenant Smedley D. Butler, joining a battalion under the command of Colonel Bower, of the British service, to capture a gun that had been annoying us considerably. The gun had been moved to the opposite side of the river and that part of the expedition was unsuccessful. Two villages were captured after sharp hand-to-hand fighting. When ordered to retire some of the Wei hei wei regiment (British Chinese) were unable to fall back because the fire was too hot. Lieutenant Butler asked and received permission to support these men. He deployed the Marines, advanced at the double to the position, opened fire by volleys, and permitted the Chinese regiment to fall back in comfort. He then fell back by sections in fine order and without casualty. I have before mentioned the fine qualities of Mr. Butler, in control of men, courage, and excellent example in his own person of all the qualities most admirable in a soldier. Colonel Bower sent his thanks for the services of our men and expressed his admiration for their training and fine discipline.

2. We were shelled each day until the 9th, the Chinese fire being very accurate; our barracks were struck three times and the adjoining houses five or six times each. The British officers who had been in the siege of Ladysmith

state that the shelling here was far more severe. On the 9th an attack was made, in two columns, on the west arsenal and along our west flank, the Chinese having spread around our flank on that side and threatened the river communication. I had command of the right column, consisting of our men and Japanese sailors, joined later by Sublieutenant Kennedy, of the royal navy, with a Maxim gun. We were shelled heavily by the Chinese, but fortunately without casualty, although frequently covered with dust and stung by stones and gravel thrown up by bursting shell. The left column hotly engaged the enemy and had them on the run in about thirty minutes. Some of the retreating enemy fled to the westward, while others retreated to the arsenal, too far out of range for me to reach them. The Chinese guns mounted in a mud fort to the westward of the arsenal were silenced by the Japanese and British mountain batteries and my column rushed into the arsenal, the action up to this point lasting about three hours. The Japanese cavalry had charged the retreating Chinese, killing many and capturing a field battery, four guns, and many standards. Proceeding to the first part of the arsenal, about 1,500 yards from the walls of Tientsin, we were met by a heavy rifle fire. My men were deployed and placed lying on the roofs of the huts, while the Japanese deployed on our right. We cleared the plane between the arsenal and city of the enemy, but the fire from the wall was very heavy. Fortunately we had no casualties. The arsenal was destroyed and our troops ordered to withdraw to the European city. Observing that the bridge crossing the canal from the arsenal to the main road would be under a heavy fire from the city walls and the outside villages, I took my men across and deployed them along the mud wall, located the enemy, and when the wounded and artillery came to the bridge opened fire by section volleys. This fire kept down the enemy, and all the wounded, the Japanese and British mountain batteries, the British, Indian, and Japanese infantry passed without casualty. The Russian infantry relieved me to protect the passage of the troops and, I learn, had three or four casualties. In the afternoon of the same day I received, through the Japanese chief of staff, a present from Major-General Fukishima of one of the field guns captured by his cavalry. The present was accompanied by the following note:

"Consulate du Japon,
"Tientsin, le 10 juillet 1900.

"Mon Cher Major: Je fais envoyer á votre quartier général un des quatre canons que l'on a pris hier sur l'ennemi. C'est un souvenir du succés que nos forces réunies ont remporté dans cette même journée.

"Agréez mon cher Major, l'expression de mes sentiments très distingués,
"Y. Fukishima."

The same evening I received a letter from Vice-Admiral Seymour, commander in chief of the British forces, and another from Lieutenant-Colonel Bower, commanding the Hongkong regiment. Copies follow:

"Tientsin, July 9, 1900.

"Sir: I beg leave to thank you for having been good enough to send out a force to co-operate in the action of to-day. The steadiness of your men and the way in which they entered the arsenal was much admired. The actual command of the expedition was, as you know, under the Japanese general, with our general, Brigadier-General Dorward, assisting; but as being myself present,

I desire to thank you for your valuable assistance, and to assure you that we are always happy to have your officers and men associated with ours.

"I have the honor to be, sir, your obedient servant,

"C. H. SEYMOUR, *Vice-Admiral.*"

"TIENTSIN, July 10, 1900.

"DEAR MAJOR WALLER: The officers who were with the detachment of my regiment that were dragging the guns yesterday have asked me to write and thank you for the support you gave them yesterday by keeping down the enemy's fire while they were crossing an exposed place. Their opinion is that had it not been for the action taken by you and your men they would have had a good many casualties. Please accept my most sincere thanks and believe me,

"Yours sincerely, "P. BOWER."

I sent the following letters in acknowledgment:

"TIENTSIN, CHINA, July 10, 1900.

"SIR: I have the honor to acknowledge the receipt of your kind and flattering letter of yesterday's date relative to the services rendered by my officers and men. While I think, sir, that you overestimated these services, the opinion of one in such a high position as Vice-Admiral Seymor is most grateful to us.

"I hope, sir, that we may be able to retain the good opinion you have so gracefully expressed.

"I have the honor to remain, sir, your obedient servant,

"LITTLETON W. T. WALLER, *Major, U. S. Marine Corps.*

"THE COMMANDER IN CHIEF, BRITISH FORCES,

*"Tientsin, China."*

"HEADQUARTERS UNITED STATES FORCES,

"TIENTSIN, July 10, 1900.

"SIR: On the part of my Government, myself, officers, and men, permit me to express to you our grateful thanks for the token of appreciation of the work done during the present expedition by the men under my command.

"I assure you, sir, that nothing could be more appreciated by us than the gun you have so gracefully presented.

"If you will permit me to say it, I should be glad to express my humble opinion of the troops under your command. I do not believe that there exists to-day a better body of men, as to discipline, training, courage, and ability to adapt themselves to surrounding conditions.

"Hoping, sir, to have the honor to serve under your command and direction, and with the assurance that for the purposes of this expedition you may call on me, my officers, and men at any time,

"I have the honor to remain, your obedient servant,

"LITTLETON W. T. WALLER,

*"Major, U. S. Marine Corps, Commanding United States Forces."*

"TIENTSIN, CHINA, July 10, 1900.

"DEAR COLONEL BOWER: I have to acknowledge the receipt of your kind note relative to the attack on the Chinese right on yesterday and to the assistance rendered by us to your regiment at the bridge.

"We did what I am sure you would have done under the circumstances. I hope you will feel that we feel toward you as comrades in arms, always ready to do any service at any time.

"Please express the thanks of my officers and men to your officers for their good opinion. It is very grateful to us.

"Sincerely yours, "LITTLETON W. T. WALLER.'

At the same time I sent the following letter to Vice-Admiral Seymour, inviting his attention to the admirable work done by Sublieutenant Kennedy with his Maxim:

"TIENTSIN, CHINA, July 10, 1900.

"SIR: Please permit me to express my thanks through you to Sublieutenant Kennedy, R. N., for the valuable assistance rendered by him and his men in yesterday's attack on the west arsenal.

"Lieutenant Kennedy volunteered and courteously placed himself under my orders, performing excellent work with his Maxim against the retreating enemy.

"I have the honor to remain, sir, your obedient servant,

"LITTLETON W. T. WALLER, *Major, U. S. Marines.*

"THE COMMANDER IN CHIEF, BRITISH FORCES, *Tientsin.*"

The shelling of the European city continued until and during the 13th. On that day we attacked the Walled City in force. The report of our part of it is, I believe, covered by the report of Colonel Meade. I wish particularly to mention, however, the conduct of Private S. D. Sugar, my orderly. On two occasions I found it necessary to send orders to Captain Long, who was about a thousand yards in our right rear protecting our flank and rear from a considerable body of the enemy. I sent Private Sugar, and each time he walked quickly over the field that was swept by bullets, delivering his message and returning promptly to the firing line. I think promotion to sergeant is not a great reward for his conduct.

On the morning of the 15th, at request of Brigadier-General Dorward, commanding the British forces, I took command of a detachment of mounted infantry composed of our own men and men of the Royal Welsh Fusileers. There were 25 men in all and 4 officers—Captain Gwynne and Lieutenant Flower, of the Fusileers, and Lieutenant A. E. Harding, my adjutant. Our orders were to scout to the west of the Walled City and look for the enemy, supposed to be in a fort near there. We made the fort and rushed it without opposition, securing eight guns, many standards, arms, carts, ammunition, etc.

I cannot speak too highly of the conduct of the officers of the Fusileers. This battalion has been by our side since June 23. They have responded to my orders with the greatest alacrity and willingness, all the officers and men ready to go anywhere. Captain Gwynne and Lieutenant Flower have been thrown more with me than the others; the first, because he commanded the battalion, with the latter as his adjutant. I have thanked General Dorward officially for their services and invited his attention to their courage and fidelity. I hope they may receive the promotion they so richly deserve.

On the 15th I took a part of my battalion to guard the American quarter of the Walled City, the place having been divided into four quarters—English, Japanese, American, and French. I found in the yamen of the salt commissioner, then under our guard and on fire, a couple of vaults of silver, amounting in all to about $800,000 silver. The estimated value is, I believe, $376,000. I think that a little low. After several days' hard work this was secured and is now in the Hongkong-Shanghai Bank under our guard, awaiting the disposition of the Washington Government. This, I believe, closes the military operations to date. I am under orders to move forward and have been ready for two days. General Chaffee arrived to-day and assumed command of the forces in China. We are therefore under his orders on shore.

Major Biddle has arrived at Tonku and reports that he expects to arrive here to-morrow. While I am glad to see Major Biddle I am sorry to lose command of the men. I have led them always until now. I suppose I must bow to the fortunes of war and a complication unforseen by anyone.

Very respectfully, L. W. T. WALLER,
*Major, U. S. Marine Corps, Commanding.*

THE BRIGADIER-GENERAL, COMMANDANT, U. S. MARINE CORPS,
*Washington, D. C.*

After the capture of the city of Tientsin, the allied commanders held a conference relative to forming a provisional government, which, according to their decision, should consist of one Russian, one Japanese and one Englishman, representing the three powers having the largest forces at hand. This body was called the Provisional Government Council, and it continued to administer affairs for a considerable period, until the affairs of the Chinese Empire were settled to the satisfaction of the Powers.

Pending the completion of arrangements for the provisional government, the following proclamation was issued by the commanders of the allied forces:

TO THE INHABITANTS OF THE CITY OF TIENTSIN:

In bombarding the city of Tientsin the allied forces only replied to the attack made by the rebels on the foreign settlements.

At present, as your authorities, forgetting their duties, have deserted their posts, the allied forces consider it their duty to establish in the city a temporary administration, which you all have to obey. This administration will protect everyone wishing to deal in a friendly manner with foreigners, but will punish without mercy everyone who causes trouble.

Let the bad people tremble, but the good people should feel reassured and quietly return to their houses and begin their usual work. Thus peace will be restored.

Respect this.

Tientsin, the 16th July, 1900.

Approved by:

Allemange: Von Usedom, Capitaine de Navire.
Autriche Hongrie: J. Tudrak, Lieutenant de Vaisseau.
Etats Unis d'Amérique: Colonel Meade, American Marines.
France: De Pelacol, Colonel.
Grande-Bretagne: Le Général Dorward, Captain Bayly.
Italie: G. Sirianni, Lieutenant de Vaisseau.
Japon: Le Général Fukushima.
Russie: Vice-Admiral Alexeieff. STESSEL, *General Major.*

As is well known, the capture of Tientsin was followed by a wholesale looting of public and private buildings, in which the soldiers of all nations were implicated. In mentioning the fact, Col. Meade remarks, "It is impossible to prevent it, unless all unite in the suppression. This cannot be effected." The operations in China furnish an eminently good object lesson on the possibilities of alien nations in co-operating for any cause worthy the name "good."

In the course of this career of pillage, a building known as the "treasury," or the "salt commissioner's yamen," was looted and burned by soldiers of the Powers, who secured considerable booty in the shape of bullion and other valuables. On the arrival of the Marines at this spot, a careful examination convinced the American officers that some of the treasure vaults had not been reached: accordingly a systematic examination of the ruins was begun, under the direction of Major Waller. The result was the discovery of a large amount of silver bullion, much of which hade been melted and fused with bricks and mortar by the burning of the building, and had to be taken out with picks.

Reporting on this find, under date July 20, Col. Meade writes:

"Have already treasure to the value of about $300,000 in gold, and about twice or three times as much more to bring into town.

"We have not yet found the gold vaults of the treasury, but are on the track of them. They may or may not contain treasure."

This treasure was first stored at the barracks of the Marine regiment, but later, "in order to remove all danger of demoralization" to the command, it was removed to the premises of the Hong Kong and Shanghai Bank.

On the 20th Colonel Meade appointed a committee of three persons, consisting of Major George Richards, Assistant Paymaster, U. S. M. C., Capt. M. J. Shaw, Acting Commissary and Quartermaster, U. S. M. C., and C. H. Christopher Moller, Agent at Tientsin of the firm of J. P. Morgan & Co., New York, "to count the bullion and estimate the value of the treasure in our possession." On the 27th this committee reported to Major Waller, as follows:

"We have counted 16,716 shoes of 'sycee' of varying condition and weight, a greater part of which is in bad condition and will require refining. We estimate the value of this to be $334,320, United States money. The remaining slag and refuse sweeping we have weighed and they approximate to 2,889 shoes of 'sycee,' which, allowing for cost of refining and probable inaccuracies, we estimate the value to be $42,000, United States money. The total value of the treasure we believe to be about $376,300, United States currency."

The Secretary of the Navy having authorized the sale of the treasure, Rear-Admiral Remey appointed a board, for the purpose of conducting the necessary negotiations, consisting of Major Richards, Captain Shaw and Capt. W. B. Lemly, U. S. M. C. Major Waller, in apprising these officers of their appointment, transmitted a copy of the agreement between Colonel Meade and Mr. Kenneth R. Campbell, of the firm of Moller & Campbell, bullion brokers and agents of J. P. Morgan & Co.:

HEADQUARTERS FIRST REGIMENT, UNITED STATES MARINES,
TIENTSIN, CHINA, July 21, 1900.

MR. KENNETH R. CAMPBELL, Present.

SIR: In the event of the captured treasure from Tientsin city, now under guard of my regiment, being sold by order of the United States Government, I agree to accept your offer at the rate of $20 gold per shoe of sycee, including the silver fused with foreign matter which is to be taken at a valuation to be determined by a fair appraisement by the commissioners appointed to count it.

Payment to be made by drafts on J. P. Morgan & Co., New York, at sixty, ninety, and one hundred and twenty days sight.

Delivery to be given at Taku in lighter.

This agreement to be subject to the approval of my commander in chief, to whom I have communicated the matter.

ROB'T L. MEADE,
*Colonel, U. S. M. C., Commanding United States Force in Tientsin.*

Confirmed: KENNETH R. CAMPBELL.

Witness to both signatures, C. H. MOLLER,
*Member of the counting board.*

The negotiations of Major Richards and Captains Shaw and Lemly, on behalf of the U. S. Government, with Mr. Campbell, were carried to a successful issue; the bullion and treasure being purchased at its full appraised value, $376,300. According to the terms of an agreement, executed Aug. 1, 1900, the payments were made in three sight drafts on J. P. Morgan & Co., of New York, one for $100,000 at 60 days, one for $100,000 at 90 days, and one for $176,300 at 120 days, payable to the order of the Secretary of the Navy.

---

## CHAPTER LI.

**1900.**—The Capture of Pekin by the China Relief Expedition, also the Work of the Marine Guard During the Long Siege.

**1900.**—Two detachments of Marines were sent to China early in 1900, going straight to Pekin, in order to serve as a guard at the U. S. Legation during the disturbances throughout the Empire, remaining there through the troublous period between their arrival and the capture of the City by the allied forces. For several months no authentic news, either of the foreign ministers or of the legation guards, reached the outside world, and numerous sensational reports were circulated regarding the destruction of both; all of which were proved untrue.

The first detachment of 1 sergeant, 2 corporals and 25 privates, under command of Capt. John T. Myers, U. S. M. C., was detached from the U. S. S. Oregon on May 24, 1900, and sent to Pekin. On May 29 another detachment, consisting of 1 sergeant, 1 corporal, 1 drummer and 23 privates, under command of Capt. Newt. H. Hall, U. S. M. C., was detached from the U. S. S. Newark, and sent to Pekin by way of Tientsin.

At Pekin, as at Tientsin, the Imperial troops seemed to co-operate perfectly with the "Boxers," in attacking the foreigners, and the regular and volunteer guards had a difficult task in preventing wholesale massacres. However, all guards, particularly the U. S. Marines, rendered exceptionally good service, deserving the highest praise for their untiring vigilance and repeated exhibitions of heroic gallantry.

The first official communications to reach the outside world were the following enclosures sent by Mr. John Fowler, U. S. Consul at Chefoo, China, to the Department of State:

No. 302. CONSULATE OF THE UNITED STATES.

CHEFOO, CHINA, August 7, 1900.

SIR: I have the honor to inclose herewith copy of a letter from Minister Conger, dated July 21, the copies of three memoranda all relating to the situation at Pekin up to July 21, obtained by me from unofficial sources.

I have the honor to be, sir, your obedient servant,

JOHN FOWLER, *Consul.*

HON. DAVID J. HILL, *Assistant Secretary of State, Washington, D. C.*

BRITISH LEGATION, PEKIN, July 21.

From June 20 to July 16 repeated attacks by Chinese troops on all sides, both rifle fire and artillery, including two 3-inch Krupp guns.

Since July 16, armistice, but cordon strictly drawn, both sides strengthening positions.

We hold at present following line: Two hundred yards wall Tartar City south of American legation, Russian and British legations half of park opposite east of latter, also French and German legations; all outside this line burnt and ruins held by Chinese, whose barricades are close to ours.

All women and children in British legation. Food sufficient for fortnight at most. Ammunition running short.

Casualties to date 62 killed, including Strouts (captain of Marines), David Oliphant, Warren, and double that number wounded in hospital, including Halliday (captain of Marines). Rest of legation all well.

Important that relief force, when near, should advance rapidly to prevent attack on legations by retreating Chinese forces.

Yesterday we refused a renewed demand to leave Pekin and proceed to Tientsin.

MACDONALD, *British Minister.*

FERNSTALK, *Boston.*

All Pekin, Tungchow Americans, also Walkers, Chapins, Smiths, Wyckoffs, Verity, Hobart, Terry, Mackay safe Pekin; wire friends. All property destroyed.

Eastern Shanghai, Record, Chicago, June 20. Kettler murdered, secretary wounded en route tsung li yamen by Chinese troops. Foreign residents besieged British legation. Since then under daily fire artillery rifles Chinese. Fortunately cowardice prevented successful rushes. Our loss 60 killed, 70 wounded. Theirs exceed thousands. No word from outside world. Food plenty—rice, horses. Yesterday flag truce message from Jung Lu requested Macdonald willing truce. Replied willing, provided Chinese came no closer. Shell firing ceased; quiet now. Hope it means relief, having defeated Chinese, are nearing. All exhausted constant watching, fighting, building barricades, digging trenches night, day.

All legations except British utterly wrecked, shell shot. Austrian, Italian, Belgian, Holland, burned ground; British also much shattered. American Marines still hold vital position city wall commanding legations after brilliant sortie July 3, Captain Myers driving back hordes Kansuli troops, he slightly wounded; captured flags, arms. Greatest credit due Secretary Squiers, whose military experience, energy, invaluable present dangers, treachery; possible entry city defeated Chinese army. Intense anxiety early relief.

COLTMAN.

Another report, which I have not been able to copy, reports Lippett badly wounded. LITTLETON W. T. WALLER.

TIENTSIN, July 27, 1900.

Messenger who left Pekin 21st arrived to-day with messages from several ministers. Minister Conger sends following to Consul Ragsdale: "Have been besieged in British legation five weeks under continual fire of Chinese troops, but since 16th by agreement there has been no firing; 50 Marines of all nationalities killed and more wounded. We have provisions for several weeks, but little ammunition. If they continue to shell us as they have done, we can't hold out long, complete massacre will follow. Hope relief can come soon; glad to hear of victory at Tientsin."

Gist of other messages missionaries uninjured, but missions destroyed, customs staff and families uninjured. Chinese approached Brtish minister under flag of truce and proposed cessation of hostilities; ministers agreed providing Chinese made no advance. Treachery feared. July 3 Captain Myers, American Marines, made wonderful sortie, capturing guns and standards. He was wounded slightly. Chinese also badly defeated when they attempted night attack. Foreigners hold Legation Street from French to American legations and British on north, all working at barricades, trenches, and fighting, and nearly worn out. Chinese seem to be short of ammunition. Our marines have fought like tigers against fearful odds; only Chinese cowardice prevented their hordes of savages massacreing our nationals.

PEKIN, July 21, 1900.

DEAR RAGSDALE: Have been besieged in British legation five weeks under continual fire from Chinese troops, but since 16th, by agreement, there has been no firing. Fifty Marines of all nations have been killed, and more wounded.

We have provisions for several weeks, but little ammunition. If they continue to shell us as they have done, we can't hold out long. Complete massacre will follow. Hope relief can come soon. Glad to hear of victory at Tientsin, but regret its terrible cost.

Hope you are all safe and will send cipher message by bearer.

E. H. CONGER.

DEPARTMENT OF STATE, WASHINGTON, September 14, 1900.

SIR: I have the honor to inclose herewith a copy of a dispatch from the consul at Chefoo, China, reporting the situation at Pekin to July 21, 1900.

I have the honor to be, sir, your obedient servant,

ALVEY A. ADEE, *Acting Secretary.*

THE SECRETARY OF THE NAVY.

(Inclosure: From Chefoo, No. 302, August 7, 1900, with inclosures.)

(First Indorsement.)

DEPARTMENT OF THE NAVY, BUREAU OF NAVIGATION,

September 26, 1900.

Respectfully referred to the Brigadier-General, Commandant, United States Marine Corps, via. the office of the Assistant Secretary, for his information and return. A. S. CROWNINSHIELD, *Chief of Bureau.*

(Second Indorsement.)

NAVY DEPARTMENT, September 28, 1900.

Referred to the Commandant of the Marine Corps for his information and return. LONG, *Secretary.*

On Aug. 19, Rear-Admiral Remey sent the following cablegram to the Navy Department from Taku:

Taku, 19th. Authentic report Pekin 15th from Latimer. Troops moving on Imperial City; clearing out Tartar City. All Americans who remained in Pekin are well. There have been no deaths among them except one child. Captain Myers has recovered from wound, has typhoid fever, crisis passed, now convalescent. Assistant Surgeon Lippett was wounded, left upper leg bone fractured, leg saved, now recovering. Following killed during siege in Pekin: Sergeant J. Fanning, Privates C. B. King, J. W. Tutcher, J. Kennedy, R. E. Thomas, A. Turner. H. Fisher. Wounded: Private J. Schroeder, elbow, severe and dangerously ill, fever; Seaman J. Mitchell, wound upper arm, severe but recovering; all other wounded and sick have returned to duty. Casualties Major Biddle's command attack San Tan Pating: First Lieutenant Butler, chest; Private Green, wrist; Private Warrell, right temple; all slight. Reported from Chinese sources Royal family have escaped and en route to Sianfu.

The thrilling history of the operations of the Marines during the trying period of the seige of Pekin is ably recounted in the following letters from Captains Myers and Hall:

UNITED STATES NAVAL HOSPITAL, TIENTSIN, CHINA, September 26, 1900.

SIR: I have the honor to submit a detailed report of the operations of the Marine guard recently on duty at the United States legation, Pekin, China, during the period in which I had active command, from May 29 to July 3, 1900.

2. The guard consisted of 25 Marines from the U. S. S. Oregon; 23 Marines, 3 bluejackets, 1 chief machinist, and 1 hospital apprentice from the U. S. S. Newark. Captain N. H. Hall, U. S. Marine Corps, and Assistant Surgeon T. M. Lippitt, U. S. N., were also attached to my command. Chief Machinist Peterson reported for duty on June 3, arriving in Pekin unarmed.

3. The guard, equipped in heavy marching order and without baggage, left the Newark on the morning of May 29 and arrived at Tientsin that night, having been compelled to come up by boat, as the railroad officials refused to sell tickets to an armed force without permission from the viceroy. We were the first troops to arrive in Tientsin, and our reception was enthusiastic; nearly all the foreign residents, with a band, were on the bund to meet us, and escorted us to the quarters which had been provided by Captain McCalla and Paymaster Jewett, who had come up by train that afternoon. We remained there, waiting for the other guards to arrive and for the necessary permission to go to Pekin, 4 miles from the station to the Chien Men gate was absolutely silent—a silence until Thursday, May 31, when we left at 4.30 P. M., reaching the railroad terminus outside the city at about 11 o'clock. Captain McCalla, Paymaster Jewett, Naval Cadet Courtney, and Warrant Machinist Molineaux accompanied us.

We headed the column on the march into the city, meeting with no opposition, although we had been warned to be on the lookout for trouble.

The dense mass of Chinese which thronged either side of the roadway for the which seemed more ominous than a demonstration of hostility would have been.

On our arrival we were assigned to quarters adjoining the legation compound and in the rear of the Russian bank.

4. The ammunition carried was as follows: 9,720 rounds in men's belts, 8,000 for Colt gun, and 10,000 in boxes, making a total of 27,720 rounds, of which I had been told 7,000 rounds remained when the relief arrived. Five days' rations and two large ship's breakers of water were also taken.

5. Between the 1st and 5th of June nothing of any importance happened. Guards were established on the legations grounds and a daily routine of drills and exercises was instituted.

Captain McCalla, Paymaster Jewett, and Naval Cadet Courtney returned to Tientsin on the 2d.

On June 6 the railroad communication was interrupted and many stations burned. The situation appeared critical, and I wired Captain McCalla that 25 officers commanding the various guards, the English marine officer, at my request, called a meeting of all the officers, and it was decided that at the first sign of an outbreak all the noncombatants together with all provisions, should more men were needed to properly protect the legation.

On June 7, there having been no plan for common defence adopted by the be sent to the English legation; and that all streets leading into the legation quarter should at once be barricaded, no Chinese being allowed to enter without a pass. It was also agreed that we should endeavor to hold all the legations as long as possible, and as a last resort, to fall back upon the English legation.

On June 8, at the request of Mr. Conger, sent 10 men, under Corporal Hunt, to guard the Methodist mission, where a number of American missionaries had assembled. The next day, this number being greatly increased by the arrival of refugees from Tungchow and outlying districts of the city, 10 more men were sent, and Captain Hall was detailed to command the detachment. The mission buildings lay about three-quarters of a mile to the eastward and outside of our lines, on a small street leading off the Hatamun.

On June 10, 11, and 12 nothing of any moment occurred. The feeling of unrest deepened in the city, and foreigners were openly insulted in the streets.

On June 13, two Chinese in full Boxer regalia, carrying naked swords, appeared on Legation street, followed by a large crowd. They were shot at and pursued by the German sentries, the German minister himself leading the chase. One of them was captured and the other escaped. As the crowd was thronging in front of the legation, the alarm was sounded and the Colt gun run into the street, upon which they all disappeared. This incident tended to greatly add to the excitement already manifested among the Chinese. At 5 P. M. the outside chapel of the Methodist mission was set fire to by the Boxers, who now openly appeared in all sections of the city. The usual crowd collected, whereupon Captain Hall promptly dispersed them, using only the bayonet. The chapel was totally destroyed. At 6 P. M. the streets in the legation quarter were cleared and rough barricades built. That night all the outlying missions and churches were burned except the Peitang Cathedral, where the French and Italians had sent guards to protect the Roman Catholic priests, nuns, and native converts there assembled.

On June 14, in the early morning, large numbers of Chinese Christians, most of them horribly burned or badly wounded, appeared at our barricade. They had come from the districts surrounding the Nantung Cathedral, which was then burning, bringing the most frightful tales of outrages committed by the Boxers. These people were allowed to enter, and after the most serious cases among the wounded had been treated by the American and Russian surgeons they were sent to the French legation. The officer commanding the Russian guards decided, in the name of humanity, to send out a party for the rescue of those Christians still in hiding among the burning buildings. At his request, I detailed 10 men to go with them. This party was led by an American—Mr. Pethick, a veteran of our Civil War, whose knowledge of China and the Chinese language and whose personal bravery while under fire rendered his services of the greatest value to the besieged. The mission of the rescuers was entirely successful, they returned escorting about 150 Christians, having shot a number of Boxers and looters. That night the excitement in the Chinese city was intense, and the shouting and cries of "Kill!" "Kill!" continued until early morning.

On June 15 sent 10 men to accompany the English on an expedition to the northeastern part of the city for the purpose of rescuing some native Christians supposed to be in hiding there. Dr. Lippitt volunteered and was given permission to go with them. They failed to find the Christians, but on the way discovered a temple where a Boxer meeting was evidently in progress. Our forces, which had been augmented by the arrival of some Japanese and Austrians, surrounded the place and after a short fight killed 45 of the Boxers.

On June 16 the Boxers set fire to Watson's drug store, in the Southern City, not far from the Chien Men entrance. The explosion of the chemicals caused the fire to spread rapidly, destroying the richest business portion of the city, and finally igniting the outer of the two gates, which was totally destroyed. The imperial troops upon the wall waved their banners energetically, but failed to fire upon the incendiaries.

On June 17, about 5 o'clock in the evening, a fire was started on Legation Street, half a mile to the westward of our barricade. The Russians promptly started down the street and succeeded in shooting the man who had applied the torch; then after two hours' hard work, aided by the municipal fire department, got the fire under control. I followed with as many men as could be spared from the legation and patrolled the streets between the burning building and our lines, causing the Chinese to tear down all the matting and inflammable materials. In this work the English assisted.

On June 18 the Tsungli Yamen, the members of which had been full of assurances of friendship and protection, suddenly changed their tone, and in a message to the ministers announced that the Taku forts had been taken by the powers; that a state of war existed, and that the ministers would be given twenty-four hours to leave the city, accompanied by their suites, protection being guaranteed as far as Tientsin.

On June 20 the ministers held a meeting in the morning, shortly after which the German minister, unattended save by his interpreter, set out to visit the officers of the Tsungli Yamen which was situated on the Hatamun Street, some distance outside of our lines. Shortly after the interpreter sought refuge at the American mission in a badly wounded condition. He stated that Baron von Ketteler had been shot in the back and killed by an Imperial soldier, scores of whom lined the road in the vicinity of the yamen. The firing had then become general, but he had managed to make his escape. On the receipt of this news the Germans at once sent out a detachment to obtain, if possible, the body of their minister, but they were fired upon and compelled to return unsuccessful. Realizing that the crisis had come, all the women and children were sent to the English legation, each foreign guard sending ten men for the greater protection of that legation. All food supplies were sent to the same place, and as we had within our lines the only foreign stores in Pekin and several Chinese stores of rice and wheat, the contents of each was appropriated for our use. In the afternoon, the English and Russians having each lent me ten men, I started with them and fifteen of my own to escort Captain Hall and his charges within our lines, which was safely accomplished. At 6 P. M. the Chinese soldiers began to fire upon us. The attack, however, did not seem to be an organized one. Any hope that the Imperial Government would put down the trouble had long been dispelled, as our spies brought us word that the Boxers were entering the city through all the gates, guarded as they were by soldiery, and in all parts of the city mingling freely with the troops, with whom they appeared to be on the best of terms.

On June 21 the Methodist mission and stores carrying goods were burned. A desultory firing by the enemy was kept up all day. About this time Sir Claude McDonald was chosen by the ministers, and this choice agreed to by the foreign officers, to act as commander in chief. Captain Strouts, R. M. L. I., was appointed chief of staff. Upon his death, on July 16, Mr. H. G. Squiers, first secretary of the United States legation took his place. This gentleman having served ten years in our army filled the post with great credit. In addition to these duties he, throughout the siege, acted as commissary for the American guard, his services being of the greatest value.

CHARLES LAURIE McCAWLEY,

MAJOR AND QUARTERMASTER U. S. M. C Plate 9

Friday, June 22, an attempt was made to burn the English legation. Later in the day, owing to some misunderstanding of orders, nearly all the foreign guards retired upon the English legation. The mistake being quickly discovered, our positions were at once reoccupied, before the Chinese knew of our absence.

On June 23 the customs building and Austrian legation were burned, the Austrians having been forced to fall back on the French. The Russian bank was also fired, and as this was very near us we spent the day alternately fighting fire and Chinese.

On June 24 the Chinese from the Chien Men gate came down the wall and, protected by the parapet, opened a harassing fire upon the legation grounds and buildings. They were driven back by the Germans, who had gone on the wall in the rear of their legation, some 500 yards below. Later in the day I made several attempts to occupy a position on the way opposite our legation, but the dense smoke which drifted across, togegther with the fact that the coolies would not build barricades in face of the fire which the Chinese directed down the wall, rendered these attempts unsuccessful. During the day Private King was shot and instantly killed while on duty as a sentry. In the evening the fire in the Russian bank spread to the quarters occupied by the guard, which were totally destroyed.

On June 25, succeeded in occupying a position on the wall. The Germans at the same time gained a foothold and threw up a barricade some 500 or 600 yards in my rear, facing the Chinese troops who occupied the Hatamun gate. During the day the building of bombproofs was begun in the English legation, as the shell fire was now incessant.

On June 26 Sergeant Fanning was killed while on duty. The Chinese busied themselves building barricades in our front.

On June 27, in broad daylight, a number of Chinese sallied from their barricades on the wall and advanced on the run toward our own. A few volleys drove them back in confusion. This was the only instance in my experience of the Chinese leaving shelter to make an attack. Firing continued at intervals during the night, being kept up incessantly during the 28th, 29th, and 30th. The nights being very dark, the Chinese managed to advance their barricades on the wall until they gained a position at the corner of the bastion opposite that occupied by my forces, at a distance of about 30 or 40 yards. The Russians were now daily furnishing ten men to assist n defending our barricade and the English six or seven men to help hold the trench crossing the wall to the United States legation, which was our only means of communication with the latter place. The Germans had also been sending men until the attack on their own barricade became so determined that every man was needed.

On June 29 Dr. Lippitt was severely wounded while crossing the courtyard of the legation. His place was taken by Dr. George Lowry, an American missionary, who took up his residence in the legation and rendered every assistance in his power in giving first and to the wounded, who were later sent to the nternational hosuital in the English compound. His services were of great value, and he at all times exposed himself to fire when the necessity arose.

On July 1 the Germans were forced to leave the wall by a heavy shell fire. As this left my rear unguarded and exposed men going up and down the ramp and across the street to a heavy fire, I also withdrew into our legation. After some consultation with the American minister and Mr. Squiers it was decided that the place must be retaken and held at any cost. I was assured that all the coolies available would be sent that night to build cross barricades on the ramp, and also to build a barricade across the wall in my rear. We thereupon, after an absence of about fifteen minutes, returned, finding that the Chinse had once more failed to discover our retreat. In the evening, by order of Sir Claude McDonald, I was relieved by Captain Hall and went below to get some rest, having had little or no sleep since the wall was occupied.

On July 2, at about dusk, relieved Captain Hall and found that during the preceding night and day the Chinese had succeeded in building a wall into and across the bastion and were then busily engaged in erecting a tower directly on my left flank, the fire from which, when completed, would reach all parts of our position. Their work had been done with such infinite caution and so little exposure that, although the fire upon them was incessant, their labor was in no way hindered. I at once reported the condition of affairs, and was asked by Mr. Squiers if it were feasible for me to make a charge, provided reinforcements were sent. I replied that we would try. Later a communication was received from Mr. Conger telling me that after a consultation with the English and Russian ministers and military officers it had been decided to instruct me to take the enemy's barricade in my front. The Russians were to send five more men with an officer and the English about twenty. The details were of course left to me. These men arrived between 2 and 3 a. m., and as the Chinese had almost finished their tower and were amusing themselves throwing stones into our barricade, I at once made the dispositions for the advance. The Russians were directed to take the inside of the wall, keeping close to the parapet, and dislodge the enemy from their position at the head of the ramp. The English and Americans, on the left and center, were to go over barricade and, once across, to break to the left and follow, if possible, the Chinese wall in the bastion around to the rear to their position. These plans were completely successful, the Chinese being routed with considerable loss, and retiring some hundred yards down the wall to the next barricade.

A number of rifles and some ammunition were captured and turned over to the commander in chief. I regret that two of the best men in the guard, Privates Turner and Thomas, were killed, and one Englishman and one Russian wounded. Private Turner's body was picked up well inside the Chinese barricade, showing that he had evidently, in the darkness, gotten ahead of our line. Being slightly wounded myself, I sent for Captain Hall to relieve me, and went below into the legation. The next day, as the wound had gotten worse, I accepted the kind offer of the first secretary of the Russian legation and was moved into his quarters, retaining the command, however, until the 21st of the month. Captain Hall's report up to this date is appended.

6. I respectfully commend to the consideration of the commander in chief the conduct of Asst. Surg. T. M. Lippitt, who not only performed all his professional duties thoroughly, but upon two occasions took a purely military part. The first of these occurred when the captain of the Colt gun crew, having misunderstood orders, was taking his piece to the wrong position. Dr. Lippitt followed under a heavy fire and assisted in bringing the piece back and guiding it to the place originally assgned; the second, when he helped to take the Austrian 1-pounder on the wall and place it in position.

I have also to recommend the following enlisted men:

Gunner's Mate (First Class) Joseph Mitchell, U. S. S. Newark, who operated the Colt gun with the utmost courage and skill under the heaviest fire;

he also, with the assistance of Mr. Squiers, put in working order and later used successfully an old brass cannon which had been dug up inside our lines; he also captured a flag under peculiarly hazardous circumstances, on which I will later make a special report. Hospital Apprentice R. H. Stanley, of the *Newark*, who volunteered and took a message to the English legation when it was necessary to use the street down which the Chinese were firing. Sergeant Walker, Corporals Hunt and Dahlgren, who at all times cheerfully performed their duties with courage and fidelity. The conduct of the guard, with one or two exceptions, was excellent.

7. The following papers are appended:

(1) Report of Captain Hall, July 3 to July 21.

(2) List of killed and wounded, compiled by Dr. Lung.

(3) Copy of letter from United States minister.

(4) Copy of resolutions passed by American missionaries.

(5) Copy of report of Dr. Velde, the civilian in charge of the hospital, of total casualties during siege.

8. I can not close this report without expressing for myself and the men under my command our appreciation of the uniform kindness shown us and the active assistance rendered by the United States minister and his staff.

Very respectfully, JOHN T. MYERS, *Captain, U. S. Marine Corps.*

THE COMMANDER IN CHIEF, *U. S. Naval Force on Asiatic Station.*

Captain Hall's letter tells much the same story with some additional details:

PEKIN, CHINA, August 30, 1900.

SIR: I have the honor to submit herewith the following daily report for your information, during the time that you were confined to your room with your wound, but retained command of the United States detachment on guard at the United States legation in this city from July 3 to July 21, 1900. I was called about 2.30 A. M., July 3, 1900, and infomed that you were wounded, and went on the wall and took command until relieved by Captain Von Strauss about 3.30 or 4 A. M. At his request I took charge of the old barricade, making a passage through it for the men, and strengthening the rear side (Ha-ta-men side) of the barricade. I remained on the wall until ordered down by you at 11.30 A. M. It was raining continuously, and heavy fire from the enemy during the entire time. The American, Russian, and British marines were at work on the barricade that you took and the old barricade, the Chinese laborers being afraid to work under the enemy's fire.

*July* 4.—Intermittent and slack fire by the enemy during the day. Occupied in building barricades. Private Woods was wounded at the head of the east ramp by a stray bullet.

*July* 5.—Very little firing; nearly all done by the enemy. Some sniping from the American barricade at the head of the west ramp.

*July* 6, 7, 8 *and* 9.—Same as 5th.

*July* 10.—Snipping by the enemy; very little firing by us.

*July* 11.—Desultory firing by the enemy.

*July* 12.—Sniping on the part of the enemy. I was ordered to build a barricade at the far end of the first bastion on the Ha-ta-men side about 100 yards from our barricade; built about half of it during the night; almost no shots fired by the enemy; enemy's barricade on the wall in the rear of the German legation. About 12.30 P. M., at the request of Baron Rahden, the Russian commanding officer, I sent one corporal and four privates to protect his men while they were occupied in tearing down some burned buildings and went with them. There was very little firing by us, as the enemy did not open fire upon the Russians until they had finnished working and were coming back to their legation.

*July* 13.—Continuing work at the barricade; enemy sniping all day at the barricade.

*July* 14.—Continuing work on the barricade; enemy firing at the barricade continuously.

*July* 15. I was informed that the barricade was not far enough along the wall and was requested to build another. At my request, Mr. H. G. Squiers, first secretary to the United States legation, went on the wall with me to show me the exact place at which it was desired to build the barricade—the far end of the bastion about 100 yards from our last barricade and 250 from the enemy's barricade. I took Private Daly with me about 9 P. M., and went out to reconnoiter the bastion with the understanding if we were not attacked the coolies were to come out with sand bags, under charge of Privates Carr and Upham ten minutes later. As the coolies did not come at the appointed time, Private Daly asked permission to remain in the bastion while I returned for the coolies. I did not wish to leave one man in the bastion as there were stray shots flying along the wall from the front and rear and the Joss temple west of the legation. Finally, feeling sure that there was a misunderstanding in regard to the coolies, I left Private Daly and returned to the barricade for the coolies and found out that the interpreter first sent up wih them could not understand English and that Mr. Dossberg, secretary of the eBlgian legation, had just arrived and was starting for the bastion in which I had left Private Daly. I sent out Privates Carr and Upham with the coolies and went back to our first barricade to send the rest of the men to the barricade on the night of the 12th.

*July* 16.—An armistice was made with the enemy. Sick. Private Fisher killed about 9.30 A. M.

*July* 17 and 18.—Sick.

*July* 19.—Quiet. A few stray shots during the day.

*July* 20.—Same as 19th.

*July* 21.—Same as 19th and 20th. I relieved you in command about 5 P. M., on account of illness.

I respectfully invite the attention of the commanding officer to the courage and fidelity of Daniel Daly, private, U. S. Marine Corps, at all times, and to his conduct on the night of the 15th of July, 1900, when he volunteered to remain alone in the bastion under the fire of the enemy while I returned to the barricades for the laborers.

The command is very much indebted to Mr. F. H. Conger, envoy extraordinary and minister plenipotentiary of the United States of America for his many kindnesses, placing his house at our disposal for shelter, and for doing everything in his power for their health and comfort.

Messrs. Pathig, Cheshire, and Dossberg rendered valuable assistance in directing the Chinese laborers at work on the barricades.

Very respectfully, N. H. HALL, *Captain, U. S. Marine Corps.*

J. T. MYERS, *U. S. Marine Corps, Commanding U. S. Detachment, guarding U. S. Legation, Pekin, China.*

HEADQUARTERS FIRST REGIMENT U. S. MARINES,
Forwarded. By command of Major Biddle. PEKIN, CHINA, Sept. 1, 1900.

DAVID D. PORTER, *First Lieutenant and Adjutant.*

According to the report of Dr. G. A. Lung, surgeon of the expedition, the casualties of the Marine detachment were, in all, 11 wounded and 7 killed.

"The following extract from the report of Minister Conger to the Secretary of State, dated August 17, 1900, is of interest as throwing additional light on the splendid work of the Marine guard in protecting the legations during the siege of Pekin:

* * * * * * * *

" 'To our marines fell the most difficult and dangerous portion of the defense by reason of our proximity to the great city wall and the main city gate, over which the large guns were planted.

" 'Our legation, with the position which we held on the wall, was the key to the whole situation. This given up, all, including many Chinese Christians, would at once be driven into the British legation, and the congestion there increased by several hundred. The United States marines acquitted themselves nobly. Twice were they driven from the wall, and once forcd to abandon the legation, but each time, reinforced, immediately retook it, and with only a handful of men, aided by ten Russian sailors, and for a few days a few British marines, held it to the last against several hundred Chinese, with at least three pieces of artillery.

" 'The bravest and most successful event of the whole siege was an attack led by Captain Myers, of our marines, and 55 men—Americans, British and Russians—which resulted in the capture of a formidable barricade on the wall, defended by several hundred Chinese soldiers, over 50 of whom were killed. Two United States marines were killed and Captain Myers and one British marine wounded. This made our position on the wall secure, and it was held to the last with the loss of only one other man.

* * * * * * * *

" 'I can not close this dispatch without gratefully mentioning the splendid service performed by the United States marines, who arrived here on May 31 under command of Captain Myers. With slight exceptions their conduct won the admiration and gratitude of all, and I beg you to kindly communicate the fact to the Navy Department.' "

Minister Conger's high appreciation of the work of Captain Myers and his men is also shown in the following letter, dated September 3, 1900:

MY DEAR CAPTAIN MYERS: Congratulating you on the recovery from your dangerous illness and the early prospect of getting away from the scenes of your sad experiences, I can not let you go without expressing to you the profound gratitude of all the saved for the incomparable part you took in their salvation.

Yours was a most trying position from the start. Our fate depended upon holding the wall. It could not have been held except for your heroic and successful charge of July 3, when you received your ugly wound. For the brave men who fell upon that and other occasions we grieve with you and shall always honor their memory.

I beg to convey to your men anexpression of our highest appreciation and sincerest gratitude for all they so nobly did and so bravely endured on our account.

For yourself, please accept my warmest personal thanks and very best wishes for all your future. E. H. CONGER,

In his report for 1900, the Brigadier-General Commandant of the U. S. Marine Corps writes as follows of the services of Captain Myers:

"It would seem by the reports that Captain Myers and his small body of Marines succeeded in holding a dangerous and almost untenable position on the city wall, in the face of overwhelming numbers, and also that he made a brilliant sortie, driving back hordes of Chinese, on which occasion he was slightly wounded. Captain Myers's courage and gallantry merit the highest commendation, and I will in a short time recommend to the Department that he be given proper recognition for his bravery in the presence of the enemy. Minister Conger's letter and its inclosures are in full, as follows:

LEGATION OF THE UNITED STATES OF AMERICA.
*Pekin, China, August* 20, 1900.

SIR: It affords me great pleasure to transmit herewith a copy of resolutions passed by the American missionaries besieged in Pekin, expressing their hearty appreciation of the courage, fidelity, and patriotism of the American marines, who so bravely and tenaciously held the key to our salvation during the whole of the trying time.

I most heartily and sincerely join in this expression, and beg to communicate to both officers and men my personal commendation of and gratitude for their heroic and faithful service.

Yours, very respectfully, E. H. CONGER.

Maj. W. P. BIDDLE, *Commanding United States Marines in Pekin.*

HEADQUARTERS FIRST REGIMENT OF MARINES.
Pekin, China, August 23, 1900.

Respectfully forwarded to the brigadier-general, commandant, United States Marine Corps headquarters, Washington, D. C. W. P. BIDDLE,
*Major, U. S. M. C., Commanding.*

PEKIN, *Saturday*, 18, 1900.

DEAR SIR: At a meeting of the American missionaries held this morning at 8.30 the following resolution was unanimously adopted; and it was further voted that the resolution be drafted and presented to you.

"The Americans who have been besieged in Pekin desire to express their hearty appreciation of the courage, fidelity, and patriotism of the American Marines, to whom we so largely owe our salvation.

"By their bravery in holding an untenable position on the city wall in the face of overwhelming numbers, and in cooperating in driving the Chinese from a position of great strength, they made all foreigners in Pekin their debtors, and have gained for themselves an honorable name among the heroes of their country." For the meeting: ARTHUR H. SMITH, *Chairman.*
CHARLES E. EWING, *Secretary.*

Hon. E. W. CONGER, *Minister of United States of America.*

Shortly after the battle at Tientsin, Colonel Meade was relieved of the command of the First Regiment of Marines, and Major Waller succeeded him, retaining the position until the arrival of Major W. P. Biddle, U. S. M. C., and the Fourth Battalion of Marines, on August 3. Major Biddle's command, which had been originally intended for service in the Philippines, was formed at the Marine Barracks, Washington, D. C., and at the Navy-Yard, New York City, and consisted of 8 officers, two companies of 80 men each, and 60 additional men "to fill vacancies in the other battalions at Cavite." Beside Major Biddle, the officers were the following: Second Lieutenant W. C. Harllee, adjutant; Captains F. J. Moses and W. C. Neville; Second Lieutenants C. C. Carpenter, S. A. W. Patterson, J. G. Muir, and D. W. Blake. Gunnery Sergeant Horace D. Heaton was detailed as sergeant-major of the battalion.

Having been conveyed to San Francisco by train, this force embarked on the transport Grant on July 1.

Almost immediately after assuming command of the Marine regiment before Tientsin, Major Biddle joined the march of the allied forces upon Pekin. His command, at this date (Aug. 4), 482 strong, was attached to the forces under Gen. Adna R. Chaffee, U. S. A., with which was a part of the 9th Infantry and the 5th Artillery. In this march the men suffered greatly from the intense heat, and the progress was necessarily slow, but the Marines endured the hardships so remarkably well, that, when the order was given at Matow, two-thirds of the way, to leave all incapacitated men behind, only four of them were found unfit to proceed.

Only two encounters of any note occurred; one at Pietsang, on Aug. 5, when the Japanese troops engaged and defeated the enemy, but no Americans came into action, and another at Yangtsun, on Aug. 6, when the Marines acted as the support to Reilly's Battery of the 5th Artillery. In this latter fight the Marines repulsed the enemy's cavalry, appearing on the right front, and later, advancing in skirmish order, took the village without serious opposition.

In his report to Major Biddle, dated Aug. 20, Major Waller gives the following account of the fight at Yangtsun:

"On the morning of the 6th the column moved in three separate columns, Russo-French, English, and Americans. Our route lay along the left side of the railroad track, the Fourteenth Infantry leading, with a battalion of the Ninth Infantry on the right, Marines in the center, and two battalions, Fourteenth on the left, with the Fifth Artillery, Reilley's battery, between advance and main column. The enemy's lines were developed near Yangtsun in concave form. The

Fourteenth and English attacked the right, and on the left side of the road enbankment. The Ninth, with Marines in support, attacked a position on the left of the enemy's position, Reilley's battery shelling the enemy's guns. The enemy's artillery fire was very accurate. The Marines, First Battalion on the line, Second in reserve, were deployed to support artillery. Numerous changes of direction and objective tired the men, and the great heat and the steady movement through high corn began to tell, the men dropping out of the line, overcome by heat. Reaching a point near the enemy's position we were subjected to a sharp rifle fire, but did not reply, moving on until the enemy's villages could be seen. A few volleys sent them out. About the same time we discovered the enemy's cavalry about 800 yards away. Several well-directed volleys put them to flight. The artillery followed the retreating cavalry with shrapnel. The villages in our front were carried with little or no opposition. Our casualties were 1 dead, 1 wounded. The death was caused by heat. One entire company failed to move forward on the last village, being nearly prostrated by the heat in the cornfield. This was Company H, Captain Bannon."

The record of the successful expedition for the relief of Pekin was most extraordinary as far as the Marine regiment is concerned. Although, owing to the intense heat and other hardships, many men were overcome and invalided, the record for both battalions of 15 missing and 1 dead seems a very small number of casualties. As given by Major Biddle, during the march on Pekin, 118 men and 2 officers were detached; 43 men and 2 officers were placed on junks; 11 men and 1 officer were sent to hospital, giving a grand total of 198 from the original 482, and leaving a total force of 284 men and officers to reach Pekin.

According to this same report of Major Waller, the remainder of the march on Pekin and the details of its capture are as follows:

"We remained at Yankystum on the 7th, bivouacked at Tsai-tsum on the 8th; Ho-shi-wu on the 9th; Thienping on the 10th, a short distance from Matow; Dshang-dshai-wan on the 11th; Tung-chou on the 12th; Thing-fu-dsh on the 13th; all being uneventful but fatiguing days. On the 14th we advanced on Pekin. At about 12.30 P. M. I was directed to send men to the wall near the north gate, Tung-chou road, to protect the artillery from the enemy's fire. A part of A and H Companies were sent, silencing the enemy's fire. Lieutenant Butler was slighty wounded in the chest while getting the men of A Company on the wall. Our casualties for the day were 3 wounded. We bivouacked for the night just outside the walls of the Tartar City, near the legations. These people had been relieved and had joined us during the afternoon.

"On the 15th we marched to attack the Imperial City, the marines leading. We took a position over the Chien-men gate, clearing the barricades to permit the artillery to come into action on the padoga. The guns fired at the enemy on the west gate. I was ordered by the commanding general to capture a number of flags mounted on the enemy's position at the west gate. This order was revoked, the enemy opening a heavy fire from the gate to the north, the second gate leading into the Forbidden City. The First Battalion, especially Companies A and C, did excellent work in this fire. Two pieces of artillery were turned on this gate and the enemy driven out after a stubborn resistence, but not until they had killed Captain Reilly. A braver soldier, a truer friend never breathed than this admirable and lamented officer. He died at my side, touching me at the moment of the blow. He died without murmur or groan.

"The marines held this position and the west gate, taken on the 16th, until the 19th, when we moved into a position in the Tartar City. The flags were captured for the commanding general on the 15th by Lieuts. A. E. Harding and David D. Porter. I wish to mention the admirable work done by Lieutenants Butler and A. E. Harding on the 14th and 15th. Lieutenant Porter does not belong to my battalilon, but served as always, with great courage. Lieutenant Butler had just recovered from his wound received at the battle of Tienstsin, but marched with his men, bearing the hardships and fatigue, and always ready to jump to any call of duty. Lieutenant Harding, on this as on all other occasions, was ever ready to expose himself to set a good example to the men."

Major Biddle gives further details regarding the battle of August 14, 15 in two reports, both dated on the 20th and addressed to Major-General Chaffee:

*August* 14, 1900.—The marines advanced to a position near the north gate of the city, under a slight fire, and halted while a platoon from two companies was sent to the top of the wall to stop "sniping" and protect the artillery, which was successfully accomplished. The casualties of the day were 3 wounded—First Lieut. S. D. Butler, slightly wounded in chest; Private G. P. Farrell, and Private F. W. Green We bivouacked for the night just outside of the Tartar City.

*August* 15, 1900.—An advance was made against the Imperial City on this date, with the marines leading. The marines took a position on the Chien Men Gate and cleared away the barricades in order that the artillery might take possession. Two companies of the First Battalion of Marines were posted in the second story of the Pagoda, while the Second Battalion took a position along the wall, both battalions firing volleys at ranges of 900 yards at the first gate of the Imperial City, where the enemy were in force. Meanwhile the marines were subjected to a heavy small-arms fire by the enemy, as well as to some artillery fire. After a stubborn resistence the enemy were driven from their position and the marines were left to hold the Chien Men gate, the artillery withdrawing.

After the capture of Pekin, by order of General Chaffee, under date, September 4, the Marine regiment was divided between two brigades of the China Relief Expedition; the second battalion, under command of Captain F. J. Mases, being assigned to the First Brigade, and the first battalion, under command of Major Waller, being assigned to the Second Brigade.

Major Waller was, on September 14, appointed provost-marshal of that section of the Tartar city of Pekin, which was under American jurisdiction, and held the office until the removal of the Marine detachments. On the same date, First-Lieut. L. M. Little, U. S. M. C., was appointed ordnance and engineer officer on the staff of the commander of the First Brigade, China Relief Expedition.

By order of the Secretary of the Navy, cabled to Rear-Admiral Remey, on September 28, the Marine battalions were withdrawn from Pekin, and returned to the post at Cavite, P. I. At the same time all army troops, except the legation guards, were withdrawn. The legation guards on this occasion were, contrary to usual custom, composed of army troops, instead of Marines.

In his report for 1901, the Brigadier General Commandant gives the following list of brevets and promotions of officers of the Marine Corps, most of whom served in China:

Col. Robert L. Meade, appointed brigadier-general by brevet.

Major George Richards, appointed lieutenant-colonel by brevet.

Major L. W. T. Waller, appointed lieutenant-colonel by brevet and advanced two numbers.

Major Charles L. McCawley, appointed major by brevet.

Lieut. Col. Allan C. Kelton, appointed major by brevet.

Capt. John T. Myers, advanced four numbers and appointed major by brevet.

Capt. Newt H. Hall, appointed major by brevet.

Capt. Smedley D. Butler, advanced two numbers and appointed captain by brevet.

Capt. Henry Leonard, advanced two numbers.

Capt. James E. Mahoney, appointed captain by brevet.

Capt. Herbert L. Draper, appointed captain by brevet.

Capt. Charles G. Long, appointed captain by brevet.

Capt. Albert S. McLemore, appointed captain by brevet.

Capt. William N. McKelvy, appointed captain by brevet.

Capt. George C. Thorpe, appointed captain by brevet.

Capt. David D. Porter, appointed captain by brevet.

Capt. Charles G. Andresen, appointed captain by brevet.

First Lieut. Wm. G. Powell, appointed captain by brevet.

First Lieut. Arthur E. Harding, appointed captain by brevet.

Capt. Melville J. Shaw, appointed first lieutenant by brevet.

First Lieut. Wade L. Jolly, appointed first lieutenant by brevet.

In addition to these a number of officers and men received medals of honor.

Commendatory letters were sent to these officers and men of the China Relief Expedition:

Capts. Charles G. Long, Ben H. Fuller, and Philip M. Bannon, and First Lieuts. John F. McGill and Robert F. Wynne, with the men under their respective commands, merit and receive the commendation of the Department for their gallant, meritorious, and courageous conduct in the battle of Tientsin, July 13, 1900.

For their work in assisting on the barricades at Pekin, China, July 21 to August 17, 1900: Privates John O. Amman, Gottleib Brosi, William F. Donovan, Harry Gold, James J. Lavin, Robert M. Barratt, Edward J. Donovan, Henry C. Galligher, Thomas F. Hall, Jacob C. E. Martin, and Frederick J. Tinkler.

# APPENDIX.

## REGISTER OF OFFICERS, U. S. MARINE CORPS,

## 1798 TO 1903.

## DATE OF ORIGINAL ENTRY AND PROGRESSIVE RANK.

| Commandants. | Second Lieutenant | First Lieutenant | Captain. | Major | Lieutenant Colonel. | Colonel. | Brigadier-General. | Brevet Rank. | Remarks. |
|---|---|---|---|---|---|---|---|---|---|
| William W. Burrows | | | | Com. Jul 12, 1798 | Comd't May 1, 1800 | | | | Resigned Mar 6, 1804. |
| Franklin Wharton | | | Aug 3, 1798 | | Comd't Mar 7, 1804 | | | | Died Sep 1 1818 |
| Anthony Gale | Sep 2, 1798 | Mar 2, 1799 | Apl 24, 1804 | | Comd't Mar 3, 1819 | | | Brevet Major Apl 24, 1814. | Cashiered Oct 8, 1820. |
| Archibald Henderson | Jun 4, 1806 | Mar 6, 1807 | Apl 1, 1811 | | Comd't Oct 17, 1820 | Comd't Jul 1, 1834 | | Major 1814; Lt. Col. Oct 17, 1820; Brig Gen. Jan 27, 1837 | Died Jan 6, 1859. |
| John Harris | Apl 23, 1814 | Jun 18, 1814 | Jun 13, 1830 | Oct 6, 1841 | | Col. Comd't Jan 7, 1859 | | Capt. Mar 3, 1825; Major Jan 27, 1837 | Died May 2, 1864. |
| Jacob Zeilin | Oct 1, 1831 | Sep 12, 1836 | Sep 14, 1847 | Jul 26, 1861 | | Comd't Jun 10, 1864 | Comd't Mar 2, 1867 | Major Jan 9, 1847 | Died Nov 18, 1880. |
| Charles G. McCawley | Mar 3, 1847 | Jan 2, 1855 | Jul 26, 1861 | Jun 10, 1864 | Dec 5, 1867 | Col. Comd't Nov 1, 1876 | | 1st Lt. Sep 13, 1847; Major Sep 8, 1863 | Retired Jan 29, 1891. Died Oct 13, 1891. |
| [1]Charles Heywood | Apl 5, 1858 | May 30, 1861 | Nov 23, 1861 | Nov 1, 1876 | Mar 9, 1888 | Col. Comd't Jan 30, 1891 | Comd't Mar 3, 1899 | Major Mar 8, 1862; Lt. Col. Aug 5, 1864 | In service. |
| Names. | | | | | | | | | |
| Adams, George | Mar 19, 1845 | | | | | | | 1st Lt. Aug 12, 1847 | Died Oct 21, 1856. |
| Adams, Samuel C. | Nov 25, 1861 | Nov 4, 1862 | | | | | | | Drowned Apl 1, 1864, Cairo, Ill. |
| Alexander, Philip | Nov 7, 1800 | Aug 1, 1802 | | | | | | | Resigned Mar 31, 1803. |
| Allen, Austin W. | Feb 10, 1838 | | | | | | | | Trans. to army Nov 26, 1838. |
| Allen, Nathaniel | Apl 3, 1810 | | | | | | | | Resigned Jul 24, 1810. |
| Allen, Samuel K. | Mar 12, 1868 | Jun 3, 1873 | | | | | | | Died Feb 18, 1884. |
| Anderson, Jeremiah | Jun 11, 1811 | | | | | | | | Dismissed in Sep, 1811. |
| Anderson, William | Feb 17, 1807 | Jan 23, 1809 | Jun 18, 1814 | | | | | Maj. Jun 1824; Lt. Col. May 24, 1828 | Died Jun 13, 1830. |
| Andresen, Charles G. | Jul 25, 1898 | Apl 13, 1899 | Jul 23, 1900 | | | | | Capt. Jul 13, 1900 | [4]Hon. dis. Feb 11, 1899. In service. |
| Appleton, Randolph M. | May 20, 1898 | | | | | | | | [5]Hon. dis. Jan 9, 1899. |
| Armistead, F. N. | Nov 13, 1830 | Jul 1, 1834 | | | | | | | Died Apl 14, 1841. |
| Armory, William | Jul 25, 1798 | Nov 10, 1799 | | | | | | | Resigned, date unknown. |
| Arrowsmith, Thomas | Apl 19, 1812 | | | | | | | | Resigned in Oct, 1812. |
| Ashton, Richard W. | Jan 28, 1817 | | | | | | | | Resigned Jan 22, 1821. |

| | | | | | | | | | |
|---|---|---|---|---|---|---|---|---|---|
| Auchmuty, Richard | Feb 28, 1815 | Apl 18, 1817 | | | | | | Capt. Apl 18, 1827 | Resigned Apl 1, 1830. |
| Babb, Macker | Dec 8, 1899 | Jul 23, 1900 | | | | | | | In service. |
| Backstrom, Theodore E. | Feb 17, 1900 | | | | | | | | In service. |
| Bacon, Samuel | Apl 14, 1812 | Jul 8, 1812 | Jun 18, 1814 | | | | | | Resigned in Nov, 1815. |
| Bacote, Thomas W. | Jun 24, 1813 | Jun 18, 1814 | | | | | | | Resigned Jun 4, 1815. |
| Bainbridge, Theodore | May 24, 1828 | | | | | | | | Resigned Jan 18, 1832. |
| Baker, Alonzo C. | Feb 5, 1902 | | | | | | | | In service. |
| Baker, Adam N. | Sep 12, 1853 | Aug 1, 1860 | | | | | | | [2]Joined C. S. A. Dis'd May 23, 1861. |
| Baker, D. D. | Oct 20, 1832 | Dec 30, 1837 | Sep 28, 1847 | | | | | Capt. Sep 13, 1847 | Died Aug 31, 1853. |
| Baker, Joseph F. | Jun 5, 1861 | Sep 1, 1861 | Jun 22, 1864 | | | | | | Died Oct 2, 1876. |
| Baldwin, Samuel | Jan 2, 1800 | Dec 10, 1800 | | | | | | | Last appearance in Register 1836. |
| Ball, Walter | Jul 1, 1894 | | | | | | | | Resigned May 7, 1895. |
| Banning, Edmund P. | Jul 2, 1864 | | | | | | | | Resigned Apl 13, 1870. |
| Bannon, Philip M. | Jul 1, 1897 | | Mar 3, 1899 | | | | | 1st Lt. Jun 13, 1898 | In service. |
| Barclay, Thomas | Aug 18, 1799 | Oct 21, 1801 | | | | | | | Resigned May 31, 1802. |
| Barnett, George | Jun 19, 1877 | Aug 31, 1890 | Aug 11, 1898 | Mar 3, 1901 | | | | | In service. |
| Barton, Thomas B. | Jun 10, 1817 | Oct 17, 1820 | | | | | | | Resigned Mar 13, 1829. |
| Bartlett, Henry A. | Nov 25, 1861 | Nov 26, 1861 | Nov 29, 1867 | Jan 30, 1891 | | | | | Retired Feb 1, 1898; died Aug 9, 1901 |
| Bates, George T. | Feb 19, 1873 | Jun 5, 1880 | May 2, 1894 | | | | | | Retired May 2, 1894; died Jun 30, 1901. |
| Bates, John S. | | Jul 1, 1899 | | | | | | | In service. |
| Baxter, Don A. | Jun 9, 1898 | | | | | | | | [5]Hon. dis. Feb 11, 1899. |
| Bayly, Robert P. | Jul 1, 1809 | | | | | | | | Died Aug 26, 1809. |
| Bearss, Hiram I. | May 26, 1898 | May 26, 1899 | Jul 23, 1900 | | | | | | [4]Hon. dis. Feb 11, 1899. In service. |
| Beaumont, John C. | Jan 26, 1900 | Jul 23, 1901 | | | | | | | In service. |
| Bell, Daniel | Nov 8, 1800 | | | | | | | | Resigned Feb 14, 1801. |
| Bellvue, F. B. de | Apl 24, 1812 | Jun 18, 1814 | Mar 3, 1819 | | | | | | Resigned Mar 6, 1824. |
| Benson, George R. | Dec 17, 1873 | Jun 5, 1880 | | | | | | | Died Apl 29, 1892. |
| Berkeley, Randolph C. | Aug 8, 1898 | Apl 13, 1899 | Jul 23, 1900 | | | | | | [4]Hon. dis. Jan 9, 1899. In service. |
| Berrett, John J. | Nov 19, 1840 | | | | | | | | Resigned Apl 5, 1843. |
| Berryman, O. C. | Jan 24, 1870 | Nov 16, 1877 | Jul 11, 1892 | Mar 3, 1899 | Aug 15, 1900 | | | | In service. |
| Betts, Charles | Mar 28, 1820 | | | | | | | | Died Jan 7, 1822. |
| Biddle, William P. | Jun 19, 1875 | Feb 24, 1884 | Feb 7, 1894 | Mar 3, 1899 | | | | | In service. |
| Bigelow, Horatio B. | Feb 6, 1865 | | | | | | | | Resigned Feb 28, 1870. |
| Bishop, Giles, Jr. | Jul 23, 1900 | | | | | | | | In service. |
| Bishop, Henry J. | Nov 25, 1861 | Apl 1, 1864 | Jun 12, 1876 | | | | | | Died Dec 22, 1884. |
| Bissett, Henry O. | Feb 15, 1899 | | Mar 3, 1899 | | | | | | In service. |
| Blake, Daniel W. | Feb 26, 1900 | | | | | | | | In service. |
| Bloodgood, W. A. | Mar 3, 1831 | | | | | | | | Appointed purser May 2, 1834. |
| Bond, Francis A. | Mar 1, 1815 | | | | | | | | Resigned in Nov. 1816. |
| Boone, Leonard J. | Jul 7, 1812 | Jun 18, 1814 | | | | | | | [3]Disbanded Apl 18, 1817. P. E. A. |
| Bootes, James T. | | Jul 1, 1899 | | | | | | | In service. |
| Borden, Thomas S. | Jul 1, 1894 | Aug 11, 1898 | Mar 3, 1899 | | | | | 1st Lt. Jul 3, 1898 | In service. |
| Bosque, Joseph | Feb 28, 1815 | | | | | | | | Died New Orleans, 1815. |

[1] Maj. Gen. Comdt. Jul 1, 1902.
[2] Resigned, but resignation not accepted.
[3] Peace Establishment Act.
[4] Appointed for S.-A. War; discharged and reappointed.
[5] Appointed for S.-A. War only.

## DATE OF ORIGINAL ENTRY AND PROGRESSIVE RANK.

| Names. | Second Lieutenant | First Lieutenant | Captain. | Major. | Lieutenant Colonel. | Colonel. | Brigadier-General. | Brevet Rank. | Remarks. |
|---|---|---|---|---|---|---|---|---|---|
| Boyd, William L. | Sep 17, 1813 | Jun 18, 1814 | | | | | | | Resigned Jul 26, 1814. |
| Boyd, William S. | Jan 12, 1848 | Dec 13, 1857 | Jul 26 1861 | | | | | | Resigned Jul 31, 1865. |
| Bourne, William T. | Aug 5, 1824 | | | | | | | | Died Mar 4, 1826. |
| Boyle, James H. | Jun 25, 1809 | Apl 27, 1810 | | | | | | | Resigned in 1812. |
| Brackett, William | Sep 11, 1900 | | | | | | | | In service. |
| Bradford, C. H. | Nov 25, 1861 | Nov 26, 1861 | | | | | | | Died Feb 15, 1864, from wounds received in battle. |
| Bradford, E. T. | Dec 18, 1868 | Aug 20, 1874 | | | | | | | Resigned Nov. 15, 1877. |
| Bradman, F. L. | Aug 2, 1898 | Apl 8, 1899 | Jul 23, 1900 | | | | | | [1]Hon. dis. Dec 31, 1898. In service. |
| Brady, Thomas A. | Feb 3, 1837 | Mar 3, 1847 | | | | | | | Died Nov 7, 1847. |
| Breckenbridge, H. B. | Apl 15, 1812 | Jun 18, 1814 | Dec 10, 1814 | | | | | | [2]Disbanded Apl 18, 1817, P.E.A. |
| Breckinridge, James C. | Jul 7, 1898 | Apl 13, 1899 | Jul 23, 1901 | | | | | | [1]Hon. dis. Feb 11, 1899. In service. |
| Breese, James B. | Mar 18, 1864 | May 1, 1868 | | | | | | | Resigned Dec 5, 1879. |
| Brevoort, A. N. | Mar 28, 1820 | Sep 26, 1823 | Mar 6, 1838 | | | | | Capt. Sep 26, 1833. | Died Nov 26, 1866. |
| Brewerton, G. D. | Mar 3, 1819 | Sep 17, 1821 | | | | | | | Died Jan 31, 1827. |
| Brewster, W. H. | Jan 24, 1838 | | | | | | | | Resigned Feb 1, 1839. |
| Brewster, Sidney W. | Feb 17, 1900 | | | | | | | | In service. |
| Britton, Abraham | Sep 9, 1818 | | | | | | | | Resigned in 1818. |
| Broatch, James W. | | Jul 1, 1899 | | | | | | | In service. |
| Brodhead, J. T. | Apr 12, 1872 | | | | | | | | Resigned Nov 14, 1879. |
| Bronson, Amon | May 6, 1898 | | | | | | | | Resigned May 20, 1898. |
| Brooke, B. E. | Jul 8, 1833 | Feb 7, 1839 | | | | | | | Died Nov 28, 1858. |
| Brooke, Edmund | Apr 15, 1814 | Jun 18, 1814 | | | | | | | Resigned Feb 13, 1817. |
| Brooks, John, Jr. | Oct 1, 1807 | Jan 30, 1809 | | | | | | | Killed in action Sep 10, 1813. |
| [3]Broom, Charles R. | Jul 27, 1813 | Jun 18, 1814 | Mar 7, 1824 | Sep 12, 1836 | | | | Lt. Col. Mar 7, 1834. | Died Nov 14, 1840. |
| Broom, James | Dec 28, 1836 | | | | | | | | Cashiered Jul 6, 1838. |
| Broome, James | Apr 21, 1810 | Apr 14, 1812 | | | | | | | Killed in action, Jun 1, 1813 |
| Broome, John L. | Jan 12, 1848 | Sep 28, 1851 | Jul 26, 1861 | Dec 8, 1864 | Mar 16, 1879 | | | Major Apl 24, 1862; Lt. Col. Mar 14, 1863. | Retired Mar 8. 1888; died Apl 12, 1898. |
| Brown, Philip S. | Jun 9, 1898 | Apr 13, 1899 | Jul 23, 1900 | | | | | | [1]Hon. dis. Jan 9, 1899. In service. |
| Brown, Rexwald | Jul 15, 1898 | | | | | | | | [4]Hon dis. Feb 11, 1899. |
| Brown, Thomas H. | Mar 19, 1900 | | | | | | | | In service. |
| Brown, William | Jan 16, 1817 | | | | | | | | Died Jun 19, 1818. |
| Brown, W. R. | Nov 25, 1861 | Aug 18, 1862 | May 18, 1871 | Jul 11, 1892 | | | | | Retired Sep 5, 1892; died Aug 2, 1894. |
| Browning, R. L. | Nov 24, 1852 | Nov 29, 1858 | | | | | | | Lost in Levant. |
| Brownlow, W. L. | Jul 6, 1812 | Jun 18, 1814 | Oct 17, 1820 | | | | | | Died Jul 17, 1821. |
| Brunzell, Albert N. | Feb 17, 1900 | | | | | | | | In service. |
| Buchanan, James A. | Mar 3, 1847 | | | | | | | | Last appearance in Register 1850. |
| Burke, Thomas P. | Nov 20, 1830 | | | | | | | | Cashiered Oct 3, 1831. |
| Burnham, James D. | Jul 1, 1825 | | | | | | | | Trans. to army Jun 30, 1826. |

| | | | | | | | | | |
|---|---|---|---|---|---|---|---|---|---|
| Burrough, Jehu A. | Jun 5, 1861 | Sep 1, 1861 | Sep 1, 1864 | | | | | | Died Nov. 28, 1867, West Indies. |
| Burton, Norman G. | Jul 1, 1899 | Jul 23, 1901 | | | | | | | In service. |
| Bush, William S. | Jul 3, 1809 | Mar 4, 1811 | | | | | | | Killed in action Aug 19, 1812. |
| Butler, George | Feb 11, 1859 | Jul 9, 1861 | Nov 4, 1862 | Mar 16, 1879 | | | | Major, Jan 14, 1865. | Died Feb 23, 1884. |
| Butler, Smedley D. | May 20, 1898 | Apr 8, 1899 | Jul 23, 1900 | | | | | Capt. Jul 13, 1900. | [1]Hon dis. Feb 11, 1899. In service. |
| Butterfield, William | Mar 3, 1847 | | | | | | | | Resigned May 1, 1858. |
| Buttrick, James T. | Jul 23, 1900 | | | | | | | | In service. |
| Caffery, St. John S. | Jan 26, 1900 | Jul 23, 1900 | | | | | | | Died Jun 4, 1901. |
| Caldwell, Henry | Sep 2, 1798 | Mar 2, 1799 | Jan 23, 1809 | | | | | | Died Mar 12, 1812. |
| Caldwell, R. C. | Oct 17, 1834 | Mar 3, 1845 | | | | | | | Died Nov 13, 1852. |
| Cammack, William | | Sep 1, 1798 | | | | | | | Resigned Dec 31, 1800. |
| Campbell, Chandler | Sep 11, 1900 | | | | | | | | In service. |
| Campbell, G. W. | Oct 14, 1852 | | | | | | | | Died Mar 23, 1855. |
| Carmich, Daniel | | | Aug 3, 1798 | Mar 7, 1809 | | | | | Died in 1816 from wounds received at New Orleans. |
| Carmody, Robert E. | May 26, 1898 | Apr 13, 1899 | | | | | | | [1]Hon. dis. Dec 31, 1898; died Oct 23, 1899—drowned. |
| Carpenter, C. C. | Jul 1, 1899 | Jul 23, 1900 | | | | | | | In service. |
| Carpenter, H. W. | May 26, 1898 | Apr 13, 1899 | Jul 23, 1900 | | | | | | [1]Hon. dis. Feb 11, 1899. In service. |
| Carter, Landon N. | Mar 26, 1824 | Apr 7, 1832 | Mar 3, 1847 | | | | | | Died Sep 26, 1847. |
| Cartter, W H | Mar 1, 1861 | Sep 1, 1861 | Jun 10, 1864 | | | | | | Dismissed Aug 12, 1865. |
| [5]Cash, John C. | Mar 16, 1845 | Sep 21, 1852 | Jun 5, 1861 | Nov 20, 1862 | | | | | Died Mar 8, 1877. |
| Catlin, Albertus W. | Jul 1, 1892 | Apr 19, 1893 | Mar 3, 1899 | | | | | | In service. |
| Chamberlin, Paul E. | Feb 17, 1900 | | | | | | | | In service. |
| Chapman, Melville D. | Aug 2, 1898 | | | | | | | | Resigned Dec 21, 1898. |
| Church, Albert H. | Aug 8, 1898 | | | | | | | | [4]Hon. dis. Dec 31, 1898. |
| Church, F. L. | Jul 12, 1862 | Aug 13, 1865 | | | | | | | Resigned Aug 8, 1868. |
| Church, Jonathan | | Sep 5, 1798 | | | | | | | Resigned Oct 26, 1801. |
| Clark, A. H. | Jul 1, 1884 | | | | | | | | Died at sea May 7, 1887. |
| Clark, Lemuel | | | Aug 3, 1798 | | | | | | Resigned Nov 30, 1801. |
| Clark, N. S. | Apr 20, 1812 | | | | | | | | Resigned, 1812. |
| Claypole, John | Mar 30, 1799 | | | | | | | | Resigned Dec 23, 1800. |
| Clements, James M. | Jan 28, 1817 | Mar 28, 1820 | | | | | | | Died May 27, 1822. |
| Clifford, W. H., Jr. | | Jul 1, 1899 | | | | | | | In service. |
| Clinch, B. | | Sep 5, 1798 | | | | | | | Resigned Jun 30, 1808. |
| Clinton, Thomas M. | Jul 23, 1900 | | | | | | | | In service. |
| Cochrane, Henry C. | Sep 7, 1861 | Aug 20, 1865 | Mar 16, 1879 | Feb 1, 1898 | Mar 3, 1899 | Jan 11, 1900 | | | In service. |
| Coejman, S. S. | Mar 28, 1820 | Mar 8, 1824 | | | | | | | Died Dec 26, 1825. |
| Coffin, H. G. | May 4, 1866 | | | | | | | | Dismissed Feb 17, 1872. |
| Cohen, D. M. | Aug 10, 1855 | 1861 | Jul 26, 1861 | Dec 5, 1867 | | | | | Retired Oct 12, 1869; died May 28, 1891. |
| Cole, Eli K. | Jul 1, 1890 | Jun 5, 1892 | Mar 3, 1899 | | | | | | In service. |
| Collier, G. W. | Sep 5, 1860 | Sep 1, 1861 | Nov 20, 1862 | Apl 18, 1880 | Aug 18, 1889 | | | | Retired Oct 23, 1889; died Dec 23, 1892. |

[1] Appointed for S. A. War, discharged and reappointed.
[2] Peace Establishment Act.
[3] Paymaster and captain.
[4] Appointed for S. A. War only.
[5] Paymaster with rank of major.

## DATE OF ORIGINAL ENTRY AND PROGRESSIVE RANK.

| Names. | Second Lieutenant | First Lieutenant | Captain. | Major. | Lieutenant Colonel. | Colonel. | Brigadier-General. | Brevet Rank. | Remarks. |
|---|---|---|---|---|---|---|---|---|---|
| [1]Collum, Richard S. | Sep 7, 1861 | Dec 30, 1862 | Mar 13, 1872 | Q. M. Jun 20, 1897. | | | | | Retired Jun 26, 1892. |
| Colton, Diamond | | Sep 5, 1798 | | | | | | | Resigned Jun 24, 1799. |
| Colvocoresses, Harold | Sep 5, 1900 | | | | | | | | In service. |
| Contee, John | Apl 17, 1812 | Jul 24, 1812 | | | | | | | Resigned Sep 13, 1815. |
| Cooper, George | Mar 28, 1820 | | | | | | | | Died Sep 15, 1823. |
| Corrie, F. H. | Nov 25, 1861 | Nov 26, 1861 | Feb 12, 1870 | | | | | Capt. Jan 14, 1865. | Retired Dec 18, 1885. |
| Coston, H. H. | Mar 9, 1865 | May 18, 1871 | Mar 9, 1888 | | | | | | Retired Jun 14, 1892; died Nov 6, 1896. |
| Cowan, William | Apl 16, 1812 | Aug 20, 1812 | | | | | | | Resigned Nov 20, 1812. |
| Coxe, Charles D. | Nov 18, 1805 | Mar 4, 1807 | | | | | | | Resigned Sep 18, 1809. |
| Coyle, William R | Jan 26, 1900 | Jul 23, 1900 | | | | | | | In service. |
| Crabb, H. N. | May 7, 1822 | Feb 23, 1830 | | | | | | | Resigned Dec 30, 1837. |
| Crabb, John | Jan 31, 1809 | Jun 28, 1809 | Jun 18, 1814 | | | | | | [2]Disbanded Apl 28, 1817, P. E. A. |
| Crane, Ichabod B. | Jan 26, 1809 | Jun 28, 1809 | | | | | | | Resigned Apl 28, 1812. |
| Cruise, Lawrence | Jul 29, 1806 | | | | | | | | Last appearance in Register 1806. |
| Cummins, E. H. | Nov 2, 1810 | Apl 14, 1812 | | | | | | | Resigned May 13, 1812. |
| Cummins, F. D. | Apl 14, 1810 | Apl 1, 1811 | | | | | | | Resigned Jan 21, 1812. |
| Curtis, Joseph W. | May 4, 1840 | Mar 16, 1847 | | | | | | | Dismissed Aug 26, 1852. |
| Cutts, Richard M. | Jul 1, 1899 | Jul 23, 1900 | | | | | | | In service. |
| Dallas, A. H. | Oct 31, 1854 | | | | | | | | Resigned Sep 24, 1855. |
| Daniels, C. L. | Jul 12, 1862 | Feb 13, 1865 | | | | | | | Resigned Jul 29, 1868. |
| Darley, John | Nov 8, 1798 | Apl 10, 1800 | | | | | | | Resigned Oct 31, 1800. |
| Davis, Austin R. | Jul 1, 1894 | Jun 2, 1898 | Mar 3, 1899 | | | | | | Died July 13, 1900. |
| Davis, Henry C. | May 20, 1898 | Apl 8, 1899 | Jul 23, 1900 | | | | | | [3]Hon. dis. Feb 11, 1899. In service. |
| Davis, Kent D. | Jul 2, 1864 | | | | | | | | Died Jan 11, 1865. |
| Dawson, L. L. | Jan 13, 1859 | 1861 | Nov 23, 1861 | May 16, 1878 | | | | Major Jan 14, 1865. | Resigned Dec 20, 1880. |
| Dawson, William C | Jul 1, 1894 | Feb 1, 1898 | Mar 3, 1899 | | | | | | In service. |
| Day, John H. A. | | Jul 1, 1899 | | | | | | | In service. |
| De Granpre, P. B. | Apl 26, 1812 | Jun 18, 1814 | | | | | | | [2]Disbanded Apl 18, 1817, P. E. A. |
| Delano, Frederick H. | Aug 2, 189 | May 26, 1899 | Jul 23, 1900 | | | | | | [3]Hon. dis Dec 31, 1898. In service. |
| [4]Denny, Frank L. | Jun 16, 1880 | Dec 23, 1884 | Asst. Q. M Feb 26, 1892 | Q. M. Jun 27, 1897. | | Q.M. Mar 3, 1899. | | | In service. |
| Desha, Robert M. | Feb 28, 1816 | Apl 12, 1817 | | | | | | | Died Nov 6, 1822. |
| Devereux, A. | Nov 25, 1861 | | | | | | | | Retired Mar 31, 1864. |
| Devlin, John S. | Feb 21, 1839 | Mar 3, 1847 | | | | | | | Cashiered Sep 20, 1852. |
| Devlin, Robert E. | Jul 25, 1898 | | | | | | | | Hon. dis. Dec 31, 1898. |
| Dewey, Rupert C. | Jul 23, 1900 | | | | | | | | In service. |
| Dickins, Randolph | May 19, 1876 | Mar 2, 1884 | May 2, 1894 | Mar 3, 1889 | | | | | In service. |
| Dietrich, J. P. | Oct 17, 1834 | | | | | | | | Cashiered Apl 27, 1838. |
| Dikeman, R. W. | Jul 23, 1900 | | | | | | | | In service. |

| | | | | | | | | | |
|---|---|---|---|---|---|---|---|---|---|
| Dix, Henry E. | Mar 13, 1816 | Apl 18, 1817 | | | | | | | Died Jan 21, 1822. |
| Doughty, Isaac T. | Sep 26, 1837 | Mar 3, 1847 | Mar 14, 1856 | Nov 23, 1861 | | | | | Died Jun 21, 1890. |
| Douglass, Richard | May 7, 1822 | Apl 26, 1825 | Mar 7, 1839 | | | | | | Died Apl 20, 1851. |
| Doyen, Charles A. | Jul 1, 1883 | Oct 21, 1889 | Jun 2, 1898 | Jan 31, 1900 | | | | | In service. |
| Draper, Herbert L. | Jul 1, 1889 | Jul 1, 1891 | Mar 3, 1899 | | | | | | Died Sep 20, 1901. |
| Dulaney, William | Jun 10, 1817 | Jun 19, 1819 | Jul 1, 1834 | Nov 17, 1847 | | Jul 26, 1861 | | Capt. Jun 19, 1829; Major Mar 3, 1843; Lt. Col. Sep 13, 1847. | Died Jul 4, 1868. |
| Duncan, John H. | Jan 16, 1817 | | | | | | | | Resigned Aug 11, 1817. |
| Dunlap, Robert H. | Aug 8, 1898 | Apl 8, 1899 | Jul 23, 1900 | | | | | | [3]Hon. dis. Feb 11, 1899. In service. |
| Dutton, R. McM. | Jul 1, 1891 | Mar 17, 1893 | Mar 3, 1899 | | | | | | In service. |
| Duval, Singleton | Mar 1, 1815 | | | | | | | | Resigned in Oct, 1816. |
| Edelin, James | Mar 1, 1815 | Apl 18, 1817 | Jul 1, 1834 | Sep 14, 1847 | Jan 7, 18 7 | | | Capt. Apl 28, 1827. | Died Jul 13, 1869. |
| Edie, John R. | July 1, 1892 | | | | | | | | Resigned Jul 20, 1892. |
| Edson, Alvin | May 7, 1822 | Mar 14, 1829 | Oct 6, 1841 | | | | | | Died Jul 15, 1847. |
| Edwards, James L. | Jun 18, 1811 | | | | | | | | Resigned in Feb, 1813. |
| Edwards, Philip | | Aug 3, 1798 | | | | | | | Killed in a duel Oct 16, 1800. |
| Egan, Harvey C. | Aug 27, 1902 | | | | | | | | In service. |
| Ela, F. P. | Jan 24, 1870 | | | | | | | | Drowned in East Indies, Feb 19, 1874. |
| Elliott, George F. | Oct 12, 1870 | Mar 30, 1878 | Jun 15, 1892 | Mar 3, 1899 | Sep 11, 1899 | | | | In service. |
| Elliott, Stephen | Jul 1, 1899 | | | | | | | | Retired Jan 13, 1903. |
| Ellis, Earl H. | Dec 6, 1901 | | | | | | | | In service. |
| Ellsworth, H. G. | Apl 4, 1870 | Nov 19, 1877 | | | | | | | Died Jun 25, 1890. |
| English, George B. | Mar 1, 1815 | | | | | | | | Resigned in 1817. |
| English, Thomas S. | Jun 10, 1817 | Aug 11, 1819 | Jun 1, 1834 | Feb 18, 1853 | | | | Capt. Aug 11, 1829 | Died Mar 26, 1871. |
| Eslick, Fred M. | Dec 8, 1899 | Jul 23, 1900 | | | | | | | In service. |
| Evans, Frank E. | Jan 26, 1900 | Jul 23, 1900 | | | | | | | In service. |
| Fagan, Louis E. | Jun 14, 1862 | Dec 8, 1864 | May 16, 1878 | | | | | 1st Lt. Sep 8, 1863; Cap. Jan 14, 1865 | Retired Jun 14, 1892; died Jun 4, 1894. |
| [5]Faunt Le Roy, R. P.[6] | Jul 7, 1898 | Apl 8, 1899 | A. Q. M., May 27, 1899 | | | | | | Resigned Oct 3, 1898; retired Dec 12, 1901. |
| Fay, William G. | Mar 21, 1900 | | | | | | | | In service. |
| Feland, Logan | | Jul 1, 1899 | | | | | | | In service. |
| Fendall, P. R., Jr. | Oct 17, 1857 | 1861 | Jul 26, 1861 | Nov 1, 1876 | | | | Maj. Jun 17, 1870 | Died Mar 21, 1879. |
| Fenwick, John R. | Nov 10, 1799 | Dec 1, 1801 | Aug 13, 1809 | | | | | | Resigned Apl 1, 1811. |
| Field, Thomas Y. | Mar 3, 1847 | Oct 15, 1854 | May 30, 1861 | Jun 10, 1864 | Nov 1 1876 | Apl 18, 1880 | | 1st Lt. Sep 13, 1847 | Retired Aug 17, 1889. |
| Fillette, T. Glover | Jun 16, 1880 | Jul 16, 1886 | | | | | | | Dismissed Jun 29, 1894. |

[1] Captain and Assistant Quartermaster May 4, 1885.
[2] Peace Establishment Act.
[3] Appointed for S. A. War, discharged and reappointed.
[4] Colonel, Quartermaster.
[5] Reappointed to Marine Corps Apl 8, 1899.
[6] Captain and Assistant Quartermaster.

## DATE OF ORIGINAL ENTRY AND PROGRESSIVE RANK.

| Names. | Second Lieutenant | First Lieutenant | Captain. | Major. | Lieutenant Colonel. | Colonel. | Brigadier-General. | Brevet Rank. | Remarks. |
|---|---|---|---|---|---|---|---|---|---|
| Fisher, H. C. | Sep 7, 1871 | Sep 27, 1879 | Mar 16, 1893 | | | | | | Retired Mar 16, 1893. |
| Flint, Kingman | 1863 | | | | | | | | Died Oct 15, 1863. |
| Floyd, C. C. | Sep 9, 1818 | Mar 24, 1821 | | | | | | | Resigned Dec 1, 1824. |
| Fontané, P. H. W. | Feb 22, 1857 | 1861 | Jul 26, 1861 | | | | | | Resigned May 19, 1864. |
| Foote, Yandell | Feb 17, 1900 | | | | | | | | Resigned Apl 28, 1902. |
| Ford, Christopher | Mar 1, 1815 | Apl 18, 1817 | | | | | | | Resigned Jun 13, 1819. |
| Ford, R. O'Neill | Nov 25, 1861 | Feb 26, 1864 | | | | | | | Resigned Apl 30, 1868. |
| Ford, Henry M. | Feb 1, 1809 | Apl 14, 1810 | Jun 18, 1814 | | | | | | Cashiered in 1816. |
| Forney, James | Mar 1, 1861 | Sep 1, 1861 | Apl 23, 1864 | Feb 24, 1884 | Jan 30, 1891 | Jul 11, 1892 | | Capt. Apl 24, 1862; Maj. Apl 15, 1869; Lt. Col Mar 15, 1870 | In service. |
| Foster, Joseph | Jun 29, 1809 | Apl 27, 1810 | | | | | | | Last appearance in Register, 1810. |
| Fowler, H. W. | May 26, 1824 | May 2, 1834 | | | | | | | Resigned Jun 8, 1836. |
| Frederick, Thomas | Jun 30, 1825 | | | | | | | | Transferred to army Jul 10, 1826. |
| Freeman, W. H. | Aug 17, 1812 | Jun 18, 1814 | Jul 17, 1821 | Jul 1, 1834 | | | | Lt. Col. Feb 20, 1832 | Died Mar 11, 1843. |
| French, L. P. | Jun 14, 1862 | Apl 23, 1864 | | | | | | | Resigned Dec. 30, 1873. |
| Frothingham, Louis A. | May 20, 1898 | | | | | | | | [1]Hon. dis. Jan 9, 1899. |
| Fryer, Eli T. | Mar 21, 1900 | | | | | | | | In service. |
| Fuller, Ben. H. | Jul 1, 1891 | Mar 16, 1893 | Mar 3, 1899 | | | | | | In service. |
| Gabaudan, E. C. | Mar 18, 1864 | | | | | | | | Died Mar 25, 1868. |
| Gamble, John M. | Jan 16, 1809 | Mar 5, 1811 | Jun 18, 1814 | Jul 1 1834 | | | | Maj. Apl 19, 1816; Lt. Col. Mar 3, 1827 | Died Sep 11, 1836. |
| Gardner, Henry W. | Jun 10, 1817 | Oct 8, 1819 | | | | | | | Died Apl 26, 1825. |
| Gardner, Robert | Feb 3, 1809 | | | | | | | | Died, date unknown. |
| Garland, Addison | Oct 17, 1834 | Nov 15, 1840 | Oct 15, 1854 | Jul 26, 1861 | | | | Capt. Mar 10, 1847 | Died Jun 20, 1864. |
| Garrard, W. C. | 1815 | | | | | | | | Resigned May 31, 1815. |
| Gassaway, John | Jun 18, 1810 | | | | | | | | Resigned Dec 29, 1810. |
| Geddis, S. W. | | Jul 26, 1798 | | | | | | | Resigned Apl 9, 1801. |
| Gibson, S. H. | Oct 5, 1869 | Oct 3, 1876 | | | | | | | Retired Jan 6, 1888. |
| Gillespie, A. H. | Oct 30, 1832 | Jan 8, 1838 | Nov, 1847 | | | | | Maj. Dec 6, 1846 | Resigned Oct 14, 1854. |
| Gilman, H. K. | Jun 16, 1880 | Dec 19, 1885 | | | | | | | Resigned Dec 22, 1890. |
| Gilson, Robert M. | Jun 9, 1898 | Apl 8, 1899 | Jul 23, 1900 | | | | | | [2]Hon. dis. Feb 11, 1899; in service |
| Glisson, H. Y. | Feb 27, 1866 | | | | | | | | Died Aug 13, 1867. |
| Goldsborough, L. M. | Mar 1, 1861 | Sep 1, 1861 | | | | | | | Died Oct 15, 1863. |
| [3]Goodloe, G. C. | Apl 21, 1869 | Jan 12, 1876 | | Mar 17, 1877 | | Mar 3, 1899 | | | In service. |
| Goodrell, M. C. | Mar 9, 1865 | Apl 16, 1870 | Jul 16, 1886 | | Mar 3, 1899 | | | | In service. |
| Graham, George R. | Jul 27, 1847 | Oct 22, 1856 | Jul 26, 1861 | Jun 21, 1864 | | | | | Retired Jan 21, 1876; died Jul 20, 1889. |
| Grant, Oscar B. | Nov 25, 1861 | Nov 26, 1861 | | | | | | | Resigned Apl 23, 1862. |

| | | | | | | | | | |
|---|---|---|---|---|---|---|---|---|---|
| Gray, Charles S. | Aug 2, 1898 | | | | | | | | Died Sep 3, 1898. |
| Grayson, Alfred | Jul 26, 1810 | Apl 14, 1812 | | | | | | Capt Jun 18, 1814 | Died Jun 28, 1823. |
| Grayson, John C. | May 4, 1840 | Mar 16, 1847 | Jan 7, 1859 | | | | | | Died Feb 3, 1888. |
| Green, Farnifold | Feb 23, 1830 | | | | | | | | Resigned Jun 6, 1831. |
| Green, Israel | Mar 3, 1847 | Jul 19, 1855 | | | | | | | [4]Dismissed May 18, 1861; joined C. S. A. |
| Green, Richard D. | Mar 1, 1815 | | | | | | | | Resigned Apl 1, 1823. |
| Greene, Edward A. | Jul 23, 1900 | | | | | | | | In service. |
| Greenleaf, Robert | Mar 16, 1801 | Apl 4, 1805 | Jan 19, 1811 | | | | | | Died in 1815. |
| Gridley, J. P. V. | Mar 26, 1900 | | | | | | | | In service. |
| Grimes, John H. | Jun 5, 1861 | Sep 1, 1861 | Aug 13, 1865 | | | | | | Dismissed Feb 11, 1870. |
| Grymes, Charles | Mar 3, 1819 | Jul 17, 1821 | | | | | | Capt. Jul 20, 1831 | Died July 25, 1834. |
| Gulick, Louis J. | Jun 22, 1874 | Dec 21, 1880 | Apl 18, 1893 | | | | | | Retired Apl 18, 1893; died Apl 20, 1896. |
| Gulick, Louis M. | May 20, 1898 | May 26, 1899 | Jul 23, 1900 | | | | | | [2]Hon. dis. Feb 4, 1899; in service. |
| Gunn, C. C. | Sep 12, 1812 | | | | | | | | Resigned in 1812. |
| Haines, Henry C. | Jul 1, 1883 | Jun 25, 1896 | Aug 11, 1898 | Aug 15, 1900 | | | | | In service. |
| Hale, W. H. | Jun 5, 1861 | Sep 1, 1861 | Dec 8, 1864 | | | | | | Died Aug 28, 1867. |
| Halford, Frank | Jul 23, 1900 | | | | | | | | In service. |
| Hall, Edward | Apl 15, 1799 | | Jan 23 1809 | | | | | | Resigned Aug 13, 1809. |
| Hall, Dickinson P. | | Jul 1, 1899 | | | | | | | In service. |
| Hall, Francis C. | Jul 5, 1825 | Jul 1, 1834 | Mar 16, 1847 | | | | | | Died Jul 13, 1853. |
| Hall, John | | Aug 2, 1798 | Dec 1, 1801 | Jun 8, 1814 | | | | | [5]Disbanded Apl 18, 1817, P. E. A. |
| Hall, Joseph C. | Jun 10, 1817 | Mar 3, 1819 | | | | | | Capt. Mar 3, 1829 | Died May 17, 1833. |
| Hall, Newt H. | Jul 1, 1897 | | Mar 3, 1899 | | | | | Major Aug 14, 1900. | In service. |
| Hall, William | Apl 18, 1812 | Sep 24, 1813 | Dec 18, 1814 | | | | | | [5]Disbanded Apl 18, 1817, P. E. A. |
| Hamersly, L. R. | Jul 23, 1866 | | | | | | | | Resigned Sep 21, 1869. |
| Hanna, Charles S. | Apl 23, 1810 | Apl 14, 1812 | Jun 18, 1814 | | | | | | [5]Disbanded Apl 18, 1817, P. E. A. |
| Harding, Arthur E. | | Jul 1, 1899 | | | | | | Capt. Jun 27, 1900. | In service. |
| Harding, L. M. | Jul 1, 1899 | Jul 23, 1900 | | | | | | | In service. |
| Hardy, J. L. C. | Mar 3, 1823 | Jun 13, 1830 | Mar 3, 1847 | | | | | | Died Nov 26, 1853. |
| Harllee, William C. | Feb 17, 1900 | Jul 23, 1900 | | | | | | | In service. |
| Harrington, F. H. | Dec 8, 1864 | Oct 13, 1869 | May 4, 1885 | Aug 10, 1898 | Mar 3, 1899 | Aug 15, 1900 | | | In service. |
| Harris, John C. | Nov 25, 1861 | Feb 16, 1864 | | | | | | 1st Lt. Apl 24, 1862. | Resigned July 31, 1869. |
| Harris, Lloyd G. | Mar 9, 1865 | | | | | | | | Resigned Apl 25, 1868. |
| Harwood, Richard | | Jul 10, 1798 | | | | | | | Resigned Nov 10, 1799. |
| Hasil, Andrew | 1808 | Jan 23, 1809 | | | | | | | Resigned Aug 2, 1809. |
| Hatch, Charles S. | May 26, 1898 | Apl 13, 1899 | Jul 23, 1900 | | | | | | [2]Hon. dis. Feb 11, 1899 In service. |
| Haverstick, John W. | Mar 18, 1864 | | | | | | | | Resigned Sep 27, 1869. |
| Haycock, George B. | Mar 10, 1863 | Jun 20, 1866 | Apl 18, 1880 | | | | | Capt. Jun 7, 1870. | Retired Mar 19, 1883. |

[1] Appointed for S.-A. War only.
[2] Appointed for S.-A. War, discharged and reappointed.
[3] Colonel and Paymaster.
[4] Resigned, but not accepted by the Department.
[5] Peace Establishment Act.

## DATE OF ORIGINAL ENTRY AND PROGRESSIVE RANK.

| Names. | Second Lieutenant | First Lieutenant | Captain. | Major. | Lieutenant Colonel. | Colonel. | Brigadier-General. | Brevet Rank. | Remarks. |
|---|---|---|---|---|---|---|---|---|---|
| Hayes, Edgar | Jul 23, 1900 | | | | | | | | In service. |
| Hays, A. J. | Dec 4, 1847 | Jul 17, 1857 | | | | | | | Resigned Mar 1, 1861. |
| Heath, John | Apl 18, 1811 | Mar 31, 1812 | Jun 18, 1814 | | | | | | [1]Disbanded Apl 18, 1817, P. E. A. |
| Hebb, C. D. | Mar 14, 1856 | May 7, 1861 | Jul 26, 1861 | Jan 12, 1876 | Apl 18, 1880 | Aug 18, 1889 | | | Retired Jul 10, 1892; died Jun 23, 1897. |
| Heisler, George | Nov 25, 1861 | | | | | | | | Died July 12, 1862. |
| Henderson, C. A. | Mar 16, 1847 | Dec 10, 1855 | | | | | | 1st Lt Sep 13, 1847. | Dismissed July 22, 1863. |
| Herbert, Hilary A., Jr. | Dec 5, 1900 | | | | | | | | In service. |
| Hervilly, D', James | Mar 5, 1872 | | | | | | | | Resigned May 1, 1880. |
| Higbee, John H. | Mar 9, 1861 | Sep 1, 1861 | Jun 10, 1864 | Aug 18, 1889 | Jul 11, 1892 | | | Capt. May 24, 1863. | Retired Jun 1, 1898. |
| Hill, Charles S. | Jun 9, 1898 | Apl 8, 1899 | Jul 23, 1900 | | | | | | [2]Hon. dis. Feb 11, 1899. In service. |
| Hine, Oliver C. | May 20, 1898 | | | | | | | | [3]Hon. dis. Dec 31, 1898. |
| Hirshinger, Herbert J. | Jan 11, 1900 | Jul 23, 1900 | | | | | | | In service. |
| Hitchcock, R. E. | Jun 5, 1861 | | | | | | | | Killed in action, Battle Bull Run, July 21, 1861. |
| Hoff, H. B. | Nov 25, 1861 | Nov 26, 1861 | | | | | | | Died Aug 17, 1862. |
| Hoist, Thomas | Jan 14, 1799 | | | | | | | | Last appearance in Register 1799. |
| Holcomb, Thomas, Jr. | Mar 21, 1900 | | | | | | | | In service. |
| Holmes, George | Mar 8, 1859 | | | | | | | | Resigned Feb 28, 1861. |
| Hooker, Richard S. | Feb 17, 1900 | Jul 23, 1900 | | | | | | | In service. |
| Hooper, Thomas W. | Feb 19, 1801 | Sep 1, 1802 | | | | | | | Resigned Mar 10, 1807. |
| Hopkins, S. G. | Apl 28, 1810 | | | | | | | | Last appearance in Register 1810. |
| Hopkins, William | | Jul 1, 1899 | | | | | | | In service. |
| Houston, George P. | Oct 23, 1860 | Sep 1, 1861 | Feb 6, 1864 | Dec 21, 1880 | Oct 24, 1889 | | | Major Aug 5, 1864. | Retired Feb 1, 1891; died Feb 7, 1897. |
| Howard, John | Jun 4, 1799 | Jul 1, 1801 | | | | | | | Resigned Feb 16, 1807. |
| Howell, Becket J. | Aug 1, 1860 | | | | | | | | Resigned Mar 1, 1861. |
| [1]Howle, Parke G. | Mar 1, 1815 | Apl 18, 1817 | Jul 1, 1834 | | | | | Capt. Apl 18, 1827. | Died July 16, 1857. |
| Hoyt, Samuel | Jul 5, 1809 | | | | | | | | Resigned Mar 14, 1810. |
| Huey, James McE. | Oct 2, 1889 | Jul 23, 1900 | | | | | | | In service. |
| Huger, Jacob M. | May 1, 1799 | | | | | | | | Died Nov 8, 1799. |
| Hughes, John A. | Dec 6, 1901 | | | | | | | | In service. |
| Humphrey, C. H. | Mar 10, 1863 | | | | | | | | Resigned in 1864. |
| Huntington, R. W. | Jun 5, 1861 | Sep 1, 1861 | Jun 21, 1864 | Oct 24, 1889 | Feb 2, 1897 | Aug 10, 1898 | | | Retired Jan 10, 1900. |
| Hyde, Benjamin | Jul 2, 1812 | 1813 | | | | | | | Died Feb 10, 1815. |
| Ingate, C. L. A. | Jul 1, 1890 | Apl 30, 1892 | Mar 3, 1899 | | | | | | Died Dec 24, 1899. |
| Ingraham, H. L. | Jul 16, 1858 | | | | | | | | Resigned Mar 8, 1861. |
| Irving, Edgar | Oct 1, 1833 | | | | | | | | Resigned Feb 27, 1835. |

| | | | | | | | | | |
|---|---|---|---|---|---|---|---|---|---|
| Jackson, S. L. | Jun 16, 1880 | Jan 7, 1888 | | | | | | | Died in Nov, 1890. |
| James, James | | Aug 9, 1798 | | | | | | | Resigned May 20, 1799. |
| Johnson, John | May 23, 1800 | Jun 1, 1802 | | | | | | | Resigned in 1809. |
| Johnson Samuel B. | Apl. 1814 | Jul 16, 1814 | | | | | | | Died May 19, 1820. |
| Jolly, Wade L. | Oct 2, 1899 | Jul 23, 1900 | | | | | | 1st Lt. Jun 27, 1900. | In service |
| [5]Jonas, Edwin A. | May 20, 1898 | May 26, 1899 | A. Q. M. Jan 13, 1902 | | | | | | [2]Hon. dis. Feb 11, 1899. In service. |
| Jones, Edward | Jan 7, 1859 | | | | | | | | Resigned Oct 22, 1860. |
| Jones, James H. | Mar 3, 1847 | Sep 1, 1853 | May 7, 1861 | | Jun 10, 1864 | Mar 16, 1879 | | | Died Apl 17, 1880. |
| Jones, Roger | Jan 26, 1809 | Jun 28, 1809 | | | | | | | Resigned Jul 7, 1812. |
| Junkin, W. D. A. | Dec 3, 1900 | | | | | | | | Resigned Mar 14, 1902. |
| Kane, T. P. | Jul 1, 1890 | Jun 15, 1892 | Mar 3, 1899 | | | | | | In service. |
| Karmany, L. | Jul 1, 1883 | Aug 18, 1889 | Feb 1, 1898 | Jan 11, 1900 | | | | | In service. |
| Keene, Newton | Dec 22, 1798 | Nov 1, 1800 | | | | | | | Resigned Nov 1, 1805. |
| Kellogg, Lyman | Jul 13, 1812 | Jun 18, 1814 | Mar 3, 1819 | | | | | | Resigned Feb 9, 1820. |
| Kelton, Allan C. | Mar 31, 1869 | Apl 1, 1875 | Jun 18, 1890 | Mar 3, 1899 | Jan 11, 1900 | | | Major Jun 11, 1898. | In service. |
| Kennedy, H. W. | Mar 1, 1815 | | | | | | | | Resigned Nov 9, 1816. |
| Kennedy, Philip C. | Feb 14, 1861 | Sep 1, 1861 | 1864 | | | | | | Died Aug 31, 1864. |
| Kidd, Robert | Nov 25, 1861 | Nov 26, 1861 | | | | | | | Dismissed Dec 20, 1862. |
| Kinne, Frank A. | Jul 25, 1898 | | | | | | | | [3]Discharged Nov 10, 1898. |
| Kintzing, M. R. | Sep 8, 1841 | Mar 16, 1847 | Aug 1, 1860 | | Jun 10, 1864 | Dec 5, 1867 | | | Retired Mar 15, 1879. |
| Kipp, Howard H. | Dec 3, 1900 | | | | | | | | In service. |
| Kirkland, W. W. | Jun 26, 1855 | | | | | | | | Resigned Aug 28, 1860. |
| Kuhn, Joseph L. | Jul 27, 1813 | Jun 18, 1814 | Jun 25, 1823 | | | | | | Dismissed April 7, 1832. |
| Lander, Frank C. | Mar 15, 1900 | | | | | | | | In service. |
| Lane, Enoch C. | Feb 27, 1801 | | | | | | | | Died Jul 16, 1804. |
| Lane, Rufus H. | Jul 1, 1833 | Jan 28, 1896 | Mar 3, 1899 | | | | | | In service. |
| Lang, William | Sep 30, 1831 | Jun 21, 1836 | Jan 16, 1847 | | | | | | Died May 6, 1850. |
| [6]Lauchheimer, C. H. | Sep 17, 1877 | Jan 9, 1890 | Aug 10, 1898 | A. A. & Ins. Mar 3 1899 | | | | | In service. |
| Lay, Harry R. | Jan 26, 1900 | Jul 23, 1900 | | | | | | | In service |
| Lee, Harry | Aug 2, 1898 | Apl 13, 1899 | Jul 23, 1900 | | | | | | [2]Hon. dis. Feb 11, 1899. In service. |
| Lee, Thomas | Nov 4, 1826 | | | | | | | | Resigned Sep 30, 1831, |
| Legge, Thomas | Aug 16, 1812 | Jun 18, 1814 | | | | | | | Resigned in 1816. |
| Lejeune, John A. | Jul 1, 1890 | Feb 26, 1892 | Mar 3, 1899 | | | | | | In service. |
| [5]Lemly, William B. | May 26, 1898 | Apl 8, 1899 | A. Q. M May 27, 1899 | | | | | | [2]Hon. dis. Feb 20, 1899. In service. |
| Leonard, Henry | Jun 9, 1898 | Apl 8, 1899 | Jul 23, 1900 | | | | | | [2]Hon. dis. Feb 11, 1899. In service. |
| Lewis, James | Sep 25, 1855 | 1861 | Jul 26, 1861 | Oct 13, 1869 | | | | | Resigned Jan 11, 1876. |
| Lewis, John L. | Aug 9, 1798 | Jun 24, 1799 | | | | | | | Resigned June 23, 1801. |
| Lilly, Reuben | | Sep 9, 1798 | | | | | | | Died Apl 9, 1800. |
| Lindsay, G. F. | Apl 1, 1823 | Sep 28, 1831 | Mar 3, 1847 | | | | | | Died Sep 27, 1857. |
| Lindsay, G. F., Jr. | Mar 16, 1847 | | | | | | | | Resigned Dec 31, 1852. |

[1] Peace Establishment Act.
[2] Appointed for S.-A. War, discharged and reappointed.
[3] Appointed for S.-A. War only.
[4] Adjutant and Inspector, with rank of Major.
[5] Captain and Assistant Quartermaster.
[6] Major and Assistant Adjutant and Inspector.

# DATE OF ORIGINAL ENTRY AND PROGRESSIVE RANK.

| Names. | Second Lieutenant | First Lieutenant | Captain. | Major. | Lieutenant Colonel. | Colonel. | Brigadier-General. | Brevet Rank. | Remarks. |
|---|---|---|---|---|---|---|---|---|---|
| Linton, Thomas A. | Feb 28, 1815 | Apl 18, 1817 | Apl 7, 1832 | Mar 12, 1845 | | | | Capt. Apl 18, 1827 | Died Feb 17, 1853. |
| Little, Louis M. | Jul 1, 1899 | Jul 23, 1900 | | | | | | | In service. |
| Little, M. M. | May 17, 1822 | | | | | | | | Resigned Feb 25, 1825. |
| Llewellin, Samuel | Jan 12, 1799 | Oct 1, 1801 | | | | | | | Resigned Aug 1, 1805. |
| Logan, Samuel J. | Apl 26, 1875 | Dec 12, 1883 | | | | | | | Retired Sep 25, 1885. |
| Long, Charles G. | Jul 1, 1891 | Sep 6, 1892 | Mar 3, 1899 | | | | | Capt. Jun 11, 1898 | In service. |
| Long, H. D. F. | Jan 26, 1900 | Jul 23, 1900 | | | | | | | In service. |
| Loomis, Erastus | Feb 26, 1816 | | | | | | | | Resigned Jul 15, 1816. |
| Lord, Charles | Sep 27, 1813 | Jun 14, 1814 | | | | | | | Died Oct 30, 1827. |
| Love, John C. | Jul 20, 1804 | Jul 4, 1805 | | | | | | | Died May 13, 1807. |
| Love, Marshall | Apl 3, 1830 | | | | | | | | Died Jul 28, 1832. |
| Lowndes, Edward R. | Jul 1, 1891 | Apl 18, 1893 | Mar 3, 1899 | | | | | | In service. |
| Low, Theodore H. | Jul 1, 1894 | Mar 3, 1899 | Mar 3, 1899 | | | | | | In service. |
| Low, William W. | | Jul 1, 1899 | | | | | | | In service |
| [1]Lowry, H B. | Nov 25, 1861 | Nov 26, 1861 | Oct 16, 1869 | May 4, 1885 | | | | Capt. Sep 8, 1863 | Retired Jun 19, 1897; died May 22, 1901. |
| Lowry, John | May 7, 1822 | | | | | | | | Resigned June 30, 1825. |
| Lucas, Lewis C. | July 1, 1891 | Jun 15, 1892 | Mar 3, 1899 | | | | | Capt. Jun 13, 1898 | In service. |
| Lucket, Lloyd | Jul 5, 1812 | | | | | | | | Last appearance in Register 1812. |
| Luckett, Nelson | 1807 | Jan 23, 1809 | | | | | | | Resigned in 1812. |
| Lyman, Charles H. | | Jul 1, 1899 | | | | | | | In service. |
| Lynch, James W. | | Jul 1, 1899 | | | | | | | Retired Oct 30, 1902. |
| Lyons, Thomas F. | | Jun 8, 1899 | | | | | | | In service. |
| McAllister, Arthur | Dec 6, 1901 | | | | | | | | In service. |
| McArdle, W. M. | Mar 3, 1835 | | | | | | | | Resigned Jan 31, 1837 |
| McCauley, Cleyburn | Aug 27, 1902 | | | | | | | | In service. |
| [2]McCauley, C. L. | | | Asst. Q. M. Jun 27, 1897 | Mar 3, 1899 | | | | Major Jun 11, 1898 | In service. |
| McCauley, Ed. | Jun 17, 1870 | | | | | | | | Resigned Apl 1, 1873. |
| McCauley, James | Mar 28, 1820 | Oct 6, 1822 | Sep 12, 1836 | | | | | Capt. Oct 6, 1832 | Died Feb 22, 1839. |
| McClaskey, John W. | Dec 5, 1900 | | | | | | | | In service. |
| McClean, James | Mar 23, 1813 | | | | | | | | Resigned Aug 30, 1813. |
| McCleary, W. | Dec 16, 1799 | | | | | | | | Died Dec. 31, 1800. |
| McConnell, Richard G. | Oct 2, 1899 | Jul 23, 1900 | | | | | | | In service. |
| McCormick, Provence, Jr. | Sep 11, 1900 | | | | | | | | In service. |
| McCreary, Wirt | Jul 27, 1899 | Jul 23, 1900 | | | | | | | In service. |
| McDonald, W. J., Jr. | May 21, 1868 | May 1, 1874 | | | | | | | Resigned Sep 1, 1879. |
| McDougal, Douglas C. | Feb 17, 1900 | | | | | | | | In service. |
| McDowell, J. E. | Jul 3, 1811 | | | | | | | | Resigned Mar 25, 1812. |
| McElrath, Thomas L. | Nov 25, 1861 | Nov 26, 1861 | | | | | | | Resigned Jun 19, 1866. |

| | | | | | | | | | |
|---|---|---|---|---|---|---|---|---|---|
| McGill, John F. | Jul 7, 1898 | May 26, 1899 | Jul 23, 1900 | | | | | | [3]Hon. dis. Feb 11, 1899. In service. |
| McKean, W. B. | Nov 25, 1861 | Nov 26, 1861 | Oct 13, 1869 | | | | | | Died Aug 30, 1879. |
| McKelvey, William N. | Jul 1, 1893 | Jun 20, 1897 | Mar 3, 1899 | | | | | Capt. Jun 11, 1898 | In service. |
| McKinnon, N. A. | Apl 21, 1812 | | | | | | | | Last appearance in Register 1812. |
| McKline, John. | Sep 9, 1818 | | | | | | | | Last appearance in Register 1818. |
| McKnight, James | | Aug 3, 1798 | Sep 1, 1798 | | | | | | Killed in a duel, Oct 14, 1802. |
| McLean, Alex. C. | Mar 11, 1829 | | | | | | | | Resigned May 8, 1833. |
| McLean, G. W. | May 8, 1833 | Mar 6, 1838 | | | | | | | Resigned Feb 7, 1839. |
| McLemore, Albert S. | Jul 1 1893 | Jun 14, 1896 | Mar 3, 1899 | | | | | Capt. Jun. 11, 1898 | In service. |
| McNeill, F. B. | Oct 17, 1834 | Apl 15, 1841 | | | | | | | Died Mar 13, 1856. |
| McSherry, W. K | Mar 31, 1869 | | | | | | | | Died Oct 13, 1872. |
| McWhorter, Jacob G. | Jul 1, 1884 | Dec 23, 1890 | | | | | | | Died Jun 27, 1891. |
| Macklin, Charles F. | Jul 1, 1894 | | | | | | | | Resigned June 30, 1896. |
| Macomber, Benjamin | Mar 28, 1820 | Apl 2, 1823 | Jan 8, 1838 | Jan 7, 1859 | | | | Capt. Apl 2, 1833 | Died in 1861. |
| Maddow, W. A. T. | Oct 14, 1837 | Mar 3, 1847 | Oct 26, 1857 | | | | | Capt. Jan 3, 1847 | Died Jan 1, 1889. |
| Madison, A. L. B. | Jun 30, 1809 | Nov 16, 1810 | | | | | | | Resigned Apl 16, 1812. |
| Madison, Ambrose. | Jul 2, 1817 | | | | | | | | Resigned Jul 21, 1818. |
| Magill, Louis J. | Jul 1, 1895 | | Mar 3, 1899 | | | | | 1st Lt. & Capt. Jun 13, 1898 | In service, |
| Maguire, James. | Feb 28, 1839 | | | | | | | | Resigned Jan 25, 1842. |
| Mahoney, J. E. | Jul 1, 1883 | Jul 1, 1890 | Mar 3, 1899 | | | | | Capt. Jun 11, 1898 | In service. |
| Maine, John. | Aug 12, 1798 | | | | | | | | Resigned Mar 14, 1799. |
| Mannix, D. P. | Dec 30, 1862 | Feb 12, 1870 | Dec 19, 1885 | | | | | | Died Feb 6, 1894. |
| Manwaring, E. B. | Jul 23, 1900 | | | | | | | | In service. |
| Marin, Arthur T. | May 26, 1898 | Apl 13, 1899 | Jul 23, 1901 | | | | | | [3]Hon. dis. Feb 11, 1899. In service. |
| Marston, Ward. | Mar 3, 1819 | Oct 30, 1821 | Jul 1, 1834 | Dec 10, 1855 | Jul 26, 1861 | | | Capt. Oct 30, 1831; Maj. Jan 2, 1847 | Died Apl 17, 1882, |
| Martin, Nicholas | Apl 26, 1810 | | | | | | | | Last appearance in Register 1810. |
| Massey, Lee. | 1807 | Jan 28, 1809 | | | | | | | Drowned Feb 7, 1812. |
| Mather, George H. | Dec 8, 1899 | Jul 23, 1900 | | | | | | | In service. |
| Mathews, S. H. | Nov 25, 1861 | Nov 26, 1861 | | | | | | | Dismissed Jul 1, 1863. |
| Matthews, Arthur J. | May 26, 1898 | Apl 13, 1899 | Jul 23, 1900 | | | | | | [3]Hon. dis. Feb 11, 1899. In service. |
| Matthews, Hugh L. | Feb 17, 1900 | May 19, 1902 | | | | | | | In service. |
| Mayson, F. G. | Nov 14, 1845 | | | | | | | | Resigned Jan 26, 1850. |
| Meade, Robert L. | Jun 14, 1862 | Apl 2, 1864 | Jan 22, 1876 | Sep 6, 1892 | Aug 10, 1898 | Mar 3, 1899 | | 1st Lt. Sep 8, 1863; Brig.-Gen. July 13, 1900 | In service. |
| Meeker, E. P. | Jun 14, 1862 | Nov 17, 1864 | Nov 19, 1877 | | | | | Capt. Jan 14, 1865 | Retired Mar 15, 1893. |
| Meiere, Julius E. | Apl 16, 1855 | 1861 | | | | | | | [4]Dis'd May 6, 1861; joined C. S. A. |
| Memminger, George. | | | Aug 3, 1798 | | | | | | Died Aug 31, 1798. |
| Mercer, Carroll. | Jun 16, 1880 | Feb 25, 1887 | | | | | | | Resigned Jul 1, 1890. |

[1] Major and Quartermaster.
[2] Major and Assistant Quartermaster.
[3] Appointed for S.-A. War, discharged and reappointed.
[4] Resignation tendered to join the South, but not accepted by the Department.

## DATE OF ORIGINAL ENTRY AND PROGRESSIVE RANK.

| Names. | Second Lieutenant | First Lieutenant | Captain. | Major. | Lieutenant Colonel. | Colonel. | Brigadier-General. | Brevet Rank. | Remarks. |
|---|---|---|---|---|---|---|---|---|---|
| Mercer, Samuel | Mar 12, 1868 | Jan 1, 1874 | Oct 24, 1889 | | | | | | Retired Jan 27, 1896; died Jul 22, 1896. |
| Miller, Ellis B. | Jul 23, 1900 | | | | | | | | In service. |
| Miller, E. R. | Feb 6, 1865 | | | | | | | | Retired Oct 12, 1869. |
| Miller, Louis G. | Feb 23, 1900 | | | | | | | | In service. |
| Miller, Samuel | Jun 1, 1808 | Mar 7, 1809 | Jun 18, 1814 | Jul 1, 1834 | Oct 6, 1841 | | | Maj. Aug 24, 1814; Lt.-Col. Mar 3, 1827 | Died Dec 9, 1855. |
| Mills, James J. | Mar 1, 1815 | Apl 18, 1817 | | | | | | | Resigned Sep 12, 1817. |
| Monell, Theodore | Jul 23, 1900 | | | | | | | | In service. |
| Montegot, T R. | Apl 15, 1812 | Jun 18, 1814 | | | | | | | [1]Disbanded Apl 18, 1817; P. E. A. |
| Morgan, John C. | Mar 10, 1865 | Jul 2, 1871 | | | | | | | Resigned Nov 28, 1877. |
| Morris, Gouverneur | Jul 2, 1864 | | | | | | | | Died Dec 25, 1865. |
| Mosby, Beverly C. | Jul 7, 1898 | | | | | | | | [2]Hon. dis. Dec 31, 1898. |
| Mosby, Robert | 1810 | Apl 12, 1812 | | | | | | | Resigned Apl 30, 1814. |
| Moseby, Joseph | Apl 20, 1810 | | | | | | | | Last appearance in Register 1810. |
| Moses, Franklin, J. | Jul 1, 1883 | Oct 27, 1890 | Mar 3, 1899 | | | | | | In service. |
| Moses, Lawrence H. | Jul 1, 1892 | Dec 26, 1893 | Mar 3, 1899 | | | | | | In service. |
| Mott, Thomas A. | Jan 26, 1900 | Jul 23, 1901 | | | | | | | In service. |
| Muir, John G. | Jan 13, 1900 | Jul 23, 1900 | | | | | | | In service. |
| Mullany, F. A. | Jul 23, 1866 | Jun 21, 1872 | | | | | | | Resigned May 11, 1880. |
| Munroe, Frank | Jun 5, 1861 | Sep 1, 1861 | Jun 10, 1864 | | | | | | Died Nov 18, 1877. |
| Murphy, P. St. C. | Feb 19, 1873 | Apl 18, 1880 | Mar 16, 1893 | Mar 3, 1899 | | | | Major Jul 3, 1898. | In service. |
| Murray, W. B. | Jul 2, 1864 | Aug 2, 1869 | | | | | | | Dismissed May 17, 1871. |
| Muse, W. S. | Mar 18, 1864 | Apl 27, 1867 | Dec 21, 1880 | Jun 2, 1898 | Mar 3, 1899 | Jan 31, 1900 | | | Retired Aug 14, 1900. |
| Myers, John T. | Feb 25, 1895 | Aug 11, 1898 | Mar 3, 1899 | | | | | Major Jul 20, 1900. | In service. |
| Neale, Joseph | May 8, 1799 | | | | | | | | Died Sep 22, 1800. |
| Neill, R. R. | Feb 6, 1865 | Oct 16, 1869 | | | | | | | Resigned June 2, 1873. |
| Neumann, B. S. | Jul 1, 1891 | Jul 11, 1892 | Mar 3, 1899 | | | | | | Dismissed May 17, 1902. |
| Neville, Francis S. | May 22, 1826 | | | | | | | | Struck off Apl 3, 1830. |
| Neville, Francis S. | Oct 30, 1830 | | | | | | | | Resigned July 15, 1833. |
| Neville, Wendell C. | Jul 1 1892 | Feb 7, 1894 | Mar 3, 1899 | | | | | Capt. Jun 13, 1898. | In service. |
| Newton, E. B. | Jan 16, 1817 | | | | | | | | Resigned Jan 1, 1819. |
| Nichols, Robert H. | Feb 27, 1801 | | | | | | | | Resigned Apl 5, 1801. |
| Nicholson, A. A. | Mar 13, 1816 | | | | | | | | Resigned July 10, 1817. |
| [3]Nicholson, A. A. | Mar 28, 1820 | May 27, 1822 | Jul 25, 1834 | | | | | Capt. May 27, 1832. | Died July 18, 1855. |
| [4]Nicholson, A. S. | Mar 16, 1847 | Mar 14, 1856 | | May 6, 1861 | | | | 1st Lt. Sep 13, 1847. | Retired May 1, 1894. |
| Nicholson, Jessup | Jun 7, 1873 | May 12, 1880 | | | | | | | Retired Feb 24, 1887; died May 22 1893. |

| | | | | | | | | | |
|---|---|---|---|---|---|---|---|---|---|
| Nicholson, Joseph S. | Mar 16, 1847 | | | | | | | 1st Lt. Sep 13, 1847. | Resigned Sep 5, 1850. |
| Nicoll, William | Dec 24, 1813 | Jun 18, 1814 | | | | | | | Died Mar 24, 1821. |
| Niven, Manchlin | May 20, 1898 | | | | | | | | [2]Hon dis. Dec. 31, 1898. |
| Noa, Walter E. | Dec 6, 1901 | | | | | | | | In service. |
| Nokes N. L. | Nov 25, 1861 | Jun 30, 1863 | Mar 20, 1872 | | | | | | Died Oct 7, 1883. |
| Norvell, Freeman | Mar 3, 1847 | | | | | | | 1st Lt. Sep 13. 1847. | Dismissed Jun 26, 1855. |
| Nowell, Ed. S. | May 9, 1815 | | | | | | | | Died July 7, 1817. |
| Nye, C. H. | Nov 25, 1861 | Nov 26, 1861 | | | | | | | Dropped Nov 16, 1864. |
| O'Bannon, P. N. | Jan 18, 1801 | Oct 15, 1802 | | | | | | | Resigned Mar 6, 1807. |
| O'Brien, A. H. | Mar 10, 1869 | Nov 29, 1874 | | | | | | | Resigned Mar 30, 1875. |
| O'Leary, Arthur J. | Jul 23, 1900 | | | | | | | | In service. |
| Olcott, Henry | Oct 19, 1812 | Jun 18, 1814 | | | | | | | Died Sep 17, 1821. |
| Osborne, W. S | Oct 9, 1800 | Jul 1, 1802 | | | | | | | Resigned Apl 26, 1806. |
| Page, John S. | Mar 13, 1816 | | | | | | | | Cashiered Jun 12, 1817. |
| Palmer, Aulick | Mar 12, 1866 | Mar 13, 1872 | | | | | | | Resigned Sep 26, 1879. |
| Parker, H. M. | Jan 23, 1809 | Jun 28, 1809 | | | | | | | Resigned Feb 28, 1812. |
| Parker, W. H | Jun 5, 1861 | Sep 1, 1861 | Dec 8, 1864 | | | | | Major Jan 15, 1864. | Died May 9, 1872. |
| Parker, William H. | Aug 8, 1898 | May 26, 1899 | Jul 23, 1901 | | | | | | [5]Hon. dis. Dec 31, 1898. In service. |
| Patterson, S. A. W. | Jan 11, 1900 | Jul 23, 1900 | | | | | | | In service. |
| Payne, John O. | Dec 2, 1853 | | | | | | | | Dismissed Oct 25, 1858. |
| Peet, F. T., Jr. | Jun 14, 1862 | Sep 1, 1864 | | | | | | | Resigned Aug 28, 1869. |
| Pendleton, Joseph H. | Jul 1, 1884 | Jun 28, 1891 | Mar 3, 1899 | | | | | | In service. |
| Perkins, C. Marrast | Jul 1, 1884 | Jan 30, 1891 | Mar 3, 1899 | | | | | | In service. |
| Perry, William F. | May 3, 1847 | | | | | | | | Resigned Nov 15, 1848. |
| Pickering, Woodell A. | Oct 21, 1900 | | | | | | | | In service. |
| Pile, Jones | Jul 12, 1862 | | | | | | | | Killed in action Dec 24, 1864. |
| Pinckney, Thomas H. | 1807 | Jan 23, 1809 | | | | | | | Resigned Aug 31, 1811. |
| Pope, Percival C. | Nov 25 1861 | Nov 26, 1861 | Apl 14, 1870 | Feb 2, 1891 | Jun 2, 1898 | Mar 3, 1899 | | Capt. Sep 8, 1863. | In service |
| Porter, Charles K. | 1817 | Mar 3, 1819 | | | | | | | Resigned Aug 12, 1819. |
| Porter, Carlisle P. | Dec 20, 1866 | Feb 1, 1873 | Aug 18, 1889 | | Mar 3, 1899 | | | | Retired Sep 10, 1899. |
| Porter, David D. | May 26, 1898 | Apl 8, 1899 | Jul 23, 1900 | | | | | Capt. Oct. 8, 1899. | [5]Hon. dis. Feb 11, 1899. In service. |
| Porter, James. | Dec 6, 1799 | | | | | | | | Resigned Jun 30, 1801. |
| Powell, S. W. | Nov 25, 1861 | Nov 20, 1862 | | | | | | | Resigned Feb 12, 1865. |
| Powell, William G. | | Jul 1 1899 | | | | | | Capt. June 21, 1900. | In service. |
| Powers, Darius C. | Jun 13, 1836 | | | | | | | | Dismissed Jul 10, 1837. |
| Prime, Joshua. | Apl 27, 1812 | | | | | | | | Died Feb 7, 1813. |
| [6]Prince, Thomas C. | Jul 1, 1884 | Feb 2, 1891 | A. Q. M. Jun 29, 1897 | Mar 3, 1899 | | | | | In service. |
| Pritchett, William H. | Mar 2, 1900 | | | | | | | | In service. |

[1] Peace Establishment Act.
[2] Appointed for S.-A. War only.
[3] Quartermaster, with the rank of Major.
[4] Adjutant and Inspector, with the rank of Major.
[5] Appointed for the S.-A. War, discharged and reappointed.
[6] Major and Assistant Quartermaster.

## DATE OF ORIGINAL ENTRY AND PROGRESSIVE RANK.

| Names. | Second Lieutenant | First Lieutenant | Captain. | Major. | Lieutenant Colonel. | Colonel. | Brigadier-General. | Brevet Rank. | Remarks. |
|---|---|---|---|---|---|---|---|---|---|
| Prochazka, Julius | Jul 1, 1891 | | | | | | | | Died Oct 10, 1892. |
| Purcell, Lee B | Oct 2, 1899 | Jul 23, 1901 | | | | | | | In service. |
| Quackenbush, S. W | Dec 4, 1869 | Sep 8, 1877 | Jun 15, 1892 | | | | | | Died Jun 13, 1896. |
| Queen, Henry W | Mar 14, 1842 | Sep 28, 1847 | | | | | | | Died at sea April 4, 1858. |
| [1]Radford, Cyrus S | Jul 1, 1892 | May 2, 1897 | Mar 8, 1899 | | | | | | In service. |
| Ragland, James | Feb 2, 1809 | | | | | | | | Resigned May 17, 1809. |
| Ramsay, Alan | Mar 1, 1857 | 1861 | Jul 26, 1861 | | | | | | Died Feb 15, 1864. |
| Ramsey, Fr-d A. | Dec 5, 1900 | | | | | | | | In service. |
| Randolph, W. A | Jul 30, 1823 | | | | | | | | Dismissed Oct 21, 1826. |
| Rankin, Robert | Sep 2, 1798 | Mar 2, 1799 | Jan 16, 1808 | | | | | | Resigned Jan 1, 1809. |
| Rask, Olof H | Oct 2, 1899 | Jul 23, 1900 | | | | | | | Died May 29, 1902. |
| Rathborne, J. H | Oct 25, 1858 | | | | | | | | Resigned Jan 16, 1862. |
| Read, Jacob | Mar 3, 1847 | Aug 19, 1855 | | | | | | | Resigned Feb 28, 1861. |
| Reber, John M | Jun 14, 1862 | | | | | | | | Dismissed Jun 20, 1863. |
| Reddick, Josiah | Sep 5, 1798 | May 25, 1799 | | | | | | | Resigned June 30, 1802. |
| Redles, William L | Sep 11, 1900 | | | | | | | | In service. |
| [2]Reid, George C | Jul 2, 1864 | Aug 29, 1869 | Apl 2, 1884 | A. & I. May 2, 1894 | | Mar 3, 1899 | | | In service. |
| Reid, George C | May 20, 1898 | Apl 8, 1899 | Jul 23, 1900 | | | | | | [7]Hon. dis. Feb 11, 1899. In service. |
| Reisinger, H. C | Jan 26, 1900 | Jul 23, 1900 | | | | | | | In service. |
| [3]Remey, W. B | Nov 25, 1861 | Feb 17, 1864 | Jun 21, 1872 | | | | | | Retired Jun 4, 1892; died Jan 20, 1895. |
| Reynolds, E. McD | Mar 3, 1847 | Nov 27, 1853 | May 24, 1861 | | | | | 1st Lt. Sep 13, 1847 | Dismissed Dec 7, 1864. |
| Reynolds, John G | May 26, 1824 | May 17, 1833 | Mar 3, 1847 | 1861 | Jul 26, 1861 | | | Maj. Sep 13, 1847 | Died Nov 2, 1865. |
| Reynolds, Michael | May 1, 1799 | Apl 9, 1801 | Jan 23, 1809 | | | | | | Resigned Jan 30. 1811. |
| Rhea, Robert Y | Mar 21, 1900 | | | | | | | | In service. |
| Rich, Jabez C | Jun 21, 1834 | Feb 24, 1839 | Nov 27, 1853 | | | | | | Dismissed May 22, 1861. |
| [4]Richards, George | Jul 1, 1893 | Feb 2, 1897 | | A. P. M. Mar 3, 1899 | | | | Lt. Col. Jul 13, 1900 | In service. |
| Richardson, Benj | Jun 5, 1813 | Jun 18, 1814 | | | | | | | Resigned Oct 1, 1824. |
| Riddle, R. B | Apl 24, 1810 | | | | | | | | Died in Sep, 1811. |
| Riggs, Myron C | Aug 18, 1857 | | | | | | | | Resigned Jan 9, 1858. |
| Rittenhouse, Benj. F | Aug 8, 1898 | May 26, 1899 | | | | | | | [7]Hon. dis. Oct 11, 1899. In service. |
| Rixey, P M., Jr | Feb 17, 1900 | | | | | | | | In service. |
| Robards, Frank F | Jul 23, 1900 | | | | | | | | In service. |
| Robins, G. W | Apl 26, 1832 | Dec 11, 1836 | | | | | | | Died Mar 1, 1845. |
| Robinson, E. R | Jul 2, 1864 | Sep 28, 1869 | Dec 23, 1884 | Aug 10, 1898 | | | | | Retired Jan 24, 1899. |
| Rogers, Austin R | Sep 11, 1900 | | | | | | | | In service. |
| Rogers, S. W | Mar 3, 1821 | | | | | | | | Died Sep 27, 1823. |
| Roosevelt, Henry L | Dec 8, 1899 | Jul 23, 1900 | | | | | | | In service. |

| | | | | | | | | | |
|---|---|---|---|---|---|---|---|---|---|
| Ross, Andrew | Mar 3, 1821 | Oct 1, 1824 | | | | | | | Died Dec 11, 1836, of wounds received in action. |
| Roumfort, A. L. | Jul, 1817 | | | | | | | | Died in 1818. |
| Russell, B. R. | Oct 16, 1869 | Nov 1, 1876 | Feb 2, 1891 | Mar 3, 1899 | Jan 31, 1900 | | | | Retired March 1, 1903. |
| Russell, John H., Jr. | Jul 1, 1894 | Aug 10, 1898 | Mar 3, 1899 | | | | | | In service. |
| [5]Russell, W. W. | Apl 5, 1843 | Nov 18, 1847 | | | | | | Capt. Dec 10, 1847 | Died Oct 31, 1862. |
| Salladay, Jas. M. | Dec 8, 1899 | Jul 23, 1900 | | | | | | | In service. |
| Saltmarch, E. G. | Jon 14, 1862 | Jun 10, 1864 | | | | | | | Resigned Jul 1, 1871. |
| Sawyer, Raymond E. | Jan 2, 1900 | | | | | | | | Resigned Oct 12, 1900. |
| Sayre, Calvin L. | Jun 3, 1858 | | | | | | | | Resigned Feb 14, 1861. |
| Schmumburgh, J. W. | Mar 14, 1829 | | | | | | | | Cashiered Oct 20, 1832. |
| [6]Schenck, W. S. | Jul 5, 1871 | Sep 2, 1879 | Jan 4, 1880 | | | | | | Retired Oct 12, 1891; died Jun 4, 1893. |
| Schermerhorn, J. | Jan 10, 1858 | Nov 16, 1861 | Nov 16, 1861 | | | | | | Died Jan 16, 1876. |
| Schwable, Frank J. | Mar 21, 1900 | | | | | | | | In service. |
| Scott, Frank | Dec 9, 1870 | May 16, 1878 | | | | | | | Resigned Jun 30, 1883. |
| Scott, Henry H. | Dec 3, 1900 | | | | | | | | Resigned Jul 31, 1901. |
| Scott, W. B. | Jul 4, 1809 | | | | | | | | Resigned Mar 13, 1810. |
| Sells, David M. | Nov 25, 1861 | | | | | | | | Resigned July 17, 1863. |
| Searcy, Lafayette | Oct 30, 1831 | | | | | | | | Resigned Oct 31, 1841. |
| Sears, Charles | Mar 28, 1820 | | | | | | | | Died May 29, 1821. |
| Sevier, Alexander | Apl 27, 1810 | Apl 17, 1812 | Jun 18, 1814 | | | | | Major, 1814 | Resigned Apl 3, 1816. |
| Shailer, Julius C. | Mar 21, 1870 | | | | | | | | Resigned Jun 15, 1877. |
| Shaw, Melville J. | Jul 1, 1896 | | Mar 3, 1899 | | | | | 1st Lt. Jun 11, 1898. | In service. |
| Sherburne, John H. | Mar 12, 1866 | Mar 20, 1872 | | | | | | | Resigned Apl 30, 1874. |
| Sherburne, W. W. | Sep 28, 1811 | | | | | | | | Resigned in 1811. |
| Sherodin, M. R. | Mar 20, 1799 | | | | | | | | Resigned July 31, 1802. |
| Sherman, C. L. | Jul 12, 1862 | Dec 9, 1864 | | | | | | 1st Lt. Aug 5, 1861. | Resigned Sep 7, 1877. |
| Shuttleworth, W. L. | Feb 28, 1839 | Mar 16, 1847 | Sep 28, 1857 | | | Jun 10, 1864 | | Capt. Mar 10, 1847. | Died Sep 27, 1871. |
| Sibley, Berton W. | Jul 23, 1900 | | | | | | | | In service. |
| Simms, John D. | Oct 7, 1841 | Sep 27, 1847 | 1861 | | | | | Capt. Sep 13, 1847. | [8]Dismissed Jul 8, 1861; joined C. S. A. |
| Singletary, J. G. | Mar 1, 1815 | | | | | | | | Last appearance in Register 1815. |
| [9]Slack, W. B. | Jun 28, 1839 | Mar 3, 1847 | Feb 22, 1857 | Aug 13, 1860 | | | | Capt. Mar 10, 1847. | Retired May 3, 1885; died Dec 3, 1895. |
| Slack, W. B., Jr. | May 12, 1869 | | | | | | | | Died Sep 27, 1874. |
| Sloan, Thomas T. | Oct 17, 1834 | Oct 31, 1840 | | | | | | | Died Feb 10, 1850. |
| Smalley, E. A. | Nov 25, 1861 | Nov 26, 1861 | | | | | | | Resigned July 26, 1866. |
| Smith, Constantine | Aug 27, 1825 | | | | | | | | Transferred to Army Nov 20, 1830. |
| Smith, Jacob G. | Sep 1, 1802 | | | | | | | | Resigned Mar 24, 1809. |
| Smith, Richard | Jul 20, 1806 | Mar 8, 1807 | Mar 13, 1812 | | | | | Major Jun 18, 1814; Lt. Col. Jun 18, 1824. | Cashiered Feb 23, 1830. |

1 Captain and Assistant Quartermaster.
2 Colonel and Adjutant and Inspector.
3 Judge Advocate-General, with the rank of Colonel.
4 Major and Assistant Paymaster.
5 Paymaster, with rank of Major.
6 Captain and Assistant Quartermaster.
7 Appointed for S.-A. War, discharged and reappointed.
8 Resignation tendered to join the South, but not accepted by the Department.
9 Quartermaster, with the rank of Major.

## DATE OF ORIGINAL ENTRY AND PROGRESSIVE RANK.

| Names. | Second Lieutenant | First Lieutenant | Captain. | Major. | Lieutenant Colonel. | Colonel. | Brigadier-General. | Brevet Rank. | Remarks. |
|---|---|---|---|---|---|---|---|---|---|
| Smith, Richard L | Apl 26, 1812 | | | | | | | | Resigned July 8, 1812. |
| Smith, William E | Jul 23, 1900 | | | | | | | | In service. |
| Smyser, John D | Mar 21, 1870 | | | | | | | | Retired May 3, 1878. |
| Snowden, Charles | Mar 1, 1815 | | | | | | | | Last appearance in Register 1815 |
| Snyder, Harold C | Jul 1, 1899 | Jul 23, 1900 | | | | | | | In service. |
| South, Hamilton D | Jul 23, 1900 | | | | | | | | In service. |
| Sparks, Alex. D | Dec 10, 1855 | | | | | | | | Resigned in 1855. |
| Spearing, Charles F | May 7, 1822 | Jan 31, 1827 | | | | | | | Cashiered Sep 23, 1841. |
| Spicer, W. F. | Mar 13, 1872 | Jan 4, 1880 | Sep 6, 1892 | Mar 3, 1899 | Mar 3, 1901 | | | | In service |
| Sprague, John T | Oct 17, 1834 | | | | | | | | Resigned July 3, 1837. |
| Squires, W. J | Nov 25, 1861 | Nov 26, 1861 | Aug 24, 1867 | | | | | | Resigned Jan 12, 1870 |
| Stafford, Leroy A | Jul 1, 1890 | | | | | | | | Resigned June 30, 1892. |
| Stark, Alex W | Jul 19, 1855 | Feb 28, 1861 | | | | | | | Dismissed Jan 9, 1862. |
| Stark, W. B. | Jul 1, 1831 | Jul 1, 1834 | Mar 16, 1847 | | | | | | Died Aug 18, 1855. |
| Stayton, William H | Jul 1, 1883 | Jun 18, 1890 | | | | | | | Resigned Jun 30, 1891. |
| Stephen, Henry | Feb 28, 1815 | | | | | | | | Resigned Mar 20, 1816. |
| Sterne, Francis W | Apl 22, 1812 | Jul 24, 1813 | Dec 10, 1814 | | | | | | [1]Disbanded Apl 18, 1817, P. E. A. |
| Stevenson, Andrew | Mar 26, 1875 | | | | | | | | Resigned Apl 25, 1879. |
| Stewart, C. G. | 1808 | 1809 | | | | | | | Last appearance in Register 1809. |
| Stewart, Richard | Apl 23, 1812 | | | | | | | | Last appearance in Register 1812 |
| Stickney, David | | Oct 31, 1798 | | | | | | | Resigned Sep 30, 1801. |
| Stillman, C. A. | Nov 25, 1861 | Nov 26, 1861 | Dec 5, 1867 | | | | | | Retired Mar 12, 1872. |
| Stoddard, George C | Jun 14, 1862 | Jun 10, 1864 | | | | | | Capt Dec 6, 1864 | Died Apl 26, 1867. |
| Strong, William | Mar 1, 1811 | Apl 16, 1812 | Jul 18, 1814 | | | | | | Resigned Oct 23, 1818. |
| Strother, Benjamin | | Oct 31, 1798 | | | | | | | Resigned Oct 31, 1800. |
| Stuart, Alex | Jun 1, 1806 | | | | | | | | Resigned Aug 6, 1806. |
| Sturgeon, E. B | Nov 25, 1861 | | | | | | | | Dismissed Jul 1, 1863. |
| Sullivan, R. B | Jul 23, 1900 | | | | | | | | In service. |
| [2]Sutherland, D. J | Mar 29, 1842 | 1847 | | | | | | Capt. Sep 18, 1847. | Dismissed Jul 23, 1860. |
| Sutton, Francis E | Jul 1, 1883 | Mar 9, 1888 | | | | | | | Drowned at Apia, Samoa, Mar 16, 1889. |
| Swift, Thomas R | 1807 | Jan 27, 1809 | Jun 18, 1814 | | | | | | [1]Disbanded Apl 18, 1817, P. E. A. |
| Swift, William F | Mar 1, 1815 | | | | | | | | Last appearance in Register 1815 |
| Tallman, James | | Sep 5, 1798 | | | | | | | Dismissed Jan 2, 1799. |
| Tansill, Robert | Nov 3, 1840 | Mar 16, 1847 | Nov 29, 1858 | | | | | Capt. Nov 17, 1847. | [3]Dismissed Aug 24, 1861; joined C. S. A. |
| Tattnall, J. R. F | Nov 3, 1847 | Feb 22, 1857 | | | | | | | [3]Dismissed Nov 22, 1861; joined C. S. A. |
| Taussig, Charles A | Jul 7, 1898 | | | | | | | | [4]Hon. dis. Jan 9, 1899. |
| Taylor, A. S. | Jul 2, 1864 | Jul 30, 1868 | Dec 12, 1883 | | | | | | Retired Jun 17, 1890. |
| Taylor, Algernon S | Feb 21, 1839 | Mar 3, 1847 | Jul 17, 1857 | | | | | Capt. Mar 27, 1847. | [3]Dismissed May 6, 1861; joined C. S. A. |

| | | | | | | | | | |
|---|---|---|---|---|---|---|---|---|---|
| Taylor, Charles B. | Jan 26, 1900 | Jul 23, 1900 | | | | | | | In service. |
| Taylor, R. D. | Mar 2, 1839 | | | | | | | | Died Nov 13, 1845. |
| Terrett, George H. | Apl 1, 1830 | Jul 1, 1834 | Mar 16, 1847 | | | | | Major Sep 13, 1847. | [3]Dismissed May 6, 1861; joined C. S. A. |
| Thayer, Cotton | Jun 25, 1799 | | | | | | | | Resigned Jul 15, 1801. |
| Theall, Elisha | Jul 1, 1893 | Feb 2, 1897 | | | | | | | Retired Feb 2, 1897. |
| Thomas, Frederick | Jul 1, 1825 | | | | | | | | Transferred to Army Jul 14, 1826. |
| Thompson, George H. | Mar 11, 1865 | | | | | | | | Died at sea May 2, 1868. |
| Thompson, Gillies | Apl 12, 1815 | | | | | | | | Resigned Dec 12, 1815. |
| Thompson, James | May 2, 1799 | Jul 1, 1800 | Jan 23, 1809 | | | | | | Resigned Jan 12, 1810. |
| Thornton, Francis | Apl 25, 1810 | Apl 3, 1811 | | | | | | | Resigned in 1817. |
| Thorpe, George C. | May 20, 1898 | Apl 8, 1899 | Jul 23, 1900 | | | | | Capt. Oct 8, 1899. | [5]Hon. dis. Feb 11, 1899. In service. |
| Tilton, McLane | Mar 2, 1861 | Sep 1, 1861 | Jun 10, 1864 | Mar 9, 1888 | Feb 2, 1891 | | | | Retired Feb 1, 1897. |
| Towns, Thomas | Aug 30, 1805 | | | | | | | | Resigned Mar 31, 1806. |
| Tracy, James K. | Jul 23, 1900 | | | | | | | | In service. |
| Treadwell, Thomas C. | Jul 1, 1892 | May 3, 1894 | Mar 3, 1899 | | | | | | In service. |
| Tupper, Charles C. | Mar 3, 1819 | Jan 21, 1822 | Jul 1, 1834 | | | | | Capt. Jan 18, 1832 | Died Jan 18, 1838. |
| Turner, George P. | Sep 27, 1856 | | | | | | | | [3]Dismissed Jun 25, 1861; joined C. S. A. |
| Turner, James A. | Jun 16, 1880 | May 4, 1885 | | | | | | | Retired Dec 25, 1893; died Jan 23, 1895. |
| Turner, Thomas C. | Aug 27, 1902 | | | | | | | | In service. |
| Turner, William C. | Mar 12, 1875 | Oct 8, 1883 | | | | | | | Died Aug 31, 1890. |
| Turrill, Julius S. | Oct 2, 1899 | Jul 23, 1900 | | | | | | | In service. |
| Twiggs, Levi | Nov 10, 1813 | Jun 18, 1814 | Feb 28, 1830 | Nov 15, 1840 | | | | Capt. Mar 3, 1825 | Killed in battle Sep 13, 1847. |
| [6]Tyler, H. B. | Mar 3, 1823 | Apl 1, 1830 | Mar 12, 1845 | | | | | | [3]Dismissed May 4, 1861; joined C. S. A. |
| Tyler, H. B., Jr. | Jan 2, 1855 | | | | | | | | [3]Dismissed June 21, 1861; joined C. S. A. |
| Udell, Fred A. | Sep 11, 1900 | | | | | | | | In service. |
| Urquehart, John | Mar 5, 1811 | Apl 30, 1812 | | | | | | | Resigned in 1812. |
| Van Orden, George | | Jul 1, 1899 | | | | | | | In service. |
| Wadleigh, John W. | Jan 26, 1900 | Jul 23, 1900 | | | | | | | In service. |
| Wainwright, R. D. | Feb 15, 1807 | Jan 23, 1809 | Sep 29, 1812 | | Jul 1, 1834 | | | Maj Mar 3, 1823; Lt. Col. Mar 3, 1827 | Died Oct 5, 1841. |
| Wainwright, R. D. | Dec 4, 1869 | Mar 17, 1877 | Jun 5, 1892 | | | | | | Retired Apl 17, 1893. |
| Wait, Marmaduke | Jan 16, 1801 | | | | | | | | Last appearance in Register 1801. |
| Waldron, N. S. | Sep 13, 1831 | Jul 25, 1834 | Mar 16, 1847 | | | | | Maj. Jul 22, 1848 | Died Feb 21, 1857. |
| Wales, Samuel | Nov 10, 1800 | | | | | | | | Resigned Feb 20, 1801. |
| [7]Walker, G. W. | Jun 16, 1817 | Mar 3, 1821 | Jul 1, 1834 | | | | | Capt. Mar 3, 1831 | Died Aug 29, 1851. |
| Walker, Ralph E. | May 20, 1898 | | Mar 3, 1899 | | | | | | Retired Feb 15, 1901. |

[1] Peace Establishment Act.
[2] Quartermaster, with the rank of Major.
[3] Resignation tendered to join the South, but not accepted by the Department.
[4] Appointed for S.-A. War only.
[5] Appointed for S.-A. War, discharged and reappointed.
[6] Adjutant and Inspector, with rank of Major.
[7] Brevet Captain and Quartermaster.

## DATE OF ORIGINAL ENTRY AND PROGRESSIVE RANK.

| Names. | Second Lieutenant | First Lieutenant | Captain. | Major. | Lieutenant Colonel. | Colonel. | Brigadier-General. | Brevet Rank. | Remarks. |
|---|---|---|---|---|---|---|---|---|---|
| Wallace, Rush R., Jr. | Dec 29, 1899 | Jul 23, 1900 | | | | | | | In service. |
| Wallace, William | Jun 14, 1862 | Jun 10, 1864 | Oct 3, 1876 | | | | | 1st Lt. Sep 8, 1863; Capt. Jan 14, 1865 | Died Dec 11, 1883. |
| Wallach, Richard | Apl 24, 1869 | Jan 22, 1876 | Jan 30, 1891 | Mar 3, 1899 | | | | | Retired Jul 24, 1899. |
| Waller, L. W. T. | Jun 16, 1880 | Sep 26, 1885 | Jun 14, 1896 | Jul 25, 1899 | | | | Lt. Col. Jul 13, 1900 | In service. |
| Ward, A. W. | Jun 14, 1862 | | | | | | | | Died Mar 17, 1867. |
| Washburn, I. H. | Mar 18, 1864 | Aug 20, 1867 | Mar 20, 1883 | | | | | | Retired Jul 15, 1886; died Feb 6, 1896. |
| Watkins, T. L. C. | May 22, 1826 | Jul 1, 1834 | | | | | | | Died Oct 1, 1840. |
| Watson, Arthur L. | Mar 9, 1865 | Mar 18, 1871 | | | | | | | Died Jun 30, 1882. |
| Watson, H. B. | Oct 5, 1836 | Mar 3, 1847 | | | | | | Capt. Nov 20, 1847 | Resigned Jan 1, 1855. |
| Watson, Josiah | Jun 21, 1835 | Mar 3, 1847 | Dec 10, 1855 | Nov 16, 1861 | | | | | Died Feb 5, 1864. |
| Watson, Samuel E. | Jul 4, 1812 | Jun 18, 1814 | Mar 28, 1820 | Jul 1, 1834 | | | | Lt. Col. Mar 28, 1830 | Died Nov 17, 1847. |
| Weaver, James | | Sep 27, 1798 | | | | | | | Resigned Aug 31, 1803. |
| Webster, F. D. | Mar 18, 1864 | Dec 5, 1867 | Oct 8, 1883 | | | | | | Retired Apl 1, 1884. |
| Webster, Le Roy C. | Jul 1, 1874 | Jul 1, 1883 | Apl 18, 1893 | | | | | | Died Jun 17, 1899. |
| [1]Weed, Elijah J. | Jan 16, 1817 | Mar 3, 1819 | Jul 1, 1834 | | | | | Capt Nov 7, 1828 | Died Mar 5, 1838. |
| Welch, Henry | Mar 3, 1847 | | | | | | | | Died Aug 27, 1847. |
| Welles, George M. | Jul 12, 1862 | Jan 11, 1865 | | | | | | | Died Oct 18, 1883. |
| West, Ernest E. | | Jul 1, 1899 | | | | | | | In service. |
| West, Edward L. | Oct 17, 1834 | Oct 6, 1841 | | | | | | | Died Mar 30, 1851. |
| Westcott, Charles T., Jr. | Feb 17, 1900 | | | | | | | | In service. |
| Wharton, Thomas | Jan 28, 1799 | Oct 17, 1800 | | | | | | | Died in 1801. |
| Whetcroft, W. W. | Feb 19, 1817 | | | | | | | | Resigned Aug 3, 1820. |
| Whipple David | May 29, 1871 | Mar 16, 1879 | | | | | | | Died Mar 1, 1884. |
| White, F. B. | Sep 21, 1813 | Jun 18, 1814 | | | | | | | Killed in a duel Sep 25, 1819. |
| White, Harry K. | Jul 1, 1883 | Mar 26, 1889 | Feb 2, 1897 | Sep 11, 1899 | | | | | In service. |
| Whiting, Henry | Jul 1, 1874 | Mar 20, 1883 | | | | | | | Died Jan 8, 1890. |
| Whitney, L. F. | Oct 17, 1834 | | | | | | | | Cashiered Mar 14, 1840. |
| Whittier, Rufus A. | Oct 22, 1856 | | | | | | | | Dismissed June 30, 1858. |
| [2]Wiley, James | Jan 9, 1847 | Sep 27, 1856 | Jul 26, 1861 | | | | | | Retired Jun 20, 1872; died Jul 7, 1895. |
| Williams, A. S. | Oct 2, 1899 | Jul 23, 1900 | | | | | | | In service. |
| Williams, Charles F. | Jun 14, 1862 | Jun 10, 1864 | Nov 1, 1876 | Feb 2, 1897 | | Mar 3, 1899 | | Capt. Jan 14, 1865 | Died Jan 30, 1900. |
| Williams, Dion | Jul 1, 1893 | Jun 30, 1894 | Mar 3, 1899 | | | | | | In service. |
| Williams, H. A. | Sep 28, 1798 | | | | | | | | Died Jul 1, 1800. |
| Williams, John | Aug 20, 1805 | Mar 2, 1807 | Jan 31, 1811 | | | | | | Died of wounds received in action, Aug, 1812. |
| Williams Job G. | May 7, 1822 | Dec 27, 1825 | Nov 15, 1840 | | | | | | Died Sep 26, 1856. |

| | | | | | | | | | |
|---|---|---|---|---|---|---|---|---|---|
| Williams, R. P. | Oct 2, 1899 | Jul 23, 1901 | | | | | | | In service. |
| Wilmer, E. P. | Jan 23, 1809 | | | | | | | | Resigned June 19, 1809. |
| Wilson, Isaac R. | May 10, 1838 | | | | | | | | Resigned May 29, 1841. |
| Wilson, Thomas S. | Dec 13, 1857 | | | | | | | | [3]Dismissed Aug 24, 1861; joined C. S. A. |
| Wilson, W. D. | Jan 23, 1809 | Jun 28, 1809 | | | | | | | Resigned Sep 14, 1809. |
| Wiltse, Frank S. | Oct 22, 1900 | | | | | | | | In service. |
| Winthrop, P. W. | 1808 | Feb 13, 1809 | | | | | | | Resigned May 6, 1811. |
| Wise, Frederic M., Jr. | Jul 1, 1899 | Jul 23, 1900 | | | | | | | In service. |
| Wood, Thomas N. | Jul 1, 1876 | Apl 2, 1884 | Jan 28, 1896 | Jun 18, 1899 | | | | | In service. |
| Woodson, Joseph | Apl 22, 1810 | Apl 2, 1811 | | | | | | | Resigned June 17, 1814. |
| Woog, Benjamin B. | | Jul 1, 1899 | | | | | | | In service. |
| Wormley, W. W. | Aug 15, 1805 | | | | | | | | Resigned June 17, 1806. |
| Wortman, Ward K. | Jul 1, 1902 | | | | | | | | Revoked Oct 10, 1902. |
| Wright, John N. | Oct 2, 1899 | Jul 23, 1900 | | | | | | | In service. |
| Wright, William | 1817 | | | | | | | | Resigned May 8, 1818. |
| Wynkoop, D. S. | | Nov 16, 1798 | | | | | | | Lost in the "Insurgente." |
| Wynne, Robert F. | Jun 9, 1898 | May 26, 1899 | Jul 23, 1900 | | | | | | [4]Hon. dis. Feb 11, 1899. In service. |
| Yates, Thomas Y. | 1817 | | | | | | | | Resigned Aug 17, 1817. |
| Young, Albert B. | Mar 18, 1864 | Nov 29, 1867 | | | | | | | Resigned Jan 31, 1873. |
| Young, J. M. T. | Jul 2, 1864 | Aug 9, 1868 | Sep 24, 1884 | Aug 10, 1898 | | | | | Retired Jan 21, 1899; died Aug 1, 1902. |
| Young, James B. | Jun 14, 1862 | Jun 22, 1864 | | | | | | | Resigned Jan 20, 1865. |
| Young, John F. | Apl 4, 1838 | | | | | | | | Died Feb 9, 1839 |
| Young, T. M. W. | Feb 20, 1822 | | | | | | | | Died July 25, 1825. |
| Young, W. L. | Feb 23, 1835 | Mar 12, 1845 | | | | | | Capt. Sep 13, 1847. | Died Dec 12, 1857. |
| Youngblood, E. H. | Mar 27, 1856 | | | | | | | | Resigned Feb 28, 1857. |
| Zeilin, W. F. | Dec 19, 1871 | | | | | | | | Died Jun 4, 1880. |

1 Quartermaster, with the rank of Major.
2 Assistant Quartermaster, with the rank of Captain.
3 Resignation tendered to join the South, but not accepted by the Department.
4 Appointed for S.-A. War, discharged and reappointed.

# LIST OF OFFICERS BREVETTED FOR SERVICES DURING THE CIVIL, SPANISH-AMERICAN, PHILIPPINE AND CHINA WARS.

The following is a complete list of officers of the Marine Corps honored by brevet titles for distinguished services during the Civil and subsequent wars, with a statement of the reasons which led to their distinction:

## CIVIL WAR.

*Lieutenant-Colonels by Brevet.*—Brevet Major John L. Broome, for gallant and meritorious services at the second battle of Vicksburg, July 15, 1863, to date from March 14, 1863; Brevet Major Charles Heywood, for gallant and meritorious services at the battle of Mobile Bay, to date from August 5, 1864; Brevet Major James Forney, for meritorious services in defeating a rebel raid at Gunpowder Bridge, in July, 1864.

*Majors by Brevet.*—Captain Charles G. McCawley, for gallant and meritorious services at the night attack upon Fort Sumter, to date from September 8, 1863; Captain John L. Broome, for gallant and meritorious services, to date from April 24, 1862; Captain P. R. Fendall, for gallant and meritorious conduct at the battles of Hatteras Inlet, Port Royal, Sewell's Point, Ocrakoke Inlet, Savannah and James Rivers, to date from June 7, 1870; Captain Charles Heywood, for distinguished gallantry in presence of the enemy, March 8, 1862; Captain Lucien L. Dawson, for gallant and meritorious services at the attack upon Fort Fisher, January 13, 14, and 15, 1865; Captain George P. Houston, for gallant and meritorious services at the battle of Mobile Bay, to date from August 5, 1864; Captain James Forney, for gallant and meritorious services in the action with the savages at Formosa, June 13, 1867; Captain George Butler, for gallant and meritorious services at the attack upon Fort Fisher, January 13, 14, and 15, 1865; Captain W. H. Parker, for gallant and meritorious services at the attack upon Fort Fisher, January 13, 14, and 15, 1865.

*Captains by Brevet.*—First Lieutenant James Forney, for gallant and meritorious services at the attack on Forts Jackson and St. Philip,

April 24, 1862; First Lieutenant John H. Higbee, for gallant and meritorious services at Port Hudson and Grand Gulf, May 25, 1863; First Lientenant F. H. Corrie, for gallant and meritorious services at the attack upon Fort Fisher, January 13, 14, and 15, 1865; First Lieutenant H. B. Lowry, for gallant and meritorious services at the night attack upon Fort Sumter, September 8, 1863; First Lieutenant P. C. Pope, for gallant and meritorious services at the night attack upon Fort Sumter, September 8, 1863; First Lieutenant William Wallace, for gallant and meritorious services at the attack upon Fort Fisher, January 13, 14, and 15, 1865; First Lieutenant George G. Stoddard, for gallant and meritorious services at the battle of Boyd's Neck, and at the battle of Tulifinney Cross-Roads, to date from December 6, 1864; First Lieutenant E. P. Meeker, for gallant and meritorious services services at the attack upon Fort Fisher, January 13, 14, and 15, 1865; First Lieutenant E, P. Meeker, for gallant and meritorious services at the attack upon Fort Fisher, January 13, 14, and 15, 1865; First Lieutenant L. E. Fagan, for gallant and meritorious services at the attack upon Fort Fisher, January 13, 14, and 15, 1865; First Lieutenant George B. Haycock, for gallant and meritorious services at Gunpowder Bridge, in July, 1864, to date from June 7, 1870.

*First Lieutenants by Brevet.*—Second Lieutenant John C. Harris, for gallant and meritorious services at the attack upon Forts Jackson and St. Philip, April 24, 1862; Second Lieutenant R. L. Meade, for gallant and meritorious services at the night attack upon Fort Sumter, September 8, 1863; Second Lieutenant William Wallace, for gallant and meritorious services at the night attack upon Fort Sumter, September 8, 1863; Second Lieutenant Louis E. Fagan, for gallant and meritorious services at the night attack upon Fort Sumter, September 8, 1863; Second Lieutenant Charles Sherman, for gallant and meritorious services at the battle of Mobile Bay, August 5, 1864.

## SPANISH-AMERICAN WAR.

*Majors by Brevet.*—Captain Allan C. Kelton, for distinguished conduct and public service in the presence of the enemy, at Guantanamo, Cuba, June 11, 1896; Captain Charles L. McCawley, for distinguished conduct and public service in the presence of the enemy, at Guantanamo, Cuba, June 11, 1898; Captain Paul St. C. Murphy, for gallant service in the naval battle of Santiago, Cuba, July 3, 1898.

*Captains by Brevet.*—First Lieutenant Charles G. Long, for distinguished conduct and public service in the presence of the enemy, at Guantanamo, Cuba, June 11, 1898; First Lieutenant Lewis C. Lucas, conspicuous conduct in the battle of Guantanamo, Cuba, June 13, 1898; First Lieutenant William N. McKelvy, for distinguished conduct and public service in the presence of the enemy, at Guantanamo, Cuba, June 11, 1898; First Lieutenant Albert S. McLemore, for distinguished conduct and public service in the presence of the enemy, at Guantanamo, Cuba, June 11, 1898; First Lieutenant James E. Mahoney, for distinguished conduct and public service in the presence of the enemy, at Guantanamo, Cuba, June 11, 1898; First Lieutenant Wendell C. Neville, for conspicuous conduct in the battle of Guantanamo, Cuba, June 13, 1898; Brevet First Lieutenant Louis J. Magill, for good judgment and gallantry in the battle of Guantanamo, Cuba, June 13, 1898.

*First Lieutenants by Brevet.*—Second Lieutenant Philip M. Bannon, for distinguished service at the battle of Guantanamo, Cuba, June 13, 1898; Second Lieutenant Thomas S. Borden, for distinguished service in the naval battle of Santiago, Cuba, July 3, 1898; Second Lieutenant Louis J. Magill, for good judgment and gallantry in the battle of Guantanamo, Cuba, June 13, 1898; Second Lieutenant Melville J. Shaw, for distinguished conduct and public service in the presence of the enemy, at Guantanamo, Cuba, June 11, 1898.

### WAR IN THE PHILIPPINES.

*Captains by Brevet.*—First Lieutenant David D. Porter, for distinguished conduct and public service in the presence of the enemy, at Novaleta, P I., October 8, 1899; First Lieutenant George C. Thorpe, for distinguished conduct and public service in the presence of the enemy, at Novaleta, P. I., October 8, 1899.

### WAR IN CHINA.

*Brigadier-General by Brevet.*—Colonel R. L. Meade, for distinguished conduct and public service in the presence of the enemy, at the battle of Tien Tsin, China, July 13, 1900.

*Lieutenant-Colonels by Brevet.*—Major George Richards, for distinguished conduct in the presence of the enemy, at the battle of Tien Tsin, China, July 13, 1900; Major L. W. T. Waller, for distinguished conduct and public service in the presence of the enemy, near Tien Tsin, China, July 13, 1900.

*Majors by Brevet.*—Captain Newt H. Hall, for distinguished conduct in the presence of the enemy, at the siege of Pekin, from June 20, 1900, to August 14, 1900, to take rank from August 14, 1900;

Captain John T. Myers, for distinguished conduct in the presence of the enemy, in defense of the legations at Pekin, China, July 20, 1900.

*Captains by Brevet.*—First Lieutenant Charles G. Andersen, for distinguished conduct and public service in the presence of the enemy, near Tien Tsin, China, July 13, 1900; First Lieutenant Smedley D. Butler, for distinguished conduct and public service in the presence of the enemy, near Tien Tsin, China, July 13, 1900; First Lieutenant William G. Powell, for distinguished conduct and public service in the presence of the enemy, at Tien Tsin, China, June 21, 1900; First Lieutenant Arthur E. Harding, for distinguished conduct and public service in the presence of the enemy, near Tien Tsin, China, June 27, 1900.

*First Lieutenant by Brevet.*—Second Lieutenant Wade L. Jolly, for distinguished conduct and public service in the presence of the enemy, near Tien Tsin, China, June 27, 1900.

THE END.

Made in the USA
Monee, IL
05 November 2026

67911908R00262